Automatic Speech Recognition and Understanding in Air Traffic Management

Automatic Speech Recognition and Understanding in Air Traffic Management

Editors

Hartmut Helmke
Oliver Ohneiser

MDPI

Basel • Beijing • Wuhan • Barcelona • Belgrade • Novi Sad • Cluj • Manchester

Editors

Hartmut Helmke
Department Controller
Assistance
Institute of Flight Guidance
German Aerospace Center (DLR)
Braunschweig
Germany

Oliver Ohneiser
Department Controller
Assistance
Institute of Flight Guidance
German Aerospace Center (DLR)
Braunschweig
Germany

Editorial Office
MDPI
St. Alban-Anlage 66
4052 Basel, Switzerland

This is a reprint of articles from the Special Issue published online in the open access journal *Aerospace* (ISSN 2226-4310) (available at: www.mdpi.com/journal/aerospace/special_issues/COQJVT00W7).

For citation purposes, cite each article independently as indicated on the article page online and as indicated below:

Lastname, A.A.; Lastname, B.B. Article Title. *Journal Name* **Year**, *Volume Number*, Page Range.

ISBN 978-3-7258-0316-3 (Hbk)
ISBN 978-3-7258-0315-6 (PDF)
doi.org/10.3390/books978-3-7258-0315-6

Cover image courtesy of Oliver Ohneiser

Contents

About the Editors

Hartmut Helmke

Hon. Prof. Dr.-Ing. Hartmut Helmke holds a Diploma degree in Computer Science from the University Karlsruhe (Germany) in 1989 and a doctor degree in Chemical Engineering from Technical University of Stuttgart (Germany) in 1999. He joined the German Aerospace Center (DLR) in Braunschweig (Germany) in 1989, working on Artificial Intelligence-based expert systems. Since 1999, he has concentrated on Controller Assistance Systems in DLR's Institute of Flight Guidance. He was responsible for the Arrival Manager *4D-CARMA*. Since 2012, he has led a row of projects on automatic speech recognition and understanding in air traffic management, such as *AcListant®*, *AcListant®-Strips*, *PJ.16-04 CWP HMI ASR*, *PJ.10-W2-96 PROSA ASR-002*, *MALORCA*, and *HAAWAII*. Prof. Helmke has been an assistant professor for Computer Science at Ostfalia, University of Applied Sciences (Wolfenbüttel, Germany) since 2001.

Oliver Ohneiser

Dr.-Ing. Oliver Ohneiser received his bachelor's degree in Information Technology from Baden-Württemberg Cooperative State University Mannheim (Germany) in 2009, his master's degree in Computer Science as well as his doctorate degree (PhD) in Aerospace Engineering from the Technical University of Braunschweig (Germany) in 2011 and 2017, respectively. He joined DLR in 2006 and is now with the department "Controller Assistance" of DLR's Institute of Flight Guidance in Braunschweig. Dr. Ohneiser investigates modern interaction technologies at controller working positions and led projects concerning automatic speech recognition and understanding in air traffic management such as *TriControl*, *PJ.16-04 CWP HMI* (Solution with multiple activities), and *PJ.05-W2-97 DTT ASR-006*. Dr. Ohneiser has been a private lecturer for Aeronautical Informatics at Clausthal University of Technology (Germany) since 2021.

Preface

Ever since the emergence of Alexa, Google Assistant, and Siri, voice recognition technologies have been seamlessly integrated into our everyday lives. This innovation not only liberates our hands when inputting a new address into a navigation system but also has the potential to reduce air traffic controllers' (ATCos) workload and enhance air traffic management (ATM) safety.

Although the idea of using data links has been around for more than 30 years, voice communication between ATCos and pilots using radio equipment is still the main communication channel used in air traffic control. ATCos issue verbal commands to the cockpit crew.

Whenever the information from voice communication has to be digitized, ATCos are burdened to manually enter the information, although it was already uttered. On the one hand, Automatic Speech Recognition (ASR) transforms the analog voice signal into a spoken sequence of words, e.g.,

"speed bird four eight six descend flight level one two zero".

On the other hand, Automatic Speech Understanding extracts the meaning from the above sequence of words, e.g., that the aircraft with the callsign BAW486 should descend to roughly twelve thousand feet. The different approaches in Europe and in the US to modeling spoken words and semantics in machine-readable form are discussed in the article by Chen et al. We can also model the above sequence of words, for example, as

"speedbird 4 8 6 descend flight level 1 2 0"

where "speed bird" is written in one word, and the numbers are transcribed as digits instead of words.

When the formal problems of representing lexical, syntactic, and semantic information are solved, the spoken callsigns still need to be extracted from a verbal transmission. This challenge is addressed in the article by Garcia et al. The authors do not only address the extraction of callsigns spoken by ATCos in a lab environment but also by pilots, even in a noisy operational environment, i.e., from the cockpit, when pilots with different accents are flying in Spanish airspace. Highlighting the displayed aircraft callsign of the flight that is currently transmitting via radio reduces the workload of air traffic controllers in searching for the flight on the radar display.

Callsign highlighting is also addressed in the article by Kasttet et al. The above-mentioned callsign BAW486 can be spoken as "speed bird four eight six", "british airways four eight six", or "speed bird forty eight six". It can be abbreviated as "speed bird six" or "four eighty six". The output of automatic speech recognition is seldom perfect. For BAW486, it could also be "speed four eight six" or "speed bird five height six". Therefore, the authors rely on the available Automatic Dependent Surveillance-Broadcast (ADS-B) data, which contain a list of callsigns currently within the area for which the ATCo is responsible. The most similar callsign is extracted by a Fuzzy string matching algorithm, allowing for callsign extraction rates better than 95%. A similar string matching algorithm was also used in three articles by Ohneiser et al., Kleinert et al., and Ahrenhold et al.

Another application of Automatic Speech Recognition and Understanding (ASRU) is the pre-filling of radar labels with information extracted from ATCo voice transmissions, for example, for the Vienna approach control. This not only means that spoken callsigns need to be extracted by the system but also that the spoken commands with the associated values, qualifiers, and conditions need to be extracted too. The article by Ahrenhold et al. quantifies the effects of ASRU support in terms of increasing safety and human performance. An implemented ASRU system was validated within a human-in-the-loop environment by ATCos in different traffic density scenarios. In the baseline condition, ATCos performed the filling of radar labels by entering the voice transmission content manually with a mouse and keyboard into the aircraft radar label. In the proposed solution condition, ATCos were supported by ASRU, which, most of the time, automatically pre-filled the information into the radar labels. The solution condition led to a reduction in the clicking time the ATCos needed to maintain the radar labels by a factor of more than 30. The application was validated by the SESAR 3 JU (Single European Sky ATM Research Programme Joint Undertaking) and achieved a technological readiness level of 6 (TRL6).

A similar application is presented in the article by Kleinert et al. Here, ASRU is integrated into an Advanced Surface Movement Guidance and Control System (A-SMGCS). ASRU provides the A-SMGCS with the ability to automatically adapt the apron controller route planning based on voice communication. This relieves the controllers of the burden of manually entering a lot of information into the A-SMGCS. Validations with the ASRU-enhanced A-SMGCS were performed in the complex apron environment of Frankfurt airport with 14 apron controllers in a human-in-the-loop simulation in the summer of 2022. This integration significantly reduced the workload of the controllers and increased safety as well as the overall performance.

The article by Ohneiser et al. addresses a lower technology readiness level. Ten ATCos from Lithuania and Austria participated in a human-in-the-loop simulation of DLR to validate ASRU support within a prototypic multiple remote tower controller working position. The ASRU supports ATCos by (1) highlighting recognized callsigns, (2) inputting recognized commands from ATCo voice transmissions in electronic flight strips, (3) offering corrections to ASRU output, (4) automatically accepting ASRU output, and (5) feeding the digital air traffic control system. The presented results motivate the technology to be brought to a higher technology readiness level, which is also confirmed by subjective feedback from questionnaires and objective measurement of workload reduction based on a performed secondary task.

Most of the results of ASRU applications still result from the lab environment, which requires simulations with ATCos and especially the integration of simulation pilots. The article by Zuluaga-Gómez and Prasad et al. presents a virtual simulation pilot agent to reduce the number of needed simulation pilots, especially in the context of ATCo training. The agent also includes a text-to-speech engine and, therefore, automatically generates the pilots' readbacks. The framework employs state-of-the-art artificial intelligence-based tools such as Wav2Vec 2.0, Conformer, Bidirectional Encoder Representations from Transformers (BERT), and the Tacotron speech synthesis model.

Voice communication between ATCos and pilots is not always split into different voice streams, especially when the push-to-talk (PTT) signal is not available. Khalil et al. present a pipeline that deploys (1) speech activity detection to identify speech segments, (2) a speech-to-text system to generate transcriptions of audio segments, (3) a text-based speaker role classification to detect if the ATCo or pilot is the speaker, and (4) unsupervised speaker clustering to create a cluster of each individual pilot speaker from the obtained speech utterances.

The development of machine learning-based ASR systems demands large-scale annotated datasets, which are currently lacking in the field. The ATCO2 project aimed to develop a unique platform to collect, preprocess, and transcribe large amounts of ATC audio data from airspace in real time. The article by Zuluaga-Gómez and Nigmatulina et al. reviews (1) robust ASR, (2) natural language processing, (3) English language identification, and (4) contextual ASR biasing with surveillance data.

The work of Park and Na investigates on-board ASR within small unmanned aerial vehicles (UAV), where computer resources are very limited. Therefore, the authors propose that variable Hidden Markov-Models (HMM) are sufficient, although it is known that Deep Neural Networks

(DNN) outperform HMM in complex application domains, e.g., noisy environments and in instances with complex unstructured grammar. They show that the recognition speed of the used HMM is 100 times faster than the speed of the used DNN.

Xu et al. show that ATC speech data can also be used for purposes in which the recognition of words or even ATC concepts is not relevant. They join speech and gaze data from a laboratory environment to detect fatigue based on the entropy weight method. The authors compare automatic fatigue state recognition with controller self-ratings on the Karolinska Sleepiness Scale and achieve an accuracy rate of 86%.

Bringing ASRU technology from the lab environment to the ops room requires a safety assessment. A safety assessment process consists of defining design requirements for ASRU technology application in normal, abnormal, and degraded modes of ATC operations. Pinska-Chauvin et al. identified eight functional hazards based on the analysis of four use cases. The safety assessment was supported by top-down and bottom-up modeling and analysis of the causes of hazards to derive system design requirements for the purpose of mitigating hazards. The assessment of how well the specified design requirements were achieved was supported by evidence generated from two real-time simulations for pre-filling radar labels and callsign highlighting with pre-industrial ASRU prototypes in approach and en-route operational environments. It was demonstrated that the use of ASRU does not increase safety risks. This article has already been selected for the cover of the November 2023 issue of the MDPI *Aerospace* journal.

Aerospace's Special Issue on "Automatic Speech Recognition and Understanding in Air Traffic Management" contains 12 articles authored by 54 different authors. Those authors work for 23 different institutions, i.e., research organizations, universities, non-profit associations, speech recognition and understanding institutions, air navigation service providers, airports, and ATM system suppliers. The authors come from 13 countries on four continents. Only the continents of Australia and Antarctica are missing. This shows the worldwide interest in deeply analyzing the benefits and drawbacks of ASRU in ATM as well as the process of transferring this technology to industry for operational use in aviation.

Hartmut Helmke and Oliver Ohneiser
Editors

aerospace

MDPI

Article

Effects of Language Ontology on Transatlantic Automatic Speech Understanding Research Collaboration in the Air Traffic Management Domain

Shuo Chen [1,*], Hartmut Helmke [2], Robert M. Tarakan [1], Oliver Ohneiser [2], Hunter Kopald [1] and Matthias Kleinert [2]

[1] The MITRE Corporation, 7515 Colshire Dr, McLean, VA 22102, USA; rtarakan@mitre.org (R.M.T.); hkopald@mitre.org (H.K.)

[2] German Aerospace Center (DLR) Braunschweig, Institute of Flight Guidance, Lilienthalplatz 7, 38108 Braunschweig, Germany; hartmut.helmke@dlr.de (H.H.); oliver.ohneiser@dlr.de (O.O.); matthias.kleinert@dlr.de (M.K.)

* Correspondence: chen@mitre.org

Abstract: As researchers around the globe develop applications for the use of Automatic Speech Recognition and Understanding (ASRU) in the Air Traffic Management (ATM) domain, Air Traffic Control (ATC) language ontologies will play a critical role in enabling research collaboration. The MITRE Corporation (MITRE) and the German Aerospace Center (DLR), having independently developed ATC language ontologies for specific applications, recently compared these ontologies to identify opportunities for improvement and harmonization. This paper extends the topic in two ways. First, this paper describes the specific ways in which ontologies facilitate the sharing of and collaboration on data, models, algorithms, metrics, and applications in the ATM domain. Second, this paper provides comparative analysis of word frequencies in ATC speech in the United States and Europe to illustrate that, whereas methods and tools for evaluating ASRU applications can be shared across researchers, the specific models would not work well between regions due to differences in the underlying corpus data.

Keywords: automatic speech recognition; natural language understanding; semantic interpretation; air traffic control; radio communications; intent representation; semantic ontology; performance metrics

Citation: Chen, S.; Helmke, H.; Tarakan, R.M.; Ohneiser, O.; Kopald, H.; Kleinert, M. Effects of Language Ontology on Transatlantic Automatic Speech Understanding Research Collaboration in the Air Traffic Management Domain. *Aerospace* **2023**, *10*, 526. https://doi.org/10.3390/aerospace10060526

Academic Editor: Judith Rosenow

Received: 14 April 2023
Revised: 16 May 2023
Accepted: 17 May 2023
Published: 1 June 2023

1. Introduction

1.1. Broad Context of the Study

For more than a decade, researchers in the United States and Europe have been developing and proving the benefit of Automatic Speech Recognition (ASR) applications in the Air Traffic Management (ATM) domain. In the United States, in support of the Federal Aviation Administration (FAA), the MITRE Corporation (MITRE) has developed capabilities to use Air Traffic Controller (ATCo)–pilot voice communication information for operational purposes, such as notifying ATCos of unsafe situations or analyzing operations to identify opportunities for safety or efficiency improvements. In Europe, as part of the Single European Sky ATM Research (SESAR) program, the German Aerospace Center (DLR) has led the development and testing of prototypic applications to enhance ATCo automation interactions, reduce ATCo workload, and identify safety issues in real time. Both MITRE [1] and DLR [2] have investigated the potential for automatic detection of readback errors, which are pilot errors in reading back ATCo instructions.

Key to most applications of ASR is the semantic meaning of the words spoken and transcribed, specifically in the context of the application in which the information will be used. Thus, we use the term Automatic Speech Recognition and Understanding (ASRU) to describe the speech-to-text and the text-to-meaning processes as one. ASRU for the Air

Traffic Control (ATC) domain needs to transcribe domain-specific words and phrases and then interpret their ATC meaning. For example, "lufthansa three twenty one one seventy knots until four contact tower eighteen four five" needs to be understood to capture the flight's callsign (DLH321) and the instructions it received (speed 170 knots until four miles from the runway; contact the tower on this radio frequency 118.450).

To represent the information contained in the speech—both the words and their semantic meaning in the ATC context—MITRE and European stakeholders, led by DLR, independently developed ATC language ontologies in support of ATM application development. A common ontology, used in both Europe and the US, could enable better sharing and reuse of data, models, algorithms, and software between the US and Europe.

In a recent paper [3], we described our collaboration to compare ontologies and identify opportunities for improvement and harmonization. This paper expands on that topic to discuss the impact of the ontology on future research and development collaboration, describing several ways that an ATC ontology is critical to facilitating collaboration between researchers and to appropriately evaluating ASRU applications in the ATM domain. This paper also examines the word-level differences between United States and European ATC speech to provide quantitative understanding of the corpus data that feed the ASRU models, informing their potential cross-use between regions. The analysis shows that whereas the methods and tools for developing and measuring ASRU performance can be shared across regions (e.g., between the US and Europe), the specific models built for the different regions would likely not work well across regions.

1.2. Structure of the Paper

This paper expands on the ontology study described in [3]. The following sections are organized as follows. Section 1.3 summarizes the uses of ASRU in ATC to date. Section 1.4 lays out the levels of an ontology in the context of the ATC domain. Section 2 presents two different concrete instantiations of ATC ontology and recalls examples presented in [3] that illustrate representations of ATC semantics using these ontologies. Section 3 describes the value of ATC ontology in facilitating collaboration between research groups and presents specific applications and the semantic representations they rely on. Section 4 presents a quantitative comparison of ATC speech at the word level between the United States and Europe. Finally, Section 5 completes the paper with our conclusions and next steps.

1.3. Background

Voice communications are an essential part of ATC because they are the primary means of communicating intention, situation awareness, and environmental context. Over the last decade, researchers have invested tremendous effort into advancing the accuracy and sophistication of in-domain ASR and Natural Language Understanding (NLU) capabilities to enable human–machine teaming that improves aviation safety and efficiency [4].

Early applications of ASR and NLU focused on simulation pilots for high-fidelity controller training simulators because these applications were in controlled environments with well-defined phraseology and a limited set of speakers [5–7]. Other examples for replacing pseudo-pilots in training environments are from the FAA [8,9], DLR [10], and DFS [11]. Later applications in lab settings expanded to simulation pilots for human-in-the-loop simulations in ATM research measuring workload [12]. With the adoption of electronic flight strips in ATC facilities, Helmke et al. [13] applied ASRU to demonstrate the effectiveness of speech assistants in reducing controller workload and improving efficiency. Prototypes demonstrating the use of ASRU to enhance safety in live operations also emerged. ASRU can support the detection of anomalous trajectories [14]. It can also support the detection of closed runway operations and wrong surface operations in the tower domain [15]. The efficacy of using ASRU to automatically detect readback discrepancies was analyzed in the US [1] and in Europe [2]. A safety monitoring framework that applied ASR and deep learning to flight conformance monitoring and conflict detection has been proposed by [16]. The growing prevalence of uncrewed aerial vehicles has also led to use cases in autonomous piloting. Text-to-speech and NLP can enable communications between human controllers and autonomous artificial intelligence pilots as advocated by [17]. Finally, the accuracy and robustness achieved by mature in-domain ASR has enabled mining of large-scale ATC communication recordings for post-operational analyses. Chen et al. [18] measured approach procedure usage across the U.S. National Airspace System using automatically transcribed radio communications in post analyses. Similarly, reference [19] assessed the quantity of pilot weather reports delivered over the radio against the quantity of pilot reports manually filed during the same time frame.

A common theme across all these applications is the use of a language understanding layer that distills and disambiguates semantic meaning from the text transcripts generated by ASR. Although there is variability in the semantic structures and concepts relevant to each use case, almost all extracted semantics relate to the representation of controller and pilot intent or situation awareness. Currently, research groups in the US and Europe create and maintain their own semantic taxonomies or ontologies to define the elements and relationships that represent intentions or situational context relevant to their specific use cases. These elements usually cover ATC concepts such as aircraft callsigns, command types, and command values in a structured human-readable and machine-readable formats.

The European ontology was defined by fourteen European partners from the ATM industry as well as by air navigation service providers (ANSPs) funded by SESAR 2020 [20]. The ontology was refined through use by different projects, such as STARFiSH [21], "HMI Interaction Modes for Airport Tower" [22,23] in the tower environment, "HMI Interaction modes for approach control" [24], and HAAWAII [25], which expanded the ontology to support pilot utterances [2].

The MITRE ontology was developed and matured over several years, with many contributing projects. Our earliest ontology was created for the simulation pilot component of an enroute ATCo trainer [5]. It was later expanded to incorporate tower domain phraseology for projects such as the Closed Runway Operations Prevention Device [15]. More recently, to support the varied use cases required of our large-scale, post-processing capability [18], the ontology was expanded to cover most of the phraseology for the standard operations documented in [26]. With each iteration we made it more robust and flexible to cover regional phraseology variations across the operational domains, i.e., tower, terminal, and enroute airspace.

1.4. What We Mean by Ontology

An ontology in the context of this paper is a collection of rules, entities, attributes, and relationships that define how language meaning is represented in a particular domain. An ontology introduces structure to ASRU by distinguishing between the four levels of language communication—lexical, syntactical, semantic, and conceptual [27]—and defining meaning representation within these levels.

The **lexical level** deals with words and distinguishes between synonyms—words with the same meaning that are spoken differently. For example, the words *nine* and *niner* denote the same numerical value in the ATC domain. Similarly, *speed bird* and *speedbird* signify the same commercial airline. The ontology rules at this level specify the universe of words (i.e., the vocabulary) that may appear in ATC radio communications.

The **syntactical level** deals with grammar and distinguishes between similar meaning phrases that are worded differently. For example, the phrases *runway two seven left cleared to land*, and *cleared to land two seven left* are syntactically different because they have different word ordering; however, they have the same meaning, which is to convey clearance to land on runway two seven left.

The **semantic representation level** deals with meaning despite differences in vocabulary or grammar that do not affect the meaning of the communication. The ontology rules at this level may deal with meaning that is explicitly spoken as well as meaning that is implied. Both phrases from the syntactical level example may be mapped to an agreed form such as CTL RWY 27L or RW27L CLEARED_TO_LAND. Later in this paper, we discuss how these semantics are represented in the European and MITRE ontologies.

The **conceptual level** deals with a higher level of understanding that goes beyond the semantic level. It captures the bigger picture, which in the ATC domain can be bigger than the sum of the individual radio transmissions. An example of an event at the conceptual level is the concept of an aircraft being in the arrival phase of flight. For some applications, this is more important than knowing the particular set of altitude and speed reductions an ATCo issued. Another example is the speech associated with a go-around, which might involve a back-and-forth discussion between an ATCo and pilot followed by a series of ATCo instructions.

In this paper, the ontology instantiations we describe primarily address the lexical and semantic level described above. However, we believe ontologies can and should expand to cover any information that is relevant to the application using language interpretation.

2. A Comparison of Two ATC Ontologies

This section recaps the comparison of US and European ATC ontology instantiations described in [3].

2.1. Lexical Level

At the lexical level, MITRE's ontology specifies that both speech and non-speech sounds during ATC radio communications should be captured in the transcription. Furthermore, the transcription should closely represent the sounds present in the audio without additional annotation or meaning inference. This means speaker hesitation sounds such as "um" and "uh", partially spoken words, foreign words such as "bonjour" or "ciao", and initialisms such as "n d b" and "i l s" are transcribed as they sound. These rules were based on best practices in automatic speech recognition training corpus creation.

The European ontology at the lexical level requires that both speech and non-speech sounds be annotated in the transcription. Special annotation is associated with non-English words spoken in a radio transmission to indicate non-English content. Domain-specific acronyms and initialisms such as "NDB" and ILS" are transcribed as words in the vocabulary. Special handling is associated with domain-specific synonyms such as "nine" and "niner", which are transcribed to a single lexical representation, "nine". Both ontologies stick to the standard 26 letters in the English alphabet, i.e., "a" to "z" in lower- and uppercase form. Diacritical marks such as the umlaut "ä" in German or the acute accent "é" in French are not supported.

The differences that we observed at the word level can be summed up as fitting into the following categories:

- Identical words with different spellings (e.g., *juliett* versus *juliet*).
- How initialisms are handled (e.g., *ILS* versus *i l s*).
- Words with similar meaning and different pronunciations and spelling (e.g., *nine* versus *niner*).
- Words absent from one ontology or the other (e.g., the word *altimeter* does not occur in European ATC communications and the corresponding ICAO term *QNH* is absent from US ATC communications) [28].
- Whether speech disfluencies and coarticulation are captured at the word level (e.g., *cleartalan* versus *cleared to land*).
- Words not represented in the US English language (e.g., the German word *wiederhoeren* for a farewell).

These differences can have an impact on ASR speed and accuracy performance and on the end user or downstream software application.

2.2. Semantic Level

At the semantic level, MITRE's ontology (SL_{US}) specifies a set of entities, attributes, and relationships that capture meaning at the command or clearance level. Figure 1 illustrates the ontology of SL_{US} in graph format. At the highest level, SL_{US} starts with a concept called *Command Interpretation* that represents an instruction, and it has a mandatory attribute called *Command Type*. The *Command Type* attribute declares the type of the instruction, such as an aircraft maneuver such as "climb" or a clearance to fly a procedure such as "cleared ILS two one approach".

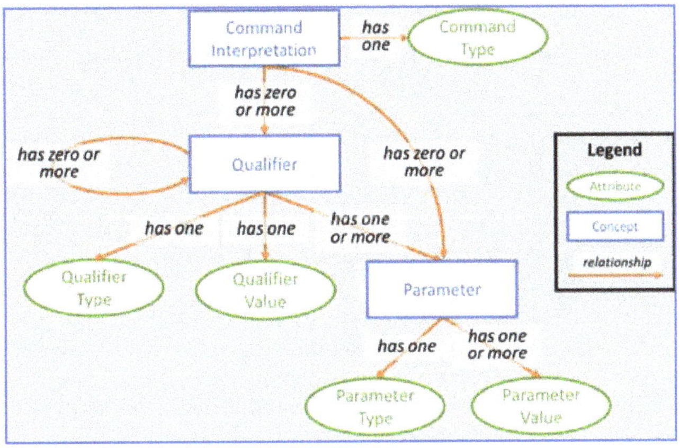

Figure 1. Graphical representation of SL_{US} ontology.

Each *Command Interpretation* can have zero or more child concepts called *Qualifiers* and *Parameters*. Both characterize, modify, and/or add values to the instruction. *Qualifiers* disambiguate or characterize *Parameters* by representing value units that are lexically present in the transcript, e.g., "flight level", "heading", "knots", etc. *Qualifiers* can be nested to represent deeper, hierarchical relationships. For example, to represent the condition "until the dulles VOR", the highest-level *Qualifier* would represent the preposition "until", its child *Qualifier* would represent the waypoint type "VOR", and its child *Parameter* would represent the name of the waypoint "dulles".

Parameters represent the value payloads for instructions that require a value, such as a heading (in degree) for a turn instruction or an altitude (in feet or flight level) for a climb instruction. A *Parameter* may exist without a *Qualifier* parent if the format of the *Parameter* value or the instruction's command type makes the *Parameter* inherently unambiguous. For example, in the instruction "climb three four zero", the command type "climb" allows us to infer that an altitude must be represented in the *Parameter* and the value format in three digits allows us to infer that the altitude is in flight level even though a unit is not explicitly stated. Figure 2 illustrates the SL_{US} ontology as a block diagram for comparison with the semantic level of the European ontology in Figure 3.

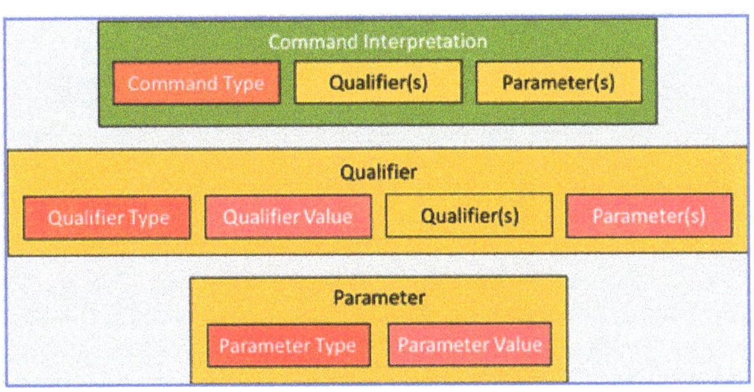

Figure 2. Block diagram of SL_{US} ontology; optional elements in orange.

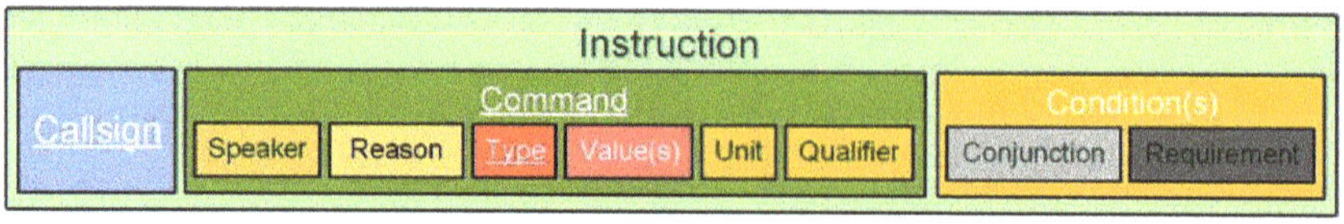

Figure 3. Block diagram of SL_{EU} ontology; optional elements in orange.

In comparison, Figure 3 illustrates the semantic level of the European ontology (SL_{EU}). At its highest level, SL_{EU} starts with a concept called *Instruction*, i.e., a mandatory *Callsign*, a mandatory *Command*, and optional *Conditions*. If the *Callsign* cannot be extracted from the transmission, the *Callsign* is "NO_CALLSIGN". A *Command* concept always has a *Type* attribute that declares the type of instruction represented. When no *Command* is found in a transcript, a *Command* concept with *Type* "NO_CONCEPT" is created. Depending on the *Type*, no *Value* or one or more *Values* can follow. If a *Value* is available, the optional attributes *Unit* and *Qualifier* are possible. The optional *Condition* concept can be present for any *Type* and more than one may be associated with one *Command*.

Type can consist of a subtype, as illustrated by the command CLEARED ILS. The Speaker attribute can have the values "ATCO" or "PILOT". If not specified, it is ATCO or can be derived from additional available context information. The *Reason* attribute is only relevant for pilot transmissions. Then the values "REQ=REQUEST", "REP=REPORTING", or an empty value are possible. The empty value, i.e., the default value, in most cases contains a pilot's readback. The *Reason* attribute is motivated by the examples in Table 1.

Table 1. Examples of ontology representations on ATC communications transcripts.

Lexical Representation	eurowings 1 3 9 alpha cleared I L S approach oh 8 right auf wiedersehen	
	MITRE Ontology Word-Level	eurowings uh one three niner alfa cleared i l s approach oh eight right auf wiedersehen
	European Ontology Word-Level	euro wings [hes] one three nine alfa cleared ILS approach [spk] O eight right [NE German] auf wiedersehen [/NE]
Controller Transmission	november three mike victor cleared I L S runway two one approach	
	SL_{US}	Callsign: {N, 3MV, GA}, Cleared: {21, ILS}
	SL_{EU}	N123MV (CLEARED ILS) 21
Callsign and Unit Inference	fedex five eighty two heavy maintain four thousand three hundred	
	SL_{US}	Callsign: {FDX, 582, H, Commercial}, Maintain: {Feet, 4300}
	SL_{EU}	FDX482 (MAINTAIN ALTITUDE) 4300 none
Transmission with Multiple Commands	good day american seven twenty six descend three thousand feet turn right heading three four zero	
	SL_{US}	Callsign: {AAL, 726, Commercial}, Courtesy, Descend: {3000, Feet}, TurnRight: {340, Heading}
	SL_{EU}	AAL726 GREETING, AAL726 DESCEND 3000 ft, AAL726 HEADING 340 RIGHT
Transmissions without Callsign	fly zero four zero cleared I L S approach	
	SL_{US}	Fly: {040, Heading}, Cleared: {ILS}
	SL_{EU}	NO_CALLSIGN HEADING 040 none, NO_CALLSIGN (CLEARED ILS) none
	lufthansa one two charlie go ahead	
	SL_{US}	Callsign: {DLH, 12C, Commercial}
	SL_{EU}	DLH12C NO_CONCEPT

Table 1. *Cont.*

Transmissions with more than one Callsign	lufthansa six alfa charlie descend one eight zero break break speed bird six nine one turn right heading zero nine five cleared I L S runway three four right	
	SL_{US}	Callsign: {DLH, 6AC, Commercial}, Descend: {180, FL}, Callsign: {BAW, 691, Commercial}, TurnRight: {95, Heading} Cleared: {34R, ILS}
	SL_{EU}	DLH6AC DESCEND 180 none, BAW691 HEADING 095 RIGHT, BAW691 (CLEARED ILS) 34R
	stand by first speed bird sixty nine thirteen turn right by ten degrees	
	SL_{US}	Callsign: {BAW, 6913, Commercial}, TurnRight: {10, Degrees}
	SL_{EU}	NO_CALLSIGN CALL_YOU_BACK, BAW6913 TURN_BY 10 RIGHT
Altitude with Limiting Condition	maintain four thousand feet until established	
	SL_{US}	Maintain: {4000, Feet}
	SL_{EU}	(MAINTAIN ALTITUDE) 4000 ft (UNTIL ESTABLISHED)
Instructions with Position-Based Conditions	at dart two you are cleared I L S runway two one left	
	SL_{US}	Cleared: {21L, ILS}
	SL_{EU}	NO_CALLSIGN (CLEARED ILS) 21L (WHEN PASSING DART2)
	leaving baggins descend and maintain one four thousand feet	
	SL_{US}	Descend: {14,000, Feet}
	SL_{EU}	NO_CALLSIGN DESCEND 14,000 ft (WHEN PASSING BGGNS)
Instructions with Advisories	maintain two fifty knots for traffic	
	SL_{US}	Maintain: {250, Knots, for traffic}
	SL_{EU}	NO_CALLSIGN (MAINTAIN SPEED) 250 kt, NO_CALLSIGN (INFORMATION TRAFFIC) none
	traffic twelve o'clock two miles same direction and let's see the helicopter	
	SL_{US}	Traffic: {Distance: 2, OClock: 12, TrafficType: helicopter}
	SL_{EU}	NO_CALLSIGN (INFORMATION TRAFFIC)
	caution wake turbulence one zero miles in trail of a heavy boeing seven eighty seven we'll be going into this [unk]	
	SL_{US}	Wake: ()
	SL_{EU}	(CAUTION WAKE_TURBULENCE)
Pilot Transmission as Readback	descend flight level one seven zero silver speed	
	SL_{US}	Descend: {170, FL}
	SL_{EU}	NO_CALLSIGN PILOT DESCEND 170 FL
Pilot Transmission as Report	speed bird two one alfa flight level two one two descend flight level one seven zero inbound dexon	
	SL_{US}	Callsign: {BAW, 21A, Commercial} Descend: {170, FL}
	SL_{EU}	BAW21A PILOT REP ALTITUDE 212 FL BAW21A PILOT REP DESCEND 170 FL BAW21A PILOT REP DIRECT_TO DEXON none
Correction of Instruction	speed bird one one descend level six correction altitude six thousand feet	
	SL_{US}	Callsign: {BAW, 11, Commercial}, Descend: {6000, Feet},
	SL_{EU}	BAW11 CORRECTION BAW11 DESCEND 6000 ft
	speed bird one one descend level six correction six thousand feet disregard turn left heading three two five degrees	
	SL_{US}	Callsign: {BAW, 11, Commercial} Descend: {6000, Feet} TurnLeft: {325, Heading}
	SL_{EU}	BAW11 DISREGARD BAW11 HEADING 325 LEFT

The differences that we observed between SL_{US} and SL_{EU} at the semantic level can be summed up as fitting into the following categories:

- How callsigns are represented.
- The extent of and representation of inferred and implied information in the semantic representations.
- The level of detail represented for advisory-type transmissions (e.g., traffic advisories, pilot call-in status information).
- Which less-common ATCo instructions have defined representations.
- How ambiguous ATCo instructions are represented.

For a detailed comparison of the semantic-level ontology overlap between the MITRE and European ontology instantiations, refer to Tables A1–A6 in the Appendix A.

2.3. Examples of Ontology Representations from ATC Communications

In reference [3], we presented several examples of word-level and semantic interpretation representations as defined by the European and MITRE ontology instantiations. We summarize them again below in Table 1 to illustrate the similarities and differences between the two ontology instantiations.

2.4. Quantifying the Differences

MITRE and DLR each exchanged 100 transmissions, with transcripts and semantic annotations, from the terminal area of a major US airport and a European hub airport. The US transcripts and annotations were manually transformed into the European format and vice versa. We assessed the word-level differences at the transcript level in terms of Levenshtein distance [29].

Out of 1554 total words in transmissions, 187 of them required modification to adhere to the other party's ontology, i.e., 12.0% of words were modified through substitution (89), deletion (35), and insertion (63). We omit uppercase to lowercase transformation from this measure. Figure 4 shows a sample transcript and its transformation.

```
all     right cleared for the ils     two five
alright       cleared for the i   l s two five
```

Figure 4. SL_{EU} structure for ambiguous instructions. Word-level difference between European (first row) and US (second row) transcripts resulting in a Levenshtein distance of 5.

In the following bullets, we list and explain some of the most often occurring cases from the 200 transcripts that are represented differently at the word level in the MITRE and European ontologies as sketched in Section 2.3:

- Separation and combination of words/letters
  - "ILS" vs. "i l s" (23 times)
  - "southwest", etc., vs. "south west", etc. (19 times)
- Different spellings
  - "nine" vs. "niner" (9 times)
  - "juliett" vs. "juliet" (6 times)
  - "OK" vs. "okay" (4 times)
- Special sounds and their notation
  - "[unk]" vs. no transcription (7 times)
  - "[hes]" vs. "uh" (7 times)

Table 2, taken from [3], shows the overlap in commands represented by the MITRE and European ontologies at the semantic level after analyzing 121 ATCo instructions from Europe and 120 from the US. DESCEND in SL_{EU} corresponds to Descend in SL_{US}. MAINTAIN ALTITUDE with Value and Unit in SL_{EU} corresponds to Maintain with the US Qualifier feet or FL. The Cleared ILS Z in SL_{US} now corresponds to CLEARED ILSZ in

SL_{EU}. GREETING and FAREWELL in SL_{EU} correspond to Courtesy in SL_{US}. SL_{US}'s Radar Service Terminated is currently not modelled in SL_{EU}. In contrast, SL_{US} does not model SL_{EU}'s CALL_ YOU_BACK command type.

Table 2. Percentage of overlap based on analysis of 241 ATCo instructions.

Type of Semantic Comparison	Overlap of Concepts
Concept present in both ontologies before adaptation	82%
Corresponding concept after small adjustments	95%
Achievable match with existing model structures	100%

3. Impact of Ontology on Collaboration

Up to this point in the paper, we have described and compared two ontology instantiations that define simplifying meaning representations for ATC communications. In the remainder of this paper, we will describe how these ontologies assist collaboration, highlighting their benefits and shortfalls. Specifically, we examine the extent to which data, models, algorithms, and applications can be shared between research groups given operational and geographic differences and how the differences manifested in ATC communications can be bridged with the help of ontologies.

3.1. *Data Sharing*

3.1.1. Text Data

In a perfect world, there would exist only one ground truth transcript for a segment of speech audio. However, as the ontology differences summarized above show, even when there is agreement on what was spoken, lexical representation of the spoken content can still differ. Although these differences in representation may seem superficial, they leave lasting impressions on models created using these lexical representations and can lead to artificially inflated error metrics if overlooked and in some cases can increase the number of actual errors.

For example, consider the two nominal examples in Table 3, where the original ground truth transcripts are transcribed according to the European ontology rules and the automatically transcribed text is generated by a speech recognizer that has modeled language following the MITRE ontology rules.

Table 3. Nominal examples of transcription error without lexical translation.

	Ground Truth Transcript	Automatically Transcribed Text
Example 1	good day american seven twenty six descend three thousand feet turn right heading three four zero	good day american seven twenty six descend four thousand feet turn right heading three zero
Example 2	cleared ILS three four	cleared i l s three five

In Example 1, when the automatically transcribed text is assessed against the ground truth transcript using word error rate (WER), a common metric for assessing speech recognition accuracy, the WER evaluates to 12.5% because of one substitution error (three by four) and one deletion error (four is missing) against a total of 16 words in the ground truth. This WER is reasonable because in this scenario the ground truth and the speech recognizer have the same lexical representation for all words in the transcript.

In contrast, in Example 2, the three errors (1 substitution and 2 insertions) resulting from differences in lexical representation ("i l s" instead of "ILS") compound the actual substitution error ("four" by "five") and results in a WER of 100%. In this scenario, lexical differences artificially inflate the true WER from 25% to 100%. Furthermore, if the semantic parser does use the same lexical representation, the difference can lead to parse errors, which in turn lead to semantic errors.

Thus, a mechanism for translation between different lexical representations is often required when sharing raw text data. By explicitly defining the rules for lexical representation, ontologies play a critical role in highlighting what is required of the translation process and facilitate its design without extensive data analysis and exploration. Because WER is an indicator of lexical representation mismatch, it can be repurposed to measure the effectiveness of the translation process.

3.1.2. Semantic Annotations

Semantic representation differences are often much more obvious than lexical representation differences but they still require the same, if not more, attention to translation. The complexities of semantic representation make ontologies even more critical to the translation design process. Though an exhaustive comparison of ontology instantiations may seem daunting, it is still much easier than an exhaustive search for syntactic and semantic samples in raw text data!

As in the case with lexical translation, a measure of semantic representation mismatch is needed to assess effectiveness of the translation process. We outline below our simple scheme for measuring semantic translation accuracy that is independent of semantic concept type or subcomponents and treats all semantic components with equal importance. These metrics can be used to compare semantic labels that have been mapped from one ontology representation to another and then back again to assess semantic content loss from the conversion. Table 4 lists definitions that are the building blocks for the accuracy metrics, and Table 5 defines the metrics and their formulas.

Table 4. Definition of basic element for accuracy calculation.

Name	Definition
True Positive (TP)	TP is the total number of True Positives: The concept is present and correctly and fully (including all subcomponents) detected
False Positive (FP)	FP is the total number of False Positives: The concept is incorrectly detected, i.e., either the concept is not present at all or one or more of its subcomponents are incorrect
True Negative (TN)	TN is the total number of True Negatives: The concept is correctly not detected
False Negative (FN)	FN is the total number of False Negatives: A concept is not detected when it should have been
Total (TA)	TA is the total number of annotated transcripts, i.e., the number of gold transcripts

Table 5. Accuracy metrics for semantic representations.

Name	Definition
Recall	$\frac{TP}{TP+FN}$
Precision	$\frac{TP}{TP+FP}$
Accuracy	$\frac{TP+TN}{TP+TN+FP+FN}$
F_1-Score	$\frac{2*Recall*Precision}{Recall+Precision}$
F_α-Score	$\frac{(1+\alpha^2)*\text{Recall}*\text{Precision}}{(\alpha^2*Precision)+Recall}$
Command Recognition Rate (RcR)	$\frac{TP+TN}{TA}$
Command Recognition Error Rate (CRER)	$\frac{FP}{TA}$
Command Rejection Rate (RjR)	$\frac{FN}{TA}$

Consider the nominal example of semantic translation for the transcript in Table 6, *"good day american seven twenty six descend three thousand feet turn right heading three four zero"*. We use this example to illustrate the metrics in action.

Table 6. Nominal example of semantic translation.

Ground Truth Semantics	Translated Semantics
AAL726 GREETING, AAL726 DESCEND 3000 ft, AAL726 HEADING 340 RIGHT	AAL726 GREETING, AAL726 DESCEND, AAL726 HEADING 340 RIGHT

In this example, there are 2 TPs (greeting and heading change) and 1 FP (due to the missing altitude in the altitude change), 0 FN, 0 TN, and TA = 3. Table 7 summarizes the accuracy metrics calculated on this nominal example.

Table 7. Accuracy metrics calculated on nominal example of semantic translation.

Name	Definition	Example
Recall	2/(2 + 0)	100%
Precision	2/(2 + 1)	66%
Accuracy	(2 + 0)/(2 + 0 + 1 + 0)	66%
F1-Score	$F_1 = \frac{2*100\%*33\%}{100\%+33\%}$	50%
Command Recognition Rate (RcR)	(2 + 0)/3	66%
Command Recognition Error Rate (CRER)	1/3	33%
Command Rejection Rate (RjR)	0/3	0%

The range for all metrics, with the exception of the Command Recognition Error Rate (CRER), is between 0 and 1. The CRER could go above 1 if (many) concepts not present in the ground truth are generated.

These metrics provide a general measure of the semantic coverage overlap between ontologies, i.e., when there is significant overlap, the CRER is low and when there is little overlap, the CRER is high. These same metrics can measure the extraction accuracy of a rules-based or deep neural network semantic parser in a general sense, but they should be modified and supplemented before use as a measure of application accuracy performance. We detail the rationale for and examples of application-specific metrics later in this section.

3.2. *Reusing Models and Algorithms*

3.2.1. Automatic Speech Recognition Models

In today's world of large pre-trained models, automatic speech recognition models are usually robust enough to transplant into new geographic regions, environments, and domains with minimal finetuning. Some models can even adapt to language changes with little to no finetuning! However, there are idiosyncrasies in the ATC language that can reduce a speech recognition model's performance if they are not addressed during transplantation between geographic regions or simply throughout prolonged use. Specifically, the quantity of airspace and region-specific, i.e., site-specific, proper nouns used during ATC radio communications requires special handling and maintenance when operating a speech recognition model in the ATC domain.

A lot of the vocabulary that appears in ATCo–pilot communications includes general purpose words such as climb, descend, cleared, to, for, and, until, one, two, three, alfa, bravo, and charlie. These are simple to document in the word level of an ontology. However, depending on the quantity of airspace that the ASRU is intended to cover, a significant percentage (90% or more) of the vocabulary could be made up of names, such as those for airline callsigns, facility identifiers, location identifiers, navigational aids, and procedure identifiers.

The site-independent, general-purpose vocabulary is relatively static and short—just a few hundred words covers most ATCo–pilot voice communications. Section 4 will show that 551 words cover 95% of the spoken words in the US data. The vocabulary of names that are used in ATCo–pilot voice communications is much larger (tens of thousands if covering the entire United States airspace) and subject to change to accommodate airspace and procedure revisions and airline and pilot callsign name additions. This name list is disproportionately large compared with the general-purpose word list but not excessively large by ASR standards. More importantly, the list of names is much more dynamic, which creates a challenge. Just as software can deteriorate over time (i.e., software rot), ASRU ontologies (and their associated models) can degrade over time if they are not maintained. For ASRU applications, an outdated word-level ontology is likely to result in out-of-vocabulary errors, which can negatively affect ASR accuracy and the accuracy of all downstream capabilities. The same applies for the sequence of words, i.e., the ICAO-phraseology and the deviation from ICAO phraseology [28]. This is a serious lifecycle maintenance issue. It is a particularly large challenge for applications that need to be scaled-up to cover multiple ATC sectors and facilities. Newer ASR models, which transcribe at the letter level, and language model tokenizers, which tokenize at the subword level, may eliminate the problem of "out-of-vocabulary" words but not the challenge of correctly recognizing and interpreting these words given their low occurrence in the training data. Furthermore, the unconstrained vocabulary in these models presents its own problems to interpretation.

Changes on the ATC operations side are made on the 28-day AIRAC (Aeronautical Information Regulation And Control) cycle. The number of changes during any one AIRAC cycle is usually small and the changes are known well in advance. Changes on the commercial airline side do not follow an official cycle but tend to be relatively uncommon. There are two subcategories of names that can present unique problems for ASRU: Military callsigns and five-letter waypoint names.

Military callsigns are a challenge because they can be introduced ad hoc and are not always known in advance of a flight's departure. The FAA ATC handbook [26] states that: *U.S. Air Force, Air National Guard, Military District of Washington priority aircraft, and USAF civil disturbance aircraft. Pronounceable words of 3 to 6 letters followed by a 1- to 5-digit number.* These pronounceable words may be pilots' names or nicknames and these words might not otherwise appear in an ATC ontology and associated ASR model. For example, "*honda five*" and "*maverick zero zero seven*" are examples of accepted military callsigns.

Five-letter waypoint names present a different challenge for ASRU. They are part of the AIRAC update cycle and are published in advance, but only the five-letter codes are published, not their pronunciations. In many cases, the waypoint codes correspond to obvious words or can be sounded out using a simple algorithm—but not always! For example: *GNDLF*, *YEBUY*, and *ISACE*. Whereas pronunciation can be handled manually on a small scale by talking to the ATC personnel for a facility for some applications, it does not easily scale to applications involving multiple ATC facilities or large amounts of airspace.

In ASRU, there is a fundamental tradeoff between a vocabulary that is too small, resulting in out-of-vocabulary errors, and a vocabulary that is too large, resulting in confusion between similar sounding words. An ASR built using a larger word-level ontology is not always better. Furthermore, it may not be possible to know and include the region-specific names in the vocabulary until you know the region where the model will be used. Thus, a word-level ontology may only specify the general-purpose vocabulary explicitly and define rules for how this vocabulary should be augmented with site-specific names before use. This issue contributes to the challenge of sharing ASR models trained and/or used between different ATC facilities or regions. Well-designed ASRU tools can simplify the adding of this site-specific information to the ontology and corresponding software.

3.2.2. Semantic Parsing Algorithms

Semantic parse algorithms translate lexical representations into semantic representations by capturing and translating the syntactical relationships between words. The mechanism for semantic parsing could be a rules-based algorithm or a machine-learning-based neural network model. Both are sensitive to lexical representation changes because they operate so closely on lexical and syntactic relationships.

Rules-based semantic parse algorithms could be considered a part of the ontology at the syntactical level because they contain rules about which relationships between lexical representations are meaningful and how they can be interpreted to construe higher-level semantic concepts. As every acceptable permutation of words must be explicitly or implicitly specified for interpretation, rules-based parse algorithms inherently document the syntactic level of the ontology; however, they can be incredibly labor intensive to create and maintain. Transplanting a rules-based semantic parse algorithm into a new region requires adapting the parse algorithm to regional lexicons, site-specific operational communications, and jargon. This inherently updates the syntactic level of the ontology as part of the model transition process.

Machine-learning-based models for semantic parsing learn the syntactic relationships from the hierarchies present in the semantic labels. In one sense, this eases the burden of rule creation, but it shifts it instead to data labeling, because the data labels must reflect the relationships between lexical entities in order for the model to learn them. Furthermore, as the syntactic rules are no longer explicitly stated as rules but hidden within the model weights, exact syntactic relationships can be difficult to discover and adjust for new model users, hampering reuse and even certification. In the absence of explicit syntactic rules, the semantic definitions of the ATC ontology become even more important as they capture and relay semantic hierarchies that might otherwise be overlooked without exhaustive data search and analysis.

3.3. *Sharing and Reusing Applications*

In ATC, there are common areas for improvement that come up again and again as possible avenues for ASRU application. As a result, the potential for application transition and reuse is high when an application is successful, even across geographic boundaries. In this section, we describe how ontologies facilitate application transition. We also discuss the importance of application-specific metrics and why they should be added to the ontology on an as-needed basis.

3.3.1. Examples of Application Specific Ontologies

Most applications incorporating ASRU are unlikely to use all the semantic concepts defined in an ATC ontology. Indeed, some of the applications prototyped between MITRE and DLR have only used a handful each. However, some semantic concepts appear across multiple applications, marking them as particularly important and worthy of focused research to improve extraction accuracy. Callsign is a recurring semantic concept that is relevant to multiple applications. Thus, both MITRE and DLR have special handling, such as context-based inference, to improve the detection accuracy of this concept.

Table 8 summarizes different applications of ASRU prototyped by DLR and MITRE. The table elucidates by an "X", which command semantics are used in each application. The applications are described in greater detail below the table and references of published reports are provided where available.

3.3.2. Closed Runway Operation Detection (CROD)

MITRE prototyped and field tested a closed runway operation clearance detection system that uses ASRU to detect landing or takeoff clearances to runways that are designated as closed. The system relies purely on manual entry of runway closures and passive listening on the local controller radio channel to detect a clearance to a closed runway and issue an alert. For more information on this application, please see [15].

3.3.3. Wrong Surface Operations Detection (WSOD)

An expansion on the closed runway operation clearance detection system, this more advanced prototype combines ASRU on radio communications with radar data in real time to detect discrepancies between the landing clearance runway issued over the radio and the projected landing runway inferred from radar track data. When a discrepancy is detected, the system generates an alert to the tower ATCo.

Table 8. Semantic representations relevant to specific applications.

Applications [1]	**CROD**	**WSOD**	**ACUA**	**PRLA**	**MRT**	**SMGCS**	**WLP**	**CPDLC**	**PWR**	**VFR**	**SPA**	**SPET**	**RB-E**	**RB-T**
Command-Type Categories	**Closed Runway Operation Detection**	**Wrong Surface Operations Detection**	**Approach Clearance Usage Analysis**	**Prefilling Radar Labels—Approach**	**Multiple Remote Tower Operations**	**Integration with A-SMGCS—Apron**	**Workload Prediction in London TMA**	**Integration with CPDLC**	**Pilot Weather Reports**	**Use of Visual Separation**	**Simulation Pilot—Apron**	**Simulation Pilot—Enroute Training**	**Readback Error Detection—Enroute**	**Readback Error Detection—Tower**
Acknowledgement												X	X	
Airspace Usage Clearance					X								X	
Altimeter/QNH Advisory				X				X				X	X	
Altitude Change				X	X		X	X				X	X	
Vertical Speed Instruction				X			X	X				X	X	
Attention All Aircraft												X		
Callsign		X	X	X	X	X	X	X	X	X	X	X	X	X
Cancel Clearance			X											
Correction/Disregard				X	X	X	X	X			X		X	
Courtesy							X							
Future Clearance Advisory				X									X	
Heading				X			X	X				X	X	
Holding				X	X		X	X				X	X	
Information (Wind, Traffic)				X	X				X					
Maintain Visual Separation										X				
Pilot Report									X				X	
Procedure Clearance			X	X	X			X				X		
Radar Advisory				X	X									
Radio Transfer				X	X	X	X	X			X	X	X	
Reporting Instruction				X	X	X	X				X	X	X	
Routing Clearance				X			X	X				X	X	
Runway Use Clearance	X	X		X	X	X					X			X
Speed Clearance				X				X				X	X	
Squawk					X							X	X	
Taxi/Ground Clearance					X	X					X			
Traffic Advisory						X				X	X			
Verify/Confirm				X	X						X		X	

[1] The gray shaded applications are MITRE applications, and the others are from DLR. An "X" indicates, which command semantics is used in which application.

3.3.4. Approach Clearance Usage Analysis (ACUA)

MITRE's voice data analytics capability was used to mine radio communications for approach clearances to inform a post-operational approach procedure utilization and conformance study [18]. The study used spoken approach clearances and radar tracks to detect trends in when and where flights received their approach clearances, correlation between aircraft equipage and approach clearance, and the effect of weather conditions on procedure utilization. The study was also able to use detected approach clearances to differentiate aircraft flying visual approaches from aircraft flying Required Navigation

Performance (RNP) procedures and then analyze RNP procedure conformance. For more information on this application, please see [18] for details.

3.3.5. Prefilling Radar Labels for Vienna Approach (PRLA)

DLR and Austro Control performed a validation exercise with 12 ATCos in DLR's ATMOS (Air Traffic Management Operations Simulator) from September 2022 to November 2022. The validations compared ATCos' workloads and safety effects with and without ASRU support. The evaluated application was inputting spoken commands into the aircraft radar labels on the radar screen. Therefore, the number of missing and wrong radar label inputs with and without ASRU support was determined. The details can be found in "Automatic Speech Recognition and Understanding for Radar Label Maintenance Support Increases Safety and Reduces Air Traffic Controllers' Workload" of Helmke et al. presented at the 15th USA/Europe Air Traffic Management Research and Development Seminar (ATM2023) in, Savannah, GA, USA, 5–9 June 2023.

3.3.6. Electronic Flight Strip in Multiple Remote Tower Environment (MRT)

In multiple remote tower operations, controllers need to maintain electronic flight strips for a number of airports. The manual controller inputs can be replaced by automatic inputs when using ASRU support. In the HMI interaction modes for the Airport Tower project, the tower/ground controller had to simultaneously take care of three remote airports. Their responsibilities included entering flight status changes triggered by issued clearances, such as pushback from gate, taxi with taxiways, line-up, runway clearances, etc., with an electronic pen into the flight strip system. When ASRU support was active, the flight status changes were automatically recognized from the controller utterances, entered into the flight strip system, and highlighted for their review. If an automatically detected flight status change was not manually corrected by the controller within ten seconds of entry, the values were accepted by the system. The prototypic system was validated with ten controllers from Lithuania and Austria in 2022. More details can be found in the presentation "*Understanding Tower Controller Communication for Support in Air Traffic Control Displays*" given at the SESAR Innovation Days in Budapest in 2022 by Ohneiser et al.

3.3.7. Integration of ASRU with A-SMGCS for Apron Control at Frankfurt and Simulation Pilots in Lab Environment (SMGCS and SPA)

In June 2022, Frankfurt Airport (Fraport), together with DLR, ATRiCS Advanced Traffic Solutions GmbH, and Idiap performed validation trials with 15 apron controllers in Fraport's tower training environment under the STARFiSH project. An A-SMGCS (Advanced Surface Movement Guidance and Control System) was supplemented with ASRU to enable integration of recognized controller commands into the A-SMGCS planning process and simultaneously improve ASRU performance with the addition of context from A-SMGCS. Together with manual input from the ATCo, the A-SMGCS is able to detect potentially hazardous situations and alert the ATCo. The addition of ASRU reduces the burden on the ATCo to manually input issued clearances over the radio into A-SMGCS. Research results showed that up to one third of the working time of controllers is spent on these manual inputs, which is detrimental to overall efficiency because ATCos spend less time on the optimization of traffic flow. More details can be found in the presentation "*Apron Controller Support by Integration of Automatic Speech Recognition with an Advanced Surface Movement Guidance and Control System*" given at the SESAR Innovation Days in Budapest in 2022 by Kleinert et al. Table 8 contains two columns for this application. The column "SMGCS" corresponds to the support of the ATCo in this application, whereas the column "SPA" corresponds to the support of the simulation pilots by ASRU.

3.3.8. Workload Prediction for London Terminal Area (WLP)

Under the Highly Automatic Air Traffic Controller Working Position with Artificial Intelligence Integration (HAAWAII) project, DLR, together with NATS (the Air Navigation

Service Provider of the United Kingdom), University of Brno, and Idiap developed a tool that determines an ATCo's workload in real-time, based on input from ASRU. The radio communications between ATCos and pilots at London TMA, for Heathrow Approach, was analyzed. Length of utterances, frequency usage rate, number of greetings, and number of miscommunications (say again, etc.) were evaluated for this purpose [30]. Callsign information is of minor importance here.

3.3.9. Integration of ASRU and CPDLC (CPDLC)

Under the HAAWAII project, DLR, together with NATS and Isavia ANS evaluated the performance of ASRU and CPDLC integration. More details can be found in the deliverable D5.3 of the HAAWAII project *"Real Time Capability and Pre-Filling Evaluation Report"*. In the future, ATCos and pilots will communicate their intentions via both data link, e.g., CPDLC (Controller Pilot Data Link Communication), and radio communications. In this envisioned state, ASRU and CPDLC are not competitors but complementary tools. Current CPDLC applications are expected to advance with the advent of data link with lower latency (LDACS). ASRU can reduce the number and complexity of mouse clicks required to create a CPDLC message.

3.3.10. Pilot Weather Reports (PWR)

MITRE performed a post-operational analysis on the quantity of weather-related pilot reports (PIREPs) that could be automatically detected and submitted as "synthetic PIREPs" by an ASRU-enabled capability [19]. One of the goals of this analysis was to see if synthetic PIREPs could supplement the manually submitted PIREPs present in the system today and better inform strategic and tactical planning of ATC operations throughout the US National Airspace System (NAS) while also easing the ATCo workload. This use case relied on the Callsign and Pilot Report semantic representations to generate a formatted synthetic PIREP. More details about the motivation, outcomes, and conclusions of this analysis can be found in [19].

3.3.11. Use of Visual Separation (VFR)

Pilot-to-pilot visual separation is an important component of NAS safety and efficiency because it allows aircraft to fly closer together with the pilot assuming responsibility for separation. However, determining whether pilots were maintaining visual separation can only be determined from the voice communications between ATCo and pilot. ASRU can be used to detect traffic advisories (when an ATCo points out traffic to a pilot), the pilot reporting the traffic in sight, and the instruction for pilots to "maintain visual separation" in post-operations analysis. This information is critical to understanding the safety of a given encounter between aircraft. The information can therefore be used to better prioritize operations for safety assurance review. Visual separation information can also be used to inform efficiency-perspective analysis of operations (e.g., what percentage of flights are visual separated), because it informs the spacing between aircraft, which informs throughput/capacity.

3.3.12. Simulation Pilots in Enroute Domain Controller Training (SPET)

MITRE designed and prototyped high-fidelity simulation training consoles to support controller training in the enroute domain [5]. To reduce training and simulation costs, these consoles included a real-time simulation pilot system that uses automatic flight management, ASRU, and text-to-speech technology to interact with controllers during training simulations. Automated simulation pilots can handle more aircraft workload, provide consistent performance and response times to controller instructions, and require less training than human simulation pilots. The success of this prototype led to other follow-up projects, such as terminal training applications, Human-In-The-Loop (HITL) simulations to support new technology prototyping, procedure and airspace design, and research studies in MITRE's Integration Demonstration and Experimentation (IDEA) Lab, and bilingual training consoles for international use.

3.3.13. Readback Error Detection for Enroute Controllers (RB-E)

Under the HAAWAII project, DLR, together with the Icelandic Air Navigation Service Provider Isavia ANS, University of Brno, and Idiap developed a readback error detection assistant and tested it on pilot and ATCo voice utterances directly recorded in the ops room environment of Isavia ANS [2].

3.3.14. Readback Error Detection for Tower Controllers (RB-T)

In 2016, MITRE conducted a feasibility study into the automatic detection of readback errors at the tower/local controller ATCo position using recorded live-operations audio [1]. The study focused on runway and taxiway use clearances and assessed the readiness of ASRU performance to support this type of application. Whereas automatic speech recognition performance was promising, the study found that more complex understanding logic was needed to differentiate acceptable readback discrepancies from alert-worthy readback errors. The study also identified the importance of detecting the nuances of dialogue between the ATCo and pilot during which the ATCo might have already taken corrective action and nullified the need for an alert.

3.4. *Application-Specific Metrics*

We previously described general semantic accuracy metrics for evaluating how well labeled concepts are extracted in general, irrespective of a downstream application. In an ideal world, we could have a single set of objective ASRU metrics that could be used to communicate accuracy and be meaningful across all applications. However, we cautioned that these general semantic metrics should be supplemented before use with a downstream application. In this section, we describe why metrics must be tailored to the application in order for it to be useful.

The first set of metrics to consider is the set that describes the accuracy performance of the application, i.e., the performance that is relevant to the end user (who could be an ATCo, pilot, data analyst, policy maker, etc.). The application accuracy is the ultimate measure of performance because the application's benefit is the ultimate measure of the utility of the capability.

However, there are situations where the application accuracy can diverge from the accuracy of the underlying ASRU. One case is when the application logic is such that an incorrect ASRU result can still produce the correct application output. Another case is when there is non-speech information used after ASRU processing that can improve wrong or missing ASRU output.

For example, consider the application described in Section 3.3.3, in which ASRU is used to detect the ATCo landing clearance and then surveillance track information is used to determine if the arrival is lined up for the correct runway. If the arrival is lined up for the wrong runway, the application issues an alert to the ATCo; if no landing clearance is detected for an arrival, the application does nothing.

Incorrect ASRU detection of the callsign will likely result in no alert because the system will not be able to compare the flight's track with a clearance. No alert will likely be the correct application response because most arrivals line up for the correct runway. Similarly, missing the landing clearance would also result in no alert. In other words, we are getting the right results but for the wrong reason.

In contrast, incorrect ASRU detection of the callsign could be corrected through use of other information, e.g., using the arrival's position in the landing sequence to fill in the gap in knowledge, resulting in correct application performance.

It is clear from these examples that although application performance is the ultimate measure of success, it obscures some detail of the ARSU accuracy. Detail of the ASRU accuracy can be critical for two reasons. One, it provides understanding of what kinds of application errors will result from ASRU errors. Two, it provides understanding of where ASRU accuracy can and should be improved.

Continuing the example of using ASRU to detect landing clearances that can be compared with arrival alignment to identify wrong surface alignment, ASRU errors in callsign recognition will result in ASRU failing to associate the landing clearance with the correct aircraft. Given that most aircraft line up correctly, this missed recognition will likely still result in a correct "no alert" response at the application level. On the other hand, ASRU errors in runway recognition could result in ASRU producing an incorrect assigned runway for the flight, which could then result in a false alert to the ATCo.

Thus, for an application that aims to detect and alert on runway misalignment, the ASRU accuracy measures should be defined corresponding to the ontology concepts that need to be detected for the application: callsign, landing clearance, and runway. For each concept, detection accuracy can be evaluated using the metrics defined in Table 5.

These metrics should then be produced for each concept separately, such that callsign, landing clearance, and runway would each have several associated accuracy measures: recall, precision, etc. These metrics can then be used to identify and measure performance improvements in the ASRU. For example, they differentiate between missed landing clearances due to missed callsign detection and those due to missed landing clearance detection.

Note that the concept detection accuracy can be rolled up into a single metric, producing an overall concept recognition error rate by combining the TP, FP, TN, and FN for all concepts. This overall concept recognition error rate provides a general measure of the ASRU accuracy, and improvement in this measure generally means better ASRU accuracy for the application, which in turn means better overall efficacy for the application. However, as the previous examples illustrate, rolling the detection of these concepts up into a single measure will obscure understanding about the effects of the errors on application performance or where ASRU improvements should be targeted. Using Figure 5, consider the following example.

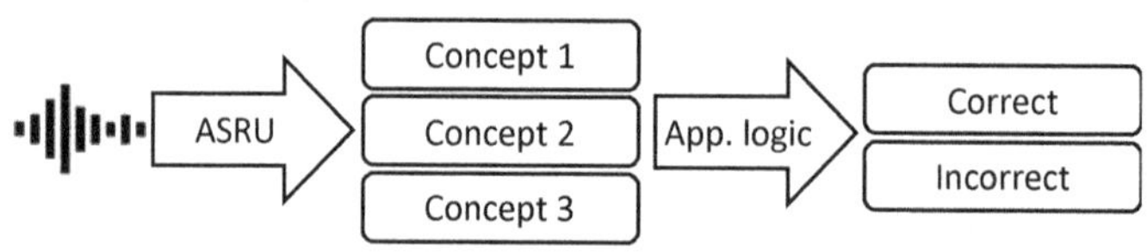

Figure 5. Example use of ASRU semantic concepts for a specific application.

Consider evaluation of ASRU performance on a set of 10 transmissions for this hypothetical application where all three basic concepts are needed to generate correct application output. A concept can be the callsign, the command type, the command value, etc. The Concept Error Rate (CER) measures the accuracy of the ASRU in detecting each concept, and a CER *should* be measured for each concept, not combined into a single metric covering the accuracy of detecting all semantic concepts. In contrast, the Command Recognition Error Rate (CRER), as defined in Table 5, measures the accuracy of the ASRU in detecting complete commands, which requires both the callsign and the instructions, which can be composed of different concepts, again.

In Case A, ASRU produces fully correct concepts for nine of the ten transmissions but zero correct concepts for one transmission. A "combined" concept error rate (CER, 3/30 = 10%) and the application error rate (1/10 = 10%) are the same. In Case B, ASRU produces fully correct output for seven of the ten transmissions but two out of three correct concepts for the remaining three transmissions. The combined CER is still 3/30 = 10% but the application error rate is now 3/10 = 30%. The CRER for Case A is 10% whereas the CRER for Case B is 30%.

The application performance for Case A is clearly better than for Case B. It is clear from this example that combined CER is obscuring important information. First, Case A will result in better application performance than Case B, despite the two having the same combined CER. Second, neither the combined CER nor the CRER tells us which concepts have room for improvement. For the example in Case A, the issue may be a systematic problem with a transmission that affects the recognition of all three concepts, such as bad audio or incorrect segmentation. For the example in Case B, did the system miss the callsign each time or one of the other concepts? Individual measures of precision and recall for each ontology concept (callsign, landing clearance, and runway in the example used above) are needed to fully assess the ASRU accuracy.

As another example, if the application only requires one concept to be detected (e.g., the closed runway operation clearance detection application described in Section 3.3.2) and does not require a callsign, then a metric such as CRER is not appropriate because it incorporates unnecessary concepts into the metric.

In summary, there is not a single metric nor type of metric that is appropriate for all applications. Practitioners should develop metrics specific to the application, covering both the application level (i.e., the performance of the application from the user's perspective) and the ASRU level (i.e., the performance of the ASRU on individual concepts needed for the application). These application-specific metrics may expand beyond accuracy measures and incorporate requirements on computing and speed performance as applications come closer to being fielded in operational settings with specific resource constraints and demands on response time.

4. Quantitative Analyses with Applied Ontologies

Thus, application-specific metrics assess overall application readiness for an operational setting and acceptability to the end-user. In this capacity, they are as important, if not more so, than the lexical and semantic level ontology when applications are transplanted into new operational environments. The general semantic accuracy metrics we described previously help researchers evaluate data, algorithms, and models; however, application-specific metrics describe the end-user experience and how he or she will be impacted by the addition of the application to the operational environment. For this reason, we recommend application-specific metrics be added to the conceptual-level definitions and rules of the ontology when an application is transitioned. These application-specific metrics can go beyond TN/TP/FN/FP and include metrics even more relevant to operations, such as false alerts per hour.

The following two subsections describe example applications and the types of ontology-related metrics needed to assess their accuracy performance.

4.1. Application-Specific Metrics for a Workload Assessment in the Lab Environment

This application is briefly described in Section 3.3.5. Table 9 summarizes the applicable semantic concepts relevant to this application. The impact on workload and safety was measured in terms of the number of missing and incorrect radar label inputs when ASRU support was present and when it was not.

Table 9. Semantic accuracy metrics for workload assessment.

WER	Total	TP	FP	FN	TN	RcR	RER	RjR	Prc	Rec	Acc	F-1	F-2	F-0.5
0.0%	17,096	16,933	71	94	11	99.1%	0.4%	0.5%	99.6%	99.4%	99.0%	99.5%	99.5%	99.6%
3.1%	17,096	15,869	368	920	10	92.9%	2.2%	5.4%	97.7%	94.5%	92.5%	96.1%	95.1%	97.1%

Table 9 summarizes the command detection accuracy when ASRU support was present during operations. Row "0.0%" shows the command detection performance with a perfect speech-to-text conversion, i.e., all incorrect detections come from errors in semantic extraction. Row "3.1%" shows the actual command detection performance during the validation trials with a speech-to-text engine that had an average WER of 3.1%.

For this use case, the application-specific metrics closely aligned with the semantic accuracy metrics described in Section 3.1.2 because the command detection accuracy translated directly into radar label entry accuracy. The number of correctly detected commands, or the command recognition rate (RcR), translated into how many entries the ATCo did not have to manually enter into the automation system. The number of incorrectly detected commands, or the command recognition error rate (CRER), translated into the number of safety risks introduced due to incorrect radar label inputs. The metric recall corresponded approximately to the command detection accuracy. They would be equal if TP+FP+FN+TN was equal to the total number of command samples (Total). The metric RER approximated 1—Prc. This correlation between RcR and Acc and the inverse correlation between RER and Prc was not present in our nominal example in Section 3.1.2 but was present in this experiment.

4.2. *Application-Specific Metrics for a Post-Operations Pilot Report Analysis*

The application itself is briefly described in Section 3.3.10. For the context of this paper, we discuss here the value of the application-level metrics used to measure the validity of this prototyped application's overall performance.

From the analyst perspective, the relevant metrics for this application were:

1. The number of correctly detected and accurately formatted pilot reports (PIREPs), i.e., correct PIREPs.
2. The number of correctly detected but incorrectly formatted PIREPs (incorrect PIREPs because they are incomplete, misleading, or both).
3. The number of PIREPs not detected or not mapped to a formatted PIREP (missed PIREPs).

The first quantity informs how much reliable supplemental information could be introduced into the US National Airspace System (NAS) by this capability. The second quantity informs how much supplemental information introduced might be misleading and potentially detrimental to planning. The final quantity informs how much potential supplemental information is being missed but would not negatively affect planning except by omission.

However, there is not a direct one-to-one correspondence between the semantic accuracy of the individual Callsign and Pilot Report concepts and the application metrics. Figure 6 illustrates the effect of different errors during the automatic PIREP detection logic within the application and their effect on the overall application performance. As the diagram shows, an error in Callsign extraction could lead to either an incorrect PIREP or a missed PIREP; an error in Pilot Report extraction could also independently lead to an incorrect PIREP or a missed PIREP, and only the combined accurate extraction of both the Callsign and Pilot Report semantics could lead to a correct PIREP.

Table 10 recaps the concept metrics of the application originally published in [19]. The final output quantities show that even when a PIREP concept is correctly detected, it may not be fully and correctly encoded (i.e., the application-level success).

Using the sample results from Table 10, we define application-specific metrics for precision and recall. We define true positive PIREPs as those that are encoded with complete information *and* PIREPs that are encoded with correct but incomplete information, on the reasoning that some information is better than none; this is an application-specific consideration. Using that definition, we calculate precision as 88% = (79 + 26)/(79 + 26 + 14). Recall is then calculated as 63% = (79 + 26)/168.

The complexity of the final application metrics is compounded by additional upstream probabilistic processes such as speech diarization, speech recognition, and text classification that could all introduce errors affecting the final result of the application. The interwoven effects of the different internal model and algorithm errors mean that no one model or algorithm is the most important and no individual model or algorithm accuracy metric could estimate overall application accuracy. Thus, the application-specific metrics are necessities invaluable for assessing the overall value of the prototype and its readiness for use in an operational setting.

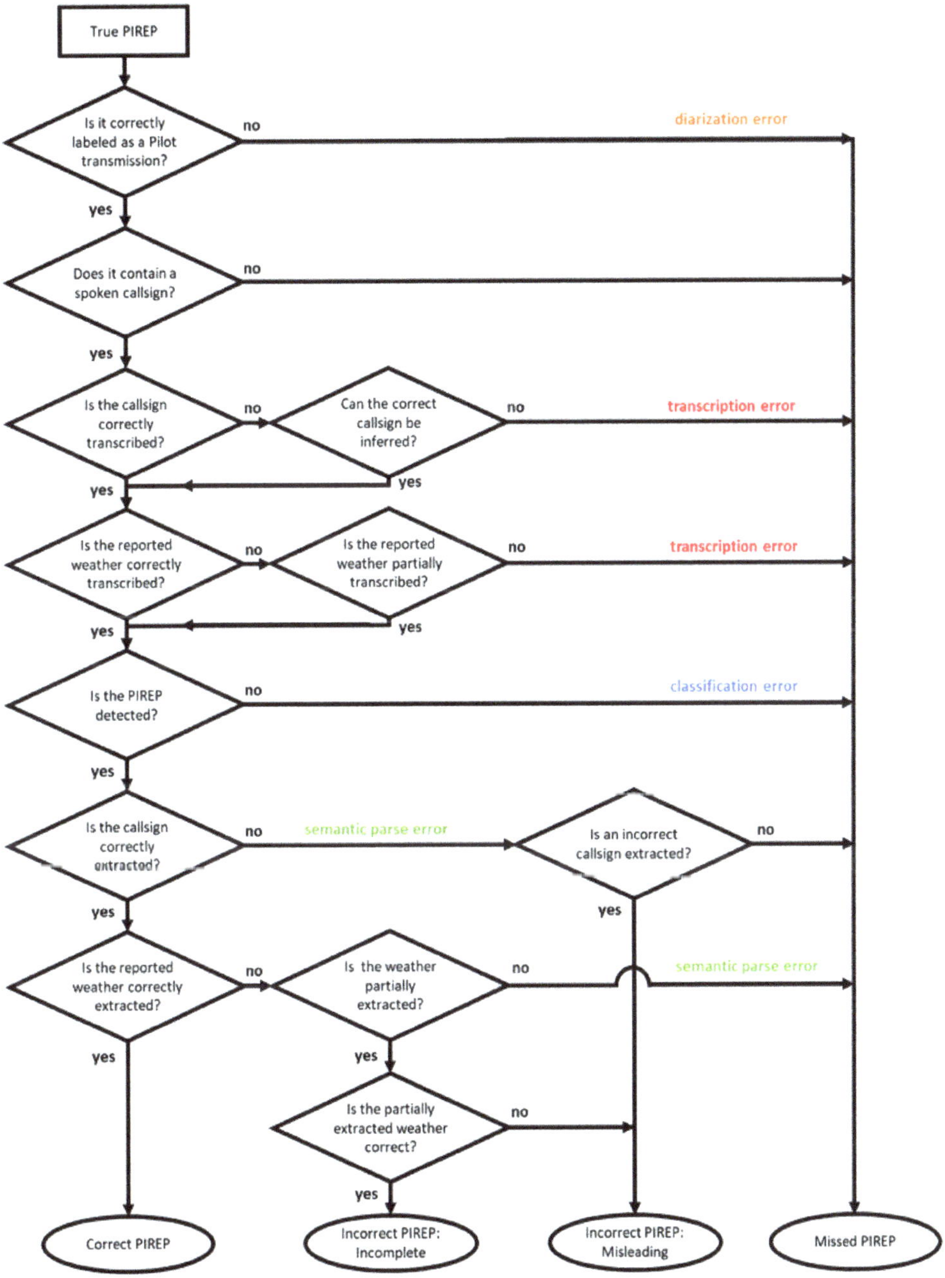

Figure 6. Effect of different ASRU errors on final PIREP application performance.

Table 10. Summary of PIREP application accuracy metrics.

Ground Truth		Detection		Encoding	
168	PIREP	161	Correct detection	79	Correct final PIREP
				34	Incorrect discard—callsign not spoken
				8	Incorrect discard—callsign not detected
				26	Incorrect final PIREP—missed details
				14	Incorrect final PIREP—incorrect flight
		7	Missed detection		
96	Not PIREP	79	Correct rejection		
		17	False detection		

4.3. European Word-Level Challenges and Statistics

As already described in Section 3.2.1, a lot of the vocabulary that appears in ATCo–pilot communications are general-purpose words such as climb, descend, cleared, etc. A large and significant percentage of the vocabulary is made up of names, e.g., airline designators, facility identifiers, location identifiers, navigational aids, and procedure identifiers. They, however, seldom occur, i.e., training data might not be available as needed.

The following Table 11 shows the results of the top 10 words in the two applications from the laboratory environment, described in Sections 3.3.5 and 3.3.7. "# Spoken" shows how often the word was really said. "Freq" shows how often this word was recognized relative to the number of all words spoken.

Table 11. Top 10 words of Vienna Approach and Frankfurt Apron Control.

Vienna Approach			Frankfurt Apron Control		
Word	**# Spoken**	**Freq**	**Word**	**# Spoken**	**Freq**
two	8841	7.4%	one	11,724	9.3%
one	8128	6.8%	november	7713	6.1%
zero	7576	6.4%	five	7100	5.6%
four	5805	4.9%	two	5520	4.4%
three	5624	4.7%	lufthansa	4994	4.0%
eight	5422	4.6%	eight	4939	3.9%
austrian	4979	4.2%	lima	4002	3.2%
six	4295	3.6%	seven	3882	3.1%
seven	4028	3.4%	four	3769	3.0%
descend	3909	3.3%	hold	3513	2.8%

Words shaded by "light blue" were not present both top 10 lists.

The Vienna data is based on 118,800 spoken words, whereas the Apron application is based on 125,800 spoken words. In "light blue", we marked the words that were only present in one or the other top 10 list but not in both. In Frankfurt, most of the taxi way names start with the letter "N", e.g., N1, N6, etc. Most of the flights to and from Vienna are from "Austrian Airlines", whereas it is "Lufthansa" for Frankfurt.

Table 12 shows the top 10 word for London TMA (Section 3.3.8) and for the enroute traffic managed by Isavia ANS (Section 3.3.9). The fact that "Reykjavik" is within the Top 10 of Icelandic traffic control is quite clear. Reykjavik is the capital of Iceland and the station name ATCos and pilots are using. "speed" being the sixth most frequent used word in London traffic might be surprising; however, knowing that "speed" is used both in speed commands and also in the callsign "speed bird" (for British Airways) explains the high occurrence. The London data is based on 102,952 spoken words, whereas the enroute application is based on 73,980 spoken words.

Table 12. Top 10 words of London TMA and Isavia Enroute Traffic.

London TMA			Isavia ANS Enroute Traffic		
Word	**# Spoken**	**Freq**	**Word**	**# Spoken**	**Freq**
one	7599	7.4%	one	4371	5.9%
zero	6284	6.1%	zero	3849	5.2%
five	5191	5.0%	three	3255	4.4%
two	5019	4.9%	five	3230	4.4%
seven	3702	3.6%	seven	3064	4.1%
speed	3677	3.6%	two	2830	3.8%
three	3536	3.4%	six	2436	3.3%
six	3198	3.1%	reykjavik	2202	3.0%
four	3113	3.0%	nine	2057	2.8%
eight	2965	2.9%	four	1962	2.7%

Words shaded by "light blue" were not present both top 10 lists.

Investigating the statistics for all four ASRU applications, we get the values shown in Table 13. The ten digits make the top 10. The digit "four" has the highest word error rate. It is often mixed with "for", a problem which can be solved afterwards at the semantic level.

Table 13. Top 10 words of four DLR applications from Tables 11 and 12.

Word	# Spoken	Freq
one	31,822	7.6%
two	22,210	5.3%
zero	19,378	4.6%
five	19,266	4.6%
three	15,346	3.7%
eight	15,085	3.6%
seven	14,676	3.5%
four	14,649	3.5%
six	13,313	3.2%
nine	9998	2.4%

Table 14 shows the "Number of Words" evaluated for each of the four applications. For Vienna, 179 words were observed more than four times, i.e., at least five times. The first 62 most occurring words for Vienna already sum up to 95% of all the spoken words. For 99% of all spoken words, we need 112 words. All in all, we have 347 different words observed for the Vienna ASRU applications (row "words for 100%").

Table 14. Statistics on word level for different ASRU applications.

	Vienna	Frankfurt	NATS	Isavia	All
Number of Words	118,794	125,810	102,952	73,980	421,536
Spoken >4 times	179	291	497	583	931
Words for 95%	62	110	205	322	256
Words for 99%	112	203	432	754	619
Words for 100%	347	520	899	1375	1972

The word statistics in Table 14 also show the difference between lab experiments and real-life data from the ops room. The number of used words is much bigger in the ops room environment than in the lab environment. This is supported by the number of words occurring more than four times and also by the 95%, 99%, and 100% thresholds. In the Icelandic enroute airspace English, Icelandic and Norwegian words are used, which explains the high number of different words.

4.4. US Word-Level Statistics

A similar analysis has been performed by MITRE. It is based on 70 ATC facilities all over the US with a corpus of 1,248,436 words. Table 15 is similar to Table 14 for the European word-level statistics.

Table 15. Statistics on word level for different MITRE data sets.

	From Corpus Partition of 99,513 Transmissions/1,248,436 Words		
	Unique Words		**Cumulative Word Count Percentage**
Spoken >1 time	4471	1st 50 words	60%
Spoken >4 times	2640	1st 100 words	74%
Words for 95%	542	1st 150 words	81%
Words for 99%	1884	1st 500 words	94%
Words for 100%	7236	1st 1000 words	98%

Table 16 shows the top 10 word occurrences from the MITRE analysis. The 10 digits are also the most frequently used words in the US.

Table 16. Occurrence of digits in MITRE data set.

Word	# Spoken	Freq	Additional Information
one	56,298	4.5%	
two	54,376	4.4%	
three, tree	45,112	3.6%	tree: 167
zero, oh	43,584	3.5%	oh: 3168
five, fife	32,038	2.6%	fife: 1
four	31,035	2.5%	
seven	27,466	2.2%	
six	26,410	2.1%	
eight	22,324	1.8%	
nine, niner	21,901	1.8%	niner: 7193
All	360,544	29.0%	

Looking into the details we observe some other interesting differences such as "one" and "two" also being the top words in US. The word "nine" would be rank only sixteen by occurrence; however, when combined with "niner", the composite moves into the top 10 in terms of occurrence frequency. One surprising observation is that "nine" is used more often than "niner", although niner is the recommended spoken form for the digit by the ICAO [28]. The European transcription ontology does not even distinguish between "nine" and "niner". Both words are mapped to "nine". Europe also does not distinguish between "five" and "fife" or "three" and "tree". Manual transcribers may not have even been able to distinguish between them.

The digit "oh" for "zero", transcribed in Europe as a capital "O", is observed in the European data only 59 times and only in the operational environment data sets from NATS and Isavia. This is a negligible percentage. However, in the US data, the more than 7000 occurrences constitute a significant percentage.

The 10 digits from "zero" to "nine" cover 42% of all words observed in the European DLR data set. In the MITRE data set, the same digits comprise 29% of all spoken words, when "niner", etc., are also considered. Our hypothesis for this is that ATCos and pilots are not limited to the ten digits, as recommended by ICAO [28]. They also use the other group-form digit words such as "ten", "twenty", "thirteen", "fourteen", "hundred", "thousand", etc. When these additional numbers are summed up together with "zero" through "nine", then numerical words comprise 40% of all words spoken, which is shown in Table 17.

Table 17. Frequency of values between 10 and 1000 in MITRE and DLR data sets.

		MITRE		DLR	
Word	**Numerical Value**	**# Spoken**	**Freq**	**# Spoken**	**Freq**
ten	10	4033	0.3%	270	0.1%
eleven	11	2788	0.2%	8	0.0%
twelve	12	3185	0.3%	5	0.0%
thirteen	13	1810	0.1%	2	0.0%
fourteen	14	2274	0.2%	4	0.0%
fifteen	15	2671	0.2%	13	0.0%
sixteen	16	2224	0.2%	5	0.0%
seventeen	17	2085	0.2%	3	0.0%
eighteen	18	2251	0.2%	62	0.0%
nineteen	19	2404	0.2%	327	0.1%
twenty	20	17,323	1.4%	972	0.2%
thirty	30	14,773	1.2%	101	0.0%
forty	40	11,961	1.0%	45	0.0%
fifty	50	11,327	0.9%	201	0.0%
sixty	60	7907	0.6%	286	0.1%
seventy	70	6882	0.6%	14	0.0%
eighty	80	7339	0.6%	317	0.1%
ninety	90	6401	0.5%	21	0.0%
hundred	100	4726	0.4%	1329	0.3%
thousand	1000	13,732	1.1%	5019	1.2%
All		128,096	10.3%	9004	2.1%

The words "hundred" and "thousand" have nearly the same frequency in the MITRE and DLR data sets. These words are recommended by ICAO. The combined occurrences of words for 11 through 90 are negligible in DLR's data set. They sum up to only 0.6% of the words spoken, whereas in the MITRE data sets they sum up to over 10%, which is significantly more. Furthermore, analysis of the US data set by speaker showed that ATCos and pilots used group-form numbers about equally, so the difference in group-form word occurrence between the US and European data sets can be attributed to differences in word usage by region, i.e., between the US and Europe, not speaker.

Moreover, very interesting is how small a percentage of the most frequently occurring words in the data set comprise in the overall data set vocabulary. Table 18 summarizes the top occurring words that comprise 95% of words in the data set and the percentage of the vocabulary they represent. This top 95% of words present in the corpus is made of 551 distinct words and includes all the numbers and letters but not most of the airline, ATC facility, and waypoint names. This 551-word set is about 7.61% of the data set's 7236 distinct word vocabulary, which means the remaining 92.39% of the distinct words in the vocabulary comprise only 5% of the data corpus in terms of occurrence.

This last statistic illustrates one of the biggest challenges for ASRU in the ATC domain. The large variety of distinct waypoint, airline, and airport names relevant to understanding is hard to recognize correctly because they have low occurrence in the data set. The reason for their low occurrence is because a training corpus for ASR or semantic parse is often deliberately varied to improve robustness and reduce overfitting, which means they are collected from many facilities and regions. However, the geographical spread of the audio data sources, while improving general robustness, dilutes the observation frequency of regional waypoint, airline, and facility names. This scarcity of a large percentage of the vocabulary in the training data subsequently leads to misrecognition of these words and misinterpretation unless deliberate action is taken to correct or improve their detection.

The findings of this analysis lead to our conclusion that although the methods and tools for developing and measuring ASRU performance can be shared across regions (e.g., between the US and Europe), the specific models built for specific regions would likely not work well across regions.

Table 18. Word classification of MITRE words.

Meaning Category	Definition	Examples	# Spoken	Percentage of Corpus Words	Percentage of Vocabulary
Other		climb, fly, contact, thanks, until	527,579	42.3%	4.78%
Numeric	Digits, other numbers, number modifiers	zero, ten, hundred, triple, point	498,066	39.9%	0.51%
Callsign Words	Airline names, aircraft types, air service types	United, Cessna, Medevac	56,265	4.5%	0.90%
Phonetic Alphabet	Phonetic alphabet words	Bravo, Charlie, Zulu	48,077	3.9%	0.39%
Place Names	ATC facilities and airport names	Atlanta, Reno	21,958	1.8%	0.50%
Initials	Letters, e.g., "i l s"	V, O, R, J, F, K, D, F	16,543	1.3%	0.35%
Filler Words	Words that fill up space but do not add substance	uh, um	10,356	0.8%	0.03%
Multiple Meanings	Words that can be all or part of airline names, airport names, or general-purpose words	Sky, Midway, Wisconsin	7084	0.6%	0.12%
Waypoint Names	Named fixes and waypoints	SAILZ, KEEEL, HUNTR, KARLA	706	0.1%	0.04%
	Total		1,186,634	95.0%	7.61%

5. Conclusions

This paper built off our comparative analysis of the two ontologies in [3] in two ways. First, this paper describes the impact of ontologies on collaboration on data, models, and applications. We described several ways that an ATC ontology is critical to facilitating collaboration between researchers and to appropriate evaluating ASRU applications in the ATM domain, using examples of specific applications to illustrate how ontology facilitates development of the metrics targeted for the application.

Second, this paper presents a word-level comparison of US and European ATC speech, specifically focusing on similarities and differences in the types of words. Although there are significant similarities (e.g., in both regions, digits make up the top 10 most spoken words), there are also significant differences (e.g., the frequency of group-form numbers). This analysis leads to our conclusion that whereas the methods and tools for developing and measuring ASRU performance can be shared across regions (e.g., between US and Europe), the specific models built for the different regions would likely not work well across regions.

Future work is needed to develop capabilities to make methods and tools more shareable between ontologies. This effort could involve modifying one or both ontologies and/or creating translation mechanisms to automatically convert data from one ontology to the other. Ultimately, research funding is critical to informing the effective and available paths forward.

Author Contributions: Conceptualization, S.C., H.K., R.M.T. and H.H.; data curation, S.C., H.K. and M.K.; formal analysis, H.H., S.C. and R.M.T.; funding acquisition, H.K. and H.H.; investigation, S.C., R.M.T., H.K. and H.H.; methodology, H.H., R.M.T. and S.C.; project administration, H.H. and H.K.; resources, H.H., O.O. and H.K.; software, H.H., S.C. and M.K.; supervision, H.K., R.M.T. and H.H.; validation, R.M.T.; visualization, S.C. and O.O.; writing—original draft, S.C., H.K. and R.M.T.; writing—review and editing: O.O., H.H., R.M.T., H.K., S.C. and M.K. All authors have read and agreed to the published version of the manuscript.

Funding: This is the copyright work of The MITRE Corporation, and was produced for the U. S. Government under Contract Number 693KA8-22-C-00001, and is subject to Federal Aviation Administration Acquisition Management System Clause 3.5-13, Rights In Data-General, Alt. III and Alt. IV

(Jan 2009). No other use other than that granted to the U. S. Government, or to those acting on behalf of the U. S. Government, under that Clause is authorized without the express written permission of The MITRE Corporation. For further information, please contact The MITRE Corporation, Contracts Management Office, 7515 Colshire Drive, McLean, VA 22102-7539, (703) 983-6000. Approved for Public Release; Distribution Unlimited. Public Release Case Number 23-1072. DLR acknowledges the support of the SESAR Joint Undertaking for developing the first version of the ontology in PJ.16-04-W2 SESAR2020 (grant agreement No 734141) and for funding the extended version in the HAAWAII project (grant agreement No 884287).

Data Availability Statement: Data can be made available by contacting the corresponding author if the data is not protected by, e.g., GDPR or other contracts.

Acknowledgments: Many thanks to Yuan-Jun Wei and Weiye Ma for their curation of the MITRE ASR corpus. MITRE also acknowledges the support from the FAA for the research funding and for allowing access to recorded ATCo–pilot live-operations radio voice communications.

Conflicts of Interest: The authors declare no conflict of interest. The funders had no role in the design of the study; in the collection, analyses, or interpretation of data; in the writing of the manuscript; or in the decision to publish the results.

Appendix A. Command Types in European and MITRE Ontology

Table A1. Altitude clearances in MITRE and European ontology.

MITRE	European	Example/Explanation
Climb	CLIMB	climb to flight level three two zero
Descend	DESCEND	descend to flight level one four zero
Tries always to derive, whether CLIMB or DESCEND	ALTITUDE	if no descend or climb keyword is provided/recognized in transmission
StopAltitude	STOP_ALTITUDE/ STOP_CLIMB/ STOP_DESCEND	stop descent at flight level one zero zero
Maintain	MAINTAIN ALTITUDE/ PRESENT_ALTITUDE	maintain flight level one eight zero; maintain present level
Cancel	NO_ALTI_RESTRICTIONS	No altitude constraints at all.

Table A2. Speed clearances in MITRE and European ontology.

MITRE	European	Example/Explanation
IncreaseSpeed	INCREASE/INCREASE_BY	increase to zero point eight four mach
ReduceSpeed	REDUCE/REDUCE_BY	reduce speed to two two zero knots
Tries always to derive, whether REDUCE or INCREASE.	SPEED	if no reduce or increase keyword is provided/recognized in transmission
	RESUME_NORMAL_SPEED	Still the published speed constraints are relevant.
Cancel SpeedRestriction	NO_SPEED_RESTRICTIONS	The speed restriction is removed
DoNotExceed	OR_LESS used as qualifier	Speed limit
Maintain	MAINTAIN SPEED/PRESENT_SPEED	maintain present speed
SpeedChange		SPEED, INCREASE, REDUCE used in Europe
	REDUCE_FINAL_APPROACH_SPEED	reduce final approach speed
	REDUCE_MIN_APPROACH_SPEED	reduce minimum approach speed
	REDUCE_MIN_CLEAN_SPEED	reduce minimum clean speed
	HIGH_SPEED_APPROVED	speed is yours

Table A3. Altitude change rate clearances in MITRE and European ontology.

MITRE	European	Example/Explanation
Climb (At)	RATE_OF_CLIMB	climb with two thousand feet per minute (or greater)/ climb at three thousand feet per minute
Descend (At)	RATE_OF_DESCENT	descend with two thousand five hundred feet per minute
Maintain		maintain three thousand in the climb/ maintain three thousand five feet per minute in the climb
	VERTICAL_RATE	if no climb or descent keyword is provided/recognized in transmission
	EXPEDITE_PASSING	expedite passing flight level three four zero

Table A4. Heading clearances in MITRE and European ontology.

MITRE	European	Example/Explanation
TurnLeft, TurnRight	HEADING/TURN/TURN_BY (Qualifier LEFT/RIGHT)	turn left heading two seven zero; turn right by one zero degrees
Turn	TURN/TURN_BY (Qualifier LEFT/RIGHT)	turn right by one zero degrees
	TURN (without a value) Qualifier LEFT/RIGHT)	turn right
Fly	HEADING (Qualifier none)	fly heading three six zero (no keyword left/right recognized)
Maintain	CONTINUE_PRESENT_HEADING/MAINTAIN HEADING	continue present heading
	MAGNETIC_TRACK	magnetic track one one five

Table A5. Routing clearances in MITRE and European ontology.

MITRE	European	Example/Explanation
Direct	DIRECT_TO/DIRECT Approach_Leg/LatLong	direct to delta lima four five five/ direct final runway three four direct six zero north zero one five west
Resume	NAVIGATION_OWN	own navigation
Cleared	CLEARED TO	cleared to london heathrow
Circle	ORBIT (Qualifier LEFT/RIGHT)	make orbits to the left

Table A6. Procedure clearances in MITRE and European ontology.

MITRE	European	Example/Explanation
Cleared (STAR/SID/Approach)	CLEARED/CLEARED VIA/MISS_APP_PROC	cleared via sorok one november
Intercept (Approach/ApproachType)	INTERCEPT_LOCALIZER	intercept localizer for runway
	INTERCEPT_GLIDEPATH	intercept glidepath
	JOIN_TRAFFIC_CIRCUIT	right traffic circuit for runway three four
Join	TRANSITION	join nerdu four november transition
Resume	NAVIGATION_OWN	resume navigation
Continue	CONTINUE_APPROACH	continue approach runway zero one
Cleared	CLEARED Approach_Type	cleared Rnav approacch zero nine center
Cancel (Approach/ SpeedRestriction/ AltitudeRestriction)	CANCEL Approach_Type	cancel approach for runway zero five
Climb (Via)		climb via the capital one departure
Descend (Via)		descend via the cavalier four arrival
	GO_AROUND	go around

References

1. Chen, S.; Kopald, H.; Chong, R.S.; Wei, Y.-J.; Levonian, Z. Readback error detection using automatic speech recognition. In Proceedings of the 12th USA/Europe Air Traffic Management Research and Development Seminar (ATM2017), Seattle, WA, USA, 26–30 June 2017.
2. Helmke, H.; Ondřej, K.; Shetty, S.; Arilíusson, H.; Simiganosch, T.S.; Kleinert, M.; Ohneiser, O.; Ehr, H.; Zuluaga, J.P. Readback error detection by automatic speech recognition and understanding: Results of HAAWAII project for Isavia's enroute airspace. In Proceedings of the 12th SESAR Innovation Days, Budapest, Hungary, 5–8 December 2022.
3. Helmke, H.; Ohneiser, O.; Kleinert, M.; Chen, S.; Kopald, H.D.; Tarakan, R.M. Transatlantic Approaches for Automatic Speech Understanding in Air Traffic Management. In Proceedings of the Submitted to 15th USA/Europe Air Traffic Management Research and Development Seminar (ATM2023), Savannah, GA, USA, 5–9 June 2023.
4. Lin, Y. Spoken Instruction Understanding in Air Traffic Control: Challenge, Technique, and Application. *Aerospace* **2021**, *8*, 65. [CrossRef]
5. Tarakan, R.; Baldwin, K.; Rozen, N. An automated simulation pilot capability to support advanced air traffic controller training. In Proceedings of the 26th Congress of ICAS and 8th AIAA ATIO, Anchorage, AK, USA, 14–19 September 2008.
6. Schultheis, S. Integrating advanced technology into air traffic controller training. In Proceedings of the 14th AIAA Aviation Technology, Integration, and Operations Conference, Atlanta, GA, USA, 16–20 June 2014.
7. Updegrove, J.A.; Jafer, S. Optimization of Air Traffic Control Training at the Federal Aviation Administration Academy. *Aerospace* **2017**, *4*, 50. [CrossRef]
8. Federal Aviation Administration. *2012 National Aviation Research Plan*; Federal Aviation Administration: Washington, DC, USA, 2012.
9. Baldwin, K. *Air Traffic Controller Training Performance Assessment and Feedback: Data Collection and Processing*; The MITRE Corporation: McLean, VA, USA, 2021.

10. Schäfer, D. Context-Sensitive Speech Recognition in the Air Traffic Control Simulation. Ph.D. Thesis, University of Armed Forces, Munich, Germany, 2001.
11. Ciupka, S. Siris big sister captures DFS (original German title: Siris große Schwester erobert die DFS. *Transmission* **2012**, *1*, 14–15.
12. Cordero, J.M.; Rodriguez, N.; de Pablo, J.M.; Dorado, M. Automated speech recognition in controller communications applied to workload measurement. In Proceedings of the 3rd SESAR Innovation Days, Stockholm, Sweden, 26–28 November 2013.
13. Helmke, H.; Ohneiser, O.; Buxbaum, J.; Kern, C. Increasing ATM efficiency with assistant-based speech recognition. In Proceedings of the 12th USA/Europe Air Traffic Management Research and Development Seminar (ATM2017), Seattle, WA, USA, 26–30 June 2017.
14. Subramanian, S.V.; Kostiuk, P.F.; Katz, G. Custom IBM Watson speech-to-text model for anomaly detection using ATC-pilot voice communication. In Proceedings of the 2018 Aviation Technology, Integration, and Operations Conference, Atlanta, GA, USA, 25–29 June 2018.
15. Kopald, H.; Chen, S. Design and evaluation of the closed runway operation prevention device. In Proceedings of the Human Factors and Ergonomics Society Annual Meeting, Chicago, IL, USA, 27–31 October 2014.
16. Lin, Y.; Deng, L.; Chen, Z.; Wu, X.; Zhang, J., Yang, B. A real-time ATC safety monitoring framework using a deep learning approach. *IEEE Trans. Intell. Transp. Syst.* **2020**, *21*, 4572–4581. [CrossRef]
17. Lowry, M.; Pressburger, T.; Dahl, D.A.; Dalal, M. Towards autonomous piloting: Communicating with air traffic control. In Proceedings of the AIAA Scitech 2019 Forum, San Diego, CA, USA, 7–11 January 2019.
18. Chen, S.; Kopald, H.; Tarakan, R.; Anand, G.; Meyer, K. Characterizing national airspace system operations using automated voice data processing. In Proceedings of the 13th USA/Europe Air Traffic Management Research and Development Seminar (ATM2019), Vienna, Austria, 17–21 June 2019.
19. Chen, S.; Kopald, H.; Avjian, B.; Fronzak, M. Automatic pilot report extraction from radio communications. In Proceedings of the 2022 IEEE/AIAA 41st Digital Avionics Systems Conference (DASC), Portsmouth, VA, USA, 18–22 September 2022.
20. Helmke, H.; Slotty, M.; Poiger, M.; Herrer, D.F.; Ohneiser, O.; Vink, N.; Cerna, A.; Hartikainen, P.; Josefsson, B.; Langr, D.; et al. Ontology for transcription of ATC speech commands of SESAR 2020 solution PJ.16-04. In Proceedings of the IEEE/AIAA 37th Digital Avionics Systems Conference (DASC), London, UK, 23–27 September 2018.
21. Bundesministerium für Bildung und Forschung, "KI-in der Praxis," Bundesministerium für Bildung und Forschung. Available online: https://www.softwaresysteme.pt-dlr.de/de/ki-in-der-praxis.php (accessed on 13 April 2023).
22. Deutsches Zentrum für Luft und Raumfahrt e.V. (DLR). Virtual/Augmented Reality Applications for Tower (SESAR Solution PJ.05-W2-97.1). Available online: https://www.remote-tower.eu/wp/project-pj05-w2/solution-97-1/ (accessed on 13 April 2023).
23. SESAR Joint Undertaking. Industrial Research Project: Digital Technologies for Tower. *SESAR 3 Joint Undertaking*. Available online: https://www.sesarju.eu/projects/DTT (accessed on 13 April 2023).
24. European Commission. PJ.10 W2 Separation Management and Controller Tools. [Online]. Available online: https://cordis.europa.eu/programme/id/H2020_SESAR-IR-VLD-WAVE2-10-2019/de (accessed on 13 April 2023).
25. Deutsches Zentrum für Luft- und Raumfahrt e.V. (DLR). HAAWAII: Highly Automated Air Traffic Controller Workstations with Artificial Intelligence Integration. Available online: https://www.haawaii.de/wp/ (accessed on 13 April 2023).
26. Federal Aviation Administration. *JO 7110.65Z—Air Traffic Control*; Federal Aviation Administration: Washington, DC, USA, 2021.
27. Foley, J.D.; Van Dam, A. *Fundamentals of Interactive Computer Graphics*, 1st ed.; Addison-Wesley Publishing Company: Reading, MA, USA, 1982.
28. International Civil Aviation Organization. *Procedures for Air Navigation Services (PANS)—Air Traffic Management (Doc 4444)*; International Civil Aviation Organization: Montreal, QC, Canada, 2016.
29. Levenshtein, V.I. Binary codes capable of correcting deletions, insertions, and reversals. In *Soviet Physics—Doklady*; American Institute of Physics: College Park, ML, USA, 1965; Volume 10, pp. 707–710.
30. Harfmann, J. D5.4 Human Performance Metrics Evaluation. Publicly available Deliverable of SESAR-2 funded HAAWAII Project, 2022-08-05, version 01.00.00. 2022. Available online: https://www.haawaii.de/wp/dissemination/references/ (accessed on 13 April 2023).

aerospace MDPI

Article

Automatic Flight Callsign Identification on a Controller Working Position: Real-Time Simulation and Analysis of Operational Recordings

Raquel García [1,*], Juan Albarrán [2], Adrián Fabio [1], Fernando Celorrio [1,2], Carlos Pinto de Oliveira [3] and Cristina Bárcena [2]

1 Centro de Referencia I+D+i ATM (CRIDA A.I.E), 28022 Madrid, Spain; afabio@e-crida.enaire.es (A.F.); fcelorrio@enaire.es (F.C.)
2 ENAIRE, 28022 Madrid, Spain; jaalbarran@enaire.es (J.A.); cpbarcena@enaire.es (C.B.)
3 EML Speech Technology GmbH, 69120 Heidelberg, Germany; carlos.pintodeoliveira@eml.org
* Correspondence: rglasheras@e-crida.enaire.es

Abstract: In the air traffic management (ATM) environment, air traffic controllers (ATCos) and flight crews, (FCs) communicate via voice to exchange different types of data such as commands, readbacks (confirmation of reception of the command) and information related to the air traffic environment. Speech recognition can be used in these voice exchanges to support ATCos in their work; each time a flight identification or callsign is mentioned by the controller or the pilot, the flight is recognised through automatic speech recognition (ASR) and the callsign is highlighted on the ATCo screen to increase their situational awareness and safety. This paper presents the work that is being performed within SESAR2020-founded solution PJ.10-W2-96 ASR in callsign recognition via voice by Enaire, Indra, and Crida using ASR models developed jointly by EML Speech Technology GmbH (EML) and Crida. The paper describes the ATCo speech environment and presents the main requirements impacting the design, the implementation performed, and the outcomes obtained using real operation communications and real-time simulations. The findings indicate a way forward incorporating partial recognition of callsigns and enriching the phonetization of company names to improve the recognition rates, currently set at 84–87% for controllers and 49–67% for flight crew.

Keywords: speech recognition; human–computer interaction; situational awareness; air traffic management; air traffic controller; flight callsign; ASR; VRS

Citation: García, R.; Albarrán, J.; Fabio, A.; Celorrio, F.; Pinto de Oliveira, C.; Bárcena, C. Automatic Flight Callsign Identification on a Controller Working Position: Real-Time Simulation and Analysis of Operational Recordings. *Aerospace* **2023**, *10*, 433. https://doi.org/10.3390/aerospace10050433

Academic Editor: Judith Rosenow

Received: 16 March 2023
Revised: 24 April 2023
Accepted: 28 April 2023
Published: 4 May 2023

1. Introduction

ATCos work with a Controller Working Position (CWP) which displays all of the information needed to support them in performing the safe, orderly, and efficient management of flights. On the CWP, flights are presented as radar tracks with an associated label indicating as a minimum the flight identification or callsign, current flight level, current speed, and the next point of the route.

While performing their tasks, ATCos must communicate with flight crews to provide them with commands and information. This communication can be performed via voice or via datalink.

Communication between ATCos and FC follows the standard established by the International Civil Aviation Organization (ICAO) [1]. This standard states that when communications are initiated by ATCos they must:

- Start by the identification or the callsign of the flight being addressed;
- Continue by issuing the command with its qualifiers or information.

Example:
Iberia three four two descend flight level two five zero

Control commands safety-related parts must always be acknowledged by the FC whose answer:

- Starts with the command with its qualifiers;
- Ends with the identification or callsign of flight.

This answer is known as readback and it is vital for ensuring mutual understanding between the FC and the ATCo of the intended plan for the aircraft. ICAO [2] requires "Flight Crew shall read back to the air traffic controller safety-related parts of ATCo clea-rances and instructions which are transmitted by voice".

The answer to the previous command would be:

Descending to flight level two five zero, Iberia three four two

When the FC initiates communication with the ATCos they will start the communication with the callsign and follow it with the necessary information. FC can initiate communications for several reasons:

- FCs always have to call air traffic control when they are about to enter a new air traffic service, ATS, unit or sector; they make a call prior to the boundary between both airspaces. The FC communicates with the ATCo to make them aware of their presence and confirm that voice communication is feasible for emergency use. In this communication the FC will typically greet the ATCo and provide some information related to the flight. Example: Good morning Ryanair nine zero three five flight level three hundred.
- The FC usually starts communications at any time with ATCos to request modifying vertical/horizontal trajectories and/or the speed to fly at the optimum performance of the aircraft.
- Another important reason to initiate a call from the FC is requesting to modify their flight level, route, speed, or any other flight condition because of adverse weather such as encountering cumulonimbus, severe turbulence, icing etc. Example: Air Europa six alfa bravo requesting flight level four zero zero due to severe turbulence.

The ATM community has investigated ASR mainly using communications from controller utterances [3–5]. This is due mainly to the fact that the ASR is seen as a means to free the controller from the necessity to manually introduce commands on the CWP, but also because of the characteristics of controllers and pilot communications.

There are some basic features in communications initiated by the controller:

- The voice signal used for speech recognition from controllers' voice utterances is extracted directly from the jack of the controller. This signal has a low degree of noise.
- Controllers' language is English or the local language of the ground station [6].
- Usually, controllers of an air navigation service provider will have similar accents when speaking.
- The percentage of women/men in air traffic control differs from one country to another. In Spain or France, the percentage is around 33% women [7,8].
- Communications from controller to flight crew can be standardised as [9]: call id + command + qualifier 1 + qualifier 2.

On the other hand, there are some basic features in communications initiated by flight crew.

- The voice signal used for speech recognition from flight crew voice utterances is extracted from radio communications. The quality of these communications is highly dependent on:
    1. The distance of the aircraft to the receiving radio station.
    2. The signal-to-noise ratio, SNR, can vary from 10 dB to −5 dB [10].
    3. The quality of the signal transportation from the radio station to the air traffic control facility where the signal is analysed.
- Flight crew language is English or the local language of the ground station [6].

- FCs have very different accents usually, but not always, relating to the flight company country. Countries that are in the routes of international flights have even higher rates of different accents.
- Communications from flight crew to controller can, similarly to the controller's ones, be decomposed as: call id + command + qualifier 1 + qualifier 2. Alternatively, if it is a readback: command + qualifier 1 + qualifier 2 + call id.

Finally, there is environmental information that can be exploited. Each ATCo has a list of flights that either are in their sector, are about to enter into it or are of interest (e.g., because they fly near the sector border). This information is provided by a flight data processor (FDP) that ensures that the list of flights is updated with new incorporations and cancellations once the flight is no longer of interest.

The level of automation is having continuous improvements and enhancements introducing new functions to assist the ATCo for better situational awareness and a reduction in workload supporting them to focus attention when and where needed. Within these new functions it is the ASR Project that requires a new Human–Machine Interface (HMI) presentation. The new methods of interaction have to be compatible with the other systems and subsystems within the CWP to benefit the controller's duties.

Identification of the key information present in the communication exchange is necessary to provide the new HMI. Information extraction from written text can follow very different approaches and several factors (language, domain, entity type) impact the selected technique [11,12]. The extraction of information in the ATM domain has mainly used knowledge-based methods and machine learning models [5,13].

The work performed in the project uses as baseline an ASR prototype that has been developed between Enaire, Crida and EML to support the quantification of the controller's workload [14,15]. The prototype follows a hybrid architecture and uses a knowledge-based method to identify the callsign. The performance of the algorithm was considered adequate in its previous use, thus the latest investigations in information extraction have not been considered in this work.

2. Materials and Methods

As presented in the introduction, the presence of the flight identification or callsign is a common feature in the communications procedures in current operations. Callsign recognition and illumination is considered as a quick win by Enaire that can be implemented in any CWP as they are equipped with a radar surveillance service that can display the radar track and callsign identification regardless of the unit where they are installed: en-route/terminal-manoeuvring area, TMA, or in tower, TWR, units. The work hypotheses are:

Hypothesis 1 (H1). *The integration of an ASR system in an operational CWP and voice communication system, VCS, can be performed without negatively impacting the other system.*

Hypothesis 2 (H2). *ASR will decrease ATCos' workload by guiding the attention of the controller to the aircraft demanding an action.*

Hypothesis 3 (H3). *ASR will increase of ATCos' situational awareness by quickly identifying new flights entering the sector or flight crews requesting actions from ATCos.*

Hypothesis 4 (H4). *ASR will increase aviation safety by illuminating the callsign coming from an ATCo's utterances ensuring they are addressing the proper aircraft.*

To test the hypothesis a prototype that meets several ATM related requirements was developed and later tested through two different and complementary approaches: a real-time simulation and a statistical analysis. Real-time simulation is a human-in-the-loop technique that tests a concept in a controlled, repeatable, and realistic environment. The technique is well established in the air traffic management environment to validate tools and concepts [16,17]. It provides qualitative and quantitative feedback. Some aspects of the technique that need to be taken into account are the level of realism of the environment, the adequacy of the test subjects, and the number of runs that takes place.

The real-time simulation was performed on Crida's premises with Enaire controllers, and a prototype developed by Indra the ASR engine provided by "EML Speech Processing Server" in November 2021. The environment realism was high as operational controller working positions and voice communication systems were used as hardware while the simulation scenarios were based on Spanish operational scenarios. The test-subjects were Enaire's operational controllers with over 10 years of experience, supported by professional pilots acting as pseudopilots. The number of runs was low as only six runs were performed.

The statistical analysis was designed to address two weaknesses of the real-time simulation, RTS. One of the weaknesses is the difference between ATCo–Pilot communications in a laboratory and operational environments. The second one is the low statistical feedback due to the low number of runs. The statistical analysis uses operational recordings from Spanish airspace. The analysis took place in February–March 2022.

2.1. Requirements to Be Met by the System

To perform the experiment several requirements have been identified on the ASR engine and on the voice recognition system, VRS.

2.1.1. Basic ASR Engine Requirements for Callsign Identification

Due to the ATM application in which voice recognition is going to be used, there are three outstanding requirements:

- The voice recognition system, VRS, shall be able to function without connection to sources external to the area control centre, ACC.
- The callsign illumination must be produced as soon as possible once the communication has started.
- The ASR engine shall be able to process the utterance in English and the local language, when local languages are allowed.

The first requirement limits the available ASR engines, as it must be autonomous. Enaire considers flight management as a strategic field, and therefore, an ACC must be able to provide its service even if it is isolated. This is a requirement set by Enaire that may not be shared by other air navigation service providers, (ANSP). This requirement may nevertheless change in the near future to align with the strategy to deliver the European commission's Digital European Sky [18] and Enaire's strategic plan [19].

The second requirement implies that the ASR engine must be able to perform in streaming and provide partial transcriptions. As in most of the use cases, the callsign is at the beginning of the phrase and the time of the initiation of these partial transcriptions is also critical. In project PJ.10-W2-96 ASR [20,21], the requirement has been established at one second after the ATCo has said the callsign. This value needs to be validated.

2.1.2. VRS Requirements for Callsign Identification

The fact that a mistake in callsign illumination can mislead the ATCo which in turn may provoke an accident puts in place a new set of requirements on the VRS:

- It is preferable not to have a callsign illumination rather than a wrong callsign illumination.
- The VRS will use the sector flight list from the CWP to improve its performance.

The callsign detection algorithm also should include the callsign rules defined by ICAO [1] and is able to detect a flight indicative independently of the method or the language used by the controller (pilot) to address it. These methods include:

- The radio callsign e.g., Beeline/Cactus.
- The company name e.g., Brussels Airlines/US Airways.
- ICAO designator using aeronautical alphabet. e.g., Bravo Echo Lima (BEL)/Alpha Whiskey Eco (AWE).

- All of the possible modes to pronounce a number. e.g., one zero zero, ten zero, one hundred.

ATCos may pronounce more than one callsign in one utterance, e.g., because they give instructions to different aircraft or because they are informing a flight about a traffic that may influence them.

ATCos and FCs are allowed to refer only to partial callsigns once the first communication is established and there is no possibility of confusion.

2.2. Real-Time Simulation

To investigate the benefits of callsign illumination through a real-time simulation, a VRS has been integrated in an Enaire's operational CWP. The VRS was developed by Indra based on Voice, a recognition system developed by Crida using EML's ASR engines. The ASR models contained have been developed jointly by EML and Crida [14] and are able to work on-the-fly processing the audio signal in real-time streaming.

Enaire's ATM system is SACTA (Air Traffic Control Automatic System) developed by Indra. The communication system that processes audio signals has recently been upgraded to COMETA (Integrated Voice IP Communication into SACTA). Figure 1 presents the architecture used. The audio is extracted by the audio extractor and sent to Voice for speech and event recognition. The delivery module then sends the event (callsign highlight) information to the SACTA CWP. The SACTA CWP also sends the environmental information to Voice via the delivery module.

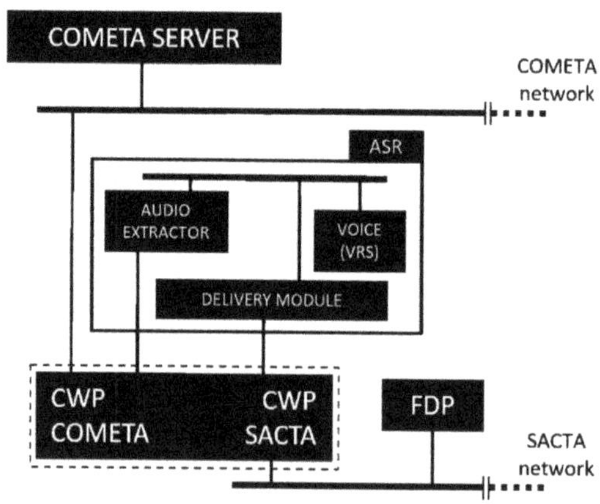

Figure 1. System Architecture.

The VRS module uses environment information to improve the recognition rate and allows the system to perform a safety check on the correct identification of the flight.

COMETA processes the audio signal following the aeronautical standard [20]. The raw audio is extracted and provided to the VRS. COMETA distinguishes between controller and FC communications. The signal is tagged with a flag indicating the source, FC (0) or ATCo (1), Figure 2. The ATCo can be in charge of one or several frequencies for radio reception (RX) and transmission (TX) depending on the sector configuration, E.g., one planner controller may listen not only to the frequency of their sector, but also to the frequency of the adjacent TMA sector to increase the situational awareness of departing flights. The system is linked to the frequency that the controller transmits (TX).

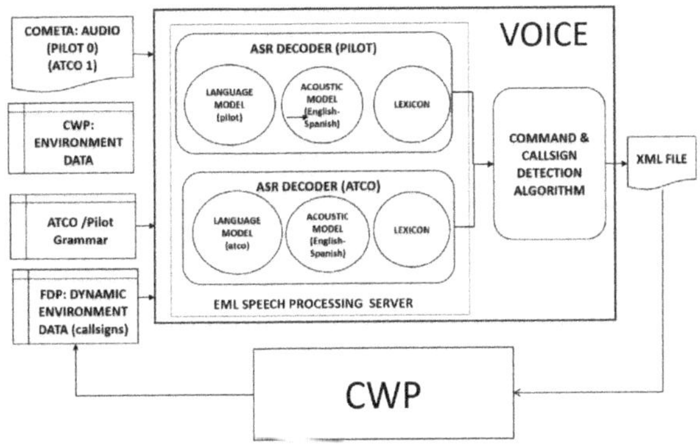

Figure 2. VRS Architecture.

The list of possible flights of interest can be provided to the VRS from two different sources.

- The FDP has the list of flights that are of interest for the ACC (composed of several sectors). The FDP ensures that the list of flights in each CWP is updated with new incorporations or cancellations once the flight is no longer of interest.
- The CWP has the list of flights that are of interest for the sector. This list is smaller than the previous one, but some flights may not be covered, for example, last minute flights deviated due to weather.

After performing a cost/benefit analysis regarding the size of each file, the system implications, and the number of flights that may be impacted, Enaire decided that the CWP would be the one providing the list. This list is provided to the dynamic lexicon update feature of "EML Speech Processing Server" (to enhance the callsign detection) and to the detection algorithm. It is updated dynamically.

ASR is provided to the "Voice" application with the "EML Speech Processing Server". The recognition engines are set for real-time streaming transcriptions with partial results, using the latest state-of-the-art technologies in ASR such as bidirectional long-short term memory (BILSTM) neural networks [22] for acoustic modelling, voice activity detection (VAD) [23] and a dynamic lexicon update feature. Due to the different characteristics of the communications, two different recognition models have been developed. Both models use the same multilingual (English and Spanish) acoustic model trained with 1000 h of recordings (about 400 h of English and 600 h of Spanish recordings out of the ATM domain) and then adapted with approximately 100 h of ATM domain recordings, which were manually transcribed from operational controller communications (these contain phrases that can be in Spanish, English, or mixing both). The difference between both recognition models is on the two class-based language models, one developed for controller communications (ATCo), and one for flight crew communications (FC). The ATCo language model is a more mature model that has been trained with operational transcriptions (approximately 90 k) gathered along several years of collaboration between Crida and EML. The FC model is a newer model that has been adapted with approximately 14 k transcriptions (from only 12 h of flight crew operational recordings).

The development of the Spanish–English acoustic model addresses some of the challenges associated with multilingual speech [24]: the simultaneous use of several languages in one sentence and the different speech rates related to each one.

The "EML Speech Processing Server" dynamic lexicon update feature enables language model class entries to be defined at runtime without having to restart the recognition engines, allowing the "Voice" application to update for a given use. The waypoint class allows the use of the same language model for different ACCs by only changing the list of entries for the waypoint class. The callsign class allows to quickly adapt the model to a specific sector, date, and time by providing the ATM-planned flights.

The command and callsign detection algorithm analyses the text recognised by the "EML Speech Processing Server" and classifies the words according to the most probable value. The algorithm follows a rule-based grammar approach [25], that was developed to support the analysis of the controller's workload calculation through a postprocessing method that included several different sources of information including flight plans, radar tracks and radio communications [14,26]. It not only classifies callsigns but also the different type of commands that are issued by controllers such as flight level or speed change. The algorithm was updated to take into account the information provided by the flight plans.

ICAO annex 10 [6] rules for callsigns are used within the algorithm to identify a possible callsign. These rules indicate that callsigns have three letters to identify the aircraft operator, followed by between one and four alphanumeric characters. The algorithm also includes the requirements listed in the VSR requirements for callsign identification paragraph. Another input to the algorithm is a list of keywords that provides possible callsign identification (radio callsign, company name, and ICAO designator) of the companies that are (or have been) authorised to operate in Spain. Finally, the algorithm increases or decreases the probability of callsign taking into account the language model; as an example, in controllers' utterances the callsign is usually at the beginning of the phrase and is followed by a command.

Once a sequence is classified as a probable callsign, it is compared against the list of possible callsigns received from the FDP. This list has the identifications of the flights that are in or near the sector and are of interest to the controller. In the implementation performed, only complete callsigns are recognised: only when the complete set of alphanumeric characters has been completely transcribed, identified by the callsign detection algorithm and are present in the FDP list, the VRS considers that there is a match. When a callsign is positively detected, a file with the information is created and sent to the Human–Machine Interface, HMI. The HMI displays a white circle around the radar track that flashes during a configurable time, currently set as 5 s, see Figure 3. The HMI is able to highlight up to 5 callsigns simultaneously, as more than one communication can be performed during this time.

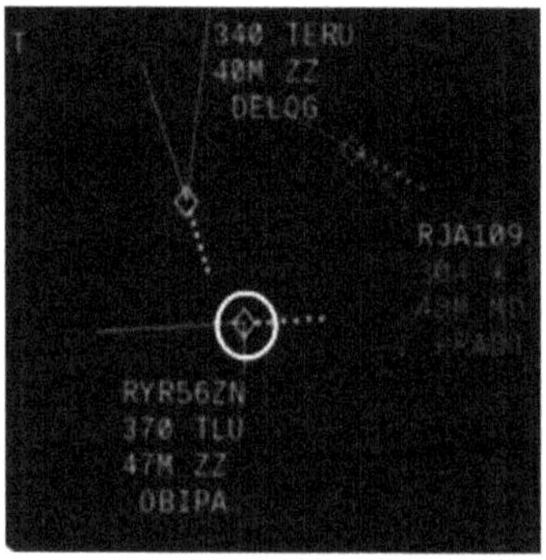

Figure 3. Radar track highlight following callsign detection (Captured from Supplementary Materials).

The real-time simulation addressed two sectors of the Madrid Flight Information Region, FIR, which has medium complexity. The sectors are Zamora–Toledo Integrated, LECMZTI, in blue in Figure 4, and Castejon–Zaragoza Integrated, LECMCZI, in red. The figure has been obtained from Enaire's Aeronautical Information web application [27].

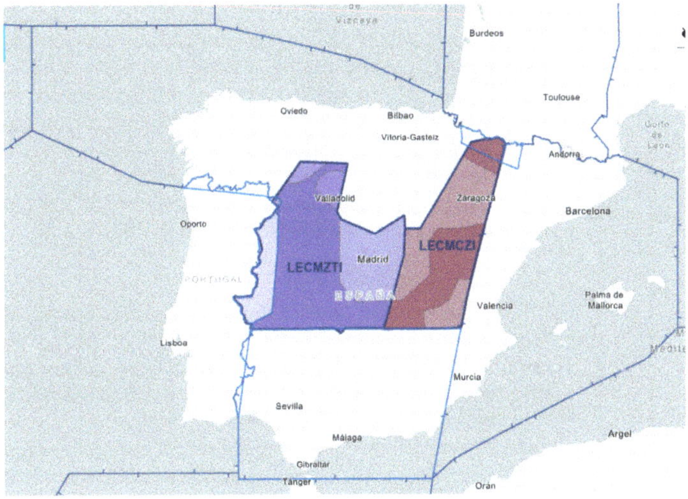

Figure 4. Simulated sectors in the real-time simulation.

This configuration is an operational configuration that is used at night. This configuration facilitates the evaluation of the validation objectives:

- The sectors have several entry points where the flight crew performs their first call (related to the highlight of callsigns on the CWP from pilot utterances).
- The sectors are quite wide and integrate nine control volumes. This implies that there are very different traffic flows that require different types of control and facilitates the creation of situations where the traffic is focused in one area or disperse along the whole sector (related to both, the highlight of callsigns on the CWP from the pilot and controller's utterances).
- There are several airports within the control volume, the main one being Madrid- Barajas airport, LEMD in its ICAO code. This airport was used in the north configuration and generated traffic flows to/from both sectors.

Controllers were operational controllers from Madrid FIR; thus, they were familiar with the scenario and the control rules. The control operation rules used were the operational ones with one simplification: the lower level to hand over traffic to TMAs and airport in all the volumes was the same, FL210.

Two types of exercise were used. Both had from medium-to-high traffic loads that supported the test of technical and operational requirements. The traffic for the exercise was created by adapting real traffic from 14 July 2019. The traffic adaptation included the modification of callsigns, flight levels and entry times. The traffic sample covered 67 different airlines plus 10 general aviation registration numbers.

The exercises were performed in an integrated controller position: one controller performs the executive and planning roles. One pseudopilot was assigned to each position. The pseudopilots have an active pilot license and have participated in previous validation activities.

The first exercise had a constant flow of aircraft that entered the controlled sectors from collateral dependencies and adjacent sectors. There were traffic peaks to concatenate calls and facilitate situations where the controller focused on one part of the sector. The traffic flight followed instrumental flight rules, IFR. There were flights from commercial airlines and general aviation.

The second exercise started in one sector configuration (one controller in charge of both sectors, Config R21 is used at nights with low traffic). The traffic steadily increases and the exercise leader, acting as supervisor, decides to split the sectors around minute 10. The traffic continues increasing until it starts to decrease. Near the end of the exercise, the traffic is low again and both sectors are grouped again. The exercise ends with one controller managing both sectors. The exercise allows the analysis of the requirements related to sectors splitting and grouping, the rapid successive communications from different pilots, and the overlap of communication from different aircraft. Traffic followed instrument flight rules (IFR), visual flight rules (VFR) and operational air traffic (OAT).

The simulation took place across two days. Each day controllers performed three runs, one with the reference scenario (without the ASR) and two with the solution one. Controllers rotated between the different sectors in each run. The results were gathered by data logs, questionnaires, debriefings, and observations.

2.3. Statistical Approach

As the RTS had some limitations, i.e., the number of scheduled runs was low, the number of controllers and pseudopilots was limited to two of each, and finally, the utterances that would be analysed were from a simulation environment which could impact the natural language of the speakers. To overcome these limitations a statistical approach was planned. The statistical approach includes the analysis of operational recordings from different types of sector and several actors, both controllers and flight crew.

The statistical test was performed between January and February 2022 with recordings from 2019. It was decided to use recordings from 2019 due to the impact of the COVID-19 pandemic on the amount and diversity of flights between 2020 and 2021.

Operational recordings fed the Voice prototype to obtain information on the callsign and event identification. The outcomes from the prototype were compared to a gold standard manually created, and rates regarding callsign and command identification have been obtained through comparison.

From the architectural point of view, this approach used part of the previous RTS prototype. It used the "Voice" prototype and callsign algorithm to transform the audios to text and identify the callsigns. Flight plans and environment data were also provided, although not dynamically, and finally the audios did not come from the COMETA system as it was not deployed at the time.

The operational environment for the statistical approach belongs to the sectors Lower Castejon (CJL), Upper Castejon (CJU) and Santiago (SAN), all of them from Madrid ACC (LECM). CJL and CJU were also used in the RTS. Figure 5 [27], presents the locations of the sectors within Spain's airspace, SAN in light blue and CJL and CJU in dark blue.

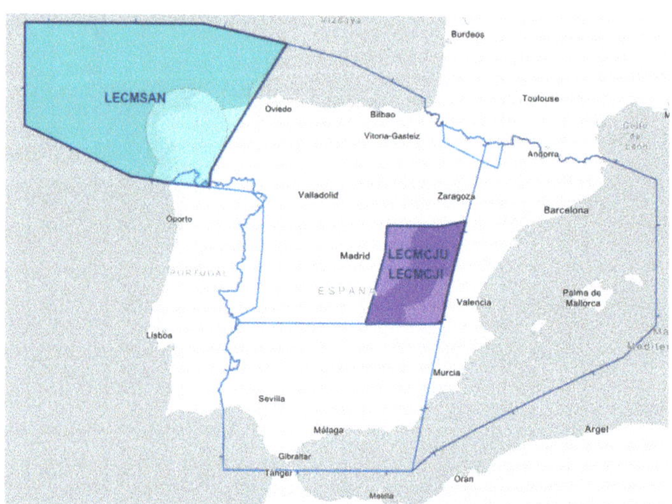

Figure 5. Simulated sectors in the statistical approach.

These sectors were selected due to their complementary characteristics that provide a wide sample of technical (i.e., signal-to-noise ratio, native speakers' origin) and operational (i.e., type of commands) characteristics:

- CJL is a sector with good radio coverage whose main traffic flows are to and from Madrid-Barajas Airport, the major Spanish airport. It limits with Madrid TMA and the surface. Control service is provided to all aircraft from FL210 to FL325. Information service is provided from SFC to FL210 outside the TMA/airport areas and airways.
- CJU is a sector with good radio coverage and quality whose main traffic flows are to and from Madrid-Barajas Airport, and over flights to the south of Spain. Control service is provided to all aircraft from FL325 to FL660.
- The SAN sector includes a large proportion of oceanic airspace that has lower radio coverage compared to CJL and CJU. Its main flows are overflights to/from the America, and to/ from United Kingdom. Free route airspace is implemented in this sector. Control service is provided to all aircraft from FL210 to FL660. Information service is provided from SFC to FL210 outside the TMA/airport and control areas.

3. Results

Technical results were gathered from system and data logs, while operational results have been collected from questionnaires, observations, and debriefings.

3.1. Technical Results

The real-time simulation produced 1139 communications from ATCo and FC. Several of the recordings were disregarded in the final analysis because they were just noise or did not contain a callsign. The traffic sample covered 67 different airlines plus 10 general aviation registration numbers that were addressed either in Spanish or English according to the airline origin.

Callsign recognition rates obtained from the RTS analysis appear in Table 1. The row speaker indicates the number of callsigns contained and detected in controllers' or flight crew utterances. No false recognition was performed.

Table 1. Callsign recognition from RTS analysis.

Speaker	Callsigns	Detected Call Sings	Detection Rate
Controller	859	721	84%
Flight Crew	687	457	67%

Regarding the identification of numbers within callsigns:

- Thousands are correctly identified (100%);
- Hundreds are correctly identified (100%);
- Numbers between 11 and 99 (e.g., 13, 18, 34) have very high recognition rates (98%);
- Numbers with triple have different success rates;
    - 111 (triple one) was transcribed correctly 88% and one time as 341;
    - 666 (triple six) had 67% success but was transcribed 326;
    - 777 (triple seven) was the least accurate of the group being transcribed as 37 in 84% of the utterances;
    - 888 (triple eight) has a success of 75%, but was transcribed as 68 two times.

One problem detected during the analysis is that if the transcription misses the ICAO name of the company, the algorithm may identify groups of four numbers and letters as possible callsigns. Due to the flight list check performed these wrong identifications were not presented to the CWP.

The statistical approach used 449 operational recordings from ATCo and FC utterances. Several of the recordings were disregarded in the final analysis because they were just noise or did not contain a callsign. The traffic sample covered 29 different airlines from 18 different countries.

Callsign recognition rates obtained from the analysis appear in Table 2. No false recognition was performed. An additional study was performed in the statistical approach. The row first call/request in the table is a subset of the pilot utterances where the FC initiates the communication, i.e., the first time a flight enters a sector or a request from the pilot not expected by the controller. These communications are of special interest as they imply a change to the controller's attention focus.

Table 2. Callsign recognition from statistical analysis.

Speaker	Callsigns	Detected Call Sings	Detection Rate
Controller	143	127	87%
Flight Crew	158	77	49%
Flight Crew First call/request	65	38	58.5%

In the statistical analysis a review taking into account the airline was also performed.

The percentage of correctly detected callsigns is higher for ATCOs than for flight crew in both cases as the algorithm is optimised for the ATCo utterances.

Regarding the comparison between simulation and operational recordings, the percentage of the ATCos are similar but the percentage of the flight crew is better in the simulation. This was already expected as the quality of the recording (ratio signal-to-noise) is better in the simulation and the accent (mother tongue) of the pseudopilots is unique (Spanish) while the one from the operational recordings is very diverse with companies from 18 different countries.

No callsign was wrongly recognised as only complete callsigns were detected. Feedback from the controllers indicated that they would like to have higher recognition rates even if some callsigns were incorrectly detected and highlighted.

In the statistical analysis detection per company was also performed. Table 3 presents the analysis. In the table the companies with less than five appearances have been removed as they have been considered that the sample is too low to infer a tendency.

Table 3. Callsign detection per company and speaker.

Airline		Controller Utterances			Flight Crew		
		Call Sings	Detected	Rate	Call Sings	Detected	Rate
American Airlines	AAL	5	4	80%	4	2	50%
Air Europa	AEA	8	8	100%	7	6	86%
Aegean airlines	AEE	9	6	67%	6	0	0%
Air Nostrum	ANE	5	5	100%	5	5	100%
Condor	CFG	9	8	89%	8	3	38%
Iberia	IBE	8	8	100%	11	11	100%
Iberia Express	IBS	7	7	100%	4	1	25%
Ryanair	RYR	18	17	94%	23	11	48%
Tap Portugal	TAP	7	7	100%	11	4	36%
Thomson	TOM	14	13	93%	18	12	67%
Emirates Airlines	UAE	11	11	100%	14	4	29%

From the analysis of the utterance according to the company, it can be inferred that in controllers' utterances, most of the airlines have over a 90% detection rate except Condor (89%), American Airlines (80%), and Aegean Airlines (67%).

In flight crew utterances the detection rate highly varies from one company to another. Spanish companies have high detection rates, i.e., Air Nostrum (100%), while for other companies the detection varies from 67% (Thomson) to 29% (Emirates), with the notable exception of Aegean Airlines that is not recognised anytime.

The callsign illumination took 3.02 s after the initialization of the utterance to the illumination in the CWP during the RTS. If the callsign was at the end of the phrase the reaction time of the prototype was lower, 0.93 s.

3.2. *Operational Results*

Operational feedback was obtained in the RTS regarding human performance (workload, accuracy, timeliness, and coherency of the information provided) and safety (situational awareness and errors induced by the prototype).

3.2.1. Human Performance

Workload was collected through Nasa-TLX [28], tailor-made questionnaires, and debriefings. The Nasa-TLX scored 9.1 (out of 20) for the reference and 7.9 (out of 20) for the solution questionnaire. The tailor-made questionnaire and debriefings indicated that workload slightly decreased in the solution scenario.

Accuracy was collected through tailor-made questionnaires, debriefings, and data logs. The feedback was that the tool needed improvement in the recognition rates to be able to effectively support them. Controllers indicated that they would prefer some occasional false positive callsign recognised if that would mean higher recognition rates.

Timeliness was collected through tailor-made questionnaires, debriefings, and data logs. The timeliness rated as adequate for the callsigns at the end of the utterance but inadequate when the callsign was at the begging of the utterance.

Feedback on the coherency of the information provided was collected through tailor-made questionnaires, and debriefings. Controllers were satisfied with the flexibility of the tool that allowed them to address the flight using very different approaches i.e., using English or Spanish, the radio name, spelling, numbers in hundreds, thousands, double, or triple. They also appreciated the HMI of the symbol on the radar track, its font, colour, and duration. A request was made to make the HMI different to be able to distinguish when the callsign highlight was from controller or flight crew utterances.

3.2.2. Safety

Situational awareness was collected through SASHA questionnaires [29] and debriefings. The overall score of the SASHA questionnaire was 4.0 (out of 6) in the reference questionnaire and 4.4 (out of 6) in the solution questionnaire. The situational awareness improved slightly with the use of VRS. During the debriefings and tailor-made questionnaires, controllers stated that situational awareness was improved but they considered that the VRS recognition rates were not high enough to allow them to completely confide and exploit the tool. They consider that higher callsign recognition rates and timeliness would further improve their situational awareness.

Errors induced by the prototype were collected through debriefings and data logs. No error resulted from the introduction of the VRS. No false recognition was performed during the simulation. This can be attributed to the requirement that indicates that only callsigns in the FDP list with the complete alphanumeric sequence provide a positive detection.

3.2.3. Additional Findings

Finally, when asked by each individual functionality, controllers appreciated especially the identification of flights from flight crew utterances. They considered that with higher robustness (meaning accuracy and timeliness) it would support to develop their tasks more efficiently, reduce workload and increase situational awareness. They considered that identification of flights from controller utterances could be especially helpful when a controller needs to understand the sector situation but is not located directly in front of the screen. On these occasions, following the performance of the controller on the radio can be difficult and having a callsign highlighting the flights from controllers' utterance will support them. At Enaire this situation happens:

- During a shift change. The entering controller may sit near the departing controller during a period of time to be able to grasp the situation before actually controlling the flights.
- When new controllers have onsite training. The new controller may be near the experienced controller following the issued commands, or a supervisor may be near the new controller.

4. Discussion

The experiment was able to connect a preindustrial VRS prototype with an ASR engine with operational systems. This connection included an operational SACTA 4 CWP that provides context information in real time to the VRS (flight plan list in this approach), receives information from the VRS and presents it to the controller in a coherent approach with the rest of the CWP information. The exercise also included connection with an operational voice communication system, COMETA, that provided the ATCo–Flight Crew communication exchange following the aeronautical standards. This integration was performed without any impact to either of the systems, demonstrating the feasibility of the technical solution. It is especially significant that no delay was introduced in the voice exchange between both actors. Hypothesis 1 is confirmed.

Although the callsign recognition was not as high as in other controller speech recognition studies [30,31], controllers' workload was reduced, and situational awareness was increased. Hypotheses 2 and 3 are confirmed. This outcome aligns with other studies [32], about the importance on high callsign recognition rates on the ATCo's perception of ASR technology support. The improvement in workload and situational awareness was not as high as expected due to the accuracy and timeliness of the VRS prototype.

The controller model has higher recognition rates than the FC model. Callsign recognition in FC utterances from the RTS was higher than from the operational ones. These outcomes were already expected due to the different maturity of both voice models; the inherent aspects of operational FC utterances with worse signal/noise relation, and higher number of different accents when compared with the controller model.

Callsigns in Spanish or with a Spanish accent have higher recognition rates than the rest. This outcome is attributable to the callsign training database which is composed mainly of Spanish controllers' utterances. The recent update with FC utterances has not been enough to cover the gap.

The bad results of some companies can be related to two different causes, training of the ASR model and phonetization of the company.

A low representation of the company in the training database would reflect in both the callsign identification from controller and flight crew utterances. Two examples of this effect can be identified in Aegean airlines in the statistical approach (67% ATCo recognition rate and 0% FC recognition rate), and Nile Air airline (Nile Bird or NIA callsign) in the RTS execution (0% FC recognition rate). Regarding this last example, Nile Air, it should be noted that, during the RTS, controllers spelt the callsign (NIA) as they were not familiarised with the airline. By doing this, they obtained a 100% recognition rate.

An incorrect or incomplete phonetization of the airline callsign would also provide low recognition rates. The airlines' callsigns are phonetised in English and Spanish, and enriched with the training database. One example of this problem is the airline Atlantic Airways whose callsign is Faroeline. Faroeline is correctly transcribed and identified in Spanish utterances but was transcribed as Flyer nine in pilot utterances. The lack of training could be enriched by enlarging the possible phonetisation with the native language of the airline country. Nevertheless, this method has its drawbacks as depending on the airline company, the nationality of pilots can be very diverse. As an example, Ryanair has a multinational group of pilots which, in 2019, counted with people from 53 different nations [32].

Future implementations where controller utterances are used to automatically implement the command on the CWP will require the recognition of partial callsigns, as already performed in some experiments [33]. This approach requests the review of the algorithm to minimise the identified problem of grouping alphanumeric characters not related to callsigns. The trade-off between false positives and recognition rate is something that needs to be investigated.

The FC usually speak differently when they start the dialogue. The FC wants to grab the attention of ATCos and are conscious that the controllers need to change the attention focus. Therefore, when initiating contact or check-in calls, they usually speak louder, slower and pronounce more clearly. The identification of these flights was one of the most appreciated by controllers, which supports the feedback provided in other ATC speech recognition experiments [34].

As already mentioned, other studies have higher callsign recognition rates [30,35]. These studies use a statistical approach for the information extraction, such as machine learning algorithms or deep neuronal networks. Comparison between both methods indicate the rule-based algorithm outperforms the statistical approach [36]. It should be noted that the statistical approach method is very dependent on the training database [35]. The method used, rule-based grammar, enhanced with the improvements identified in database training and company phonetisation, should be compared with a statistical approach that takes advantage of the Enaire's database.

Regarding the timeliness, the difference in the recognition time depending on the position of the callsigns in the utterance impacts greatly the support perceived by the controller. Further investigation in the ASR engine, ASR models, and lexicon is being conducted in order to reduce the time of partial results sent by the "EML Speech Processing Server" to then be processed by the "Command & callsign detection algorithm" within the "Voice" application.

Although situational awareness improved and no error was induced by the VRS, controllers did not consider the use of the identification of the callsign from a controller utterance as a safety tool. They considered this use of VRS as helpful to provide context information to other controllers. Hypothesis 4 is rejected.

As the feasibility of the integration of the VRS system has been demonstrated and the identification of callsigns from FC utterances provides benefits in terms of workload and situational awareness, the way forward is the improvement of recognition rates and timelines at the beginning of the utterance following the approaches previously identified. The identification of callsigns from controller utterances is also of interest to support controllers in handover and on-hand training. Improvements in recognition rates and timeliness are less critical in this case, although also necessary.

Supplementary Materials: The following supporting information can be downloaded at: Video: https://www.youtube.com/watch?v=qLQVoS4Qkms (accessed on 28 February 2023).

Author Contributions: R.G.: conceptualization, formal analysis, methodology, investigation, validation, writing—original draft preparation, visualization; A.F. and F.C.: data curation, formal analyses, methodology, software, writing—review and editing; J.A.: investigation, methodology, validation, writing—original draft preparation; C.P.d.O.: methodology, software, writing—original draft preparation; C.B. project administration, conceptualization, supervision, resources, writing—review and editing. All authors have read and agreed to the published version of the manuscript.

Funding: This research was funded by SESAR Joint Undertaking under the European Union's Horizon 2020 research and innovation programme under grant agreement No. 874464.

Data Availability Statement: Data are not available for public consultation following Spanish law for Air Traffic Regulations, BOE-A-2018-15406 and General telecommunications regarding controller–pilot communication exchange.

Acknowledgments: To Nadal Ceñal, Julian Chaves and Mhamed Fillal from Indra that contributed to the prototype development, integration, supported the exercises and reviewed this article.

Conflicts of Interest: The authors declare no conflict of interest.

References

1. ICAO. *Procedures for Air Navigation Services (PANS)—Air Traffic Management Doc 4444*, 16th ed.; ICAO: Montreal, QC, Canada, 2016.
2. ICAO. *Annex 11—Air Traffic Services*, 15th ed.; ICAO: Montreal, QC, Canada, 2018; Para 3.7.3.1.
3. Helmke, H.; Rataj, J.; Mühlhausen, T.; Ohneiser, O.; Ehr, H.; Kleinert, M.; Oualil, Y.; Schulder, M.; Klakow, D. Assistant-based speech recognition for ATM applications. In Proceedings of the 11th USA/Europe Air Traffic Management Research and Development Seminar (ATM 2015), Lisbon, Portugal, 23–26 June 2015.
4. Nguyen, V.N.; Holone, H. Possibilities, challenges and the state of the art of automatic speech recognition in air traffic control. *Int. J. Comput. Inf. Eng.* **2015**, *9*, 1940–1949.
5. Pellegrini, T.; Farinas, J.; Delpech, E.; Lancelot, F. The Airbus Air Traffic Control Speech Recognition 2018 Challenge: Towards ATC Automatic Transcription and Call Sign Detection. *Interspeech* **2019**, *2019*, 2993–2997.
6. ICAO. *Annex 10. Aeronautical Telecommunications*, 6th ed.; ICAO: Montreal, QC, Canada, 2001; Para 5.21.2.
7. Apesteguia, E. Empleo real de las mujeres en el sector aeronáutico. *Fly News N° 4* **2017**, 80–85.
8. Delpech, E.; Laignelet, M.; Pimm, C.; Raynal, C.; Trzos, M.; Arnold, A.; Pronto, D. A Real-life, French-accented Corpus of Air Traffic Control Communications. In Proceedings of the Eleventh International Conference on Language Resources and Evaluation (LREC 2018), Miyazaki, Japan, 7–12 May 2018.
9. Helmke, H.; Slotty, M.; Poiger, M.; Herrer, D.F.; Ohneiser, O.; Vink, N.; Cerna, A.; Hartikainen, P.; Josefsson, B.; Langr, D.; et al. Ontology for transcription of ATC speech commands of SESAR 2020 solution PJ.16-04. In Proceedings of the IEEE/AIAA 37th Digital Avionics Systems Conference (DASC), London, UK, 23–27 September 2018.
10. Segura, J.C.; Ehrette, T.; Potamianos, A.; Fohr, D.; Illina, I.; Breton, P.A.; Clot, V.; Gemello, R.; Matassoni, M.; Maragos, P. *The HIWIRE Database, A Noisy and Non-Native English Speech Corpus for Cockpit Communication*; 2007; Available online: https://catalogue.elra.info/en-us/repository/browse/ELRA-S0293 (accessed on 28 February 2023).
11. Goyal, A.; Gupta, V.; Kumar, M. Recent named entity recognition and classification techniques: A systematic review. *Comput. Sci. Rev.* **2018**, *29*, 21–43. [CrossRef]
12. Altinel, B.; Can Ganiz, M. Semantic text classification: A survey of past and recent advances. *Inf. Process. Manag.* **2018**, *54*, 1129–1153. [CrossRef]
13. Nigmatulina, I.; Braun, R.A.; Zuluaga-Gomez, J.; Motlicek, P. Improving callsign recognition with air-surveillance data in air-traffic communication. In Proceedings of the Interspeech 2021 Satellite Workshop, Brno, Czech Republic, 30 August–3 September 2021.
14. Cordero, J.M.; Dorado, M.; de Pablo, J.M. Automated speech recognition in ATC environment. In Proceedings of the 2nd International Conference on Application and Theory of Automation in Command and Control Systems (ATACCS'12), London, UK, 29–31 May 2012; IRIT Press: Toulouse, France; pp. 46–53.
15. SESAR. *SESAR 2020 D3.2.020 PJ.16-04 TRL4 TVALR—Automatic Speech Recognition. V02.00.00*; SESAR: Brussels, Belgium, 2019.
16. Eurocontrol. *European Operational Concept Validation Methodology, E-OCVM Volume I—Version 3.0*; Eurocontrol: Brussels, Belgium, 2010.
17. Madson, M. Air Traffic Controllers and Real-time Simulation: A Powerful Combination. In *Journal of Air Traffic Control*; ATCA: Alexandría, VA, USA, 2004; pp. 24–27.
18. Publications Office. Single European Sky ATM Research 3 Joint Undertaking, Digital European Sky: Strategic Research and Innovation Agenda. 2020. Available online: https://data.europa.eu/doi/10.2829/117092 (accessed on 28 February 2023).
19. ENAIRE. *Plan Estratégico de Enaire 2021–2025. Plan de Vuelo 2025*; ENAIRE: Madrid, Spain, 2021.
20. SESAR. *SESAR PJ.10-W2-96 ASR Initial Technical Specification. V00.01.00*; SESAR: Brussels, Belgium, 2020.
21. EUROCAE. *EUROCAE ED-137 Interoperability Standards for VOIP ATM Components*; EUROCAE: Saint-Denis, France, 2012.
22. Fischer, V.; Ghahabi, O.; Kunzmann, S. Recent improvements to neural network based acoustic modelling in the EML real-time transcription platform. In Proceedings of the ESSV, Ulm, Germany, 7–9 March 2018; pp. 38–45.
23. Ghahabi, O.; Zhou, W.; Fischer, V. A robust voice activity detection for real-time automatic speech recognition. In Proceedings of the ESSV, Ulm, Germany, 7–9 March 2018; pp. 85–91.
24. Lin, Y. Spoken Instruction Understanding in Air Traffic Control: Challenge, Technique and Application. *Aerospace* **2021**, *8*, 65. [CrossRef]
25. Cordero, J.M.; Rodríguez, N.; De Pablo, J.M.; Dorado, M. Automated speech recognition in controllers communications applied to workload measurement. In Proceedings of the 3rd SESAR Innovation Days, Stockholm, Sweden, 26–28 November 2013.
26. Raja, J.B.; Cordero, J.M.; De Pablo, J.M. Reconocimiento y síntesis de voz en sistemas automatizados de control de tráfico aéreo. In Proceedings of the VIII Congreso Ingeniería del Transporte, A Coruña, Spain, 2–4 July 2008.
27. Insignia. Available online: https://insignia.enaire.es (accessed on 28 February 2023).
28. Arnegard, R.; Comstock, J., Jr.; Raimond, J.R. *The Multi-Attribute Task Battery for Human Operator Workload and Strategic Behaviour Research*; Nasa Technical Memorandum 104174; NASA: Washington, DC, USA, 1992.

29. Dehn, D.M. Assessing the Impact of Automation on the Air Traffic Controller: The SHAPE Questionnaires. *Air Traffic Control. Q.* **2008**, *16*, 127–146. [CrossRef]
30. Ohneiser, O.; Sarfjoo, S.; Helmke, H.; Shetty, S.; Motlicek, P.; Kleinert, M.; Her, H.; Murauskas, S. Robust Command Recognition for Lithuanian Air Traffic Controller Tower Utterances. In Proceedings of the Interspeech 2021, Brno, Czech Republic, 30 August–3 September, 2021; pp. 3291–3295.
31. Helmke, H.; Shetty, S.; Kleinert, M.; Ohneiser, O.; Ehr, H.; Prasad, A.; Motlicek, P.; Cerna, A.; Windisch, C. Measuring Speech Recognition And Understanding Performance in Air Traffic Control Domain Beyond Word Error Rates. In Proceedings of the 11th SESAR Innovation Days, Online Conference, 7–9 December 2021.
32. Efthymiou, M.; Usher, D.; O'connell, J.F.; Warnock-Smith, D.; Conyngham, G. The factors influencing entry level airline pilot retention: An empirical study of Ryanair. *J. Air Transp. Manag.* **2021**, *91*, 101997. [CrossRef]
33. Shade, W.; Reynolds, D. A comparison of speaker Clustering and Speech Recognition Techniques for Air Situational Awareness. In Proceedings of the Interspeech 2007, Antwerp, Belgium, 27–31 August 2007; pp. 2421–2424.
34. Ohneiser, O.; Balogh, G.; Rinaldi, W.; Murauskas, S.; Kis-Pál, G.; Usanovic, H.; Helmke, H.; Shetty, S.; Kleinert, M.; Ehr, H.; et al. Understanding tower controller communication for support in Air Traffic Control displays. In Proceedings of the 12th SESAR Innovation Days, Budapest, Hungary, 5–8 December 2022.
35. Badrinath, S.; Balakrishnan, H. Automatic Speech Recognition for Air Traffic Control Communications. *Transp. Res. Rec. J. Transp. Res. Board* **2021**, *2676*, 798–810.
36. Helmke, H.; Ondrej, K.; Shetty, S.; Arilíusson, H.; Simiganoschi, T.; Kleinert, M.; Ohneiser, O.; Her, H.; Zuluaga-Gomez, J.; Smrz, P. Readback error detection by automatic speech recognition and understanding. In Proceedings of the 12th SESAR Innovation Days, Budapest, Hungary, 5–8 December 2022.

aerospace

MDPI

Article

Toward Effective Aircraft Call Sign Detection Using Fuzzy String-Matching between ASR and ADS-B Data

Mohammed Saïd Kasttet [1,*], Abdelouahid Lyhyaoui [1], Douae Zbakh [1], Adil Aramja [1] and Abderazzek Kachkari [2]

1 Laboratory of Innovative Technologies (LTI), National Schools of Applied Sciences of Tangier (ENSAT), Tangier 90063, Morocco; lyhyaoui@ensat.ac.ma (A.L.); douae.zbakh@gmail.com (D.Z.); adil.aramja@gmail.com (A.A.)

2 The Moroccan Airports Authority (ONDA) Tangier-Ibn Battouta Intl. Airport, Tangier 90032, Morocco; kachkari@gmail.com

* Correspondence: mohammedsaid.kasttet@etu.uae.ac.ma

Abstract: Recently, artificial intelligence and data science have witnessed dramatic progress and rapid growth, especially Automatic Speech Recognition (ASR) technology based on Hidden Markov Models (HMMs) and Deep Neural Networks (DNNs). Consequently, new end-to-end Recurrent Neural Network (RNN) toolkits were developed with higher speed and accuracy that can often achieve a Word Error Rate (WER) below 10%. These toolkits can nowadays be deployed, for instance, within aircraft cockpits and Air Traffic Control (ATC) systems in order to identify aircraft and display recognized voice messages related to flight data, especially for airports not equipped with radar. Hence, the performance of air traffic controllers and pilots can ultimately be improved by reducing workload and stress and enforcing safety standards. Our experiment conducted at Tangier's International Airport ATC aimed to build an ASR model that is able to recognize aircraft call signs in a fast and accurate way. The acoustic and linguistic models were trained on the Ibn Battouta Speech Corpus (IBSC), resulting in an unprecedented speech dataset with approved transcription that includes real weather aerodrome observation data and flight information with a call sign captured by an ADS-B receiver. All of these data were synchronized with voice recordings in a structured format. We calculated the WER to evaluate the model's accuracy and compared different methods of dataset training for model building and adaptation. Despite the high interference in the VHF radio communication channel and fast-speaking conditions that increased the WER level to 20%, our standalone and low-cost ASR system with a trained RNN model, supported by the Deep Speech toolkit, was able to achieve call sign detection rate scores up to 96% in air traffic controller messages and 90% in pilot messages while displaying related flight information from ADS-B data using the Fuzzy string-matching algorithm.

Keywords: ATC; ASR; HMM; DNN; RNN; WER; VHF; ADS-B; METAR; GMTT; speech corpus; deep speech; call sign detection; levenshtein distance; fuzzy string matching

Citation: Kasttet, M.S.; Lyhyaoui, A.; Zbakh, D.; Aramja, A.; Kachkari, A. Toward Effective Aircraft Call Sign Detection Using Fuzzy String-Matching between ASR and ADS-B Data. *Aerospace* **2024**, *11*, 32. https://doi.org/10.3390/aerospace11010032

Academic Editors: Hartmut Helmke and Oliver Ohneiser

Received: 29 October 2023
Revised: 13 December 2023
Accepted: 15 December 2023
Published: 29 December 2023

1. Introduction

The purpose of Air Traffic Control (ATC) is to ensure the safe and efficient movement of aircraft within a specific controlled airspace. It helps prevent collisions between different aircraft and between aircraft and the surrounding obstacles, maintaining the order of air traffic and allowing quick support and collaboration in case an aircraft declares an emergency [1].

Air traffic controllers monitor the position of any aircraft assigned to their airspace and ensure aircraft separation and distancing using primary or secondary radars. The communication with pilots is ensured via Very High Frequency (VHF) radio equipment. Any change in the aircraft's heading or assigned flight level is subject to ATC approval, which ensures the appropriate horizontal and vertical separation between aircraft on the ground or in the controlled airspace is thoroughly respected.

L. Rabiner [2] defined ASR as "as a technology that involves the conversion of speech signals into a sequence of words by a computer program". Every ASR system should consider the type of speech recognizer, which can be speaker-dependent or speaker-independent. The first type requires prior training for each user to create voice patterns for hypothesis comparison. This kind of system is more accurate and has better performance. It can be designed for voice command solutions with limited vocabulary pronounced in the flight cockpit. Our application, dedicated to Air Traffic Control Officers (ATCOs), aims to recognize the pilot's spoken message and display flight data captured by the ADS-B receiver. It is a multi-user system or a speaker-independent recognizer, where the implementation is more complex considering the variety of accents and mispronunciations. It thus requires more hardware capabilities, such as memory and processor speed. In the given conditions, such systems could not achieve an accuracy lower than 10% word error rate (WER) [3].

In this paper, we introduce an ASR based on DNN, a new end-to-end RNN, and the Fuzzy string-matching algorithm to enhance ATC efficiency by reducing cognitive workload in dense traffic situations, especially in airports not equipped with radar. We use an Automatic Dependent Surveillance-Broadcast (ADS-B) receiver to provide captured flight data of all surrounding aircraft synchronized with recorded VHF voice communication. After training the ATC datasets and generating both acoustic and language models, the ASR system was able to recognize, with a reasonable WER, the spoken pilot message by matching it with the decoded call sign from ADS-B data. This threshold rate matching enables the call sign detection and display of flight-related information for ATCOs, such as speed, heading, altitude, distance, and bearing to the airport.

2. Related Work

D. Becks [4] briefly reviewed the state of the art of automatic speech recognition systems with types and modes of operation. Additionally, Georgescu [5] provides a comparison study between ASR performance and hardware requirements.

Recently, the FAA's (Federal Aviation Administration) final report on ASR methodologies [6] concluded that transformers have had a significant impact on audio and NLP fields, and their innovative architecture has been successfully integrated into various algorithms [7].

Since 1980, considerable progress has been made in ASR and applied to the ATC domain. A good description of the state-of-the-art ASR systems and their application for ATC was provided by Van Nhan Nguyen [8].

2.1. ASR in ATC

Van Nhan Nguyen [8] described three ASR systems. The Hidden Markov Model (HMM) approach has been the most widely used technique for the last two decades. It is a simple and efficient solution with automatic training, but its main weakness lies in discarding information about time dependencies. A hybrid approach was introduced to overcome this weakness of HMM. This approach combines an Artificial Neural Network and a HMM. A recognition accuracy rate of 94.2% was achieved by Wroniszewska [9] using the K-Nearest Neighbor (KNN) classifier and Genetic Algorithms (GAs). Finally, an interesting approach was proposed by Beritilli [10] using Dynamic Time Warping (DTW) and Vector Quantization-Weighted Hit Rate (VQWHR), which is a robust solution for noisy environments such as ATC.

Although the hybrid approach combines different algorithms and techniques, challenges in ASR systems still exist. To address issues such as poor signal quality from VHF communication, ambiguity in commands and instruction values, or the use of non-standard phraseology and mispronunciation in different accents (native and non-native speakers) [11], a new approach based on utilizing contextual information is introduced to improve the performance and accuracy of ASR in ATC as a post-processing approach based on syntactic, semantic and pragmatic analysis.

2.2. Contextual Knowledge in ATC

Syntactic and semantic analyses [12,13] consist of parsing the result of recognized words from ASR systems and eliminating invalid sentences or words by respecting grammatical rules highly inspired by ICAO standard phraseology. It helps correct misrecognized out-of-vocabulary words with similar ones from valid words of the ATC vocabulary. Semantic analysis is the process of testing the meaning of sentences. It can help resolve ambiguity and recognize words despite background noises [14].

2.3. Call Sign Detection (CSD)

The ability of an ASR system to detect accurate call signs in ATC communication is measured by the CSD rate. In 2018, in collaboration with IRIT (Institute for Research in Informatics of Toulouse) and Safety Data-CFH, Airbus organized a challenge for 22 teams for automatic speech recognition in ATC and call sign detection [15]. The Airbus dataset consisted of 40 h of manually transcribed voice communication with various accents and a high speech rate over noisy radio channels. The best result achieved was a 7.62% WER and 82.44% CSD rate, scored by the VOCAPIA-LMSI team. In 2020, the ATCO2 project [16] added an NLP module to extract the call sign from a recognized spoken utterance matched with surveillance data (ADS-B and radar) and improved the WER from 33% to 30%. The results showcased in [17,18] reported up to 60.4% relative improvement in call sign recognition by boosting call sign n-grams with the combination of ASR and NLP methods to use surveillance data. Finally, by leveraging surveillance information, Blatt, A et al. [14] significantly improved the accuracy of call-sign recognition in noisy air traffic control environments. The model showed a 20% improvement compared to existing methods. The study by Shetty et al. 2022 [19] focused on command extraction, including the recognition of call signs as part of the semantic meanings of ATCo utterances. Their study emphasized the importance of correctly interpreting various command components, showing that call sign recognition can be achieved within 20 ms after full call sign has been uttered, making it feasible for live data use. The research used gold transcriptions to achieve call sign recognition rates above 95% and error rates below 2.5%. With automatic transcriptions, they obtained recognition rates between 92 and 98% and error rates below 5% for most datasets. Finally, Garcia et al. 2023 [20] focus on how ASR can assist air traffic controllers (ATCos) and flight crews (FCs) in their communication. It describes a project under the SESAR2020 solution for ASR in call sign recognition, which was a collaboration between Enaire, Indra, CRIDA, and EML Speech Technology GmbH. The ASR highlights call signs on the ATCo screen to improve situational awareness and safety. The recognition rates for this system were around 84–87% for controllers and 49–67% for flight crews.

3. Automatic Speech Recognition Pipelines

3.1. Conventional Automatic Speech Recognition

Automatic Speech Recognition (ASR) is the assignment of transducing raw audio signals of spoken language into text transcriptions. It is based on statistical pattern-matching using a combination of acoustic and language models, which depends on the complexity of the application. This discussion covers the history of ASR models, from Gaussian Mixtures (GMMs) and Hidden Markov Models (HMMs) to attention-augmented DNNs. The ASR architecture is represented in Figure 1.

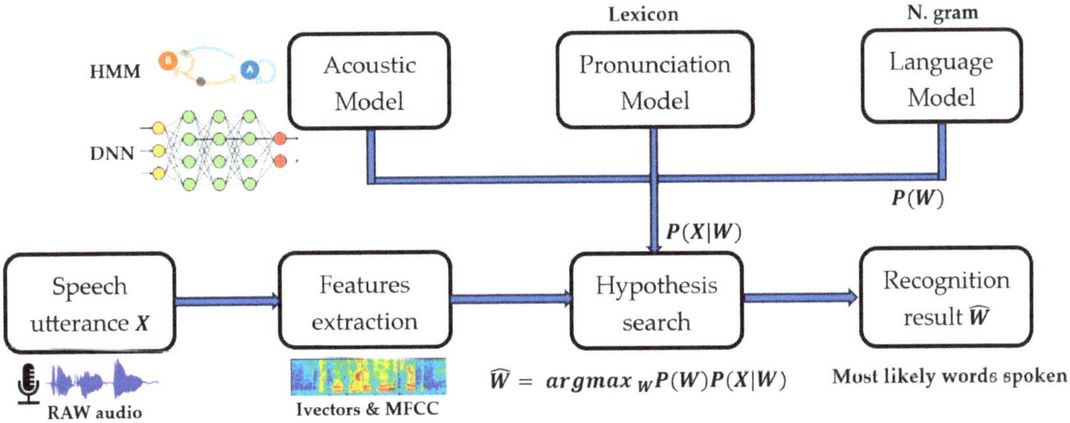

Figure 1. ASR architecture.

3.1.1. Acoustic Model

With reduced vocabulary, the acoustic model converts pronounced words into phonemes as minimal digital units. The speech processor compares the latter with stored word patterns until it matches the spoken utterance. However, in a complex situation, as in connected or continuous speech recognition, the analog voice signal is converted to digital format, typically using a 16 kHz sampling frequency. For feature extraction, the digital signal is transformed into the frequency domain using the Fast Fourier Transform (FFT). Subsequently, standard techniques [21], such as Linear Predictive Coding (LPC) and Mel Frequency Cepstral Coefficients (MFCCs), are applied. The feature numbers are determined by comparing the resulting frequency graph with stored known sounds, which allows the referencing of each phoneme found.

However, in circumstances involving a speaker with a specific accent and the noisy environment of flight cockpits and radio communications, those feature numbers cannot identify a unique sound to become a particular phoneme. The solution is to use probability techniques such as Hidden Markov Models (HMMs) that represent each phoneme and use feature numbers' probabilities to calculate the transition state's likelihood (high probability).

Recently, many techniques [22] based on neural networks (NNs) have been deployed to replace the GMM and HMM by combining recurrent and convolutional neural networks to predict states efficiently [21].

3.1.2. Language Model

The English language contains 44 phonemes; every word is a sequence of phonemes with a large number of phonetic spelling possibilities. To overcome this problem, we generated a pronunciation dictionary of 907 unique words vocabulary from all ATC datasets, known as a lexicon. All probable words delivered by the acoustic model are compared in a second N-gram model [23] or an NN called a language model [24], which can predict the next word from a set of preceding words by following standard grammatical rules. Finally, a search engine combining all models can decode and continually recognize the most likely word sequence.

The aim of the speech recognizer engine is to find the most probable word $\hat{W}$ given an acoustic signal X as input.

$$\hat{W} = argmax\ _W P(W|X) \tag{1}$$

$P(W \mid X)$ is the probability that the word W was uttered, knowing that the evidence X was observed.

Equation (1) can be rewritten using Bayes' law, as shown in Equation (2):

$$P(W|X) = \frac{P(X|W) \cdot P(W)}{P(X)} \tag{2}$$

$P(W)$ is the probability that the word W will be uttered, $P(X \mid W)$ is the probability that the acoustic evidence X will be observed when the speaker speaks the word W, and $P(X)$ is the probability that X will be observed.

So, $P(X)$ can be ignored as $P(X)$ is not dependent on the selected word string. Consequently, Equation (1) can be written as Equation (3):

$$\hat{W} = argmax\ _W P(W)P(X|W) \tag{3}$$

where $P(W)$ is determined by the language model, and $P(X \mid W)$ is determined by the acoustic model.

3.2. End-to-End Speech Speech Recognition

For optimization purposes and simplification of the training process of different models, new end-to-end models are deployed for ASR [25]. It typically uses a type of neural network called deep neural network (DNN) or recurrent neural network (RNN) architecture. It is trained on large amounts of audio data with corresponding transcriptions. It has been proven effective at transcription in many cases, especially in a noisy environment, and can potentially simplify the ASR pipeline. The end-to-end model can directly decode a feature-extracted X from spoken utterance to a sequence of words Y+ by integrating the acoustic and the language model in one process, as shown in Figure 2; it is most often used in a reduced and noisy dataset. Moreover, there is a possibility of including an optional language model called a scorer to perform the best results.

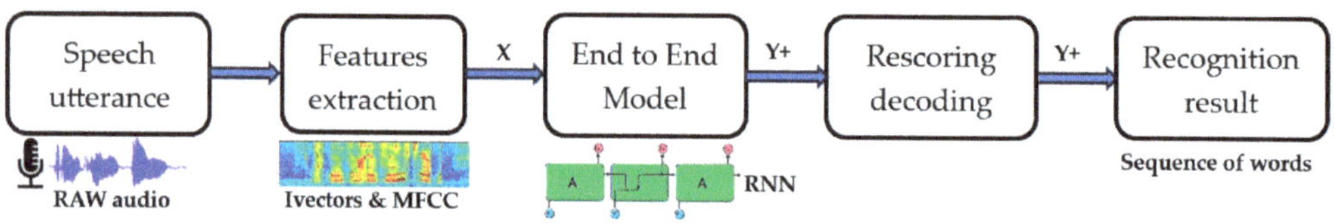

Figure 2. End-to-End ASR.

We notice that Connectionist Temporal Classification (CTC) [26] is the most popular training approach.

4. ATC Speech and Contextual Data Specification

4.1. ATC Communication

The standard communication, known as International Civil Aviation Organization (ICAO) Standard Phraseology, specifies all exchanged messages in Radio Telephony Communication (RTF) between air traffic controllers and pilots in controlled airspace, as well as in face-to-face communication between pilots and aerodrome staff in addition to the communication between pilots in the cockpit [27]. Primarily based on English or the national language, the pronunciation will be distinct between native and non-native English speakers. In some high-traffic situations, ATCOs must speak quickly to provide information and instructions for all aircraft in their allocated airspace [28]. Consequently, any recognizer system will return some broken or missing words due to the high speech rate and noisy radio signals from VHF transmission [29]. However, in some cases, it is possible to compensate for incorrect words by using the standard phraseology, as shown in Table 1, in addition to the structured contextual data such as a Meteorological Airport Report (METAR) and ADS-B flight information.

Table 1. Example of ICAO phraseology.

	Message			
	Tower	**Aircraft (Call Sign)**	**Information**	**Request Instruction**
Pilot	Tangier Approach	ARABIA six four niner	Descending flight level seven zero	Request visual approach runway 10
	Aircraft (call sign)	**Tower**	**Information**	**Instruction**
ATCo	ARABIA six four niner	Tangier Approach	Negative, last wind two seven zero degrees 25 knots	Report established for ILS approach runway 28

Table 1 shows the structured and precise nature of aviation communication exchange between the pilot and ATCO. Initially, the pilot, communicating with the Tangier Approach, identifies their aircraft as ARABIA six four niner and informs the tower that they are descending to flight level seventy FL70. The pilot then requests permission for a visual approach to runway 10. This request is part of standard aviation protocol, where pilots provide their current status and express their intended maneuvers. In response, the ATCO addresses the aircraft with the call sign ARABIA six four niner, indicating that the request is denied, possibly due to wind conditions, which are reported as two seven zero degrees at 25 knots. Instead of the requested visual approach, the ATCO instructs the pilot to prepare for an Instrument Landing System (ILS) approach for runway 28 and to report back once established on this approach. This exchange highlights the dynamic and responsive nature of air traffic communications, where ATCOs provide critical instructions and adjustments based on real-time conditions and operational requirements, ensuring the safety and efficiency of aircraft operations. The dialogue reflects the essential characteristics of air traffic communication: clarity, conciseness, and the conveyance of necessary information for the safe conduct of flights.

4.2. ADS-B Data and Call Sign

Automatic Dependent Surveillance Broadcast (ADS-B) is a technology for monitoring aircraft via satellite information. It improves the efficiency and safety of aircraft on the ground as well as in the air. It contains the flight call sign decoded in 3 letters and numbers for commercial flight, or equal to the aircraft registration number for private and general flights as shown in Table 2, speed, altitude, vertical speed, heading, and GPS latitude and longitude, as shown in Table 3. It is becoming the preferred method of real-time surveillance for ATC. Because of its reduced cost and valuable information on the call sign code, it is well suited for our application concept of ASR systems as the primary key for pilot message identification.

Table 2. Call Sign annotation.

Call Sign Annotation	Designator	Transcription
RAM982	RAM	royal air maroc niner eight two/air maroc niner height two
MAC146T	MAC	arabia maroc one four six tango/arabia one four six tango
CNTAV		charlie november tango alfa victor/charlie alfa victor

In Table 2, the provided call sign annotation data showcase the intricate and standardized method of communication in air traffic control, particularly in articulating aircraft call signs. For instance, the call sign RAM982 is designated as "RAM", and its transcription unfolds as "Royal Air Maroc Niner Eight Two". This transcription method, where numbers are spoken phonetically, is crucial for clarity, particularly in initial radio communication, where precision is paramount. Later, the call sign transcription can be reduced for more straightforward pronunciation. Similarly, MAC146T, designated "MAC", is transcribed as "Arabia Maroc One Four Six Tango". Each number and the letter 'T' (Tango) are pronounced individually, denoting a specific flight or route. The third example, CNTAV, despite lacking a clear designator, is transcribed using the phonetic alphabet as "Charlie November Tango Alpha Victor". Each letter is articulated using a corresponding word from the phonetic alphabet, ensuring each character is unmistakably understood in potentially noisy or disrupted communication environments. These examples highlight the critical importance of standardized and clear communication in aviation, especially in identifying aircraft, where even minor miscommunications can have significant implications for air traffic safety and efficiency.

Table 3. Example of ADS-B data.

Date	Time	Call Sign	Radar	Alt	Speed	Head	Vertical	Lat	Lon
6 July 2021	08:42:51	RAM982	5320	9000 ft	580 kt	320°	80 ft/min	35.46	−7.48

4.3. METAR

METAR is a weather observation report for an aerodrome and is periodically generated every 30 min. It contains wind direction and speed data, temperature, dew point, cloud cover and heights, visibility, and barometric pressure. Aircraft pilots and controllers primarily use it to determine runway-in-use and flight rules during takeoffs or landing operations.

Table 4: Example of METAR report shows an example of a weather report of Tangier Aerodrome made on 10/06/2021 at 10:30 UTC. The conditions were 15 kt wind from the west with gusts up to 30 kt, temperature of 14 °C, 84% humidity, a pressure of 1012 hPa, visibility of 7000 m, and few clouds at a height of 3000 ft. No significant changes occurred in the next two hours.

Table 4. Example of METAR report.

Aero-Drome	Day/Time	Wind Direction/Speed	Visibility	Clouds	Temp/Dew	Pressure	
GMTT	101,330 Z	27015G30KT	7000	FEW020	14/12	Q1012	NOSIG

5. Methodology and Materials

Our methodology using ASR in the specific domain of ATC involves several steps; the process begins with dataset collection for training and recording new actual speech corpus IBSC for testing; this includes communication between pilots and ATCOs, such as those with ground control and tower control, under different conditions, including varying levels of clarity, background noise, and accents. Once collected, the audio data need to be preprocessed. This stage involves cleaning the audio by reducing noise, normalizing audio levels, and segmenting it into smaller, manageable parts for easier processing. The next step is the accurate transcription of these audio files. This process is crucial and should include not only verbal communication but also annotations for non-verbal elements like flight and metrological information from ADS-B and METAR data, which is especially important in the context of ATC communications. The core of our research will involve training our chosen ASR model using the annotated data with two different toolkits based on DNN and RNN architecture. This process might require substantial computational resources and time. It is crucial to regularly validate and test the model with a separate dataset to ensure its accuracy. Special attention will be paid to how the model performs under various challenging conditions, like heavy accents, rapid speech, and noisy environments. In terms of evaluation and based on the results of our tests and validations, the model may need to be refined; this could involve adapting it with ATC data, tweaking the model parameters, or experimenting with different sets of features. Finally, the goal in the context of ATC is to achieve the highest possible accuracy and reliability, particularly under challenging conditions, due to the critical nature of ATC communications.

For call sign detection, we will implement fuzzy string matching, which is particularly important in fields like automatic speech recognition using the Levenshtein algorithm. This method centers on calculating the number of edits-insertions, deletions, or substitutions needed to transform one string into another. This algorithm is readily available in many programming languages; Python offers libraries like Fuzzy Wuzzy for this purpose [30]. We set a matching threshold based on our accuracy needs—a lower threshold means more lenient matching, while a higher one requires a closer match. We applied the algorithm to our dataset, compared each string to our target string ASR hypothesis, and calculated the similarity score. The results were evaluated and adjusted to finetune the threshold parameter of the algorithm as necessary.

5.1. Data Collection: The Ibn Battouta Speech Corpus

The Ibn Battouta Speech Corpus is a synchronized dataset of voice communication between pilots and ATCOs with weather observation data originated from Tangier's airport and current activated aircraft flight information [31], which has a very rich pronunciation accents of native and nonnative speakers thanks to its vital geographic position linking different airspaces from Morocco (GMMM, GMTT), Spain (LEZL), and Gibraltar (LXGB). The purpose is to detect and record audio speech with various accents and related captured ADS-B data plus METAR report provided by the NWS Server [32], as described in Figure 3.

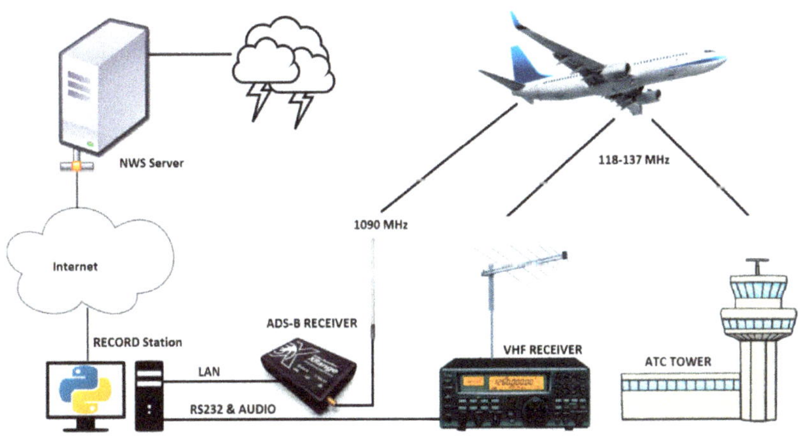

Figure 3. Ibn Battouta dataset architecture.

The voice recording was obtained with Voice Activity Detection (VAD) [33] at a rate of 16 kHz from the VHF receiver [34] tuned to the airport frequency connected to the workstation's audio input. At the same time, the ADS-B receiver provided by AirNav System [35] logs flight data, as shown in Figure 4, including the call sign code of activated aircraft with approximately 200 Nm circumference. In addition, a weather report is saved separately after downloading updated data from the US National Weather Service (NWS) Server.

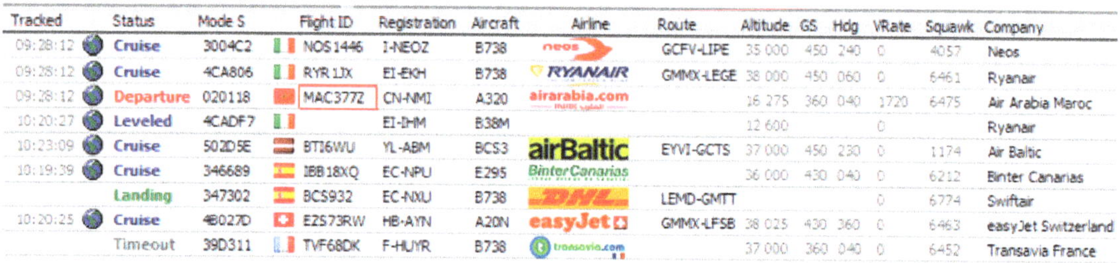

Tracked	Status	Mode S	Flight ID	Registration	Aircraft	Airline	Route	Altitude	GS	Hdg	VRate	Squawk	Company
09:28:12	Cruise	3004C2	NOS1446	I-NEOZ	B738	neos	GCFV-LIPE	35 000	450	240	0	4057	Neos
09:28:12	Cruise	4CA806	RYR1JX	EI-EKH	B738	RYANAIR	GMMX-LEGE	38 000	450	060	0	6461	Ryanair
09:28:12	Departure	020118	MAC377Z	CN-NMI	A320	airarabia.com		16 275	360	040	1720	6475	Air Arabia Maroc
10:20:27	Leveled	4CADF7		EI-IHM	B38M			12 600			0		Ryanair
10:23:09	Cruise	502D5E	BTI6WU	YL-ABM	BCS3	airBaltic	EYVI-GCTS	37 000	450	230	0	1174	Air Baltic
10:19:39	Cruise	346689	IBB18XQ	EC-NPU	E295	BinterCanarias		36 000	430	040	0	6212	Binter Canarias
	Landing	347302	BCS932	EC-NXU	B738	DHL	LEMD-GMTT				0	6774	Swiftair
10:20:25	Cruise	4B027D	EZS73RW	HB-AYN	A20N	easyJet	GMMX-LFSB	38 025	430	360	0	6463	easyJet Switzerland
	Timeout	39D311	TVF68DK	F-HUYR	B738	transavia.com		37 000	360	040	0	6452	Transavia France

Figure 4. ADS−B flight data.

5.2. Transcription and Logging

All utterances were transcribed and manually annotated by real pilots and ATCOs who authored this paper [36]. It is a time-intensive task that requires ten man-hours to transcribe one hour of speech. In order to estimate the distance and radial information, which are frequently requested by ATCOs from pilots, we integrated a Python 3 code-based program into the flight data from the ADS-B receiver using the Haversine formula [37] to calculate and save the distance and bearing between the aircraft GPS position and D-VOR installed on the Tangier airport runway. The call sign, date, and time are tagged in transcription files to enhance context-free grammar. Table 5 summarizes the dataset characteristics.

Table 5. The Ibn Battouta dataset characteristics.

	Speaker		Gender		Total
	Pilot	ATCO	Female	Male	
Number of utterances	992	1500	544	1948	2492
Duration (sec)	5416	10,040	3720	11,736	15,456
Number of words	12,936	22,224	8084	27,076	35,160
Signal Average (dB)	106	95	90	102	101
Aircraft Call Sign	832	1180	440	1572	2012

The Ibn Battouta dataset is a rich and complex collection of communications in the air traffic control context, encapsulating a wide array of spoken interactions between pilots and air traffic control officers (ATCOs). It comprises a total of 2492 utterances, divided between 992 from pilots and 1500 from ATCOs, indicating the more extensive communicative role of ATCOs in managing airspace. Notably, the dataset reveals a gender imbalance in communication, with female speakers contributing 544 utterances against 1948 from male speakers, highlighting the male predominance in this sector. The total duration of these communications is 15,456 s (4 h 20 min), with pilots accounting for 5416 s and ATCOs for a larger share of 10,040 s, reflecting the extensive and detailed nature of ATCO communications. Regarding word usage, the dataset records 35,160 words, with a significant portion (22,224 words) used by ATCOs, further emphasizing the complexity of their verbal exchanges. The signal strength, measured in decibels, averages 101 dB across the dataset, with a higher average for pilots (106 dB) compared to ATCOs (95 dB), possibly due to different communication environments or equipment. The dataset also includes a diverse array of 2012 aircraft call signs, with pilots using 832 and ATCOs 1180, adding to the complexity of speech recognition challenges in this domain. Overall, the Ibn Battouta dataset offers invaluable insights into linguistic characteristics and communicative dynamics in air traffic control.

5.3. Datasets of Training and Adapting Models

We collected the following available ATC datasets with different accents and environments (real operational and laboratory simulation) for model training phases, as summarized in Table 6.

Table 6. Dataset splitting for model training.

	Data Set	Accent	Environment	Utterance	Duration	Call Sign Annotation
Training (103 h) 46,732 utterances	LDC94S14A [38]	USA	Operational	25,120	60 h	No
	ZCU_CZ [39]	Czech	Operational	6435	15 h	No
	ATCOSIM [40]	FR/DE/CH	Simulation	8078	8 h	No
	HIWARE [41]	FR/GK/ES/IT	Simulation	7099	25 h	No
Validation (21 h) 10,024 utterance	Mixed/Unseen	Mixed	Mixed	10,024	21 h	No
Test (11 h) 5382 utterances	ATCO2 [42]	CZ/DE/CH/AU	Operational	2890	6 h	2817
	IBSC	MAR/ES/FR/EN	Operational	2492	5 h	2012

This dataset is specifically tailored for research in automatic speech recognition within the air traffic control sector, comprising a diverse range of accents, environments, and operational scenarios. It is segmented into three primary sections: training, validation, and testing, cumulatively spanning 135 h. The training set, with a substantial 103 h of audio, incorporates a wide array of utterances from the USA, Czech Republic, and a mix of countries like France, Germany, Switzerland, Greece, Spain, and Italy, covering both operational and simulation environments. The validation set offers a 21 h mixed compilation from unseen sources in varied environments. Lastly, the testing segment, totaling 11 h, includes specific datasets like ATCO2 and IBSC, representing a range of Morocco and Spain airspace in operational settings like En route and approach flight situations. This section is unique as it includes call sign annotation.

5.4. Vocabulary and Accuracy

A limited or medium ATC vocabulary size, estimated at around 500 words, and the standard phraseology of ATC grammar with its substantial semantic restrictions both allow better accuracy by increasing the probabilities of valid words and their sequences despite the noisy environment and high speech rate.

The word error rate (WER) is the standard metric for measuring the accuracy of any ASR system [43,44]. It is calculated using the formula given in Equation (4):

$$WER = \frac{I + D + S}{N} \tag{4}$$

where I is the number of insertions, D is the number of deletions, S is the number of substitutions, and N is the number of words in the sentence.

The Real-Time Factor (RTF) is included to measure the speed of ASR. It can be computed using the ratio expressed in Equation (5):

$$RTF = {}^{P}/_{I} \tag{5}$$

where P is the necessary time to process an input of duration I.

5.5. Fuzzy String-Matching

For call sign detection, we applied a string-matching algorithm called the Fuzzy, which determines the closeness of two strings. It is a technique used to identify two elements of text strings that match partially but not precisely. This algorithm is based on the

Levenshtein distance [45], a metric that evaluate the dissimilarity between two sequences of words. This measure calculates the least number of modifications required to transform one sequence of words into another.

Mathematically, the Levenshtein distance between two strings a, b is given by $lev_{a,b}(|a|,\ |b|)$ in Equation (6), where:

$$lev_{a,b}\ (i,j) = \begin{cases} \max(i,j)\ if\ \min(i,j) = 0, \\ min \begin{cases} lev_{a,b}\ (i-1,j)+1 \\ lev_{a,b}\ (i,j-1)+1 \\ lev_{a,b}\ (i-1,j-1)+1_{(a_i \neq b_j)} \end{cases} & otherwise \end{cases} \tag{6}$$

where $1_{(a_i \neq b_j)}$ is the indicator function equal to 0 when $a_i = b_j$, and equal to 1 otherwise.

A threshold value will be determined during the experimentation to assess the similarity ratio of the string matching between ADS-B Call Signs and Hypothesis transcription.

6. Experimentation

6.1. Overview

In our experiment employing the IBSC, we searched for the call sign in recognized messages in all ADS-B line data stored and synchronized with voice utterances. The call sign code in ADS-B data was parsed using an airline call sign designator database [46] from the International Air Transport Association (IATA) and phonetic transcription. The highest score of string matching allows the appropriate call sign to be identified.

We used the Pocketsphinx toolkit [47] with HMM-DNN topologies and the Deep Speech recognition toolkit [48] based on a Recurrent Neural Network (RNN) for training, adapting, and testing the IBSC dataset on an HP Z4 workstation equipped with an Nvidia GeForce RTX 2070 GPU for training acceleration, and Ubuntu 18.04 as the OS.

Precisely, we implemented the Mozilla Deep speech version 0.9.3; the RNN is fully connected and has bidirectional layers with 512 hidden units per layer. Initially, it contains three layers with clipped rectified-linear (ReLU) activation, a Long Short-Term Memory (LSTM) layer, followed by another layer with ReLU activation. Lastly, it is capped by a softmax classifier to predict the most likely alphabet letter at each point in an audio utterance.

6.2. Experimental Setup

First, we trained [49,50] two new acoustic models with five hidden layers using the Pocketsphinx toolkit and the Deep speech toolkit with TensorFlow. We then adapted [51,52] each toolkit's default English model with all ATC datasets, as indicated in Table 6. For audio data representation, we computed spectrograms of 80 linearly spaced log filter banks and an energy term. The filter banks were computed over 20 ms windows with strides of 10 ms each. The language model was a 3 g model with a 907 unique words vocabulary from all ATC datasets, which contained 55,338 utterances with a total duration of 128 h, and from the AirNav Systems database history, from which the call sign was extracted and decoded. To enhance and train this language model using the SRLIM toolkit [53], we added all air waypoints from Morocco, Spain, and Gibraltar's nearby airspaces and decoded meteorological reports, in addition to all existing commercial and private company call sign designator [54]. Finally, it took about 32 h to train each new acoustic model for 200 epochs.

For call sign detection, we used a pragmatic analysis based on ADS-B data, including flight information, to detect the call sign in a recognized pilot message that represents essential information for ATCOs to identify the aircraft; we implemented the fuzzy Wuzzy Python function using the token_set_ratio () method [55]. It returned the highest similarity ratio score for fuzzy string matching in all ADS-B data lines stored in the dataset for each recognized utterance.

6.3. Results and Discussion

After training, adapting, and following the assumptions given above, our ASR model was tested on a different corpus from Morocco, Spain, and Gibraltar airspaces to cover different accents, airspace information, and role speakers, as shown in Figure 5.

In an analysis of two prominent ASR models, PocketSphinx and Deep Speech, applied to a dataset of air traffic control communications, distinct trends emerge in their performance across various metrics. Both models were evaluated under three distinct training conditions: pretrained English, trained on ATC only, and adapted pretrained English + ATC on three key metrics:

- word error rate (WER) as defined in (4).
- fuzzy string-matching ratio given by the percentage of similarity between two strings.
- and call sign detection rate which gives the percentage of correct call signs detected among total transcribed utterances.

The baseline results of model training and adapting ATC data sets are shown in Table 7.

Table 7. Model training and adaptation.

Model		Word Error Rate			Fuzzy String-Matching Ratio			Call Sign Detection Rate		
		ATCO	PILOT	Both	ATCO	PILOT	Both	ATCO	PILOT	Both
Pocket sphinx HMM-DNN	Pretrained English	83%	91%	87%	17%	09%	13%	0%	0%	0%
	Trained on ATC only	14%	20%	17%	68%	60%	64%	94%	84%	89%
	Adapted Pretrained English + ATC	11%	13%	12%	78%	68%	73%	95%	87%	91%
Deep Speech RNN	Pretrained English	81%	89%	85%	27%	19%	23%	0%	0%	0%
	Trained on ATC only	10%	12%	11%	80%	66%	73%	93%	89%	91%
	Adapted Pretrained English + ATC	08%	10%	09%	85%	77%	81%	96%	90%	93%

For PocketSphinx, the pretrained English model trained on approximately 6500 h of data not related to the ATC condition showed in Table 7 high WERs (83% for ATCO, 91% for the pilot, and 87% overall) and low fuzzy string-matching ratios (17% for ATCO, 9% for the pilot, and 13% overall), along with a 0% call sign detection rate across all categories. However, when trained exclusively on ATC data, there was a substantial improvement in all metrics, with the call sign detection rate reaching as high as 94% for ATCO, 84% for the pilot, and 89% overall. The adaptation of pretrained English with ATC data further enhanced performance, reducing WERs to 11–13% and increasing the fuzzy string-matching ratio to around 70–78%.

Deep speech mirrored these trends but with consistently better outcomes. Under the pretrained English condition, it had slightly lower WERs and higher string-matching ratios than PocketSphinx but still had no call sign detection. Training on ATC data alone introduced significant enhancements, especially in call sign detection, reaching up to 93%. The adaptation of pretrained models with ATC data yielded the best results, with WERs dropping to as low as 8–10%, fuzzy string-matching ratios climbing to 81–85%, and call sign detection rates peaking at 93–96%.

Overall, these results clearly demonstrate that both PocketSphinx and Deep Speech significantly improve accuracy and reliability when trained on ATC-specific data, with Deep Speech showing slightly superior performance in all tested scenarios.

The results show that, in general, for low-resource data, adapting the pretrained default English model offers better performance [56] than training a new model, and using the RNN Deep Speech toolkit achieved better results in noisy environments, especially in-flight cockpit pilot transmission. Because the ATC dataset has a short duration, the model does not need to increase the depth parameter, as it may lead to overfitting. For call sign detection, the similarity ratio of the fuzzy string matching was improved when the WER was low. This means the better the message recognition accuracy, the better the fuzzy string-matching score between the decoding call sign in ADS-B data and the recognized message.

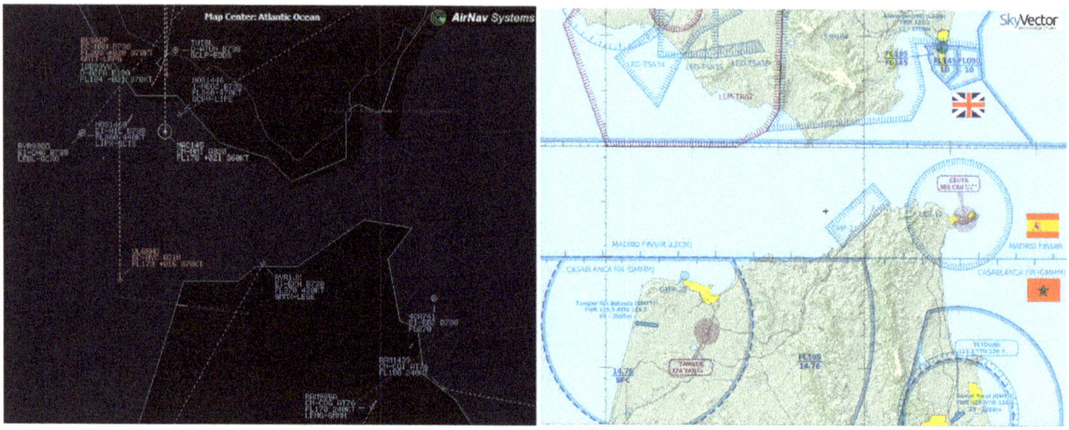

Figure 5. Test flight information region.

Table 8 details an example of a voice message by a Royal Air Maroc company pilot during the approach. The ASR hypothesis was confirmed with a WER of 12.5% by deleting the unknown word "um".

Table 8. ASR hypothesis example.

File Name	12120_20200319_170603_170606.trs
Tracking Date Time	19 March 2020 17:06:03
Real transcription	Tangier AIR MAROC zero seven four roger um continue approach
ASR hypothesis	Tangier AIR MAROC zero seven four roger continue approach
Word Error Rate WER	12.5%

By employing the data captured by the ADS-B receiver, we can assess the similarity between each detected call sign shown in Table 9; compared to the result of the ASR hypothesis from Table 8, the fuzzy string-matching score was calculated for each candidate. The model returned the call sign leading to the highest score, i.e., 89%. It corresponds to the RAM074 flight phonetic transcription (Tangier AIR MAROC zero seven four roger continue approach). A threshold of 80% is fixed to avoid the situation when all call signs have the same designator or do not concern the ASR hypothesis.

Table 9. Call sign candidates from ADS-B data.

Call Sign	Phonetic Transcription	Score	Type	Altitude	Speed	DME	Radial
BEL271	Beeline two seven one	45%	A333	35,000	470	506	225°
RAM075	Royal Air Maroc zero seven five	79%	B738	41,000	430	780	310°
BAW669	Speed bird six six niner	34%	A21N	36,000	460	380	198°
RAM074	Royal Air Maroc zero seven four	89%	B738	3325	210	20	98°
RYR8073	Ryanair eight zero seven tree	51%	B738	30,375	390	240	254°

6.4. Limitations

For private and general flights, the aircraft registration number is used as call sign instead of flight number in ADS-B data. We can apply the same Fuzzy algorithm based to search and match the phonetic transcription of aircraft registration number with the ASR hypothesis, since it is mandatory to exist in every ADS-B information and pronounced in every standard communication.

For light aircraft not equipped with ADS-B transmitters e.g., in VFR flight, we can only rely on ASR performance to detect the registration number based on phonetic transcription.

7. Conclusions and Further Work

Although the application of ASR in ATC is more challenging due to the high-security level required for air traffic management in the aviation domain, it remains possible to benefit from standard communication, a small vocabulary, and contextual information to implement simple and low-cost ASR solutions using ADS-B data to minimize the workload of ATCOs in high-traffic situations located in an airport not equipped with radar.

In our experiment, after training and adapting the ATC dataset using the Deep Speech toolkit and building the acoustic and language models based on a vocabulary dataset, we were able to demonstrate the successful detection of multiple aircraft call signs in recognized voice messages at a string-matching similarity rate starting from 60%. For safety obligations, we recommend a threshold of 80% for the fuzzy string-matching rate.

Further work in the ATC domain can present us a chance to try Whisper, the new advanced ASR system developed by OpenAI trained on 680,000 h of multilingual and multitask supervised data collected from the web, known for its high accuracy in transcribing speech, even in challenging conditions such as noisy environments or with speakers having different accents. It supports multiple languages, making it versatile for global applications. Whisper is designed to understand the conversation's context, which helps provide more accurate transcriptions.

An NLP module investigates the string position between the call sign and airport entity name; in addition, grammar rules such as gerund and key verbs like "request" and "report" in a recognized transmission would allow the detection of the speaker's role during standard ATCO and pilot communication

Airport meteorological information and the runway are usually delivered to the pilot before takeoff and landing. Decoding the METAR report, extracting the wind direction, and calculating the runway in use will help confirm the acknowledgment between the pilot and ATCO communication.

Author Contributions: Conceptualization; M.S.K.; Methodology; A.L.; Software; A.A.; Validation; D.Z.; Formal Analysis; A.L.; Investigation; A.L.; Resources; M.S.K.; Data Curation; A.K.; Writing—Original Draft Preparation; M.S.K. and A.L.; Writing—Review & Editing: A.L., A.A., A.K. and D.Z.; Visualization; Supervision; Project Administration; Funding Acquisition: M.S.K. and A.L. All authors have read and agreed to the published version of the manuscript.

Funding: This research was funded by ULTRA CONTROL Technical Center and DETROIT TELECOM Company under agreement number DTUC-1923-ACRT.

Informed Consent Statement: Informed consent was obtained from all subjects involved in the study.

Data Availability Statement: Restrictions apply to the availability of these data. Data were obtained from the European Language Resources Association (ELRA) and Aero Club Royal of Tangier; data are available under request.

Acknowledgments: The authors are grateful for the cooperation of pilots flying in Aero Club Royal of Tangier and air traffic controllers working at Tangier's Airport for their valuable help and guidance. The authors thank Khalid Choukri for providing the Datasets resources and Captain Nourddine Mountassir for his expertise and assistance in flight procedures. We would also like to thank the research team members in the Laboratory of Innovative Technologies (LTI) based at the National School of Applied Sciences of Tangier.

Conflicts of Interest: The authors declare that this study received funding from ULTRA CONTROL Technical Center and DETROIT TELECOM Company under agreement number DTUC-1923-ACRT. The funder was not involved in the study design, collection, analysis, interpretation of data, the writing of this article or the decision to submit it for publication.

References

1. Emergency Response Guidance for Aircraft Incidents Involving Dangerous Goods 2023–2024 (Doc 9481). Available online: https://store.icao.int/en/emergency-response-guidance-for-aircraft-incidents-involving-dangerous-goods-doc-9481 (accessed on 29 November 2023).
2. Rabiner, L.R.; Juang, B.H. *Fundamentals of Speech Recognition*; PTR Prentice Hall: Upper Saddle River, NJ, USA, 1993; ISBN 978-0-13-015157-5.
3. 3Play Media Study Finds Artificial Intelligence Innovation Has Led to Significant Improvements in Automatic Speech Recognition (ASR). Available online: https://www.businesswire.com/news/home/20230503005160/en/3Play-Media-Study-Finds-Artificial-Intelligence-Innovation-Has-Led-to-Significant-Improvements-in-Automatic-Speech-Recognition-ASR (accessed on 30 November 2023).
4. Beeks, D.W. Speech Recognition and Synthesis. In *Digital Avionics Handbook*; CRC Press: Boca Raton, FL, USA, 2015; ISBN 978-1-315-21698-0.
5. Georgescu, A.-L.; Pappalardo, A.; Cucu, H.; Blott, M. Performance vs. Hardware Requirements in State-of-the-Art Automatic Speech Recognition. *EURASIP J. Audio Speech Music Process.* **2021**, *2021*, 28. [CrossRef]
6. Achour, G.; Salunke, O.; Payan, A.P.; Harrison, E.; Sahbani, C.; Carannante, G.; Ditzler, G.; Bouaynaya, N.; Georgia Institute of Technology; Aerospace Systems Design Laboratory; et al. *Review of Automatic Speech Recognition Methodologies*; Federal Aviation Administration; William J. Hughes Technical Center: Egg Harbor Township, NJ, USA, 2023.
7. Wang, Y.; Mohamed, A.; Le, D.; Liu, C.; Xiao, A.; Mahadeokar, J.; Huang, H.; Tjandra, A.; Zhang, X.; Zhang, F.; et al. Transformer-Based Acoustic Modeling for Hybrid Speech Recognition. In Proceedings of the ICASSP 2020—2020 IEEE International Conference on Acoustics, Speech and Signal Processing (ICASSP), Barcelona, Spain, 4–8 May 2020; IEEE: Barcelona, Spain, 2020; pp. 6874–6878.
8. Nguyen, V.N.; Holone, H. Possibilities, Challenges and the State of the Art of Automatic Speech Recognition in Air Traffic Control. *World Acad. Sci. Eng. Technol. Int. J. Comput. Electr. Autom. Control. Inf. Eng.* **2015**, *9*, 10.
9. Wroniszewska, M.; Dziedzic, J. Voice command recognition using hybrid genetic algorithm. *TASK Q.* **2010**, *14*, 377–396.
10. Beritelli, F.; Serrano, S. A Robust Low-Complexity Algorithm for Voice Command Recognition in Adverse Acoustic Environments. In Proceedings of the 2006 8th International Conference on Signal Processing, Guilin, China, 16–20 November 2006; IEEE: Guilin, China, 2006; p. 4129154.
11. Jahchan, N.; Barbier, F.; Gita, A.D.; Khelif, K.; Delpech, E. Towards an Accent-Robust Approach for ATC Communications Transcription. In Proceedings of the Interspeech 2021, ISCA, Brno, Czech Republic, 30 August–3 September 2021; pp. 3281–3285.
12. Joakim, K. *The Integration of Automatic Speech Recognition into the Air Traffic Control System*; Flight Transportation Laboratory, Dept. of Aeronautics and Astronautics, Massachusetts Institute of Technology: Cambridge, MA, USA, 1990.
13. Schmidt, A.; Oualil, Y.; Ohneiser, O.; Kleinert, M.; Schulder, M.; Khan, A.; Helmke, H.; Klakow, D. Context-Based Recognition Network Adaptation for Improving on-Line ASR in Air Traffic Control. In Proceedings of the 2014 IEEE Spoken Language Technology Workshop (SLT), South Lake Tahoe, NV, USA, 7–10 December 2014; pp. 13–18.
14. Blatt, A.; Kocour, M.; Veselý, K.; Szöke, I.; Klakow, D. Call-Sign Recognition and Understanding for Noisy Air-Traffic Transcripts Using Surveillance Information. *arXiv* **2022**, arXiv:2204.06309.
15. Pellegrini, T.; Farinas, J.; Delpech, E.; Lancelot, F. The airbus air traffic control speech recognition 2018 challenge: Towards atc automatic transcription and call sign detection. *arXiv* **2018**, arXiv:1810.12614.
16. Zuluaga-Gomez, J.; Veselý, K.; Blatt, A.; Motlicek, P.; Klakow, D.; Tart, A.; Szöke, I.; Prasad, A.; Sarfjoo, S.; Kolčárek, P.; et al. Automatic Call Sign Detection: Matching Air Surveillance Data with Air Traffic Spoken Communications. In Proceedings of the 8th OpenSky Symposium 2020, Online, 3 December 2020; p. 14.
17. Nigmatulina, I.; Braun, R.; Zuluaga-Gomez, J.; Motlicek, P. Improving Call sign Recognition with Air-Surveillance Data in Air-Traffic Communication. *arXiv* **2021**, arXiv:2108.12156.
18. Nigmatulina, I.; Zuluaga-Gomez, J.; Prasad, A.; Sarfjoo, S.S.; Motlicek, P. A Two-Step Approach to Leverage Contextual Data: Speech Recognition in Air-Traffic Communications. *arXiv* **2022**, arXiv:2202.03725.
19. Shetty, S.; Helmke, H.; Kleinert, M.; Ohneiser, O. Early Call sign Highlighting Using Automatic Speech Recognition to Reduce Air Traffic Controller Workload. In Proceedings of the 13th International Conference on Applied Human Factors and Ergonomics (AHFE 2022), New York, NY, USA, 24–28 July 2022.
20. García, R.; Albarrán, J.; Fabio, A.; Celorrio, F.; Pinto De Oliveira, C.; Bárcena, C. Automatic Flight Callsign Identification on a Controller Working Position: Real-Time Simulation and Analysis of Operational Recordings. *Aerospace* **2023**, *10*, 433. [CrossRef]
21. Hasan, M.R.; Hasan, M.M.; Hossain, M.Z. How Many Mel-frequency Cepstral Coefficients to Be Utilized in Speech Recognition? A Study with the Bengali Language. *J. Eng.* **2021**, *2021*, 817–827. [CrossRef]
22. Deshmukh, A.M. Comparison of Hidden Markov Model and Recurrent Neural Network in Automatic Speech Recognition. *Eur. J. Eng. Res. Sci.* **2020**, *5*, 958–965. [CrossRef]

23. Pauls, A.; Klein, D. Faster and Smaller N-Gram Language Models. In Proceedings of the 49th Annual Meeting of the Association for Computational Linguistics: Human Language Technologies, Portland, OR, USA, 19–24 June 2011; Lin, D., Matsumoto, Y., Mihalcea, R., Eds.; Association for Computational Linguistics: Stroudsburg, PA, USA, 2011; pp. 258–267.
24. Song, Y.; Jiang, D.; Zhao, W.; Xu, Q.; Wong, R.C.-W.; Yang, Q. Chameleon: A Language Model Adaptation Toolkit for Automatic Speech Recognition of Conversational Speech. In Proceedings of the 2019 Conference on Empirical Methods in Natural Language Processing and the 9th International Joint Conference on Natural Language Processing (EMNLP-IJCNLP): System Demonstrations, Hong Kong, China, 3–7 November 2019; Association for Computational Linguistics: Hong Kong, China, 2019; pp. 37–42.
25. Xue, B.; Hu, S.; Xu, J.; Geng, M.; Liu, X.; Meng, H. Bayesian Neural Network Language Modeling for Speech Recognition. *arXiv* **2022**, arXiv:2208.13259. [CrossRef]
26. Graves, A.; Jaitly, N. Towards End-to-End Speech Recognition with Recurrent Neural Networks. *arXiv* **2017**, arXiv:1701.02720.
27. Kovtun, O.; Khaidari, N.; Harmash, T.; Melnyk, N.; Gnatyuk, S. Communication in Civil Aviation: Linguistic Analysis for Educational Purposes. 2020. Available online: https://www.researchgate.net/publication/344876536_Communication_in_Civil_Aviation_Linguistic_Analysis_for_Educational_Purposes (accessed on 28 October 2023).
28. Ohneiser, O.; Helmke, H.; Kleinert, M.; Ehr, H.; Balogh, G.; Tønnesen, A.; Rinaldi, W.; Mansi, S.; Piazzolla, G.; Murauskas, Š.; et al. Understanding Tower Controller Communication for Support in Air Traffic Control Displays. In Proceedings of the 12th SESAR Innovation Days, SESASR Innovation Days 2022, Budapest, Hungary, 5–8 December 2022.
29. Bollmann, S.; Fullgraf, J.; Roxlau, C.; Feuerle, T.; Hecker, P.; Krishnan, A.; Ostermann, S.; Klakow, D.; Nicolas, G.; Stefan, M.-D. Automatic Speech Recognition in Noise Polluted Cockpit Environments for Monitoring the Approach Briefing in Commercial Aviation. *Proc. Int. Workshop ATM/CNS* **2022**, *1*, 170–175. [CrossRef]
30. Fuzzy String Matching. Available online: https://pypi.org/project/fuzzywuzzy/ (accessed on 20 November 2023).
31. Saïd, K.M.; Abdelouahid, L. The IBN BATTOUTA Air Traffic Control Corpus with Real Life ADS-B and METAR Data. In *Artificial Intelligence and Industrial Applications*; Masrour, T., Cherrafi, A., El Hassani, I., Eds.; Advances in Intelligent Systems and Computing; Springer International Publishing: Cham, Switzerland, 2021; Volume 1193, pp. 371–384. ISBN 978-3-030-51185-2.
32. USA NWS Server. Available online: https://www.aviationweather.gov/metar (accessed on 30 April 2022).
33. Burileanu, D.; Pascalin, L.; Burileanu, C.; Puchiu, M. An Adaptive and Fast Speech Detection Algorithm. In *Text, Speech and Dialogue*; Sojka, P., Kopeček, I., Pala, K., Eds.; Lecture Notes in Computer Science; Springer: Berlin/Heidelberg, Germany, 2000; Volume 1902, pp. 177–182. ISBN 978-3-540-41042-3.
34. ICOM ICOM 8500. Available online: https://www.icomeurope.com/files/IC-R8500_E_20100108.pdf (accessed on 10 February 2021).
35. AirNav RadarBox. Available online: https://www.radarbox.com/presenting-the-radarbox-xrange-receiver (accessed on 18 February 2021).
36. Helmke, H.; Slotty, M.; Poiger, M.; Herrer, D.F.; Ohneiser, O.; Vink, N.; Cerna, A.; Hartikainen, P.; Josefsson, B.; Langr, D.; et al. Ontology for Transcription of ATC Speech Commands of SESAR 2020 Solution PJ.16-04. In Proceedings of the 2018 IEEE/AIAA 37th Digital Avionics Systems Conference (DASC), London, UK, 23–27 September 2018; pp. 1–10.
37. Diyasa, I.G.S.M.; Prasetya, D.A.; Idhom, M.; Sari, A.P.; Kassim, A.M. Implementation of Haversine Algorithm and Geolocation for Travel Recommendations on Smart Applications for Backpackers in Bali. In Proceedings of the 2022 International Conference on Informatics, Multimedia, Cyber and Information System (ICIMCIS), Jakarta, Indonesia, 16–17 November 2022; pp. 504–508.
38. Godfrey, J.J. *Air Traffic Control Complete 1994, 4170704 KB*; Linguistic Data Consortium: Philadelphia, PA, USA, 1994.
39. Šmídl, L. *Air Traffic Control Communication*; University of West Bohemia: Plzen, Czech Republic, 2011.
40. Hofbauer, K.; Petrik, S.; Hering, H. The ATCOSIM Corpus of Non-Prompted Clean Air Traffic Control Speech. In *Proceedings of the Sixth International Conference on Language Resources and Evaluation (LREC'08)*; Calzolari, N., Choukri, K., Maegaard, B., Mariani, J., Odijk, J., Piperidis, S., Tapias, D., Eds.; European Language Resources Association (ELRA): Marrakech, Morocco, 2008.
41. Segura, J.C.; Ehrette, T.; Potamianos, A.; Fohr, D.; Illina, I.; Breton, P.-A.; Clot, V.; Gemello, R.; Matassoni, M.; Maragos, P. The HIWIRE Database, a Noisy and Non-Native English Speech Corpus for Cockpit Communication. 2007. Available online: https://www.academia.edu/24112615/The_HIWIRE_database_a_noisy_and_non_native_english_speech_corpus_for_cockpit_communication (accessed on 28 October 2023).
42. Zuluaga-Gomez, J.; Veselý, K.; Szöke, I.; Blatt, A.; Motlicek, P.; Kocour, M.; Rigault, M.; Choukri, K.; Prasad, A.; Sarfjoo, S.S.; et al. ATCO2 Corpus: A Large-Scale Dataset for Research on Automatic Speech Recognition and Natural Language Understanding of Air Traffic Control Communications. *arXiv* **2023**, arXiv:2211.04054.
43. How to Evaluate Speech Recognition Models. Available online: https://www.assemblyai.com/blog/how-to-evaluate-speech-recognition-models/ (accessed on 28 October 2023).
44. Chowdhury, S.A.; Ali, A. Multilingual Word Error Rate Estimation: E-WER3. *arXiv* **2023**, arXiv:2304.00649.
45. Yujian, L.; Bo, L. A Normalized Levenshtein Distance Metric. *IEEE Trans. Pattern Anal. Mach. Intell.* **2007**, *29*, 1091–1095. [CrossRef] [PubMed]
46. IATA Airline and Airport Code. Available online: https://www.iata.org/en/publications/directories/code-search/ (accessed on 7 March 2022).
47. Nguyen, V.N.; Holone, H. N-Best List Re-Ranking Using Syntactic Score: A Solution for Improving Speech Recognition Accuracy in Air Traffic Control. In Proceedings of the 2016 16th International Conference on Control, Automation and Systems (ICCAS), Gyeongju, Republic of Korea, 16–19 October 2016; pp. 1309–1314.

48. Hannun, A.; Case, C.; Casper, J.; Catanzaro, B.; Diamos, G.; Elsen, E.; Prenger, R.; Satheesh, S.; Sengupta, S.; Coates, A.; et al. Deep Speech: Scaling up End-to-End Speech Recognition. *arXiv* **2014**, arXiv:14125567.
49. CMUSphinx Training an Acoustic Model. Available online: https://cmusphinx.github.io/wiki/tutorialam/ (accessed on 15 January 2023).
50. Mozilla Training Your Own Model. Available online: https://deepspeech.readthedocs.io/en/r0.9/TRAINING.html# (accessed on 14 June 2022).
51. CMUSphinx Adapting the Default Acoustic Model. Available online: https://cmusphinx.github.io/wiki/tutorialadapt/ (accessed on 15 June 2022).
52. Mozilla Deep Speech Fine-Tuning. Available online: https://deepspeech.readthedocs.io/en/r0.9/TRAINING.html#fine-tuning-same-alphabet (accessed on 19 July 2022).
53. Stolcke, A. SRILM - an Extensible Language Modeling Toolkit. In Proceedings of the 7th International Conference on Spoken Language Processing (ICSLP 2002), Denver, CO, USA, 16 September 2002; pp. 901–904.
54. Aircraft Company/Telephony/Three–Letter Designator Encode. Available online: https://www.faa.gov/air_traffic/publications/atpubs/cnt_html/chap3_section_1.html (accessed on 22 September 2023).
55. Wong, J. String Matching with FuzzyWuzzy. Available online: https://towardsdatascience.com/string-matching-with-fuzzywuzzy-e982c61f8a84 (accessed on 2 December 2023).
56. Kleinert, M.; Venkatarathinam, N.; Helmke, H.; Ohneiser, O.; Strake, M.; Fingscheidt, T. Easy Adaptation of Speech Recognition to Different Air Traffic Control Environments Using the DeepSpeech Engine; Virtual. 2021. Available online: https://elib.dlr.de/145397/ (accessed on 28 October 2023).

aerospace MDPI

Article

Validating Automatic Speech Recognition and Understanding for Pre-Filling Radar Labels—Increasing Safety While Reducing Air Traffic Controllers' Workload

Nils Ahrenhold [1,*], Hartmut Helmke [1], Thorsten Mühlhausen [1], Oliver Ohneiser [1], Matthias Kleinert [1], Heiko Ehr [1], Lucas Klamert [2] and Juan Zuluaga-Gómez [3,4]

1 German Aerospace Center (DLR), Institute of Flight Guidance, Lilienthalplatz 7, 38108 Braunschweig, Germany; hartmut.helmke@dlr.de (H.H.); thorsten.muehlhausen@dlr.de (T.M.); oliver.ohneiser@dlr.de (O.O.); matthias.kleinert@dlr.de (M.K.); heiko.ehr@dlr.de (H.E.)
2 Austro Control, 1030 Vienna, Austria; lucas.klamert@austrocontrol.at
3 Idiap Research Institute, 1920 Martigny, Switzerland; juan-pablo.zuluaga@idiap.ch
4 École Polytechnique Fédérale de Lausanne (EPFL), 1015 Lausanne, Switzerland
* Correspondence: nils.ahrenhold@dlr.de; Tel.: +49-531-295-1184

Abstract: Automatic speech recognition and understanding (ASRU) for air traffic control (ATC) has been investigated in different ATC environments and applications. The objective of this study was to quantify the effect of ASRU support for air traffic controllers (ATCos) radar label maintenance in terms of safety and human performance. Therefore, an implemented ASRU system was validated within a human-in-the-loop environment by ATCos in different traffic-density scenarios. In the baseline condition, ATCos performed radar label maintenance by entering verbally instructed ATC commands with a mouse and keyboard. In the proposed solution, ATCos were supported by ASRU, which achieved a command recognition rate of 92.5% with a command error rate of 2.4%. ASRU support reduced the number of wrong or missing inputs from ATCos into the radar label by a factor of two, which contemporaneously improved their situational awareness. Furthermore, ATCos where able to perform more successful secondary tasks when using ASRU support, indicating a greater capacity to handle unexpected events. The results from NASA TLX showed that the perceived workload decreased with a statistical significance of 4.3% across all scenarios. In conclusion, this study provides evidence that using ASRU for radar label maintenance can significantly reduce workload and improve flight safety.

Keywords: automatic speech recognition; automatic speech understanding; air traffic management; air traffic controller; radar label; human factors; assistant system; human-in-the-loop simulation

Citation: Ahrenhold, N.; Helmke, H.; Mühlhausen, T.; Ohneiser, O.; Kleinert, M.; Ehr, H.; Klamert, L.; Zuluaga-Gómez, J. Validating Automatic Speech Recognition and Understanding for Pre-Filling Radar Labels—Increasing Safety While Reducing Air Traffic Controllers' Workload. *Aerospace* **2023**, *10*, 538. https://doi.org/10.3390/aerospace10060538

Academic Editor: Judith Rosenow

Received: 21 March 2023
Revised: 15 May 2023
Accepted: 22 May 2023
Published: 5 June 2023

1. Introduction on Speech Recognition and Understanding

Speech recognition technology has made significant progress since its inception in the 1950s, with advancements in machine learning and artificial intelligence leading to increasingly sophisticated systems. Today, speech recognition technology has become an integral part of everyday life, with applications ranging from virtual assistants such as Siri and Alexa to transcription services for the hard of hearing and support functions for air traffic controllers (ATCos). However, recognizing word sequences is not the final step in creating good assistance functionality. It is crucial to understand the meaning behind word sequences. By incorporating the extracted meaning of recognized word sequences into assistance functionalities, one can measure the benefit for human operators using these support systems. The higher the technology readiness level of these support systems, which encompass the entire chain from word recognition to meaning comprehension and assistant system integration, the more realistic experiments can be conducted to assess the expected impact on human factors such as usability, workload, and errors.

1.1. Study on Speech Recognition and Understanding in a Broad Air Traffic Control Context

In Section 1.1.1, the concept of automatic speech recognition and understanding (ASRU) will be explained. Thereafter, the main research fields will be addressed and ASRU will be placed within the context of the air traffic management (ATM) domain. This section will conclude with a summary on the findings of human factors in the air traffic control context.

1.1.1. What Is Automatic Speech Recognition and Understanding in Air Traffic Control?

The concept behind automatic speech recognition and understanding in air traffic control will be outlined in the following example: An ATCo utters the following in radiotelephony communication with a flight crew: *"air serbia seven echo lima descend four thousand feet, QNH one zero one one"* with a hesitation after the QNH. The definition of common rules for transcribing such utterances may seem straightforward, but it is necessary to facilitate data exchange and minimize potential information-lossy conversion methods. In the above transcription, there might be a few cases where different notations regarding word combinations, upper case letters, and verbal hesitations are reasonable, such as:

- "air serbia" versus "airserbia"
- "~q~n~h" versus "q n h" versus. "QNH" for spelled letters
- "[hes]" for hesitations and thinking louds versus. "_aeh_", "hmm", "umm"

The HAAWAII project introduced transcription rules to normalize ATC communications [1].

The interpretation of the example transcription meaning could then be extracted as: "ASL7EL DESCEND 4000 ft, ASL7EL INFORMATION QNH 1011". Furthermore, discussions regarding the rules for annotating utterances can help to maximize the use of automatic extraction algorithms, as can be understood from alternative suggestions on how the above ATCo utterance can be annotated:

- ASL7EL descend 4000, qnh 1011
- ASL 7EL DESCEND 4000 feet, QNH 1011
- Asl7el desc 4000, Asl7el qnh 1011

The annotation rules, i.e., an ontology, for the above example in quotation marks, were first defined and agreed upon by the major European ATM stakeholders [2].

While the above transcription and annotation rules may appear straightforward, the challenge arises when dealing with greater variation in the utterances and especially deviations of ATC communications from the International Civil Aviation Organization (ICAO) phraseology [3]. Hence, the transcription of a potential pilot readback "serbia echo lima going down four thousand on the QNH one zero double one" should result in the same annotation as shown above for the ATCo utterance—with speaker "pilot" instead of "ATCo" and reason "readback"—in order to perform simple readback error checks. The complexity increases when numbers and terms are omitted or callsigns are abbreviated without adhering to defined abbreviation rules. Thus, ASRU in ATC is understood as the chain from receiving an audio signal via speech-to-text to extracting text-to-concepts in a defined format. These formats do not consider the storage method for communications inside software modules, i.e., using a structured data exchange format such as JavaScript Object Notation (JSON), and incorporating the information from different interpretation layers and recognition hypotheses.

1.1.2. What Are the Main Research Fields for Application of ASRU in ATM Domain?

The main research fields for applying ASRU in the ATM domain have varied over the last decades. Early applications focused on support in ATC training and reducing the workload of pseudo-pilots in simulations [4]. Subsequently, there was a shift towards offline analyses that assessed workload [5]. In this research, command types were extracted and counted over specific time periods [6].

In recent years, research has concentrated on quantifying the impacts of ASRU support on human performance and safety. Additionally, studies have analyzed the effects of support systems for radar label and flight strip maintenance on ATCo workload [7], resulting in changes in flight efficiency [8]. ATCos have historically used paper flight strips to manage their aircraft and corresponding clearances. These flight strips usually contained static information about the aircraft such as callsign, wake vortex categories, and designated route, as well as the aircraft's dynamic clearances such as altitude, speed, and course. Currently, electronic flight strips or interactive on-screen aircraft labels are used. However, these systems still require manual input from ATCos to update clearances. Although ATCos are used to communicate and input clearances simultaneously, the manual input requires additional workload from ATCos. Conversely, ASRU systems can be employed for automatic radar label input to remedy this effect. ATCos only need to verify the automatic input and make corrections in rare cases. This can offset the workload increase while allowing mental capacity to be directed to other tasks.

In addition, the advancement of automatic readback error detection in ATC communication—one of the most complex tasks of ASRU in ATC—is progressing. However, despite efforts to group communication threads and compare command content based on real-life operational data, the progress remains limited [9]. Throughout these developments, the need for metrics to measure the ASRU performance arose. Different metrics, such as recognition rates and error rates for complete commands and callsigns, have emerged, which are related to precision and recall [10].

In general, the benefit of more recent applications is the greater amount of usable data and the utilization of machine learning techniques [11]. However, ASRU is still not yet in operational use in ATC.

1.1.3. Human Factors in an Air Traffic Control Context

The effect of ASRU support, with command error rates below 2%, on ATCo workload has already been validated in the project AcListant®–Strips (Active Listening Assistant) project [12]. To validate the effects of ASRU support on ATCo workload, it is important to understand the different dimensions of workload. In this context, the mental workload of ATCos is connected to the task performed and related requirements, as detailed in [13]. The time for maintaining radar labels with similar user interfaces with and without ASRU support was measured using the Dusseldorf approach on a single runway as early as 2015 [8]. The time for clicking and maintaining the radar labels was reduced by a factor of three for the eight German and Austrian ATCos when they were supported by ASRU. This resulted in a better use of their mental capacity, leading to more efficient aircraft trajectories and a reduction of 60 L of fuel consumption per aircraft [8]. Furthermore, the project "Digital Tower Technologies—HMI Interaction modes for Airport Tower" investigated the impact of ASRU support for electronic flight strip maintenance. The ten Lithuanian and Austrian ATCos experienced a reduction of workload based on a secondary task measurement when ASRU [7] achieved command recognition rates of 90% and callsign recognition rates of 94%.

1.2. Research Question

Based on above-mentioned conditions, the main research question of this paper can be summarized as follows: "how can we quantify the effect of ASRU support for ATCos in radar label maintenance in terms of safety and ATCos' workload?"

To answer this main question, the following research questions are discussed in this paper:

- How accurately does ASRU extract commands that influence radar label entries?
- How many incorrect or missing radar label entries exist with and without ASRU support?

In addition to the previous derived questions, we also considered whether the ATCos corrected missing or wrong ASRU outputs.

Both questions were addressed using objective measurements, such as command recognition error rates.

- What are the consequences of over-trust or under-trust in the ASRU system?

This question is addressed through subjective data based on post-run and post-validation questionnaires, as well as objective data recorded during the simulation runs.

- How can we compensate for sequence effects induced by multiple runs under similar conditions?

The presented statistical approach allows us to obtain results with higher statistical significance without increasing the number of participants involved.

1.3. Structure of the Paper

In Section 2, the human-in-the-loop validation trials for radar label maintenance are described, including the study parameters, simulation framework, and the methods and techniques used. Additionally, the ASRU architecture is explained. Section 3 begins by presenting the subjective and objective measurements for workload, safety, and situational awareness. It then discusses the applied method to compensate for sequence effects within the objective results, considering both the ATCos' performance and ASRU performance. This is followed by the presentation of the results and discussion in Section 4. Finally, Section 5 draws a conclusion based on the results.

2. Validation Trials and Methods

The aim of the validation trials was to provide a final assessment of the ASRU by quantifying its benefits on the ATCos' performance, perceived workload, and flight safety. The difference between the baseline runs and the solution runs was the absence or presence of ASRU support for radar label maintenance in approach control. The ATC approach position was chosen as it is a highly dynamic area and is usually crowded with aircrafts at hub airports such as Vienna. The aircrafts that the ATCos needed to handle in the approach sector had already been handed over from en-route sectors and required guidance towards their final destination for a transfer of control to the tower. Therefore, it was expected that using ASRU for radar label maintenance would have an impact. In the following the general setup for the validation trials, the simulation scenarios, and the collection of results are discussed.

2.1. Infrastructure and Exercises

The ASRU validation trials were performed using the Air Traffic Management and Operation Simulator (ATMOS) at DLR's Air Traffic Validation Center. The ATMOS provided a human-in-the-loop simulation environment [14,15], which EUROCONTROL recognizes as a suitable validation method [16] for systems in the pre-industrial development phase [17]. Furthermore, the ATMOS has been previously used in several validation campaigns including mental workload analysis for air traffic controllers [18], analysis of air traffic management security [19], and assessing the impact of spaceflights on air traffic management [20]. The NARSIM software (version 8.3) [21] was deployed as a generic real-time software. Aircraft performance was modeled using the Base of Aircraft Data (BADA) model, version 3.15, from EUROCONTROL [22]. Two controller working positions (CWP) and six simulation-pilot working positions (SWP) were configured. During a simulation run, only one CWP and a maximum five SWP were used, depending on the traffic load. The second CWP and SWPs served as backups to provide redundancy in case of system failure. Furthermore, during an active simulation run, the unused CWP was already prepared for the next simulation run, which saved time and reduced procedural errors.

The ATCos' verbal utterances served as the voice input signal for the ASRU system, specifically the speech recognition engine, which was the first part of the chain. The communication between the ATCo and simulation-pilot was carried out according to the defined phraseology by ICAO [23]. Figure 1 shows the CWP setup, which included the main radar screen, a voice-over-IP (VoIP) headset, an ASRU log, and a secondary screen. These components will be explained in detail later on.

Figure 1. CWP setup including headset for radio telephony, radar screen, ASRU log, and secondary screen.

Simulation-pilots were responsible for steer their aircrafts and implementing aircraft clearances received from Approach ATCos. A DLR-developed human–machine interface (HMI) was used for the SWP. The HMI included a radar screen and a flight strip section. The flight strip section displayed aircraft performance data (such as velocity, flight level, and route) and contained a command line to enter clearances. CWP and SWPs, located in different rooms, were connected via VoIP.

The validation trials were structured in three iterative campaigns. The first two campaigns for preparing the main trials took place in autumn 2021 and spring 2022. The main campaign was carried out from 14 September until 3 November 2022. In total, twelve ATCos from the Austrian air navigation service provider (ANSP) Austro Control participated in the main campaign, consisting of eleven male ATCos and one female ATCo. The age of participants ranged from 25 to 44 years, with a mean age of 32 years and a standard deviation (SD) of $SD = 7.3$. Their work experience ranged from one to 20 years, with a mean work experience of eight years ($SD = 6.8$). During the preparation campaigns (autumn 2021, spring 2022) the ATCos wore face masks in accordance with COVID-19 pandemic rules. During the main campaign, no face masks were required for ATCos. Wearing face masks had no significant effect on ASRU performance.

2.2. Simulation Components

The simulation was implemented for the Vienna airport (LOWW) terminal maneuvering area (TMA), as shown in the approach chart of Vienna in Figure 2. Figure 3 shows the main radar screen for ATCos with LOWW airspace. The ATCos' area of responsibility encompassed a combined pickup/feeder sector in Europe (feeder/final in the U.S.). LOWW consists of two dependent runways. During the simulation, runway 34 (RWY34) with a length of 3600 m was utilized. The ATCo was in charge of guiding the arrival streams to RWY34. No departures, overflights, or other types of traffic, such as visual flight rule traffic, were integrated into the simulation runs. There were no limitations regarding the aircraft type or simulated weather restrictions. Furthermore, no emergency situations or non-nominal situations such as bird strikes or runway closures were analyzed.

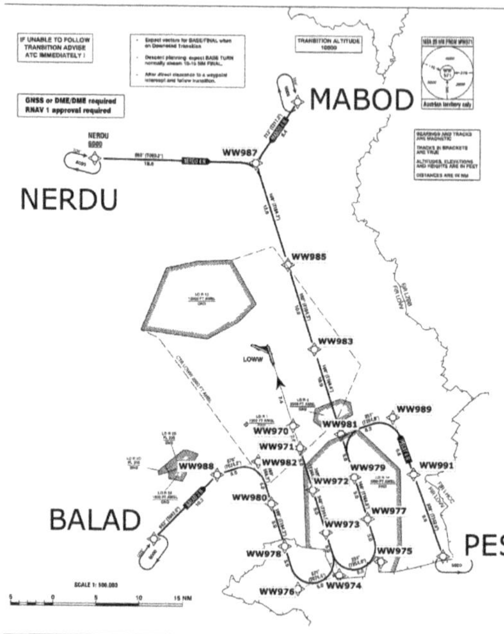

Figure 2. Approach chart Vienna (LOWW) TMA adapted from aeronautical information publication (AIP) [24] with adding the four metering fixes BALAD, NERDU, PESAT, MABOD and the names WW* of selected waypoints.

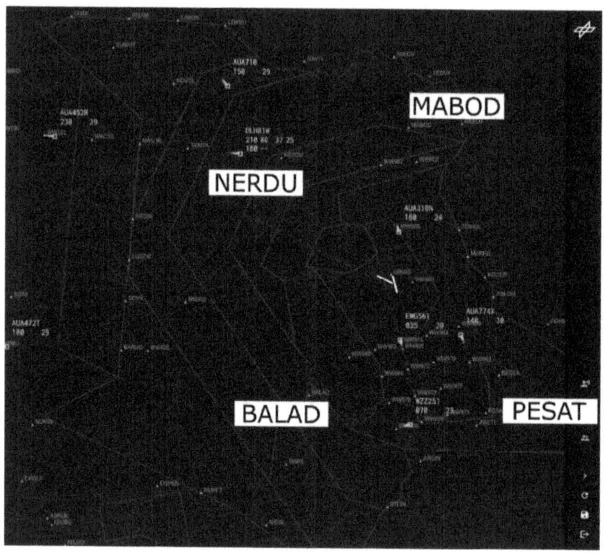

Figure 3. Main radar screen with LOWW airspace: waypoints (grey), start of transitions (NERDU, MABOD, BALAD, PESAT), as well as aircraft symbols and labels (green). This is a screen shot, how the screen is shown to the ATCo with the exception that we added the four metering fixes NERDU, MABOD, BALAD and PESAT. Overlapping of information often happens and the ATCo knows how to deal with these challenges.

A within-subject design [25,26] with the two factors "traffic flow" and "use of ASRU" was used to examine the dependent variables including mental workload, safety, situational awareness, time for maintaining radar label cells, and number of remaining incorrect inputs. Although the present experiment was conducted with a limited number of participants in the specific airspace of Vienna using a prototypic non-operation user interface, the results yielded statistical significance and were consistent with the findings from the AcListant®-Strips project conducted using the Dusseldorf approach [8,12].

Two different simulation scenarios were developed: a medium-density traffic scenario (M) with 30 arrivals per hour and a high-density traffic scenario (H) with 42 arrivals per hour. Additionally, a training traffic scenario (T) with 20 arrivals per hour was used as introductory exercise. These scenarios were developed based on recorded operational data from LOWW. For that reason, traffic flow corresponded to typical numbers and types at LOWW. All simulation scenarios (M, H) lasted for 35 min. At the beginning of the simulation scenarios, the aircrafts were already located inside the TMA under the responsibility of the ATCo. The other aircrafts were initiated outside the TMA and followed standard arrival routes (STAR)s towards the area navigation (RNAV) routes, as depicted in Figures 2 and 3. Before entering the TMA, simulation-pilots set the standard initial call according to the currently applied radio telephony procedures.

Each validation day consisted of five simulation sections per ATCo. These sections were distinguished by the following design factors: traffic flow (T, M, H) and use of ASRU. Simulation sections without ASRU support for the ATCos were referred as baseline runs. These runs represented the typical manual mouse-only input approach for radar label maintenance in the ATCo HMI. Simulation sections with ASRU support were labelled as solution runs. During the solution runs, ATCos were able to use the ASRU inputs and manual mouse input for potential corrections. Regarding the traffic flow, each ATCo started with a training run to familiarize themselves with the setup and input modalities. The support of ASRU was activated and deactivated during the training run. Afterwards, the ATCos started with an M run followed by an H run. Then, another M run was followed by the final H run. As for second decisive variable, seven ATCos started with a solution run. Among them, five of the ATCos had no ASRU support during the first simulation run. The indention was for a 50% distribution between starting with and without ASRU support. However, due to problems with the surveillance data, one runs was cancelled and needed to be repeated at the end. If an ATCo began with a solution run, she/he also ended with a solution run, with two baseline runs in between. The same procedure was applied in reverse if starting and ending with a baseline run.

Two ATCos were available on each of the six validation days. They started at 08:30 a.m. and finished the validation day at approximately 04:30 p.m. Following the previously described procedure, while one ATCo executed a simulation run, the other ATCo filled in questionnaires and rested. As a result, the ATCos did not work in parallel. All the questionnaires used in the study are described in Section 3.1. By alternating the order of simulation runs for approximately half of the ATCos, it was expected that any sequence effects or effects of exhaustion would be averaged out. While this approach helped to some degree, it still had undesired negative effects on the statistical significance of the results. In Section 3.3 a technique to compensate for these sequence effects will be presented.

2.3. HMI for Radar Label Maintenance

The ATCos were provided with three different screens, as shown in Figure 1. Figure 4a–d shows the aircraft labels in detail. Figure 4a displays the reduced data block with the following structure: the first row contained the callsign, the second row included flight level, cleared flight level, ground speed, and cleared ground speed, while the third row showed heading, waypoint, and RNAV route. In Figure 4b, the aircraft label is shown as a full data block in which the fourth label line was visibly activated through a mouse-over hovering function. The fourth label line additionally provided the present heading, remarks, and rate of climb/descent. The nine cells framed in white (highlighted for visualization purposes in the paper) could be interactively selected via a mouse click. Figure 4c shows the full data block with recognized clearances from ASRU displayed in purple. It also offers two checkmarks in the first label line. If an interactive cell was clicked, a correspondent drop-down menu appeared for value selection (see Figure 4d).

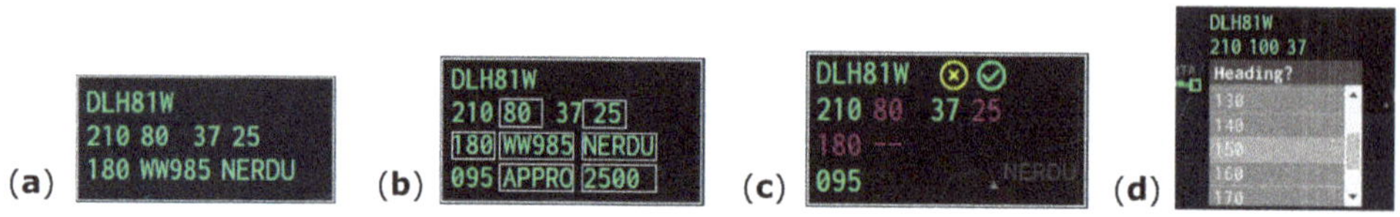

Figure 4. (**a**) Reduced (standard) aircraft data block, (**b**) full aircraft data block, (**c**) full aircraft data block with ASRU output, (**d**) drop-down menu to manipulate radar label cells.

If ATCos were provided with ASRU support in solution runs, the content of their verbal utterances was automatically extracted and transformed into relevant ASRU output values. These recognized values appeared in purple, as shown in the label cells of Figure 4c. Additionally, a yellow *Reject* and green *Accept* checkmark appeared in the top line of the label. Thus, ATCos could confirm all of the proposed values or reject them by clicking on the buttons. The accepted values turned the light green, matching the color of the rest of the aircraft clearances. In rare cases of misrecognitions, the ATCos needed to correct values manually via drop-down menus. If the ATCo did not interact with the label, the recognized values were automatically accepted after ten seconds. This time parameter was determined based on [8,27].

2.4. Additional Components

Figure 5 shows the SpeechLog provided at the CWP, as shown on the left side of Figure 1. The SpeechLog was not essential for the ATCos to operate the setup. Instead, it served as a showcase and was not considered part of the experimental setup. Nevertheless, the SpeechLog displayed an overview of the recognized ATCo utterances (word level transcriptions) and meanings (annotations with ATC concepts following the defined ontology).

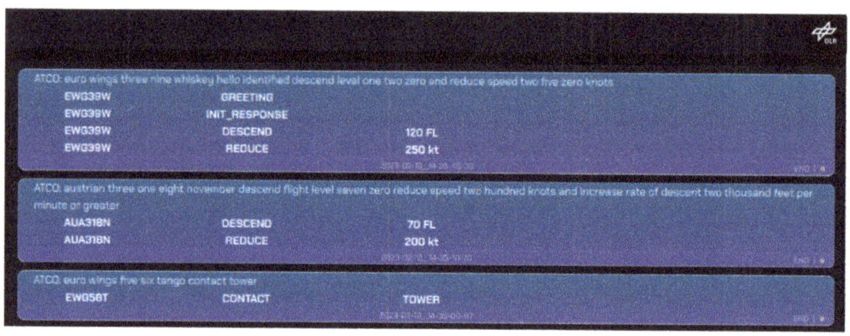

Figure 5. SpeechLog at CWP.

Finally, Figure 6 shows the SWP HMI of the simulation pilots. It provided flight strips for all aircraft within the airspace, displayed as purple strips, indicating upcoming aircrafts (left side) or those already in progress (middle). It also included a radar screen for an overview of the traffic situation (right side). Additionally, the simulation-pilots saw the ASRU word level output for comparison with or support of the self-recognized utterances. The ASRU word level output was provided in the lower left part.

Figure 6. SWP with the following three parts from left to right: strip view, workspace, radar.

2.5. Automatic Speech Recognition and Understanding

The validation software implemented the ASRU system as defined by the HAAWAII project (Highly Automated Air traffic controller Workstations with Artificial Intelligence Integration) [28]. The ASRU core mainly relied on four modules, as shown in Figure 7. The modules have already been described in detail in [29]. Here, we provide a brief summary of their functionality. In addition to the information provided in [29], we quantified the effect of the main modules to the final performance in Section 4.1 of the results.

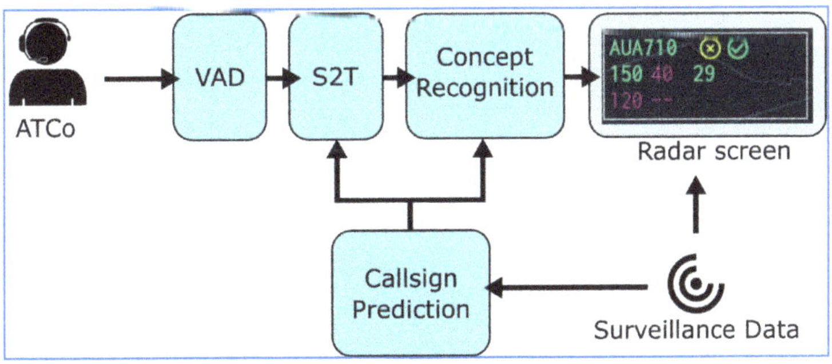

Figure 7. ASRU component setup during validation trials.

Voice Activity Detection (VAD): The VAD process is relatively straightforward, as the push-to-talk (PTT) signal is readily available. It did not have a significant impact on the performance.

Speech-to-text (S2T): Whenever the VAD detects a transmission, the signal is forwarded to S2T, and the recognition process starts in real time. S2T delivers with a minimum latency when an ATCo starts speaking and updates the recognized words continuously until the end of the transmission when the PTT button is released.

Concept Recognition: Each time a recognized word sequence is forwarded, it is analyzed by the Concept Recognition module. The analysis result is then transformed into relevant ATC concepts as defined by SESAR project PJ.16-04 CWP HMI [2] and extended by the HAAWAII project [30], as shown in Figure 8.

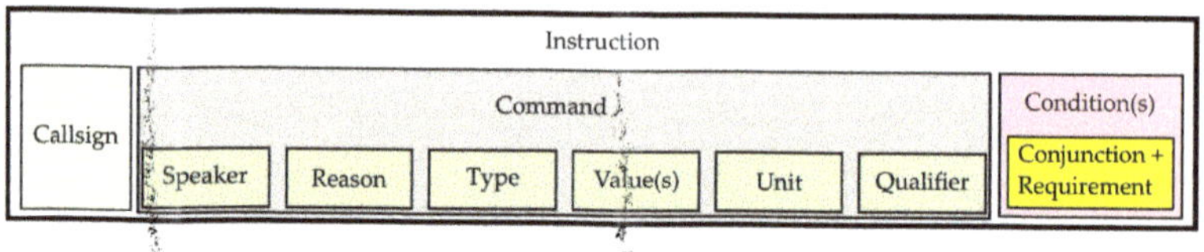

Figure 8. Elements of instructions consisting of callsigns, commands, etc.

A command at the semantic level consists of a type, values, unit, qualifier, etc., as shown in Figure 8. All these parts must be correctly extracted at the semantic level to be counted as a correct recognition. Otherwise, it is counted as an error or a rejection.

Callsign Prediction: This module considers surveillance data to determine if any recognized callsign could reasonably be part of an ATCo utterance. The output is used by S2T and Concept Recognition to enhance the recognition quality for both modules. Further details on callsign prediction and callsign extraction can be found in [31]. This model is highly critical, as indicated by the results in Section 4.1.3. It considers the callsigns of aircrafts that are currently in the relevant airspace.

3. Methods and Techniques

3.1. *Subjective ATCo Feedback Techniques*

This section describes the methods and experiments used to obtain subjective feedback from the participating ATCos during the validation trials. The subjective rating measures encompassed various aspects of mental workload, situation awareness, usability, and acceptance. An instantaneous self-assessment of workload (ISA) measurement was integrated into the radar screen of the CWP and had to be answered during the simulation runs. The answers on several questionnaires were captured after each simulation run. The set of questionnaires included the NASA-TLX (National Aeronautics and Space Administration Task Load Index) [32], Bedford Workload Scale [33], SUS (System Usability Scale) [34], CARS (Controller Acceptance Rating Scale) [35], and the three SHAPE questionnaires (Solutions for Human Automation Partnerships in European ATM) [36]—SASHA (Situation Awareness for SHAPE) ATCo, SATI (SHAPE Automation Trust Index), and AIM (Assessing the Impact on Mental Workload). All methods and experiments are explained in the following sections.

3.1.1. NASA Task Load Index (NASA TLX)

The NASA TLX was used to assess different dimensions of the workload [32]. The questionnaire includes subscales of mental demand, physical demand, temporal demand, performance, effort, and frustration. In total, the questionnaire consisted of six questions with ten answer possibilities from (1) Low to (10) High. The adapted questions can be found in Appendix A.1.

The unweighted NASA TLX was used instead of the weighted version, as previous ATC projects showed no further benefit from weighting parameters and it often caused confusion of the ATCos.

The ISA measures were used to obtain the ATCos' perceived mental workload within a defined time period [37]. Each ATCo was prompted to rate their perceived mental workload during the simulation run every five minutes for the last five minutes [38,39]. Therefore, after five minutes of simulation time, the ATCos heard a sound signal on their headset as a five-point Likert scale [40] appeared on the lower part of the radar screen. They were able to rate their perceived mental workload on a scale of one [underutilized] to five [excessively busy]. The ISA data were used afterwards to examine the mental workload. It is worth noting that in previous projects and the current study, ATCos usually did not rate their workload as a five, even during extremely busy traffic hours. Thus, there is a need for a more objective measure. Therefore, a secondary task was implemented, as described in Section 3.2.

3.1.2. Bedford Workload Scale

The Bedford Workload Scale consists of two questions that inquire about the average workload and the peak workload, with ten possible answers ranging from (1) *"Workload insignificant"* to (10) *"Task Unsustainable due to Workload"*, as shown in Figure 9. Additionally, the applied Bedford Workload Scale questionnaire had the following final open-ended question: *"Which factors/events/conditions have contributed to potentially high workload?"*.

Task Unsustainable due to Workload	10	HARDER
Workload Extremenly High	9	
Workload Very High	8	
Workload High	7	
Workload Moderate to High	6	MODERATE
Workload Moderate	5	
Workload Low to Moderate	4	
Workload Low	3	EASIER
Workload Very Low	2	
Workload Insignificant	1	

Figure 9. Screenshot of the Bedford Workload Scale questionnaire interface.

3.1.3. System Usability Scale (SUS)

The System Usability Scale, initially proposed by John Brooke [34], was used to assess the general usability with and without the ASRU support. This questionnaire consists of ten statements to be rated on a five-point scale, ranging from (1) *"fully disagree"* to (5) *"fully agree"*. The adapted questions can be found in Appendix A.2.

3.1.4. Controller Acceptance Rating Scale (CARS)

The CARS questionnaire, developed by NASA Ames [35], measures the operational acceptability and serves as an indicator for the satisfaction of human-system performance. CARS consist of a single question: *"Please read the descriptors and score your overall level of user acceptance experienced during the run. Please check the appropriate number between 1 and 10"*. The different answer options were color-coded with red, orange, and green, as shown in Table 1.

3.1.5. Solutions for Human Automation Partnerships in European ATM (SHAPE)

The SHAPE questionnaire was developed to evaluate the effects of automation on different human factors for ATCos, such as workload, situation awareness, and trust in the system [36]. It consists of three parts, as described in the following sections. Each question had seven answer possibilities, ranging from *"never"* to *"always"* or from *"none"* to *"extreme"*. The questions can be found in Appendices A.3–A.5.

For quantitative analysis, the answers were mapped to numerical values between one and seven. A numerical value of one corresponded to a good system, whereas seven corresponded to a bad system.

Table 1. Colored-coded answer options of the CARS questionnaire.

1. Improvement mandatory. Safe operation could not be maintained.
2. Major Deficiencies. Safety not compromised, but system is barely controllable and only with extreme controller compensation.
3. Major Deficiencies. Safety not compromised but system is marginally controllable. Considerable compensation is needed by the controller.
4. Major Deficiencies. System is controllable. Some compensation is needed to maintain safe operations
5. Very Objectionable Deficiencies. Maintaining adequate performance requires extensive controller compensation
6. Moderately Objectionable Deficiencies. Considerable controller compensation to achieve adequate performance.
7. Minor but Annoying Deficiencies. Desired performance requires moderate controller compensation
8. Mildly unpleasant Deficiencies. System is acceptable and minimal compensation is needed to meet desired performance.
9. Negligible Deficiencies. System is acceptable and compensation is not a factor to achieve desired performance.
10. Deficiencies are rare. System is acceptable and controller doesn't have to compensate to achieve desired performance.

Color code: Pink: Not acceptable, Yellow: Changes necessary, Green: Acceptable.

3.1.6. Situation Awareness for SHAPE (SASHA)

The SASHA questionnaire is part of the SHAPE questionnaire and addresses different dimensions of situation awareness [36]. This questionnaire consisted of six statements with the seven answer possibilities: *"never"*, *"seldom"*, *"sometimes"*, *"often"*, *"more often"*, *"very often"*, and *"always"*. The questions are provided in Appendix A.3.

3.1.7. SHAPE Automation Trust Index (SATI)

The SATI questionnaire is also part of the SHAPE questionnaire. SATI provides questions to measure the human trust in ATC systems [36]. This questionnaire consisted of six statements with the seven answer possibilities: *"never"*, *"seldom"*, *"sometimes"*, *"often"*, *"more often"*, *"very often"*, and *"always"*. The questions are provided in Appendix A.4.

3.1.8. Assessing the Impact on Mental Workload (AIM)

The AIM questionnaire, developed by Doris M. Dehn [36], assesses the impact of changes in the ATM system on the mental workload of ATCos. This questionnaire consisted of 15 questions with seven answer possibilities: *"none"*, *"very little"*, *"little"*, *"some"*, *"much"*, *"very much"*, and *"extreme"*. The questions are provided in Appendix A.5.

3.2. Objective Secondary Task for Workload Assessment: "Stroop Test"

As previously addressed, a more objective task to gather the mental workload of ATCos was needed. Therefore, a secondary task [41] based on the *Stroop Test* was integrated into the touch device located on the right side at the CWP [42] (see secondary screen in Figure 1). The central idea is that ATCos have the mental capacity to fulfil a secondary task in addition to their primary ATC task if they are underutilized. Evaluating the results of the secondary task indicates the amount of mental workload an ATCo experienced during the different simulation runs.

Figure 10 displays the interface of the secondary task. The secondary task started ten minutes after the simulation onset and provided a ten-minute execution window. The ATCos were able to begin the tasks within this time period whenever they felt comfortable with respect to handling the primary task, as shown in Figure 10a. After pressing the *START* button, the name of a color was displayed in a different color than the name itself in the upper display part, as shown in Figure 10b. In a next step, the ATCos had to select the color that was used for printing from the available options. In the example shown in Figure 10b, the correct solution was *GREEN* because the word "RED" was printed in green. After submitting their choice, the ATCo could proceed to the next task by clicking the START button. A higher number of correct responses indicated greater mental spare capacity and less mental workload occupied by the primary task [43].

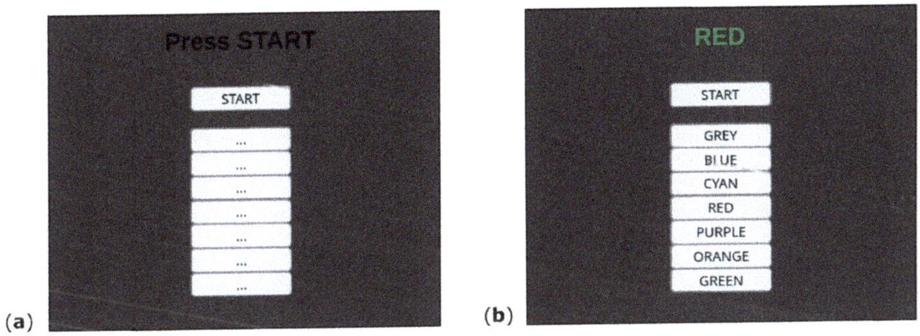

Figure 10. (**a**) Secondary task before start; (**b**) secondary task after start.

3.3. Compensation of Sequence Effects

As an example, Table 2 shows the answers to the question *"How hard did you have to work to accomplish your level of performance?"* from the NASA TLX questionnaire completed by the 12 ATCos. The answers ranged from one (low effort) to ten (high effort) for the medium traffic scenario. Columns *"ATCo Id"* show the identifier of the participant. Columns *"Sol"* and *"Base"* show the chosen answer value. The number "1/2" indicates whether the participant started with a baseline or solution run ("1") and ended with a baseline or solution run ("2").

Table 2. ATCos' answers to the NASA TLX questions to determine the effort needed to accomplish the task.

ATCo Id	Sol 1	Base 2	Diff	ATCo Id	Base 1	Sol 2	Diff
1	7	5	2	3	8	4	−4
2	7	8	−1	5	7	4	−3
4	5	5	0	7	3	2	−1
6	3	1	2	9	7	3	−4
8	4	1	3	11	6	2	−4
10	6	3	3				
12	3	3	0				
Average	5.0	3.7	1.3	Average	6.2	3.0	−3.2
Average Run 1		5.50		Average Run 2		3.42	

Light yellow color shows the first run of an ATCo, light green shows his/her second run.

Due to sequence effects, the results in the second run generally showed improvements compared to the first run for the ATCos. This would have averaged out, if 50% of the participants would have started with the baseline and 50% would have started with the solution run, but with a high standard deviation. Therefore, we decided to filter out the sequence effects not only on average, but for each participant. Furthermore, a run with ATCo ID 2 failed and was repeated, resulting in seven ATCos starting the medium scenario with solution runs and only five with baseline runs. Therefore, sequence effects do not even compensate on average. The following approach adapted from [8], was used to compensate for these sequence effects:

The average values of all 12 ATCos for the first run and the second run were calculated, as shown in the last and second-to-last rows of Table 2. The averages of the last row were used to correct the feedback values for the question. The average value of all ATCos' answers was 5.50 for the first run in the medium traffic scenario (the averages of all values in column "*Sol* 1" and column "*Base* 1" are marked in light yellow). In the second run with medium traffic, the ATCos answered with an average value of 3.42 (the averages of columns "*Base* 2" and "*Sol* 2" are marked in light green), and the ATCos performed 2.08 units better on the scale with a maximum value of 10 units. Therefore, we corrected all entries of Table 2 by 1.04 units. As shown in Table 3, the first runs, marked in light yellow, were corrected by subtracting 1.04, and second runs, marked in light green, were corrected by adding 1.04.

Table 3. ATCos' answers to NASA TLX questions to evaluate the effort needed to accomplish the task after compensating for sequence effects.

ATCo Id	Sol 1	Base 2	Diff	ATCo Id	Base 1	Sol 2	Diff
1	6	6	−0.1	3	7	5	−0.9
2	6	9	−3.1	5	6	5	−0.9
4	4	6	−2.1	7	2	3	1.1
6	2	2	−0.1	9	6	4	−1.9
8	3	2	0.9	11	5	3	−1.9
10	5	4	0.9				
12	2	4	−2.1				
Average	4.0	4.8	−0.8	Average	5.2	4.0	−1.1
Average Run 1		4.46		Average Run 2		4.46	

Light yellow color shows the first run of an ATCo, which are decremented by 1.04 compared to Table 2, light green shows his/her second run, which are incremented by 1.04. The values are rounded.

The described sequence effect compensation technique (SECT) could result in values below one or even negative values, which could not be provided by the ATCos. However, this is not important for the performed *t*-tests, as described in the following section. Only the differences between runs with and without ASRU support for the same ATCo are important.

The difference between runs with ASRU support and those without ASRU support was minus 0.58 (calculated as $7 \times 1.3 - 5 \times 3.2)/12$) in Table 2. After compensating for sequence effects as shown in Table 3, the average difference becomes minus 0.93 ($(-7 \times 0.8 - 5 \times 1.1)/12)$). If we had an equal number of runs starting with ASRU support and without ASRU support, the average value would not change. This will be the case for all heavy traffic scenarios. As demonstrated in the results section, even for medium traffic, the differences changed only slightly with and without compensating for sequence effects. More importantly, the standard deviation of the differences decreased in most cases when SECT was applied. It decreased from 2.6 in Table 2 to 1.37 in Table 3. After applying SECT, the average results of all first runs always matched the average values of all second runs.

To conclude, the experiments involved two independent variables: (i) with or without ASRU support and (ii) whether the ATCos received ASRU support in first runs or in the second runs. The presented technique can compensate for sequence effects, enabling clearer observation of the results of the ASRU support. In the presented example, the sequence effect influenced the result by 2.08 units out of 10, whereas ASRU support had only an effect of 0.58 units. By using the described technique, both effects can be separated, resulting in an ASRU effect of 0.93.

3.4. Paired T-Test to Evaluate Statiscal Significance

The differences between runs of the same ATCo in baseline and solution runs smaller now, indicating a decrease in the standard deviation sigma. This observation was also supported by the performed paired *t*-test, which was previously applied during the AcListant®–Strips project mentioned earlier to assess workload reduction benefits [12] and improvements in flight efficiency [8].

The null hypothesis H_0 states, "*ASRU support does **not** decrease the amount of work, how hard the ATCo needs to work to accomplish the required level of performance*". The test value is defined as follows:

$$T = (M - \mu_0)\frac{\sqrt{n}}{SD}, \tag{1}$$

The differences in how the ATCo needs to work (solution minus baseline runs) for each run in Table 3 were calculated, for example, for ATCo ID 3 as seven minus five. The number of differences (ATCos) is denoted as n (12 in the present case). M represents the mean value of the questionnaire answer differences "Diff" in Table 3, which in the present case is minus 0.93. SD refers to the standard deviation of the differences, which was 1.37 based on the values in Table 3. It is only important if the ASRU input results in lower value answers to the questions. Therefore, μ_0 was set to 0. With these values, we calculated T as minus 2.35.

The value T follows a t-distribution with $n - 1$ degrees of freedom. The null hypothesis H_0 can be rejected with probability of α (also known as p-value) if the calculated value for T is less than the value of the inverse t-distribution at position $t_{n-1,1-\alpha}$ with $n - 1$ degrees of freedom (in our case, -1.80 for $\alpha = 0.05$). Therefore, the hypothesis H_0 is rejected because $T = -2.35 < -1.80$. Even the minimal α can be calculated, such that $T < t_{n-1,1-\alpha}$ still holds. In this case, $\alpha = 3.8\%$. If we repeated our experiments with the 12 ATCos 1000 times, we could expect that the H_0 would not be rejected only in 38 cases. The results invalidate the negatively formulated null hypothesis, indicating that *ASRU support does decrease the amount of work required for the ATCO to accomplish the required level of performance with a probability of* $\alpha = 3.8\%$.

The probabilities of rejecting the null hypotheses for the heavy traffic scenario and for both scenarios combined were also calculated. These calculations are presented later in Table 10 in the following results section.

The effects of SECT in distinguishing between the effects of ASRU support and sequence effects are evident. Without SECT, the α value was 45.3%, compared to 3.8% after applying SECT. Without SECT, there was the need for strongly different ratings from ATCos to achieve statistical significance, even if the results were significant with only slightly different ratings. Further details are provided in [29].

4. Results and Discussion

In this section, the performance of ASRU is first presented. Secondly, the subjective feedback from questionnaires is explained and discussed. Finally, the section concludes with the objective results from performance measurements.

4.1. Results of Speech Recognition and Understanding Performance

4.1.1. Performance at the Word Level

Table 4 shows the performance on a word level, which is based on the word error rate (WER), which represents the percentage of words that were not correctly recognized. The WER is calculated using the Levenshtein distance [44]. The table also includes the number of uttered words, as well as the number of substitutions (*Subst*), deletions (*Del*), and insertions (*Ins*), which indicate the differences between the recognized words and the actual uttered words. The best performance, i.e., lowest WER, for a single ATCo on all his/her four runs was 0.7%, while the worst performance was 8.2%.

Table 4. Performance at the word level quantified as the word error rate (WER).

	# Words	Levenshtein Distance	# Subst	# Del	# Ins	WER
Total	118,816	3712	1853	1324	535	3.12%
Heavy	64,441	2148	1066	729	353	3.33%
Medium	54,375	1564	787	595	182	2.88%
Solution	59,180	1805	881	686	238	3.05%
Baseline	59,636	1907	972	638	297	3.20%

means "Number of".

It should be highlighted that there was a difference between the solution and baseline runs, with the ATCos speaking more clearly when supported by ASRU. In the baseline runs, the ATCos did not benefit from ASRU. Nevertheless, we recorded and evaluated their performance. It was also interesting to observe that the performance decreased as the number of aircrafts increased, although it did not result in a break-down of performance.

Table 5 shows the *"top words"* that were most often misrecognized, sorted by the number of absolute occurrences. The word *"two"* was recognized 32 times as a different word. It was recognized as another word (*"substituted to"*) 261 times and inserted 91 times, i.e., it was recognized, but no word was actually spoken. A total of 71 times, it was said, but no word was recognized at all. The sum of these four errors was 455. The word *"two"* was actually spoken 8841 times, and the number of recognitions, correct and wrong, was 9090 times. The word *"two"* was involved in incorrect word recognition in 5% of the instances compared to the times it was actually spoken. The word *"to"* was very often involved in these problems, more than one-third of the instances. The table also shows that many important ATC-related words were often involved in recognition problems.

Table 5. Top 10 words with challenges on word level ordered by total recognitions.

Word	Subst by	Subst to	Ins	Del	Sum	Said	Recogn	%
two	32	261	91	71	455	8841	9090	5%
one	32	130	33	44	239	8128	8215	3%
zero	20	101	8	24	153	7576	7641	2%
four	209	104	9	54	376	5804	5654	6%
three	10	73	9	25	117	5624	5671	2%
eight	99	78	37	200	414	5422	5238	8%
austrian	134	3	5	25	167	4979	4828	3%
five	77	74	10	9	170	3745	3743	5%
ILS	70	0	0	41	111	1988	1877	6%
air	24	5	23	48	100	1309	1265	8%

4.1.2. Performance at the Semantic Level

Table 6 summarizes the performance of the Concept Recognition module on the semantic level. The table distinguishes between the medium and heavy traffic scenarios, as well as between the baseline and solution runs. We did not compensate for sequence effects in this analysis, as statistical significance is not provided.

The columns *"Cmd-Recog-Rate"* and *"Cmd-Error-Rate"* show the percentage of correctly and incorrectly recognized commands, respectively. The difference between the sum of these two columns and 100% corresponds to the percentage of rejected commands. The last two columns show the same metrics for the callsign only. Details regarding the metrics used can be found in [10].

Table 6. Performance at the semantic level for different traffic complexities and for baseline and solution runs.

	Cmd-Recog-Rate	Cmd-Error-Rate	Csgn-Recog-Rate	Csgn-Error-Rate
All Scenarios	92.1%	2.8%	97.8%	0.6%
Medium	92.7%	2.7%	97.9%	0.5%
Heavy	91.7%	2.9%	97.8%	0.6%
Baseline	91.8%	2.8%	97.5%	0.6%
Solution	92.4%	2.8%	98.1%	0.5%

The differences between baseline and solution runs, as observed in in Table 4 on the word level, also occurred on the semantic level. This is not surprising, as problems at the word level cannot be fully compensated for on the semantic level. The Concept Recognition module is robust, which is shown by a command recognition rate for full commands of 92.1% and a callsign recognition rate of 97.8% in Table 6, with a word error rate of over 3%. When the ATCos were supported by ASRU in solution runs, the understanding performance increased on both the command level and the callsign level. Table 7 shows the performance on the semantic level, considering different WER and analyzing the influence of different parts of the full command on the extraction performance.

Table 7. Performance at the semantic level quantified as recognition and error rates.

Level of Evaluation	WER	Cmd-Recog-Rate	Cmd-Error-Rate	Csgn-Recog-Rate	Csgn-Error-Rate
Full Command	3.1%	92.1%	2.8%	97.8%	0.6%
Only Label		92.5%	2.4%	97.8%	0.6%
Only Label, offline		93.4%	1.7%	97.9%	0.5%
Only Label, gold	0.0%	99.3%	0.3%	99.9%	0.1%
Full Command, gold		99.1%	0.4%	99.9%	0.1%

The "*Full Command*" row represents the quality of all instruction elements, even those that were never shown in the radar label of this application, such as GREETING, CALL_YOU_BACK, DISREGARD. As our application only considered callsign, type, and value as important, the "*Only Label*" row shows the rates when unit, qualifier etc., are ignored, but still for all command types, independent of whether they were shown in the radar label. After the validation exercise, the rates were recalculated offline, considering the elimination of certain obvious software bugs. The recalculated rates for the "*Only Label*" row, based on the same word sequence inputs, are shown in the "*Only Label, offline*" row. These reported rates for all three rows received the same word sequences with an average WER of 3.1% as input. The output of the S2T block was the same, with a 3.1% WER in the offline row. Assuming a perfect S2T block with a word error rate of 0%, a command recognition rate of 99.3% is achieved. When considering the full command in "*Full Command, gold*" row, including also conditions, qualifiers etc., a command recognition rate of 99.1% is obtained. Both rows show that the Concept Recognition module effectively models the utilized phraseology, suggesting that improved S2T performance would further improve semantic-level performance.

4.1.3. Effects of Callsign Prediction on Semantic Extraction Performance

The two rows labeled "*No Context*" in Table 8 show the performance when context information is disregarded. The callsign recognition rate decreases from 99.9% to 81.6%, even with a perfect S2T engine ("gold"). For a detailed explanation, please refer to Section 4.2.2. The command recognition performance is only slightly lower at 80.6%. The row labeled "*No Context, S2T*" also shows the performance without using the available callsign information, but instead using a real S2T engine with a WER of 3.1%. In this case, the command recognition rate considerably decreases from 92.1% to 66.9%, and the callsign recognition rate decreases from 97.8% to 71.6%.

Table 8. Performance without using callsign prediction.

Full Command	WER	Cmd-Recog-Rate	Cmd-Error-Rate	Csgn-Recog-Rate	Csgn-Error-Rate
No Context, gold	0.0%	80.6%	14.4%	81.6%	14.1%
No Context, S2T	3.1%	66.9%	22.1%	71.6%	20.9%

The results of this section, presented in Table 8, demonstrate the impact of using callsign prediction. Without callsign prediction, the system performance is insufficient. However, with command prediction, we generate benefits for the ATCo with respect to workload reduction, which is shown in the next two sections.

4.2. *Subjective Results from ATCo Feedback*

This section describes the results provided by the subjective ATCo feedback, which includes ISA, NASA-TLX, Bedford Workload Scale, SUS, CARS, and the three SHAPE questionnaires: (i) SASHA ATCo, (ii) SATI, and (iii) AIM, as described in the previous section.

4.2.1. Instantaneous Self-Assessment Measure

The results from ISA provide a retrospective self-assessment of the perceived mental workload by the ATCos. Table 9 shows the ISA results based on the paired *t*-test from the validation trials. The ISA mean values were calculated for both scenarios (M and H) under conditions with and without ASRU support. Delta ISA and min α were calculated with and without considering sequence effects. A negative delta ISA value indicates that the mean ISA value was lower in the solution run compared to the baseline run.

Table 9. ISA value results and significance analysis.

Value	Composition	Medium Scenario	Heavy Scenario	Combined
ISA mean	with ASRU	2.39	2.87	2.63
	without ASRU	2.48	3.26	2.87
ISA delta	SE	−0.09	−0.39	−0.25
min α		10.6%	0.5%	0.3%
ISA delta	NSE	−0.03	−0.39	−0.21
min α		42.6%	1.1%	3.1%

SE = sequence effect; NSE = no sequence effect; minimal α values, shaded in green for 0% ≤ α < 5%, and in yellow for (|α| ≥ 10%).

It can be observed that all delta ISA values are negative, independent of whether sequence effects are considered. This indicates that solution runs received lower mean ISA values, suggesting that using ASRU support reduces the perceived mental workload of ATCos. Furthermore, the impact of considering sequence effects can be seen. The consideration influences the mean ISA value, reduces sigma, and improves the statistical significance. The minimal alpha value (α min) decreases from 0.7% to 0.3%. Examining the ISA mean values reveals that supporting ATCos with ASRU lowers the mean ISA value in both the M and H scenarios. However, the greatest impact can be seen for the H scenario. Here, the mean ISA value over all simulation runs was almost 15% lower. This result indicates that ASRU support is particularly effective in reducing the ATCos' perceived mental workload during high traffic hours, corresponding to the H scenario.

4.2.2. NASA TLX

Table 10 shows the differences in the six NASA TLX question ratings, which was calculated as the mean solution value minus the mean baseline value. The last row provides a summary by displaying the arithmetic average of all six ratings. Weights between 1 (low workload) and 10 (high workload) were possible, as described in Section 3.

Table 10. Results of NASA TLX questionnaires for the different traffic scenarios with and without compensating for sequence effects.

Hypotheses	Medium				Heavy				Both			
	Diff		α		Diff		α		Diff		α	
	SE	NSE	SE	NSE	SE	NSE	SE	NSE	SE	NSE	SE	NSE
MD	−0.08	−0.35	44.2%	14.0%	−0.50	−0.50	17.9%	17.5%	−0.29	−0.42	22.8%	8.5%
PD	−0.42	−0.54	14.9%	4.8%	−1.08	−1.08	2.5%	2.0%	−0.75	−0.81	1.4%	0.4%
TD	−0.08	−0.26	43.2%	23.6%	−0.33	−0.33	24.3%	20.8%	−0.21	−0.30	26.9%	13.6%
OP	0.42	0.38	−6.1%	−7.6%	−0.08	−0.08	44.8%	44.7%	0.17	0.15	−31.5%	−33.1%
EF	−0.58	−0.93	22.6%	1.9%	−0.75	−0.75	4.1%	2.9%	−0.67	0.84	6.5%	0.2%
FR	−0.33	−0.50	29.6%	17.9%	0.08	0.08	−45.4%	−44.7%	−0.13	−0.21	39.6%	30.7%
Summary	−0.18	−0.37	34.0%	9.2%	−0.44	−0.44	15.9%	13.0%	−0.31	−0.41	15.6%	4.4%

Minimal α values are shaded in green for 0% ≤ α < 5%, in light green for 5% ≤ α < 10%, in light red for light evidence (−10% ≤ α < −5%) that results were worse with ASRU support, and in yellow for the rest (|α| ≥ 10%). MD = mental demand, PD = physical demand, TD = temporal demand, OP = operational performance, EF = effort, FR = frustration. The blue color of "OP" is explained in the text below.

The four columns labeled *"Medium"* show the results of the performed *t*-test for the medium traffic scenarios. The columns labeled *"Heavy"* show the results for the heavy scenario, and the columns below *"Both"* combine the *"Medium"* and *"Heavy"* columns. The six columns labeled *"Diff"* show the average differences in the answers between the runs with and without ASRU support. Negative values indicate a lower workload in the solutions runs with ASRU support. The *"SE"* columns contain the values before eliminating the sequence effects, while the *"NSE"* columns show the values afterward elimination. The six columns under *"α"* show the *p*-value, which indicates the statistical significance or the probability of the null hypothesis (see Section 3.3) being valid. In the following discussion, we will focus only on the values in the *"NSE"* columns. However, we also include the values in the *"SE"* columns to show the effectiveness of our SECT approach.

The differences in the *"Diff"* columns for the *"Heavy"* scenarios did not change when considering sequence effects. In the case of the medium traffic scenarios, the differences slightly vary because there were more solution runs as the first runs of the day compared to the baseline runs (with a ratio of seven to five). Therefore, the differences in columns *"Both"* also change.

In the majority of cases, the application of SECT led to an improvement in statistical significance, resulting in a decrease in the *p*-value. This shows the value of SECT in compensating for sequence effects. For the medium runs, statistically significant results (α < 5%) were obtained in two out of the six cases when the sequence effects were eliminated. Without eliminating the sequence effects, the results are not statistically significant. For the heavy traffic scenarios, the (color of the) statistical significance did not change, but in all cases, α decreased or did not change. For the combined scenarios, the statistical significance improved in five out of two cases, and in two cases, it "improves" to a different statistically significant range, transitioning from a yellow color code to light green or from light green to green.

Question "1" (MD) addresses the mental demand, which decreased by 0.5 units out of 10 in the heavy traffic scenarios, but with a high standard deviation. Question "2" (PD) addresses the physical demand, which showed a statistically significant decrease in all runs. The same trend was observed for the related question (EF): *"How hard did you have to work to accomplish your level of performance?"*. The answers to (TD), *"How hurried or rushed was the pace of the task"*, did not exhibit statistically significant changes. The same applies to (FR) *"discouraged, irritated, stressed"* and (OP) *"successfully accomplishing the task"*. For the latter, there was even a tendency for the ATCos to subjectively believe that they performed better, at least in the heavy traffic runs without ASRU support. Later sections show that this was only a subjective feeling.

Question 4 (OP) *"How successful were you in accomplishing, what you were asked to do?"* is the only question for which the answer *"low"* corresponded to a poor performance. An explanation could be that some ATCos did not always recognize this when answering the questions. We mark these questions in the following tables in blue as the blue "OP" indicates in Table 10. It should be pointed out again, that we have transformed the answers already before presenting them in the table, so that negative differences mean "better with ASRU".

4.2.3. Bedford Workload Scale

Table 11 displays the results from the Bedford Workload Scale after performing the *t*-test, as described previously. *Medium*, *Heavy*, and *Both* represent the corresponding results for the scenarios used during the validation trials. Columns indicated by *SE* show the results with sequence effects. Columns with *NSE* show the results after compensating for the sequence effects.

Table 11. Results of the Bedford Workload Scale for the different traffic scenarios with and without compensating for the sequence effects.

Hypotheses	Medium				Heavy				Both			
	Diff		α		Diff		α		Diff		α	
	SE	NSE	SE	NSE	SE	NSE	SE	NSE	SE	NSE	SE	NSE
Average	−0.42	−0.49	11.1%	6.1%	−0.33	−0.33	18.7%	16.7%	−0.38	−0.41	6.7%	3.8%
Peak	−0.33	−0.44	15.9%	4.6%	−0.17	−0.17	34.4%	34.3%	−0.25	−0.31	17.2%	10.4%
Summary	−0.38	−0.47	11.9%	4.2%	−0.25	−0.25	23.6%	22.7%	−0.31	−0.36	9.0%	4.6%

Minimal α values are shaded in green for 0% ≤ α < 5%, in light green for 5% ≤ α < 10%, and in yellow for the rest (|α| ≥ 10%). The columns were previously explained in the NASA TLX results Section 4.1.2.

The average and peak workload change for ATCos in the M scenario ranged from [−0.33 to −0.49]. Thus, the average and peak workloads were lower with ASRU support. In the H scenario, the differences ranged from [−0.17 to −0.33]. The highest value for the difference was calculated for the average workload. The statistical significance for the H scenario remained largely unchanged with or without considering sequence effects, ranging from 16.2% to 34.1%. For both scenarios together, the differences between with and without ASRU support fall within the interval of [−0.25 to −0.41], indicating an overall improved perceived workload (lower) when using ASRU support. Statistical significance mostly improved after compensating for sequence effects. The results for the peak workload are not statistically significant, because the α values are still greater than 10%.

Overall, the results from the Bedford Workload Scale demonstrate that applying ASRU support for ATCos improved the results by lowering the perceived workload. Greater effects were recorded for the M scenario compared to the H scenario. Nevertheless, the relative change was minor. Compensating for the sequence effects significantly improved the statistical significance in all cases. In addition to the results from the Bedford Workload Scale, direct feedback was also gathered from ATCos. This feedback is summarized below.

There are three areas of feedback regarding the factors contributing to high workload for ATCos: (1) HMI aspects that were related to ASRU, (2) HMI aspects that were not related to ASRU, and (3) simulation aspects such as the amount of traffic, the simulation-pilots, and the requirement to enter all clearances into the system.

Regarding *"HMI aspects that were related to ASRU"*, the ATCos identified areas for improvement in the radar label interaction, such as reduced scrolling, using drop-down menus for inputs, and addressing issues with incorrect system inputs, especially if the callsign was wrongly recognized. However, some ATCos also acknowledged the potential usefulness of ASRU if they were more familiar with the new HMI. The aspect of "getting used to the HMI" was also the main criticism for the second feedback area, "HMI aspects that were not related to ASRU". The differences between the TopSky system used in Vienna and the prototypic CWP in Braunschweig caused some difficulties, such as the unavailability of distance measuring or the number of required clicks for system input. Most of the feedback concerned the third area of simulation aspects, where ATCos faced a high traffic load in the high-density traffic scenario. This included radio frequency congestion due to many transmissions, different speed handling, sometimes uncommon flight profiles, a few inaccurate simulation-pilot inputs, and more traffic than they were accustomed to handling alone. The main difference may have been the requirement to enter all instructed commands into the ATC system, which the ATCos do not need to do in their usual system.

4.2.4. System Usability Scale (SUS)

Table 12 displays the results of the SUS after performing a *t*-test, as described previously. *Medium*, *Heavy*, and *Both* represent the corresponding results for the scenarios used during the validation trials. Columns indicated by *SE* show the results with the sequence effects. Columns with *NSE* show the results after compensating for the sequence effects.

Table 12. Results of system usability scale for the different traffic scenarios with and without compensating for the sequence effects.

Hypotheses	Medium				Heavy				Both			
	Diff		α		Diff		α		Diff		α	
	SE	NSE	SE	NSE	SE	NSE	SE	NSE	SE	NSE	SE	NSE
1	−1.10	−1.06	2.4%	2.7%	−1.00	−1.00	0.9%	0.8%	−1.05	−1.03	0.1%	0.1%
2	−1.20	−1.18	0.3%	0.4%	−0.80	−0.80	1.5%	1.5%	−1.00	−0.99	2×10^{-4}	2×10^{-4}
3	−1.50	−1.46	3×10^{-6}	6×10^{-6}	−1.00	−1.00	0.3%	0.3%	−1.25	−1.23	3×10^{-7}	3×10^{-7}
4	0.20	0.22	−24.3%	−21.9%	0.70	0.70	−1.8%	−1.8%	0.45	0.46	−2.1%	−1.8%
5	−0.80	−0.84	2.2%	1.5%	−1.00	−0.94	1.6%	1.5%	−0.89	−0.89	0.1%	0.1%
6	−0.40	−0.35	7.4%	8.5%	0.00	0.00	NR	−50.0%	−0.20	−0.17	15.9%	16.9%
7	−0.20	−0.22	24.3%	22.5%	−0.30	−0.30	16.0%	15.9%	−0.25	−0.26	11.1%	10.3%
8	−1.29	−1.28	0.1%	0.1%	−1.40	−1.40	0.6%	0.7%	−1.35	−1.35	1×10^{-4}	1×10^{-4}
9	−0.80	−0.76	0.9%	1.1%	−0.70	−0.70	4.8%	3.6%	−0.75	0.73	0.2%	0.2%
10	−0.30	−0.39	22.2%	15.2%	−0.10	−0.10	37.3%	36.6%	−0.20	−0.25	20.8%	15.0%
Summary	−0.75	−0.74	0.2%	0.2%	−0.55	−0.55	0.8%	0.8%	−0.65	−0.64	8×10^{-5}	8×10^{-5}

Minimal α values are shaded in green for $0\% \leq \alpha < 5\%$, in light green for $5\% \leq \alpha < 10\%$, in orange if we had high evidence ($-5\% \leq \alpha < 0\%$) that the results were worse with ASRU support, and in yellow for the rest ($|\alpha| \geq 10\%$). "NR" means "no result", i.e., the average deviations are 0.0%. The columns were previously explained in the NASA TLX results section. Sometimes, not all ATCos answered all questions. In rare cases, a different number of answer pairs were obtained for starting with baseline or starting with solutions runs. Therefore, the entries in columns SE and NSE for the heavy traffic scenarios can be different, as shown for question 5. The blue colors in column 1 are already explained at the end of Section 4.2.2.

The results from the SUS assessment show the highest changes in the M runs when comparing runs with and without ASRU support, which range between 0.22 and −1.46. Thus, in most reported cases, the ASRU support enabled a better usability of the system. Statistical significance (*p*-value) ranged between $3 \times 10^{-8}\%$ and −50%. For the M runs, in three cases, a *p*-value larger than $|20\%|$ was reported (bold framed cells) after compensating for the sequence effects, which indicates no statistical significance. For the H scenario, the differences ranged from [−1.4 to 0.70]. In one case (question 4), the *p*-value of −1.8% indicated that the results were statistically significance and indicate a better performance without ASRU support. Row 4 indicates that *"I think that I would need the support of a technical person to be able to use this system"*. The same effect can be seen for that question, when analyzing the results of the *t*-tests for both scenarios combined, since the experience with the given system was relatively low compared to their general working experience with the TopSky system. However, when all 10 questions (row "summary") were combined, the results indicated that the overall system had a higher usability while using the ASRU support during the common ATC task. The statistical significance was very high, with an average value of 8×10^{-5}.

4.2.5. Controller Acceptance Rating Scale (CARS)

Table 13 shows the results of the CARS analysis. *Medium*, *Heavy*, and *Both* represent the corresponding results for the scenarios used during the validation trials. Columns indicated by *SE* show the results with sequence effects. Columns with *NSE* show the results after compensating the sequence effects.

Table 13. Results of the controller acceptance rating scale (CARS) for the different traffic scenarios with and without compensating for sequence effects.

Hypotheses	Medium				Heavy				Both			
	Diff		α		Diff		α		Diff		α	
	SE	NSE	SE	NSE	SE	NSE	SE	NSE	SE	NSE	SE	NSE
Maturity	−1.90	−1.72	2.3%	3.7%	−1.22	−1.12	2.7%	4.6%	−1.50	−1.36	0.3%	0.7%

Minimal α values are shaded in green for 0% ≤ α < 5%, see the NASA TLX results Section 4.2.2 for column names.

The CARS results show that for each scenario (M and H) as well as when combining both scenarios (*Both*), the differences were between −1.12 (Heavy) and −1.36 (Both) on the 10-point scale after compensating for sequence effects. This suggests that the ATCo acceptance increased with the usage of ASRU support compared to simulation runs without ASRU support. The *p*-values for all three cases were below 5%, indicating that the null hypothesis is invalid and there is statistical significance with the usage of ASRU support.

4.2.6. Situation Awareness for SHAPE (SASHA)

Table 14 shows the results of the SASHA analysis. SASHA is the first of three assessments from the SHAPE questionnaire, which analyses the situational awareness of ATCos. *Medium*, *Heavy*, and *Both* represent the corresponding results for the scenarios used during the validation trials. Columns indicated by *SE* show the results with sequence effects. Columns with *NSE* show the results after compensating for the sequence effects.

Table 14. Results of situation awareness using SHAPE for the different traffic scenarios with and without compensating for sequence effects.

Hypotheses	Medium				Heavy				Both			
	Diff		α		Diff		α		Diff		α	
	SE	NSE	SE	NSE	SE	NSE	SE	NSE	SE	NSE	SE	NSE
1	−0.09	−0.13	35.6%	25.4%	−0.50	−0.50	1.1%	0.9%	−0.30	−0.32	3.2%	1.4%
2	0.18	0.16	−21.0%	−22.6%	−0.17	−0.17	20.9%	20.9%	0.00	−0.01	NR	47.8%
3	−0.55	−0.58	1.0%	0.2%	−0.42	−0.42	12.5%	9.5%	−0.48	−0.49	1.4%	0.5%
4	−0.27	−0.28	12.8%	11.7%	−0.33	−0.33	14.2%	11.2%	−0.30	−0.31	6.1%	4.3%
5	0.00	−0.01	NR	47.2%	0.00	0.00	NR	NR	0.00	−0.01	NR	49.1%
6	−0.09	−0.18	42.8%	32.0%	−0.58	−0.58	7.7%	2.8%	−0.35	−0.39	13.8%	5.5%
Summary	−0.14	−0.17	21.5%	8.6%	−0.33	−0.33	8.2%	5.3%	−0.24	−0.26	5.4%	1.8%

Minimal α values are shaded in green for 0% ≤ α < 5%, in light green for 5% ≤ α < 10%, and in yellow for |α| ≥ 10%, "NR" means "no result", i.e., the average deviations are 0.0%. The columns itself and the blue color in column 1 were previously explained in the NASA TLX results Section 4.2.2.

The SASHA results show that for the M, H, and Both scenarios, the average differences were between −0.17 (Medium) and −0.33 (Heavy) after compensating for the sequence effects. This suggests that the situational awareness of the ATCos slightly increased across all scenarios when using the ASRU support during the simulation runs. The greatest positive impact on the ATCos' situational awareness was recorded during the H scenario. The p-values after compensating for the sequence effects reduce for the *Both* scenarios combined (α = 1.8%) to below 5%. This indicates that the null hypothesis was invalid and the statistical significance improved with the use of the ASRU support. For the M scenario, the p-value after compensating for sequence effects was 8.6%, and for the H scenario, it was 5.3%. Here, the statistical significance was slightly improved with SECT. One possible explanation is that during the M scenarios, the ATCos had more time to verify their current planning process (situational awareness) and thus did not feel the need for any support system. However, during the H scenario, there was less time between different verbal ATC instructions to check their own planning process. In this case, the spare time obtained through the ASRU radar label input was valued even more, which improved the ATCos' situational awareness.

4.2.7. SHAPE Automation Trust Index (SATI)

Table 15 shows the results of the SATI analysis. SATI was the second of three assessments from the SHAPE questionnaire, which analyzed the ATCos' trust in the automated functions or systems. *Medium*, *Heavy*, and *Both* represent the corresponding results for the scenarios used during the validation trials. Columns indicated by *SE* show the results with sequence effects. Columns with *NSE* show the results after compensating for the sequence effects.

Table 15. Results of the SHAPE Automation Trust Index (SATI) for the different traffic scenarios with and without compensating for sequence effects.

Hypotheses	Medium				Heavy				Both			
	Diff		α		Diff		α		Diff		α	
	SE	NSE	SE	NSE	SE	NSE	SE	NSE	SE	NSE	SE	NSE
1	−1.10	−1.21	4.4%	2.7%	−2.09	−2.07	2×10^{-4}	2×10^{-4}	−1.62	−1.66	1×10^{-4}	5×10^{-5}
2	0.10	0.14	−42.6%	−39.8%	−0.45	−0.45	12.5%	12.7%	−0.19	−0.17	27.9%	29.8%
3	0.10	0.21	−42.6%	−34.0%	−0.64	−0.60	5.8%	5.9%	−0.29	−0.21	19.4%	24.8%
4	−0.70	−0.66	9.4%	10.4%	−1.09	−1.05	0.8%	0.6%	−0.90	−0.87	0.4%	0.4%
5	−0.10	−0.12	43.3%	41.9%	−1.45	−1.45	3×10^{-4}	3×10^{-4}	−0.81	−0.82	1.5%	1.4%
6	−1.10	−1.23	6.8%	3.7%	−1.10	1.11	6.2%	5.9%	−1.14	−1.17	1.5%	0.8%
Summary	−0.47	−0.48	17.7%	17.0%	−1.15	−1.12	0.2%	0.2%	−0.83	−0.82	0.5%	0.5%

Minimal α values are shaded in green for $0\% \leq \alpha < 5\%$, in light green for $5\% \leq \alpha < 10\%$, and in yellow for $|\alpha| \geq 10\%$. The columns were previously explained in the NASA TLX results Section 4.2.2.

The SATI results show that for the M scenario, H scenario, and the combined Both scenarios, the average difference ranged from −0.48 (Medium) to −1.12 (Heavy) after compensating for sequence effects. This suggests that he ATCos' trust in the system increased when using the ASRU support compared to the simulation runs without ASRU support. The highest average difference was recorded during the H scenario. The p-value ranged below 5% for the H scenario (α = 0.2%) and Both scenarios (α = 0.5%), indicating that the null hypothesis was invalid and the usage of ASRU support increased the statistical significance. For the M scenario, the average p-value was greater than 10% (α = 17.0%), which indicates that no increase in trust could be achieved by using ASRU support. This applies before and after compensating for sequence effects. During the M scenario, the ATCos might have had enough time to explore the system and were not dependent on ASRU support. This effect could have decreased the statistical significance compared to the H scenario, where there was less time to create doubts and the system had to be used as implemented.

4.2.8. Assessing the Impact on Mental Workload (AIM)

Table 16 shows the results of the AIM analysis. AIM was the third of three assessments from the SHAPE questionnaire used in this study, which analyzed the ATCos' mental workload experienced. *Medium*, *Heavy*, and *Both* represent the corresponding results for the scenarios used during the validation trials. Columns indicated by *SE* show the results with sequence effects. Columns with *NSE* show the results after compensating for the sequence effects.

Table 16. Results of assessing the impact on mental workload for the different traffic scenarios with and without compensating for sequence effects.

Hypotheses	Medium				Heavy				Both			
	Diff		α		Diff		α		Diff		α	
	SE	NSE	SE	NSE	SE	NSE	SE	NSE	SE	NSE	SE	NSE
1	0.33	0.22	−22.7%	−29.1%	−0.08	−0.08	39.3%	37.1%	0.13	0.07	−32.1%	−38.5%
2	0.42	0.29	−20.8%	−26.5%	−0.67	−0.67	2.3%	1.8%	−0.13	−0.19	34.4%	25.7%
3	0.00	−0.14	NR	36.9%	−0.75	−0.75	3.5%	3.3%	−0.38	−0.44	12.0%	6.3%
4	0.25	0.15	−28.5%	−35.4%	−0.50	−0.50	6.2%	6.0%	−0.13	−0.17	32.5%	25.3%
5	0.75	0.60	−4.7%	−5.3%	0.00	0.00	NR	NR	0.38	0.30	−8.3%	−10.0%
6	0.82	0.63	−4.1%	−7.8%	0.18	0.21	−31.2%	−28.0%	0.50	0.42	−4.7%	−6.9%
7	0.08	−0.07	−43.2%	43.2%	−0.42	−0.42	13.7%	13.3%	−0.17	−0.24	29.5%	18.8%
8	0.58	0.54	−8.6%	−10.0%	−0.17	−0.17	32.2%	29.0%	0.21	0.19	−22.9%	−23.7%
9	−0.17	−0.39	37.5%	12.6%	−0.67	−0.67	8.2%	8.1%	−0.42	−0.53	11.9%	3.4%
10	0.17	−0.03	−37.2%	47.1%	−0.25	−0.25	22.2%	21.6%	−0.04	−0.14	44.5%	28.5%
11	0.50	0.40	−11.1%	−14.7%	−0.30	−0.28	20.5%	22.1%	0.10	0.06	−35.9%	−41.0%
12	0.58	0.54	−4.9%	−5.9%	−0.42	−0.42	4.2%	4.1%	0.08	0.06	−35.4%	−38.8%
13	0.58	0.46	−9.4%	−12.1%	−0.58	−0.58	2.9%	2.5%	0.00	−0.06	NR	40.3%
14	0.08	−0.10	−43.0%	39.1%	−0.50	−0.50	1.1%	1.1%	−0.21	−0.30	21.3%	7.3%
15	0.17	0.00	−35.6%	NR	−0.42	−0.42	2.3%	2.2%	−0.13	−0.21	30.8%	14.8%
Summary	0.32	0.19	−20.4%	−27.3%	−0.38	−0.38	2.6%	2.5%	−0.03	−0.10	44.3%	30.2%

Minimal α values are shaded in green for 0% ≤ α < 5%, in light green for 5% ≤ α < 10%, in orange, if we have high −5% ≤ α < 0% or in light red for (−10% ≤ α < −5%), and yellow is used for the rest (∣α∣ ≥ 10%), when we have no statistical significance in any direction. "NR" means "no result", i.e., the average deviation was 0.0%. The columns were previously explained in the NASA TLX results Section 4.2.2.

The AIM results show that for the M scenario, the average difference was 0.19 after compensating for sequence effects (row "*Summary*", column "*NSE*"). This suggests that using the ASRU support increased the mental workload. For the H scenario and when combining both scenarios, the average difference ranged from −0.10 (Both) to −0.38 (Heavy) after compensating for sequence effects. These results indicate that the mental workload decreased during the simulation runs when using the ASRU support compared to the simulation runs without ASRU support.

The average *p*-value results for the M scenario (α = −27%) and when combining *Both* scenarios (α = 30%) were greater than ∣10%∣. This indicates that no statistical significance could be achieved when using the ASRU support. For the H scenario, the average *p*-value was 2.5% after compensating for sequence effects. This indicates that for the H scenario, the null hypothesis was invalid and the mental workload was improved by using the ASRU support.

4.3. Objective Results

In this section, the results from the secondary task (Stroop test) and the performance measurements are analyzed and discussed.

4.3.1. Results from Secondary Task—Stroop Test

Table 17 shows the result when the ATCos' successfully performed Stroop tests for the different traffic scenarios without and with compensating for the sequence effects. The results are obtained from [29].

Table 17. Number of successfully performed Stroop Tests.

Medium				Heavy				Both			
Diff		α		Diff		α		Diff		α	
SE	NSE	SE	NSE	SE	NSE	SE	NSE	SE	NSE	SE	NSE
11.3	13.5	9.3%	3.6%	14.3	14.3	3.6%	2.3%	12.8	13.9	1.3%	0.3%

Minimal α values are shaded in green for $0\% \leq \alpha < 5\%$, see the NASA TLX results Section 4.2.2 for column names.

The Stroop test results show that for each scenario (Medium and Heavy) as well as when combining both scenarios, the difference ranged from 13.9 (both) to 14.5 (Medium). This indicates that the ATCos were able to perform more successful Stroop tests using the ASRU support during the simulation runs. When supported by ASRU, the ASRU performs many tasks that the ATCos otherwise need to do manually. The average of successfully solved Stroop tests in the H scenario was 19.9 without ASRU support and 34.2 with ASRU support (not shown in Table 17). It is important to note that we do not plan to occupy the ATCo with an additional task. This observation demonstrates that in certain situations, such as an incident, the ASRU support provides some additional safety buffers in terms of workload capacity. In the M scenarios, the average number of successfully solved tests increased from 34.3 without ASRU support to 47.8 with ASRU support. The increase in both traffic situations was nearly the same, although the command recognition was slightly worse in the H scenarios, as shown in Table 6. The p-value results were below 5% across all scenarios. Thus, the null hypothesis was invalid and the number of possible additional tasks was improved by using the ASRU support.

4.3.2. Missing and Wrong Radar Label Cell Entries

Knowing now that ASRU support reduces ATCo workload, it is important to ensure the accuracy of the radar label contents after the ATCo has checked the ASRU output, i.e., if all the given commands show the actual situation in digital form. How often do we have missing or even wrong inputs? In theory, a person would need to count how often the radar label contents were different from the spoken commands. However, this approach is impractical. It is nearly impossible for someone to listen to ATCo utterances, completely understand them on a word level, transform to the meaning to a semantic level, and check the radar label contents with the required accuracy. A deviation of approximately 1% is expected. Transcription experiments with humans transcribing voice utterances has already shown that a word error rate of approximately 4% to 11% can be expected, especially when a person can only listen once [45]. A computer-based solution is required, which is described below.

During the experiments, all mouse clicks that changed the radar label cell contents were recorded, and Table 18 shows the results of these recordings. The correct contents of each cell for each callsign at any point in time is indirectly given. All ATCo voice transmissions were transcribed and annotated, creating what are known as gold annotations. These gold annotations were replayed and sent to the software that generated the contents of the radar label cells. As a result, the clicks are recorded again, but giving us this time the correct and complete contents of each cell for each callsign at any point in time. The cell contents during the experiments can then be compared to the correct/gold contents. The comparison of the label cell contents during the experiments to the correct contents can be done automatically, and the calculation can be automatically rerun whenever inconsistencies in the gold annotations are identified.

Table 18, taken from [29], shows the results for the baseline and the solution runs. The first column shows the number of clearances given for each cell. We did not count commands which cleared a value in a field, such as "*own navigation*" or "*no speed restrictions*", but we considered them when a calculation was missing or when wrong cell entries were present. The "*Gold*" column contains the number of commands of this type, resulting from the replay of the manual annotations. "*Clicks*" counts the number of clicks in this cell, which changed the value of the cell, ignoring clicks that cleared the value.

Table 18. Number of errors in radar-label cells after compensating for sequence effects for the heavy and medium scenarios.

	Baseline					Solution				
Type	**Gold**	**Clicks**	**Miss**	**Add**	**RR**	**Gold**	**Clicks**	**Miss**	**Add**	**RR**
Alti *	1950	1906	62	20	95%	1978	28	19	16	95%
Spd *	1102	1074	70	35	89%	1183	34	17	3	89%
Head *	936	572	351	8	94%	894	7	30	11	94%
WP *	598	589	29	14	85%	604	20	18	25	87%
Tran *	301	216	89	12	85%	289	7	23	1	88%
Rate *	63	74	13	4	67%	64	11	6	1	74%
Spec *	1367	936	14	15	93%	1372	19	34	15	92%

* Row "*Alti*" shows the number of commands, which were spoken and would require an input into the altitude cell in the radar label. "*Spd*" denotes the speed cell, "*Head*" denotes the heading cell, "*WP*" denotes the waypoint cell, "*Tran*" denotes the "Transition/Route" cell, "*Rate*" denotes the descent rate cell, and "*Spec*" denotes the ILS/approach clearance and the change frequency command type cell. Cells marked in *orange*, are analyzed in more detail in [29].

The "*Miss*" column counts the number of cell values that were missing, and the "*Add*" column represents the number of cell values which were in the cells but not spoken at that time. "*RR*" is the command recognition rate for each type. The entries in the cells "*Miss*" and "*Add*" were corrected for sequence effects as described in Section 3.3. However, the compensation effects were much smaller than for the Stroop test. The greatest change was by 1.3 in absolute numbers. Some cells in Table 18 are marked in *orange*, which require a deeper analysis or additional explanations which are provided in [29]. The sum of clicks and missing commands did not correspond to the gold column. Sometimes, the same command was repeated with the same value, due to "*say again*" or a lack of response from the simulation-pilots. Additionally, a gold command may sometimes require two entries, such as when using DIRECT_TO to a waypoint, which should also delete the heading value.

The analysis of the results in Table 18 in [29] has given insights to improve the comparison between the correct and actual label values to better reflect how ATCos of Austro Control work in daily life. The improvements, which were not part of [29], are presented in Table 19.

Table 19. Number of errors in radar label compensating sequence effects and considering the special situation of the Vienna Approach Control.

	Baseline					Solution				
Type *	**Gold**	**Clicks**	**Miss**	**Add**	**RR**	**Gold**	**Clicks**	**Miss**	**Add**	**RR**
Alti	1950	1906	58	11	95%	1978	28	19	15	95%
Spd	1102	1074	70	15	89%	1183	34	17	2	89%
Head	936	572	108	4	94%	894	7	13	10	94%
WP	598	589	29	11	85%	604	20	18	24	87%
Tran	301	216	90	11	85%	289	7	22	1	88%
Rate	63	74	13	2	67%	64	11	6	0	74%
Spec	1367	936	13	12	93%	1372	19	34	15	92%
Sum	6317	5367	380	65		6384	126	130	68	

* The rows are already explained in the footer of Table 18. The blue and orange color coding of the cells is explained in the text below.

- A missing value was not considered twice. For example, if 250 knots were intended/said and the value 240 was accidently entered or wrongly recognized, and therefore incorrect, Table 18 counted this as missing 250 and as an additional value of 240 in the "*Spd*" row. In Table 19, missing label cell values were not counted twice if they already have an entry in the "miss" column for the same given command value. This and other corrections reduced the sum of column "*Add*" from 145 to 65 in the baseline runs. The effect in solution runs was a minor reduction from 73 to 68.
- Missing entries for 2600 feet were not counted as "*Miss*" or "*Add*" because this was the interception altitude at the final approach fix, which was not instructed by ATCo after the "cleared ILS" instruction.
- If a heading value was given together with an ILS clearance, we did not expect an entry in the heading cell. This reduced the number of missing heading values from 351 in Table 18 to 107 in Table 19 for the baseline runs.

The values in Table 19 shaded in blue were previously explained in [29]. The cells marked in orange require a deeper analysis.

- In 70 instances, the ATCo missed entering the given speed value or entered a wrong value. No systematic reason was observed. CAS (calibrated air speed) speed values between 160 and 300 were involved, and the majority were between CAS 160 and CAS 220. It was assumed that the high workload prevented the ATCo from inputting the given speed value in some situations, accounting for 6.5% of the cases.
- TRANSITION commands were not deemed as important for the ATCos of Austro Control. They did not input a given transition command in their operational TopSky system. Nevertheless, Austro Control had designed the experiment and the HMI in such a way that the ATCos should input the cleared transitions, which is a benefit of using ASRU support with respect to situation awareness.
- The same applies for the given heading commands that were not input manually.

The initial question was how to verify the number of missing ATCo commands in the label cells with and without ASRU support. This was done by replaying the annotated utterances. The next question was if all the missing commands were corrected by the ATCo. The numbers in Table 19 show that this was not the case. However, it was also not the case when the ATCo manually inputs all commands. We performed paired *t*-tests as described in Section 3.3 to validate whether the differences were statistically significant. We used the data after compensating for the sequence effects from Table 19. The results without and with compensating sequence effects are shown in Table 20.

Table 20. Minimum alpha values for the hypothesis that ASRU improves the correctness of radar label cell contents without and with compensating for sequence effects.

Hypotheses	Medium		Heavy		Both	
	SE	NSE	SE	NSE	SE	SNE
Alti.	5.0×10^{-4}	5.0×10^{-4}	9.2%	8.1%	2.2×10^{-4}	4.7×10^{-4}
Spd	2.0%	2.0%	0.6%	5.4×10^{-3}	2.8×10^{-4}	5.7×10^{-4}
Head	6.2%	6.0%	1.1%	5.6×10^{-3}	1.9×10^{-3}	2.8×10^{-3}
WP	18.7%	17.0%	8.6%	8.5%	4.8%	5.1%
Tran	2.8%	1.3%	8.2×10^{-4}	4.7×10^{-4}	5.1×10^{-5}	1.5×10^{-4}
Rate	−24.3%	−20.7%	3.5%	2.9%	11.4%	11.8%
Spec	−6.8%	−7.1%	−9.3%	−9.0%	−2.6%	−2.9%
Sum	2.9×10^{-3}	2.7×10^{-3}	2.3×10^{-4}	2.3×10^{-4}	3.5×10^{-7}	3.8×10^{-6}

Minimal α values are shaded in green for $0\% \leq \alpha < 5\%$, in light green for $5\% \leq \alpha < 10\%$, in orange if we have evidence that results were worse with ASRU support ($-5\% \leq \alpha < 0\%$), in light red for ($-10\% \leq \alpha < 5\%$), and in yellow for the rest ($|\alpha| \geq 10\%$).

Table 20 shows that the results were highly statistically significant for almost all radar label cells, indicating that the correctness of radar label cells was much better if ATCos were supported by ASRU. However, statistical significance was not observed for the waypoint, rate and special radar label cells across all scenarios. The deviation for the "Spec" was intended by the design of experiment, as described in [29]. The entries of the "Spec" commands appeared in the forth label line, which was only visible for the ATCo, when the mouse was hovered over the corresponding radar label.

The results highlight that even with an ASRU command recognition rate of only 92%, which is already very good compared to other results reported in the context of SESAR-2 ASRU validation exercises [7], the ATCos workload and their human performance was not negatively impacted. Furthermore, there are an abundance of safety nets such as Monitoring Aids (MONA), Cleared Level Adherence Monitoring (CLAM), Route Adherence Monitoring (RAM), Short-term Conflict Alert (STCA), and Medium-term Conflict Detection (MTCD), which would prevent any critical safety event. Additionally, the use of eye-tracking to verify whether the ATCo visually scanned the ASRU output can help to further reduce the negative effect of ASRU errors. In case the ATCo did not check the ASRU output within a certain time period, auditory or visual attention guidance could be a possible solution.

5. Conclusions

The main research question was to quantify of the benefits of Automatic Speech Recognition and Understanding (ASRU) support for ATCos performing radar label maintenance in terms of safety and human performance. Therefore, an extensive human-in-the-loop study with twelve Austro Control ATCos was carried out at DLR Braunschweig. A method to compensate for sequence effects was introduced, which improved the statistical significance by a factor of two on average, thus reducing the number of required ATCos. Furthermore, for the first time, we were able to analyze how many radar label inputs were incorrect when ASRU support was provided and when it was not available.

The measured accuracy of speech-to-text and text-to-concept has shown that the ASRU technology functions reliably and robustly. For all radar cells, a command recognition rate of 92.5% with an error rate of 2.4% was achieved.

In terms of flight safety, the number of wrong or missing inputs from ATCos into the radar label was reduced by a factor of more than two through ASRU support usage (from 11% to 4%). Hence, ATCos had more mental spare capacity when using ASRU support for radar label maintenance, which is crucial for safety in unforeseen events such as an incident. This was demonstrated through a secondary task, where occupying less mental capacity in the primary task (air traffic control) increased the situational awareness among ATCos, which can be beneficial in safety-critical situations. These findings were confirmed by the results of the SASHA questionnaire, with a statistical significance of $\alpha = 1.8\%$. The reduction in workload was measured using NASA TLX, ISA, Bedford, SHAPE, and AIM questionnaires.

In addition to the impact of ASRU support on flight safety and workload, the ATCos reported an increased satisfactory and trust level in human-system performance when using the ASRU support. The results from the CARS and SATI questionnaire showed that ATCos acceptance increased, with $\alpha = 0.7\%$, and improved trust with high statistical significance ($\alpha = 0.5\%$). Overall, flight safety and human performance was significantly improved when ATCos use ASRU support for radar label maintenance.

The ANSP involved in the study designed the user interface in a way that required the ATCos to input all commands into the ATC system, which is not done by ATCos in their current operational system. In the future, ANSPs are strongly recommended to have all commands in digitized form, ensuring that the CWP offers a way to enter the commands without significantly increasing ATCo workload. The presented ASRU technology is a lightweight method to support this transition and increase situational awareness as an additional benefit when all commands are integrated into the ATC system.

Author Contributions: Conceptualization, H.H.; methodology, H.H.; software, H.H., M.K., O.O., J.Z.-G., N.A. and H.E.; validation, N.A., T.M. and H.H.; formal analysis, H.H.; investigation, H.H. and N.A.; resources, N.A., H.H. and L.K.; data curation, H.E., O.O., M.K., H.H. and J.Z.-G.; writing—original draft preparation, N.A., H.H. and O.O.; writing—review and editing, O.O., T.M., H.E., M.K, J.Z.-G. and L.K.; visualization, N.A., H.H., M.K., H.E., O.O. and T.M.; supervision, H.H.; project administration, H.H.; funding acquisition, H.H. All authors have read and agreed to the published version of the manuscript.

Funding: This project received funding from the SESAR Joint Undertaking under the European Union's Horizon 2020 research and innovation program under grant agreement No. 874464.

Institutional Review Board Statement: Not applicable.

Informed Consent Statement: Informed consent was obtained from all subjects involved in the study.

Data Availability Statement: Not applicable.

Acknowledgments: The authors would like to thank all the ATCos from Austro Control who supported the validations with their feedback and their voice utterances. The authors would also like to thank the employees from DLR for completing tedious transcription and annotation work.

Conflicts of Interest: The authors declare no conflict of interest. The funders had no role in the design of the study; in the collection, analyses, or interpretation of data; in the writing of the manuscript; or in the decision to publish the results.

Appendix A

For the following questionnaires, different scales were applied to answer or rate the corresponding question or statement. These scales can be found in the correspondent section.

Appendix A.1. Questions Used for NASA TLX

1. How mentally demanding was the task?
2. How physically demanding was the task?
3. How hurried or rushed was the pace of the task?
4. How successful were you in accomplishing what you were asked to do?
5. How hard did you have to work to accomplish your level of performance?
6. How insecure, discouraged, irritated, stressed, and annoyed were you?

Appendix A.2. Statements Used for SUS Questionnaire

1. I think that I would like to use this system frequently.
2. I found the system unnecessarily complex.
3. I thought the system was easy to use.
4. I think that I would need the support of a technical person to be able to use this system.
5. I found the various functions in this system were well integrated.
6. I thought there was too much inconsistency in this system.
7. I would imagine that most people would learn to use this system very quickly.
8. I found the system very cumbersome to use.
9. I felt very confident using the system.
10. I needed to learn a lot of things before I could get going with this system.

Appendix A.3. Statements Used for SASHA Questionnaire

1. In the previous run, I was ahead of the traffic.
2. In the previous run, I started to focus on a single problem or a specific aircraft.
3. In the previous run, there was a risk of forgetting something important (such as inputting the spoken command values into the labels).
4. In the previous run I was able to plan and organize my work as wanted.
5. In the previous run I was surprised by an event I did not expect (such as an aircraft call).
6. In the previous run I had to search for an item of information.

Appendix A.4. Statements Used for SATI Questionnaire

1. In the previous working period, I felt that the system was useful.
2. In the previous working period, I felt that the system was reliable.
3. In the previous working period, I felt that the system worked accurately.
4. In the previous working period, I felt that the system was understandable.
5. In the previous working period, I felt that the system worked robustly (in difficult situations, with invalid inputs, etc.).
6. In the previous working period, I felt that I was confident when working with the system.

Appendix A.5. Questions Used for AIM Questionnaire

1. In the previous run, how much effort did it take to prioritize tasks?
2. In the previous run, how much effort did it take to identify potential conflicts?
3. In the previous run, how much effort did it take to scan radar or any display?
4. In the previous run, how much effort did it take to evaluate conflict resolution options against the traffic situation and conditions?
5. In the previous run, how much effort did it take to anticipate the future traffic situation?
6. In the previous run, how much effort did it take to recognize a mismatch of available data with the traffic picture?
7. In the previous run, how much effort did it take to issue timely commands?
8. In the previous run, how much effort did it take to evaluate the consequences of a plan?
9. In the previous run, how much effort did it take to manage flight data information?
10. In the previous run, how much effort did it take to recall necessary information?
11. In the previous run, how much effort did it take to anticipate team members' needs?
12. In the previous run, how much effort did it take to prioritize requests?
13. In the previous run, how much effort did it take to scan flight progress data?
14. In the previous run, how much effort did it take to access relevant aircraft or flight information?
15. In the previous run, how much effort did it take to gather and interpret information?

References

1. Shetty, S.; Ohneiser, O.; Grezl, F.; Helmke, H.; Motlicek, P. *Transcription and Annotation Handbook. HAAWAII Deliverable D3*; HAAWAII Project: Cologne, Germany, 2020.
2. Helmke, H.; Slotty, M.; Poiger, M.; Herrer, D.F.; Ohneiser, O.; Vink, N.; Cerna, A.; Hartikainen, P.; Josefsson, B.; Langr, D. Ontology for transcription of ATC speech commands of SESAR 2020 solution PJ. 16-04. In Proceedings of the IEEE/AIAA 37th Digital Avionics Systems Conference (DASC), London, UK, 23–27 September 2018.
3. International Civil Aviation Organization (ICAO). *Procedures for Air Navigation Services (PANS)-Air Traffic Management (Doc 4444)*; International Civil Aviation Organization: Montreal, QC, Canada, 2001.
4. Schäfer, D. Context-sensitive Speech Recognition in the Air Traffic Control Simulation. Ph.D. Thesis, University of Armed Forces, Neubiberg, Germany, 2001.
5. Cordero, J.M.; Dorado, M.; de Pablo, J.M. Automated speech recognition in ATC environment. In Proceedings of the 2nd International Conference on Application and Theory of Automation in Command and Control Systems, London, UK, 29–31 May 2012.
6. Cordero, J.M.; Rodriguez, N.; de Pablo, J.M.; Dorado, M. *Automated Speech Recognition in Controller Communications Applied to Workload Measurement*; 3rd SESAR Innovation Days: Stockholm, Sweden, 2013.
7. Ohneiser, O.; Helmke, H.; Shetty, S.; Kleinert, M.; Ehr, H.; Murauskas, S.; Pagirys, T.; Balogh, G.; Tönnes, A.; Kis-Pál, G. Understanding Tower Controller Communication for Support in Air Traffic Control Display. In Proceedings of the 12th SESAR Innovation Days, Budapest, Hungary, 5–8 December 2022.
8. Helmke, H.; Ohneiser, O.; Buxbaum, J.; Kern, C. Increasing ATM efficiency with assistant-based speech recognition. In Proceedings of the 12th USA/Europe Air Traffic Management Research and Development Seminar (ATM2017), Seattle, WA, USA, 27–30 June 2017.
9. Helmke, H.; Ondrej, K.; Shetty, S.; Arilíusson, H.; Simiganosch, T.S.; Kleinert, M.; Ohneiser, O.; Ehr, H.; Zuluaga, J.-P. Readback Error Detection by Automatic Speech Recognition and Understanding-Results of HAAWAII project for Isavia's Enroute Airspace. In Proceedings of the 12th SESAR Innovation Days, Budapest, Hungary, 5–8 December 2022.

10. Helmke, H.; Shetty, S.; Kleinert, M.; Ohneiser, O.; Ehr, H.; Prasad, A.; Motlicek, P.; Cerna, A.; Windisch, C. Measuring Speech Recognition and Understanding Performance in Air Traffic Control Domain Beyond Word Error Rates. In Proceedings of the 11th SESAR Innovation Days, Virtual, 7–9 December 2021.
11. Kleinert, M.; Helmke, H.; Siol, G.; Ehr, H.; Finke, M.; Srinivasamurthy, A.; Oualil, Y. Machine learning of controller command prediction models from recorded radar data and controller speech utterances. In Proceedings of the 7th SESAR Innovation Days, Belgrade, Serbia, 28–30 November 2017.
12. Helmke, H.; Ohneiser, O.; Mühlhausen, T.; Wies, M. Reducing controller workload with automatic speech recognition. In Proceedings of the 35th Digital Avionics Systems Conference (DASC), Sacramento, CA, USA, 25–29 September 2016; pp. 1–10.
13. Eggemeier, F.T.; O'Donnell, R.D. *A Conceptual Framework for Development of a Workload Assessment Methodology*; Wright State University: Dayton, OH, USA, 1982.
14. Speelmann, V. Air Traffic Management and Operations Simulator (ATMOS). Deutsches Zentrum für Luft und Raumfahrt e.V. Available online: https://www.dlr.de/content/en/research-facilities/air-traffic-management-and-operations-simulator-atmos.html (accessed on 3 March 2023).
15. Morlang, F. Validation Facilities in the Area of ATM Bottleneck Investigation. In Proceedings of the IEEE/AIAA 25th Digital Avionics Systems Conference, Portland, OR, USA, 15–19 October 2006.
16. EUROCONTRL. *European Operational Concept Validation Methodology*; Version 3; EUROCONTROL: Brussels, Belgium, 2010.
17. Mankins, J. *Technology Readiness Level-A White Paper*; Advanced Concept Office, Office of Space Access and Technology NASA: Washington, DC, USA, 6 April 1995.
18. Fürstenau, N.; Radüntz, T. Power law model for subjective mental workload and validation through air traffic control human-in-the-loop simulation. *Cogn. Technol. Work* **2021**, *24*, 291–315. [CrossRef]
19. Milan, R.; Michael, F. Using speech analysis in voice communication: A new approach to improve air traffic management security. In Proceedings of the 7th IEEE International Conference on Cognitive Infocommunications (CogInfoCom), Wroclaw, Poland, 16–18 October 2016.
20. Kluenker, C.S. Enhanced Controller Working Position for Integrating Spaceflight into Air Traffic Management. In *Advances in Human Aspects of Transportation, Proceedings of the AHFE 2021 Virtual Conference on Human Aspects of Transportation (Online), 25–29 July 2021*; Springer Lecture Notes in Networks and Systems 270; Springer: Berlin/Heidelberg, Germany; pp. 543–550. [CrossRef]
21. Have, J.T. The development of the NLR ATC Research Simulator (Narsim): Design philosophy and potential for ATM research. *Simul. Pr. Theory* **1993**, *1*, 31–39. [CrossRef]
22. Nuic, A. *Base of Aircraft Data (BADA) Product Management Document*; EEC Technical Report No. 2009-008; EUROCONTROL: Brussels Belgium, March 2009.
23. ICAO. *ICAO Standard Phraseology: A Quick Reference Guide for Commercial Air Transport Pilots, Safety Initiative*; EUROCONTROL: Brussels, Belgium, 2011.
24. Aeronautical Information Publication. Austro Control, LOWW AD 2 MAP 11-2-4, Austro Control, Vienna Austria. Available online: https://eaip.austrocontrol.at/ (accessed on 22 April 2021).
25. Charness, G.; Gneezy, U.; Kuhn, M.A. Experimental methods: Between-subject and within-subject design. *J. Econ. Behav. Organ.* **2012**, *81*, 1–8. [CrossRef]
26. Greenwald, A.G. Within-subject designs: To use or not to use? *Psychological Bull.* **1976**, *83*, 314. [CrossRef]
27. Kleinert, M.; Helmke, H.; Moos, S.; Hlousek, P.; Windisch, C.; Ohneiser, O.; Ehr, H.; Labreuil, A. Reducing Controller Workload by Automatic Speech Recognition Assisted Radar Label Maintenance. In Proceedings of the 9th SESAR Innovation Days, Athens, Greece, 2–5 December 2019.
28. Helmke, H. The Horizon 2020 Funded HAAWAII Project. Deutsches Zentrum fuer Luft- und Raumfahrt e.V. Available online: https://www.haawaii.de/wp/ (accessed on 3 March 2023).
29. Helmke, H.; Kleinert, M.; Ahrenhold, N.; Ehr, H.; Mühlhausen, T.; Ohneiser, O.; Klamert, L.; Motlicek, P.; Prasad, A.; Zuluaga Gomez, J.; et al. Automatic Speech Recognition and Understanding for Radar Label Maintenance Support Increases Safety and Reduces Air Traffic Controllers' Workload. In Proceedings of the 15th USA/Europe Air Traffic Management Research and Development Seminar (ATM2023), Savannah, GA, USA, 17 May 2023.
30. Kleinert, M.; Helmke, H.; Shetty, S.; Ohneiser, O.; Ehr, H.; Prasad, A.; Motlicek, P.; Harfmann, J. Automated Interpretation of Air Traffic Control Communication: The Journey from Spoken Words to a Deeper Understanding of the Meaning. In Proceedings of the IEEE/AIAA 40th Digital Avionics Systems Conference (DASC), San Antonia, TX, USA, 3–7 October 2021.
31. Kleinert, M.; Shetty, S.; Helmke, H.; Ohneiser, O.; Wiese, H.; Maier, M.; Schacht, S.; Nigmatulina, I.; Saeed, S.; Motlicek, P. Apron Controller Support by Integration of Automatic Speech Recognition with an Advanced Surface Movement Guidance and Control System. In Proceedings of the 12th SESAR Innovation Days, Budapest, Hungary, 5–8 December 2022.
32. Hart, S. NASA-task load index (NASA-TLX); 20 years later. In Proceedings of the Human Factors and Ergonomics Society Annual Meeting, Los Angeles, CA, USA, 16–20 October 2006; pp. 904–908.
33. Roscoe, A. *Assessing Pilot Workload in Flight*; Royal Aircraft Establishment Bedford (United Kingdom): Bedford, UK, 1984.
34. Brooke, J. SUS—A 'Quick and Dirty' Usability Scale. In *Usability Evaluation in Industry*; Jordan, P.W., Thomas, B., McClelland, I.L., Weerdmeester, B.A., Eds.; Taylor and Francis: London, UK, 1996; pp. 189–194.

35. Lee, K.; Kerns, K.; Bone, R.; Nickelson, M. Development and validation of the controller acceptance rating scale (CARS): Results of empirical research. In Proceedings of the 4th USA/Europe Air Traffic Management R&D Seminar, Santa Fe, New Mexico, 4–7 December 2001.
36. Dehn, D.M. Assessing the Impact of Automation on the Air Traffic Controller: The SHAPE Questionnaires. *Air Traffic Control. Q.* **2008**, *16*, 127–146. [CrossRef]
37. Kirwan, B.; Evans, A.; Donohoe, L.; Kilner, A.; Lamoureux, T.; Atikinson, T.; MacKendrick, H. Human factors in the ATM system design life cycle. In Proceedings of the FAA/EUROCONTROL ATM R&D Seminar, Saclay, France, 16–20 June 1997.
38. Tattersall, A.J.; Foord, P.S. An experimental evaluation of instantaneous self-assessment as a measure of workload. *Ergonomics* **1996**, *39*, 740–748. [CrossRef] [PubMed]
39. Brennan, S.D. *An Experimental Report on Rating Scale Descriptor Sets for the Instantaneous Self-Assessment (ISA) Recorder;* Technical Report; DRA Maritime Command and Control Divison: Portsmouth, UK, 1992; Volume 92017.
40. Joshi, A.; Kale, S.; Chandel, S.; Pal, D.K. Likert scale: Explored and explained. *Br. J. Appl. Sci. Technol.* **2015**, *7*, 396. [CrossRef]
41. Kaber, D.B.; Riley, J.M. Adaptive Automation of a Dynamic Control Task Based on Secondary Task Workload Measurement. *Int. J. Cogn. Ergon.* **1999**, *3*, 169–187. [CrossRef]
42. Stroop, J.R. Studies of interference in serial verbal reactions. *J. Exp. Psychol.* **1935**, *18*, 643–662. [CrossRef]
43. Casner, S.M.; Gore, B.F. Measuring and evaluating workload: A primer. In *NASA Technical Memorandum;* 2010-216395; NASA Ames: Moffett Field, CA, USA, 2010.
44. Levenshtein, V.I. *Binary Codes Capable of Correcting Deletions, Insertions and Reversals;* Doklady Akademiii Nauk SSSR, Translator; USSR Academy of Science, Leningrad Soviet Union: Moscow, Russia, 1966; Volume 163, pp. 845–848.
45. Stolcke, A.; Droppo, J. Comparing Human and Machine Errors in Conversational Speech Transcription. *Proc. Interspeech* **2017**, 137–141. [CrossRef]

aerospace

Article

Safety Aspects of Supporting Apron Controllers with Automatic Speech Recognition and Understanding Integrated into an Advanced Surface Movement Guidance and Control System

Matthias Kleinert [1,*], Oliver Ohneiser [1], Hartmut Helmke [1], Shruthi Shetty [1], Heiko Ehr [1], Mathias Maier [2], Susanne Schacht [2] and Hanno Wiese [3]

1 German Aerospace Center (DLR), Institute of Flight Guidance, Lilienthalplatz 7, 38108 Braunschweig, Germany; oliver.ohneiser@dlr.de (O.O.); hartmut.helmke@dlr.de (H.H.); shruthi.shetty@dlr.de (S.S.); heiko.ehr@dlr.de (H.E.)
2 ATRiCS Advanced Traffic Solutions GmbH, Am Flughafen 7, 79108 Freiburg im Breisgau, Germany; mathias.maier@atrics.com (M.M.); susanne.schacht@atrics.com (S.S.)
3 Fraport AG, Frankfurt Airport Services Worldwide, 60547 Frankfurt am Main, Germany; h.wiese@fraport.de
* Correspondence: matthias.kleinert@dlr.de; Tel.: +49-531-295-2567

Abstract: The information air traffic controllers (ATCos) communicate via radio telephony is valuable for digital assistants to provide additional safety. Yet, ATCos have to enter this information manually. Assistant-based speech recognition (ABSR) has proven to be a lightweight technology that automatically extracts and successfully feeds the content of ATC communication into digital systems without additional human effort. This article explains how ABSR can be integrated into an advanced surface movement guidance and control system (A-SMGCS). The described validations were performed in the complex apron simulation training environment of Frankfurt Airport with 14 apron controllers in a human-in-the-loop simulation in summer 2022. The integration significantly reduces the workload of controllers and increases safety as well as overall performance. Based on a word error rate of 3.1%, the command recognition rate was 91.8% with a callsign recognition rate of 97.4%. This performance was enabled by the integration of A-SMGCS and ABSR: the command recognition rate improves by more than 15% absolute by considering A-SMGCS data in ABSR.

Keywords: air traffic controller; simulation pilot; workload; assistant-based speech recognition; automatic speech recognition and understanding; apron control; STARFiSH

check for updates

Citation: Kleinert, M.; Ohneiser, O.; Helmke, H.; Shetty, S.; Ehr, H.; Maier, M.; Schacht, S.; Wiese, H. Safety Aspects of Supporting Apron Controllers with Automatic Speech Recognition and Understanding Integrated into an Advanced Surface Movement Guidance and Control System. *Aerospace* **2023**, *10*, 596. https://doi.org/10.3390/aerospace10070596

Academic Editor: Michael Schultz

Received: 16 May 2023
Accepted: 19 June 2023
Published: 29 June 2023

1. Introduction

This article is an extended version of [1].

In air traffic control (ATC), there is a permanent need to increase the efficiency of handling air and ground traffic. This need exists especially at highly frequented airports such as Frankfurt. However, increasing efficiency must never come at the expense of safety. An important approach to increase efficiency and safety on the ground is by supporting ground traffic controllers' decision making through digitization and automation with new digital assistant systems that are integrated in or interoperate with advanced surface movement guidance and control systems (A-SMGCS) [2]. Currently, A-SMGCS already monitor data from different sensors and are designed to enable controllers to guide traffic more safely without reducing the capacity of traffic guidance.

1.1. Motivation

The most advanced digital assistants in apron control today already have access to a large number of sensors. Together with manual inputs from the controller, the digital assistants are able to detect potentially dangerous situations and warn the controller about them. In contrast, voice communication between controllers and pilots, one of the most important sources of information in ATC, has not yet been used by these assistants. Whenever information from voice communication needs to be digitized, controllers are burdened with the additional task of entering this information into the ATC system. Research results show that up to one third of controllers' working time is spent on these manual entries [3]. This leads to a reduction in overall efficiency as controllers spend less time optimizing traffic flow [4]. The amount of time spent on manual inputs will even increase in the coming years as future regulations require more alerting functions to be implemented, hence more inputs are needed, especially in apron control, e.g., Commission Implementing Regulation (EU) 2021/116 [5].

Assistant-based speech recognition (ABSR) has already shown in the past that it is possible to significantly reduce manual input from controllers by recognizing and understanding the controller–pilot communication and automatically providing the required inputs into digital assistants [6]. ABSR technology has been continuously developed in several projects, and possible fields of application have been identified in the context of research prototypes. So far, there has been no initial integration into a commercial system to demonstrate that ABSR can also meet the corresponding requirements for safety and usability in a system network commonly used in aviation. In the German Federal Ministry of Education and Research funded project STARFiSH (Safety and Artificial Intelligence Speech Recognition), a powerful artificial intelligence (AI)-based speech recognition system was integrated into a modern A-SMGCS for apron control [1]. The solution was supposed to reduce the additional workload of controllers as much as possible by using speech recognition and understanding capabilities. At the same time, the solution was supposed to be objectively safe and be rated as reliable, easy to use, and safe by the controllers.

1.2. Related Work

This section outlines related work for automatic speech recognition (Section 1.2.1) and understanding applications (Section 1.2.2) in air traffic control and closes with related work on how both are used to automatically fill digital flight strips or radar labels with voice information from the controller–pilot communication in Section 1.2.3.

1.2.1. Early Work on Speech Recognition in Air Traffic Control

Speech Recognition in general has a long history of development. It started in 1952 when Davis, Biddulph, and Balashek of Bell Laboratories built a digit recognition system for a single speaker called "Audrey" [7]. Over the last 70 years, technological advances have led to dramatic improvements in the field of speech recognition. An overview of the first four decades is provided by, e.g., Juang and Rabiner [8]. Connolly from FAA was one of the first to describe the steps of using automatic speech recognition (ASR) in the air traffic management (ATM) domain [9]. In the late 1980s, a first approach to incorporate speech technologies in ATC training was reported [10]. Such developments led to enhanced ASR systems, which are used in ATC training simulators to replace expensive simulation pilots, e.g., FAA [11,12], DLR [13], MITRE [14], and DFS [15].

The challenges with ASR in ATC today go beyond basic training scenarios, where often standard procedures and ICAO phraseology [16] are followed very closely. Modern ASR applications have to recognize experienced controllers, who more often make deviations from the mentioned standards. Furthermore, applications with ASR also go beyond just the scope of simulation and training. ASR is for example used to obtain more objective feedback concerning controllers' workload [17,18]. A good overview of the integration of ASR in ATC is provided in the two papers of Nguyen and Holone [19,20].

1.2.2. Speech Recognition and Understanding Applications in Air Traffic Control

In the recent past, research projects developed prototypical applications with speech recognition and understanding for all ATC domains from en route [21], via approach [4], to tower and ground [22]. These prototypes support controllers in maintaining aircraft radar labels [23] and flight strips [22] to reduce workload, recognize and highlight aircraft callsigns [24], build safety nets for tower control [25], or even offer automatic readback error detection with reports to controllers [26,27]. The systems have matured to recognize words and meanings of real-life controller and pilot utterances even beyond the simulated environments [28]. The rules of how a speech understanding system can annotate the meaning conveyed with ATC radio transmissions are defined in an ontology that was agreed between major European air traffic management stakeholders in 2018 [29]. With this ontology, different word sequences can be mapped to unique word sequence meanings, e.g., the word sequences "lufthansa zero seven tango taxi via november eight and november to stand victor one five eight" and "zero seven tango via november eight november victor one five eight" both correspond to the same annotation in the ontology "DLH07T TAXI VIA N8 N, DLH07T TAXI TO V158".

1.2.3. Related Work for Pre-Filling Flight Strips and Radar Labels

The information which air traffic controllers communicate via radio telephony is valuable for digital assistants to provide additional safety. Yet, controllers are usually burdened with entering this information manually. Assistant-based speech recognition (ABSR) has been shown to be a lightweight technology that automatically extracts ATC communication content without additional human workload and that successfully feeds digital systems [6]. DLR, together with Austro Control, DFS, and other European air navigation service providers, has demonstrated that pre-filling radar labels supported by automatic speech recognition and understanding reduces air traffic controllers' workload [3] and increases flight efficiency with respect to flight time and kerosene consumption. Fuel burn can be reduced by 60 L per aircraft in the approach phase [4]. DLR, Austro Control, Thales, and the air navigation service provider of Czech Republic have redesigned this exercise with a commercial off-the-shelf speech recognizer and an industrial radar screen. The exercise results clearly showed that speech recognition, i.e., obtaining the sequence of words from a voice signal, is not enough [30]. Speech understanding is needed for providing information for flight strips and radar labels.

Recently DLR and Austro Control analyzed the safety aspects of using speech recognition and understanding for pre-filling radar label contents. They investigated how many of the verbally spoken approach controller commands, with and without speech recognition, were finally entered into the ATC system and how many errors were made, not recognized, or not corrected by the air traffic controllers. Despite manual corrections of commands even with speech recognition and understanding support, about 4% of the spoken commands were still not correctly entered into the system. However, this result, which is initially alarming from a safety point of view, is quickly put into perspective, when considering that roughly 10% of the verbally spoken commands are incorrectly or not entered at all into the system, if no speech recognition and understanding support is available. More details are provided in [23]. The results show that speech recognition and understanding [31] is far from being perfect, but a system without speech recognition and understanding seems to be even further away.

One of the main input sources for this paper, which describes the results or transforming the support tool for approach controllers to apron controllers, were two studies of DLR: one from 2015 for the Dusseldorf approach [4] and a recent one for the Vienna approach control [23]. It was expected that the good results of command recognition in the approach area will translate one-to-one to the correctness and completeness of inputs in the apron area. Previous projects have already taken first steps towards using speech recognition and understanding in a tower or apron environment, which included for example the prediction of potential controller commands [32]. The actual use of speech recognition and understanding was then further investigated in a multiple remote tower setup [33]. In the process, relevant information for digital flight strips was automatically derived and entered from the given verbal commands. The tower environment already covered many of the command types relevant for ground/apron traffic such as taxi, hold short, and pushback instructions.

1.3. Paper Structure

Section 2 summarizes the use case of supporting apron controllers, the iterative software development approach and introduces the Software Failure Modes, Effects, and Criticality Analysis. Section 3 describes the final version of the evaluation system. Section 4 explains the validation of the developed application. Section 5 presents the validation results before Sections 6 and 7 finalize the paper with discussions and conclusions.

2. Materials and Methods

2.1. Application Use Case of Supporting Apron Controllers

Initially, the apron controller, shown as "ATCo" in Figure 1, issues a command to the pilot by radio. Without ABSR (Figure 1, left), the controller enters this command into the A-SMGCS manually either before, afterwards, or in parallel to the radio call so that the system can provide automation functions. With ABSR (Figure 1, right), an ABSR-system automatically generates, based on the radio call, a data packet including metadata from the command, which is sent to the A-SMGCS. The A-SMGCS executes valid commands and highlights the changes together with the associated aircraft symbol. No system interaction by the controller is required unless an error has occurred. If the automatic speech recognition fails, the controller needs to manually correct or enter the command in the same way as without the ABSR system.

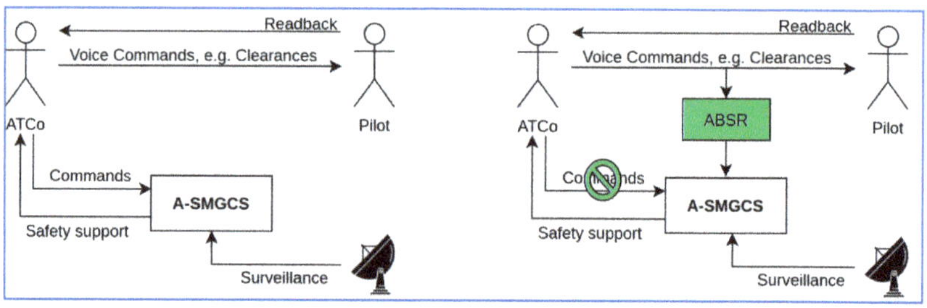

Figure 1. Interaction between human operators and digital systems in ground control without an ABSR component (**left**) and with an ABSR component (**right**).

The automatic recognition of voice commands issued by controllers to pilots should provide a solution to the problem that controllers are less able to keep an eye on traffic when they enter the information of the radio call that is necessary for modern A-SMGCS support functions. This includes, for example, entering taxi routes so that compliance can be monitored automatically.

From a technical point of view, the sequence of actions is:

1. Commands given via voice by the controller to the pilot are recorded as an audio data stream (A/D conversion of utterances).
2. The audio stream is divided into sections by detecting individual transmissions in the audio data.
3. Speech-to-text (S2T) transformation is applied on the resulting audio sections. S2T is based on neural networks trained with freely available data as well as with domain-specific recorded audio data for the target environment.
4. Relevant ATC concepts are automatically extracted from the S2T transcription using rule-based algorithms on a previously defined ontology and traffic data fed from the A-SMGCS.
5. High-level system commands are generated from the extracted ATC instructions using rules algorithmically interpreted from operational necessities according to the current traffic situation and fed into the system.

6. The changes to the system state resulting from the high-level system commands are presented to the human operators.
7. Human operators can correct or undo the automatic inputs.

We explain these steps by an example:

1. The apron controller is continuously speaking to the pilots with some gaps in between, e.g., " ... to seven five seven from the left ... lufthansa four two two good morning behind opposite air france three twenty one continue november eight lima hold short lima six ... austrian one foxtrot behind the passing". The gaps occur either because no further action is required or due to the verbal response of the (simulation) pilot, which is not available to the ABSR.
2. The audio stream sections are detected, and one continuous transmission could then be "lufthansa four two two good morning behind opposite air france three twenty one continue november eight lima hold short lima six".
3. Let us assume that the result of S2T contains some errors and results in the word sequence: "lufthansa four **to** two good morning behind opposite air **frans** three twenty one continue november eight lima **holding** short lima six" (errors marked in bold).
4. The relevant ATC instruction, being extracted by ABSR even with the errors from S2T, would be:
    a. DLH422 GREETING;
    b. DLH422 GIVE_WAY AFR A321 OPPOSITE;
    c. DLH422 CONTINUE TAXI;
    d. DLH422 TAXI VIA N8 L;
    e. DLH422 HOLD_SHORT L6.
5. The GREETING is ignored by the A-SMGCS. For the GIVE_WAY instruction the A-SMGCS may find out that the A321 from the opposite is the callsign AFR2AD. A symbol is generated in the human machine interface (HMI) of the apron controllers (and the simulation pilots), showing that DLH422 is waiting until the AFR2AD has passed. The continue statement is executed after the give way situation is resolved. The route along the taxiways N8 and L is shown. A hold short (stop) is displayed before taxiway L6.
6. In summary, the following visual output is shown to the apron controller:
    a. The aircraft symbol of DLH422 is highlighted;
    b. A GIVE_WAY symbol between the two aircrafts;
    c. The taxi route via N8 and L;
    d. A HOLD_SHORT symbol (stop) at L6.
7. The apron controller can accept or reject all three above options or can change some or all of them.

For the controller, almost all processing steps are invisible. Technology remains in the background. From the human operator's point of view, the sequence of actions is like this:

1. The callsign addressed in the controller's radio call is highlighted at the corresponding aircraft symbol in the A-SMGCS (DLH422 in the above example).
2. Once the commands to the pilot are fully uttered, they are converted into corresponding system commands that would otherwise have to be manually entered, e.g., a taxi route.
3. The result of the command input is displayed to the controller (and/or simulation pilot) in the A-SMGCS. Wherever possible, the visualization corresponds to the same visualization that would have resulted from a manual entry.
4. Special case: If an error in the data processing causes the wrong command to be sent and therefore the wrong effects (or none) to be displayed, the human operator must manually correct the command or enter it into the system. Depending on the type of command, dedicated buttons are offered for this purpose.

2.2. *Application Development*

The solution was created in four main iterations following the spiral model of software development [34]. For this purpose, an ABSR system was integrated with the A-SMGCS system TowerPad™ (see Figure 1) by iterating the following steps:

- Technical and operational requirements were determined;
- Software and interfaces were developed, implemented, and tested;
- Progress was validated by users in realistic operational scenarios in Fraport's training simulator;
- Results were analyzed to derive new requirements.

In the end, the system was intensively validated in realistic simulations with apron controllers and evaluated based on recorded data and defined metrics. The safety aspects detailed in the next subsection were addressed and focused on in the third iteration.

2.3. *Safety Considerations*

In aviation, a system can only be approved for operation if its impact on safety has been thoroughly assessed. This is even more important if it uses technology that is new and for which the currently available safety assessments are not necessarily suitable. For the use of artificial intelligence-based methods, discussions are taking place in the community regarding how the safety of AI methods can be verified or demonstrated. These discussions happen independent from air traffic management application areas.

However, there is a way out of the dilemma of the lack of approved testing and verification methods, which we saw in the STARFiSH project as a possibility to safely operate a system with AI-based speech recognition and understanding. If the AI system can be encapsulated in such a way that safety-critical outputs cannot have an immediate impact on real-world operation and must always be approved by the user of the system prior to implementation, safety will be verified during operation. However, manual checking of commands means additional effort that one does not want to impose on users for system inputs that cannot have any safety-critical consequences. Thus, it was important to identify which commands have effects that are safety-critical from an operational view.

In order to determine, which system inputs are safety-critical in this sense and which are not, a safety analysis based on the classification in Figure 2, must be performed that first determines, independently of the solution, which system inputs are potentially safety-relevant because they can endanger operational safety. For this analysis, we followed the EUROCONTROL "safety assessment methodology" (SAM) and applied the SFMECA methodology that is at the core of the "functional hazard analysis" (FHA).

Rule-based classification and execution based on safety criticality:

Rules	**Recognition good**	**Recognition bad**
Command non-critical	Implement command	Implement but offer undo
Command critical	Implement but warn	Ask ATCO for help

Figure 2. Naïve safety classification of ATC commands regarding safety criticality and recognition quality as suggested before execution of the project.

SFMECA (Software Failure Modes, Effects, and Criticality Analysis) [35] is a formalized method of risk assessment and subsequent identification of mitigation measures. It is a bottom-up method that analyzes so-called failure modes and their effects to identify (hidden) hazards at the system level. Using the standardized structure and presentation of the process and results specified by the SFMECA, a team of experts used predefined and individual "failure modes" to analyze which safety-relevant effects could be caused by the software and what their causes were for the functional requirements in the project. The requirements were grouped according to features and pre-filtered according to the evaluation dimensions:

- Safety-criticality;
- Criticality for the work of the controller;
- Risk due to potential software development errors.

Then the error cases (in categories "functionality", "timing", "sequencing", and "data, error handling") were quantitatively evaluated with their respective "root cause", checking for 26 common causes plus specific functional errors, e.g., "misdetection of callsign under own jurisdiction".

The evaluation criteria were the severity of the effects, their probability of occurrence, and the probability of timely detection of the error case. Each of these criteria was evaluated in ten gradations for each failure mode and its cause, and a risk priority number (RPN) was calculated. For sufficiently high RPNs, the SFMECA provides steps to be defined on how to reduce the risk with mitigation actions. In the project, the mitigation action envisaged was to let the controller decide on such commands instead of executing the commands directly.

Section 5.7.1 presents the results of the SFMECA and even shows that in our use case, the distinction into good and bad recognitions is not needed.

3. Description of Evaluation System

The final validation trials were conducted on five consecutive days in the apron simulator in Frankfurt in summer 2022. All necessary data were recorded, subsequently processed, evaluated, and documented along the agreed validation concept.

For the trials, an evaluation system was created that allowed us to test the hypotheses set from the project description and to adapt them to the experience gained. The following section describes the final evaluation system as it was integrated into the Fraport apron simulator.

3.1. Technical Integration into the Simulator

While the validation system was necessary to perform the final validation trials, its design was developed in iterations. From the start of the project and as a very basic technical integration, it was used to test the planned system's architecture and functionality. It was also used to record speech and validation data necessary for the iterative improvement of artificial intelligence (AI)-based speech recognition and understanding during each training session. The actual speech data were recorded, and the A-SMGCS position data, flight plan data, and commands entered by the simulation pilot were logged. Additionally, the recorded speech data were transcribed (a word-for-word transcript of the uttered speech) and annotated (information on the contained commands in the defined ontology). The recorded data were used to train the speech recognition models and adapt the algorithms for speech understanding and callsign prediction. The data used for training and adaptation contained 19 h of audio data without silence from 14,567 single utterances, aligned with corresponding transcriptions. Furthermore, around 8.5 h with respect to 7132 utterances of the transcribed data were annotated in the defined ontology. Both the transcription and annotation processes are based on automatic pre-transcription/pre-annotations generated by the speech recognition and understanding components in the quality available within the different iterations. The manual verification and correction of the pre-transcripts were executed by a human expert from Fraport who is familiar with the airport layout, the procedures, and so on. The pre-annotations were verified and corrected by experts from the DLR, which are familiar with the defined ontology and its components.

The part of the Fraport simulator that was used in the project consists of a simulation room for the apron controllers and a control room for the simulation pilots (see Figure 3).

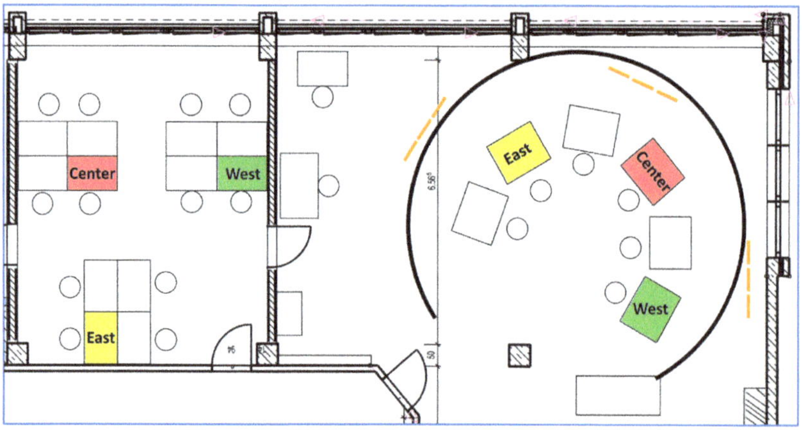

Figure 3. Simulation rooms for controllers and simulation pilots at Fraport. The left part shows the working positions of the three simulation pilots. The right part shows the simulation environment for the three apron controllers.

In operational mode, there are two different workstations. The Movement Controller (MC) workstation guides the aircraft. English is spoken on the flight frequency. In Frankfurt, three Movement workstations are usually manned, named East, Center, and West (see Figures 3 and 4). In addition to the Movement workstations, there are Operational Safety Controllers (OSC). These workstations guide the tugs and assign the follow-me vehicles. German is spoken on these frequencies. For training, usually the MC and two OSC are assigned and split in two different rooms. It was decided to not use OSC during simulation, so that five instead of three simulation days were possible with the same effort of the involved apron controllers. Everything was located in one simulation room.

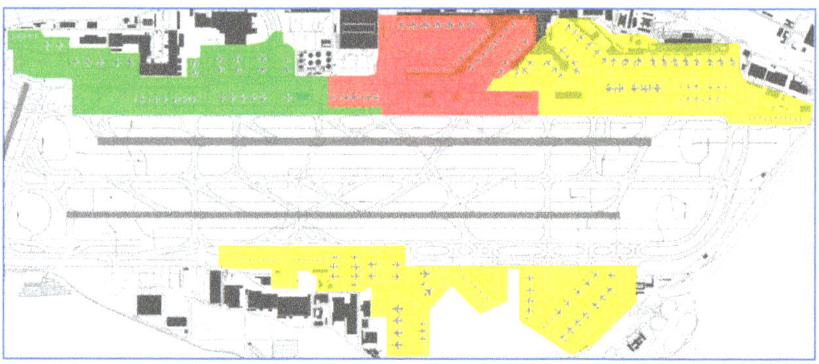

Figure 4. Areas of responsibility for the East (yellow), Center (red), and West (green) workstations for the Frankfurt apron control.

In the simulator environment, simulation pilots act as counterparts for the controllers. They sit in the simulation pilot room (left part in Figure 3). The task of the simulation pilots is to move the aircraft as instructed by the controller and to provide readback of uttered commands. The simulation pilot is in control of the same aircraft as the controller, and, therefore, controls several aircraft. The simulation pilots, like the controllers, are assigned to designated work areas (East, Center, and West). Thus, the controller always talks to the same simulation pilot during a simulation session and vice versa. Three MC workstations and three active simulation pilot positions were evaluated through ABSR support.

To use ABSR in the simulator like in real operations, various data had to be exchanged between the simulator software (ATRiCS AVATOR™), the A-SMGCS (ATRiCS TowerPad™), and DLR's ABSR system (see Figure 5).

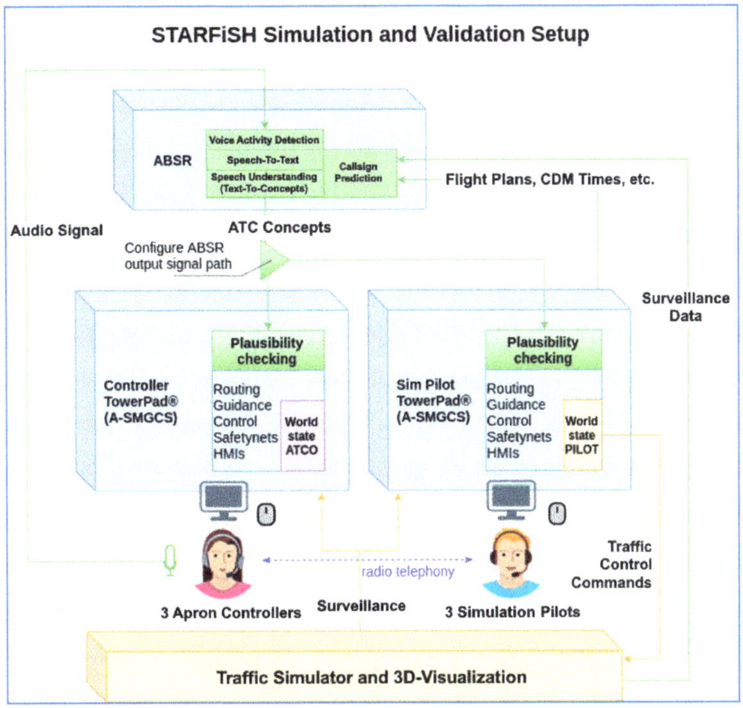

Figure 5. Simulation setup for validation trials with ABSR in an A-SGMCS system.

The surveillance data in Asterix CAT 20 format as well as flight plans and Collaborative Decision Making (CMD) Times were sent to the ABSR system and evaluated by the callsign prediction module. The audio recordings were first processed by the voice activity detection, then by the speech-to-text component, and finally by the speech understanding component. These results (ATC concepts) were forwarded as commands to the A-SMGCS and simulation pilot workstations. The latter control the traffic and radar simulator and visualization. This was performed by means of a transmission control protocol/internet protocol (TCP-IP) connection. Another interface was used to transmit flight plan data from the simulator to the ABSR system. The main interface was from the ABSR system to the A-SMGCS. Here, the recognized commands were passed to the A-SMGCS for visual display on the simulation pilot workstation and on the apron controller workstation, respectively. The interfaces and software programs as well as traffic scenarios were successfully tested in the first iteration of the project.

After the first iteration, new functions were integrated and tested in the simulator and the ABSR system in short intervals. In this way, it was possible to quickly check whether the interaction of the software worked and whether the adjustments represented an improvement or had no significant or even negative effects and should, therefore, be removed again.

Similar to the development process, an iterative approach was also taken to evaluate the results. An evaluation basis was defined and tested during the iterations and continuously improved. During these tests, it was determined that an objective measure of the cognitive load placed on controllers by system inputs was needed. Using eye-tracking sensors would have been one way to measure how often the controller's gaze is on the implemented ABSR output, e.g., to provide manual input and to determine how often the simulated traffic can be observed from an outside view. However, after further experiments, the decision was made to use a much less complex measurement method by means of secondary tasks for the participating controllers, see Section 4.4.

In the simulations, different traffic situations were used as scenarios. Scenarios from 30 min to 60 min were tested, as well as high, medium, and low traffic. After various tests, the length of 30 min and very high traffic density seemed to be most suitable to validate or falsify the validation hypotheses in the final validation trials. Two scenarios were created for the final validation trial. One scenario included runway operating direction 25, another one, operating direction 07. The two different operating directions indicate the direction in which the parallel runway system in Frankfurt is used, i.e., the direction in which aircraft take off and land. The direction depends on the weather, in particular on the wind, since landings should be made against the wind direction if possible. During operating direction 25, the runways 25 left (25 L) and 25 right (25 R) were used for inbounds/arrivals. The runways 18 and 25 center (25 C) were used for outbounds/departures. During operating direction 07, inbounds used 07 L and 07 R, whereas outbounds used 07 C and 18, i.e., in both scenarios, two inbound and two outbound runways were in use. On the ground, the operating directions affect the taxi guidance since the aircraft are then guided on other taxiways to the stand or runway. Accordingly, the ABSR and the integration of the systems could be tested in different situations. Consequently, the results should be more general and transferable to other traffic scenarios and other airports.

3.2. Assistant-Based Speech Recognition

The core of the ABSR system implemented in the STARFiSH project mainly consists of three modules (see Figure 6), which perform the conversion of the audio signal into recognized word sequences (speech recognition), the prediction of the relevant callsigns (callsign prediction), and the extraction of the semantic meaning of apron controller commands (speech understanding).

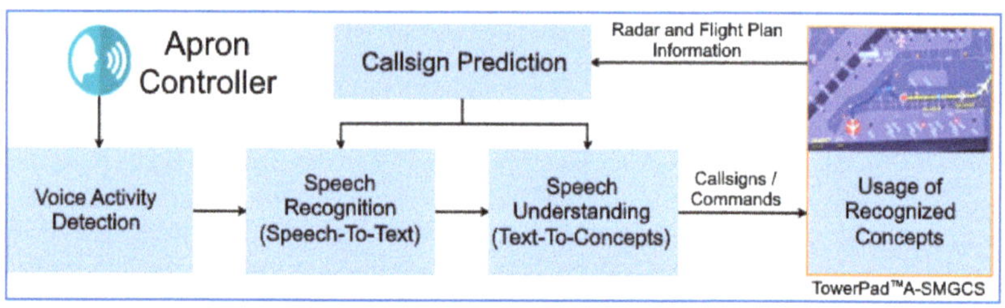

Figure 6. Modules of assistant-based speech recognition.

The only mandatory input signal to the system is the voice radio of the apron controller. To improve the recognition quality of the ABSR system, radar and flight plan information is also provided by the A-SMGCS. These data allow the generation of relevant contextual information, such as the list of aircraft callsigns that are currently relevant for operations per area of responsibility, that can be directly integrated into the recognition process of the ABSR system. In addition to the three central modules, for technical reasons, see Section 3.2.1, the project also had to implement and integrate a voice activity detection for the ABSR system, which determines when a controller's radio transmission starts and when it ends. Figure 6 provides an overview of the interfaces between the core modules. The following sections describe each module in more detail.

3.2.1. Voice Activity Detection

The goal of the ABSR system in STARFiSH is to recognize and understand the uttered commands of apron controllers. Since the audio signal is transmitted as a continuous stream of data from the voice communication system to the ABSR system, even when there is no speech at all, the system needs a way to detect the points in time a dedicated radio transmission has started and ended. Therefore, a signal is required that indicates the beginning and end of a radio message to the ABSR system. The most precise signal for this purpose would be the so-called push-to-talk (PTT) signal. This signal is triggered by controllers each time they push or release the button on the microphone they are using to start or end a radio transmission to a pilot. However, for technical reasons, the PTT signal could not be accessed for use in this project. To compensate for this problem, STARFiSH uses voice activity detection, i.e., the acoustic signal is analyzed to determine when a transmission begins and ends. The start and end of radio transmissions are detected based on the duration of previously detected silence states and a probability of reaching the end of the voice signal. Five predefined rules for detecting the end of a segment online from Kaldi have been considered without further adaptations [36].

3.2.2. Speech Recognition, i.e., Speech-to-Text (Transcriptions)

As soon as the voice activity detection detects the beginning of a radio transmission, the audio signal is forwarded to the S2T component, and the recognition process immediately starts converting the audio signal into word sequences. This means that the speech recognition system starts the recognition process as soon as the apron controller begins the radio transmission. The system then continuously provides intermediate recognitions until the controller reaches the end of the radio transmission. For example, a controller might say the following:

> *"lufthansa three charlie foxtrot taxi alfa six two alfa via november one one november november eight at november eight give way to the company A three twenty from the right".*

Let us assume that this sentence could be recognized and output by the S2T component in the following increments:

1. *"lufthansa three charlie";*
2. *"lufthansa three charlie foxtrot taxi alfa six two alfa via";*
3. *"lufthansa three charlie foxtrot taxi alfa six two alfa via november one one november november eight";*
4. *"lufthansa three charlie foxtrot taxi alfa six two alfa via november one one november november eight at november eight give way to";*
5. *"lufthansa three charlie foxtrot taxi alfa six two alfa via november one one november november eight at november eight give way to the company A three twenty from the right".*

The speech recognition engine is implemented as a hybrid deep neural network combined with a hidden Markov model (HMM). It is combined with a convolutional neural network factorized time delayed neural network (CNN-TDNNF) with six convolution layers and fifteen factorized time-delay neural networks. Overall, the model has around 31 M trainable parameters. The whole model is trained with a so-called "lattice-free maximum mutual information" as an objective function. The system follows the standard chain LF-MMI training recipe [37] of Kaldi [38], which uses high-resolution "Mel frequency cepstral coefficients" and i-vectors as input features. A typical 3-gram language model was trained and adapted using domain-specific data.

Starting from a base model, the speech recognition engine was continuously improved with new training data during the course of this project. An integration of context knowledge from callsign predictions was also implemented and contributes to the improvement of the recognition performance.

3.2.3. Speech Understanding

When a word sequence is transmitted from the S2T component, it is analyzed by the speech understanding module and converted into relevant ATC concepts, as originally defined in an ontology [29]. According to this ontology, the above word sequence *"lufthansa three charlie foxtrot taxi alfa six two alfa* via *november one one november november eight at november eight give way to company A three twenty from the right"* is transformed into the following commands:

- DLH3CF TAXI TO A62A;
- DLH3CF TAXI VIA N11 N N8;
- DLH3CF GIVE_WAY DLH A320 RIGHT WHEN AT N8.

In total, this radio transmission contains three commands. Here, the pilot of the aircraft with the callsign DLH3CF was instructed to taxi to parking position A62A via the taxiways N11, N, and N8. When arriving at taxiway N8, the pilot of the aircraft must give way to a Lufthansa (DLH), which is from the same company as the pilot addressed, has the aircraft type A320, and is coming from the right (RIGHT), before being allowed to continue taxiing.

In the case of intermediate detections from the speech recognition engine, the speech understanding module is able to provide early recognition of the callsign or, if required, early recognition of subsequent commands. The speech understanding implementation is based on a rule-based algorithm that identifies the relevant parts step by step and converts them into ATC commands. For more information, see [39].

The speech understanding module does not only convert the word sequences into ATC concepts but also makes an initial decision as to whether the extracted commands might be erroneous. Potentially erroneous commands are caused either by an erroneous interpretation of the rule-based algorithm, by an already erroneous word sequence due to a misrecognition of the speech recognition engine, or by misleading formulations of the controller. The decision, whether a command could be erroneous, is based on simple heuristic rules that determine which commands can occur together in a radio transmission. Here are some of the rules, explained by examples:

- It is logically not possible that an aircraft is instructed in a single radio transmission to taxi to two different target positions, e.g., a "TAXI TO" to two different parking positions, runways, or both in one transmission is impossible. Therefore, the module would automatically discard all "TAXI TO" commands within the transmission. Of course, with more information, it might be possible in some cases to determine which of the target positions is the correct one and only neglect one of the "TAXI TO" clearances, but that would require quite complex knowledge about the airport infrastructure to be implemented within the speech understanding component. The target application, on the other hand, which receives information from speech understanding, usually already has the required knowledge about the airport and therefore is more suitable to handle this task.
- A similar example would be a "TURN LEFT" and a "TURN RIGHT" command within one transmission and no other command in between, which is also impossible and would therefore be neglected for the same reasons.
- A less obvious example is the recognition of a "PUSHBACK" and a "TAXI TO" command in one transmission. Theoretically this might seem possible, but also these commands do not appear together and if they do, the error is usually a wrongly extracted "TAXI TO". Therefore, the heuristic says to always neglect the "TAXI TO" in this case.

However, the examples above show also that erroneous commands can only be detected if the error case is predefined. Therefore, confidence measures have furthermore been implemented for speech understanding output and are used to reduce possible false recognitions. These confidence measures can also be applied to the error cases listed above instead of neglecting the erroneous commands, but this requires that the application receiving the information is able to implement it. This means that the application then has to determine which command to neglect or not. In the end, all errors that are not detected, either by speech understanding or by the application, have to be handled manually by the apron controller in charge.

Analogous to speech recognition, speech understanding was also continuously developed and adapted based on new information. Just as in speech recognition, an integration of context knowledge from callsign prediction takes place and contributes to the improvement of recognition performance.

3.2.4. Callsign Prediction

Callsign prediction receives both radar and flight plan information from the A-SMGCS. The module uses these data to determine which callsigns may be part of a radio transmission in the near future. The radar information is used in the first step to obtain an overview of the available callsigns in the airport area. However, since many aircraft are in the airport area, but not all will be actively participating in taxiing traffic in the near future, the module also uses flight plan information dynamically provided by the A-SMGCS to determine more precisely which of the available callsigns will be addressed in the near future. For this purpose, the responsible controller position, the target startup approval time (TSAT), the actual take off time (ATOT), the actual landing time (ALDT), and the actual in block time (AIBT) are extracted from the flight plan. All relevant callsigns are forwarded to the speech recognition module (callsign boosting) and the speech understanding module to include the callsigns in the process of recognition and understanding. More information on the technique of callsign boosting, used within the speech recognition module to enhance recognition, can be found here [40,41]. The integration of callsign predictions in the speech understanding module transforms the callsigns into possible word sequences and calculates the closest match to the recognized word sequence based on the Levenshtein distance [42] to determine the correct callsign.

3.2.5. Concept Interpretation

The final stage of integrating an ABSR system into an A-SMGCS represents the testing for operational plausibility, interpretation, and implementation of the extracted concepts or commands. Figure 7 shows the running integration of ABSR into the A-SMGCS at one of the simulation pilot stations.

Figure 7. ABSR output log on the left screen and airport map for simulation pilot on the right (photo © Fraport).

Testing and interpretation are necessary prior to implementation for two reasons:

- Controller instructions via voice convey exactly the information that is necessary and sufficient for the addressed pilot in the current traffic situation. Globally, however, these instructions can be ambiguous. It is, therefore, necessary for an information technology system to unambiguously identify the addressed pilot and to make assumptions about his/her contextual knowledge in order to be able to exclude ambiguities from this perspective. A GIVE_WAY command from the right could identify several aircraft that approach from the right at the same time or consecutive taxiway crossings. The system has to determine the correct one that is implied from the traffic context.
- The extracted concepts may be erroneous. Either the controller has made a mistake, so that the verbal instruction does not correspond to what would be advised in the current traffic situation, or errors have occurred in the recording of the speech, the pause recognition, the conversion to text, or the speech understanding, so that the extracted concept is erroneous and should not be implemented.

These two sources of error cannot be distinguished. The task of this module is to admit only those commands into the assistance system that are plausible and fit into the current traffic context. If inappropriate commands are delivered by the ABSR system, the user must be given the opportunity to manually correct the error. This is technically implemented by the following steps, which are detailed in Appendix B:

1. Preprocessing;
2. Highlight the aircraft symbol on the basis of the recognized callsign;
3. Trigger multiple actions based on a single command;
4. Discard commands incompatible with the traffic situation;
5. Correctly interpret context-dependent commands;
6. Complete incomplete commands from the current traffic situation;
7. Convert commands;
8. Deal with detected errors;
9. Deal with undetected errors and identify error sources.

3.3. *Usability Considerations*

A key element to the successful implementation of automation features is the user interface. Since automation reduces the necessary interactions with the system, users may miss automatically executed actions. It is thus essential that users are still able to see all system states that are relevant for safe operations. In addition, it must be possible to quickly analyze and correct errors in the event of automation failure. This is a necessary requirement of operational safety, especially in aviation, where errors can lead to accidents.

In the STARFiSH project, the automation functions as well as the user interface were first implemented in a purely functional way and then analyzed in operation with the end users to iteratively overcome the challenges. This involved asking questions such as: Is the right information available, and is the right information being perceived? Is the user interface not overloaded with information, i.e., is important information also perceived more easily than less important information? Are delays sufficiently low?

Using various elicitation techniques (observation, brainstorming, and interviews), user requirements were thus determined in an iterative manner, and new target formulations were achieved. In total, 30 users participated in 29 evaluation days, distributed over the four iterations in the course of this project.

3.3.1. Visualization of the Automation Actions (Feedback)

In iteration 1, the main focus was to be able to check the technical integration of the systems, i.e., to investigate the question of whether recognized commands reach the human operator and are available in time from an operational point of view. Therefore, the following were implemented first:

- Display the recognized transcriptions and the resulting annotations (ATC commands) on the side of the ABSR output log, in order to be able to compare the output of the ABSR system with the received data on the TowerPad™.
- Log data at the interfaces of the ABSR system and on the working position computers of the controllers and simulation pilots, in order to be able to analyze, after the simulation runs, whether the correct commands arrive in time.
- Log the commands provided by the ABSR system in chronological order on the working position computers to give users and researchers a way to observe and verify the results of the speech inputs independently of the implementation of the commands.

During preliminary testing, it became apparent that displaying each recognized command in a table to support easy troubleshooting did not add value to the operational users of the system but was distracting. Therefore, the user interface was designed so that commands generate specific visual feedback which is integrated into the workflow. In terms of position and design, this resulted in symbolic displays specifically adapted to the command or very compact dialogs. Although the display as a table was still extended for troubleshooting, it was no longer visible at the controllers' working positions during the trials and was positioned on a second screen outside the focus of the users at the simulation pilot workstations. It was only used for evaluation and development.

Starting with iteration 3, commands were directly translated into visible actions of the system. For some actions, it was possible to use the same visual feedback to the human operator that is used for manual input, for example:

- Change a route;
- HOLD_SHORT command;
- GIVE_WAY command.

There are fundamental advantages to displaying the same feedback in the HMI regardless of the input method (by speech recognition and understanding, or by mouse or touch gesture), as there is less need for training. On the other hand, users should be able to identify if the source of a change in the user interface is the speech recognition and understanding component. This was implemented by the following user interface features:

- Highlighting of the addressed aircraft symbols without disturbing user touch or mouse input, executed in parallel, additionally multi-highlighting when several commands are executed in quick succession.
- Feedback for changes, which are scarcely visible when executed manually, such as the transfer of an aircraft to another working position.

3.3.2. Manual Error Correction

The simulation experiments showed that for some actions, an undo is ambiguous and not without side effects, e.g., when changing a taxi route. For these actions, it was easier for human operators to select the desired function directly without prior "undo", thereby implicitly overriding the wrong action.

4. Validation Trials

This section presents the preparation and results of the validation trials. All simulations took place in Fraport's training simulator, which had been retrofitted for the experiments and tests.

4.1. Pre-Simulations

During the pre-simulations, the individual parts of the system and their integration were tested, and exemplary evaluations of the simulation runs were carried out in order to determine methods for the final validation trial. The basic structure, i.e., the architecture, remained constant after the initial integration tests.

Controllers and pilots, in their corresponding positions, speak to each other on the same radio frequency. The ABSR system operates on the voice recordings of the controllers, and the command implementation takes place independently in the two instances of the A-SMGCS for the controllers and simulation pilots, respectively (see Figure 5). For both groups, ABSR support is enabled either for all working positions or for none.

The simulations in the first iterations served the dual purpose of obtaining feedback from the human operators and testing the technical integration. In later iterations, the validation methods themselves were tested as well, i.e., exemplary evaluations of the simulation runs were performed. For example, the hypotheses regarding the reduction of taxi times were discarded, since they did not differ significantly.

It was also explored what the scope of traffic should/should not be, and which additional tasks are suitable to challenge the attention of the users without tying up the support team too much.

4.2. Validation Plan

Four different combinations of the ABSR support were investigated, as shown in Table 1:

Table 1. Different combinations of ABSR support investigated during validation trials.

Condition Name	Operational Conditions
NO	No ABSR support; manual input, i.e., the baseline scenario. The controllers manually enter the spoken commands via mouse into the controller's HMI of the TowerPad™. Simulation pilots control taxi traffic at their working positions by manual input via mouse and keyboard. This corresponds to the established mode of operation without ABSR.
JC	Use of ABSR support just for controllers, i.e., automatic command recognition support for controllers plus manual correction, if ABSR fails. Commands spoken by the controller are processed by the ABSR system and transmitted to the controller's working position, where they are automatically entered for the controller. The controller receives feedback on the recognized commands via the controller's HMI and can correct errors via a mouse. No support by ABSR for the simulation pilots.
JP	Use of ABSR support just for simulation pilots, i.e., automatic command recognition and control for pilots. The commands spoken by the controllers are processed by the ABSR system, transmitted to the working position of the responsible simulation pilot, and automatically executed as control commands for the simulation pilot. The simulation pilot receives feedback on the recognized commands via the simulation pilot's HMI and can correct errors via a mouse and keyboard. No support by ABSR for the controllers.
CP	Use of ABSR support for both controllers and simulation pilots, as described individually for JC and JP conditions.

4.2.1. Validation Hypotheses

The following hypotheses were tested during the final validation trials:

H1. *(H-C-less_input): Automatic documentation (conditions JC and CP) reduces the total number of manual inputs to guide taxiing traffic at the controller's working position compared to full manual input (conditions NO and JP).*

H2. *(H-P-less_input): Automatic command recognition for simulation pilots (conditions JP and CP) reduces the total number of manual inputs to guide the taxiing traffic of simulation pilots compared to full manual input (conditions NO and JC).*

H3. *(H-C-more_cog_res): Automatic documentation (conditions JC and CP) increases the controller's free cognitive resources compared to full manual input (conditions JP and NO).*

H4. *(H-C-less_workload): Automatic documentation (conditions JC and CP) reduces the workload of the controller compared to full manual input (conditions JP and NO).*

H5. *(H-C-sit_aw_ok): Automatic documentation (conditions JC and CP) does not limit the controller's situational awareness compared to full manual input (conditions JP and NO).*

H6. *(H-C-conf): The controller's confidence in command entry automation (conditions JC and CP) is above average.*

H7. *(H-P-conf): The simulation pilot's confidence in command entry automation (conditions JP and CP) is above average.*

H8. *(H-E-CmdRR): The command extraction rate (JC, JP, and CP) in the apron environment is comparable to the quality already achieved in the approach domain by ABSR (command extraction rate for simulation-relevant commands >90%).*

H9. *(H-E-CmdER): The command extraction error rate (conditions JC, JP, and CP) in the apron environment is comparable to the quality already achieved in the approach domain by ABSR (command extraction error rate for simulation-relevant commands <5%).*

H10. *(H-E-CsgRR): The callsign extraction rate (conditions JC, JP, and CP) in the apron environment is comparable to the quality already achieved in the approach domain by ABSR (>97%).*

H11. *(H-E-CsgER): The callsign extraction error rate (conditions JC, JP, and CP) in the apron environment is comparable to the quality already achieved in the approach domain by ABSR* **H11.** *(callsign extraction error rate <2%.)*

4.2.2. Independent Variables

The independent variables (IV) of the final validation trials were as follows:

1. (IV-Input): Documentation on the controller's HMI by ABSR vs. manual input (JC and CP vs. JP and NO).
2. (IV-Control): Control of the simulation by ABSR for the controller's utterances vs. full manual input (JP and CP vs. JC and NO).

4.2.3. Dependent Variables

The dependent variables of the final validation trials are listed in Appendix A. The respective results are each compared between the different operational conditions within a scenario.

4.3. *Execution of the Final Validation Trials*

The final validation trials took place from 27th of June to 1st of July 2022 in the apron simulator in Frankfurt. The number of simultaneously active users was three controllers and three simulation pilots. For the final trials, 14 controllers were recruited who had enough experience with the A-SMGCS system (see Figure 8). On each day, a new team of controllers was on site (one controller participated twice). Half of the participants already had their first experience with the system at one of the many pre-simulations. The other half had their first contact with the ABSR system during the final trials.

Two different traffic scenarios were prepared for the final validation trial: one for runway operating direction (OD) 25 and one for OD 07. The simulation scenarios generated from these were 30 min long each. Table 2 shows the number of aircraft movements in total and the projected number of aircraft at each of the three working areas: East, Center, and West.

Table 2. Number of (#) aircraft in certain areas per operating direction.

Traffic Scenario	# Aircraft	# Arriving Aircraft	# Departing Aircraft	# Expected Aircraft East	# Expected Aircraft Center	# Expected Aircraft West
OD25	106	46	60	59	61	63
OD07	106	46	60	57	45	59

Figure 8. Photo (© Fraport) of final validation trial setup with A-SMGCS in front of the controller.

In order to generate a heavy workload, the amount of traffic in the scenarios were increased compared to usual traffic at Frankfurt Airport so that the controllers were as busy as possible all the time. The heaviest workload with respect to radio frequency usage and number of commands was expected at the Center position, followed by the East working position. At the West working position, the load was expected to be lower, even if the numbers in the Table 2 suggest otherwise. This is partially due to the type of movement (pushback-aircraft must be pushed from the parking position with the tug, etc.) and the size of the area. In the West, there were significantly fewer pushbacks, because there were less nose-in positions (so most aircraft could leave the parking stand forward under their own power). The number of commands and utterances per position for all runs are shown in Table 3.

Table 3. Number of commands (# Cmds) and utterances (# Utterances) at each of the three working positions.

	# Cmds	% of All	# Utterances	% of All
Center	5858	38%	2437	38%
West	4376	28%	1654	26%
East	5235	34%	2262	36%

The realistic maximum traffic volume in real operations was 106 per 60 min for 2019. This amount was used in the simulation trials for half an hour. This increase compensated for the fact that there were no tows on the apron and other secondary activities that would otherwise occur in reality and that could not all be represented in the simulator.

Each simulation day began with a briefing of the controllers and simulation pilots involved. In this briefing, the controllers and simulation pilots were educated on the concept of ABSR and its interaction with the controller input interface and the simulation pilot's working station. They were explained how to make manual entries in the systems and when these are required (when no ABSR support is active for the respective station, or a correction is necessary).

In addition, the controllers and simulation pilots were informed about the schedule, and the questionnaires were introduced. Afterwards, training for controllers and simulation pilots took place, in which all operational conditions were explained and tried out. During the training run, the controllers also exercised the secondary task for measuring mental load after a short introduction on how to perform it. This secondary task is discussed in more detail in Section 4.4.

The teams of three controllers and three simulation pilots remained at their working positions throughout the different simulation runs and operational conditions (OC). On each day, there were different runs for each of the two operating directions OD07 and OD25. The teams were always the same for the same OD.

When evaluating the influence of ABSR support on the work of the controller, the ABSR system was always active for the simulation pilots, i.e., only altered from on to off and vice versa for the controllers. In addition, two simulation runs were carried out in OD25, in which the ABSR system was always active on the controller's side and a change from on to off and vice versa for the simulation pilots. The influence of the ABSR system on the simulation pilot's activity was analyzed, too. Thus, six runs were performed per day. After each run, the controllers filled out a questionnaire. At the end of the day, an additional questionnaire was filled out, and the impressions, comments, and hints of the controllers and simulation pilots were recorded in a non-formal debriefing session with all participants.

In the runs, in which the ABSR system was alternately on or off for the controller, the secondary task was performed by the controllers. The task started 10 min after the start of the run and stopped 10 min later. This was performed simultaneously for all three working positions. The simulation runs with the different operational conditions and simulation scenarios were determined as follows in Table 4.

Table 4. Simulation runs with different operational conditions and simulation scenarios.

Simulation Run Name with OD	Operational Conditions and Traffic Scenario
T	Training of fully manual input and ABSR-supported input with manual corrections at controllers' and simulation pilots' working positions.
CP25	ABSR support for controllers and simulation pilots.
JC25	ABSR support just for controllers.
JP25	ABSR support just for simulation pilots.
NO25	No ABSR support.
CP07	ABSR support for controllers and simulation pilots.
JP07	ABSR support just for simulation pilots.

Table 5 below shows the order of the training and experimental runs for each controller-team, consisting of three persons. The order of the runs was changed each day to reduce (in the mean of the evaluation) learning effects that may occur during the day.

Table 5. Simulation runs per controller team and day.

Team 1 Day 1	Team 2 Day 2	Team 3 Day 3	Team 4 Day 4	Team 5 Day 5
T	T	T	T	T
NO25	CP25	JP25	JP25	JP07
JC25	JP25	NO25	CP25	CP07
JP25	JC25	CP25	JP07	CP25
CP25	NO25	JC25	CP07	JP25
JP07	CP07	JP07	NO25	JC25
CP07	JP07	CP07	JC25	NO25

4.4. Objective Workload Measurement by a Secondary Task

The questionnaires reflect the subjective experiences of the controllers, which one might argue to be the most important measure for most operationally deployed systems.

Nevertheless, we wanted to obtain more objective data that would confirm or reject our hypothesis that the proposed system reduces workload.

To measure mental load, we used a secondary task that required similar skills to the main task (controlling traffic), namely mental focus, English language proficiency, color recognition, and quick orientation on the user interface, and yet that was simple enough to be performed in parallel with the main task.

In the pre-simulations, subjects were asked to sort decks of playing cards as a secondary task to measure free mental capacity and were then made to answer questions about missing cards ("which 1–4 cards were missing?"), as was described in [3]. However, the use of this task required too much manual effort and, therefore, ran the risk of introducing errors in data recording and execution, so we chose a largely automated approach for the final validation trials using the application described below. This greatly reduced the physical and mental workload of the simulation support team and the susceptibility to errors.

For the secondary task, 10 min after the start of each simulation run, each controller (and additionally once in parallel with the simulation pilots) was asked to complete as many Stroop tasks [43] as possible in the following 10 min in addition to their main task. For this purpose, a tablet PC (6x Samsung A8) was provided to the controllers that ran an application for executing consecutive Stroop tasks [44]. The application recorded the execution time and duration of each task as well as its correctness. A high number of correctly executed Stroop tasks in the application suggests an available mental capacity that is not needed for the main task.

The atomic Stroop task is the following: when the start button is pressed, a word for a color is displayed, but in a different color to the color that the word stands for. The task for the user is to select the right button with the color word that matches the display color from a set of seven buttons, all labelled with a color word in black. The order of these buttons changes in a pseudo-random way at each repetition of the task. In Figure 9, the color word "ORANGE" is displayed in blue, so the button to press is the one labelled "BLUE".

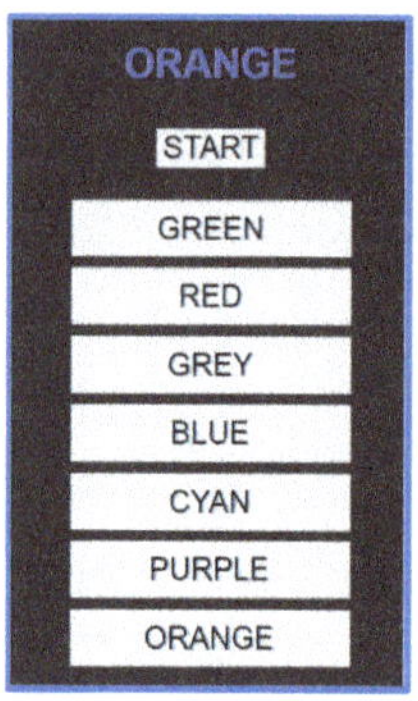

Figure 9. Example Stroop task.

5. Validation Results

5.1. Speech Recognition and Understanding Performance

Section 5.1.1 focuses on speech recognition, and Section 5.1.2 focuses on speech understanding performance results.

5.1.1. Speech-to-Text Accuracy (Speech Recognition)

A first indication for the quality of the ABSR system is provided by the so-called word error rate (WER) of the S2T component. The WER is calculated based on the Levenshtein distance [42] between the word sequence recognized by the S2T component and the actual spoken word sequence (gold transcription). This involves counting, in the recognized word sequence, how many words of the actual word sequence have been substituted (S), deleted (D), or additionally inserted (I). All three components are then added and divided by the number of words of the actual word sequence (N). Table 6 shows the WER of the developed S2T component in the final validation trials based on the verbal utterances of 14 apron controllers.

Table 6. Word error rate of recognized word sequences from the S2T component.

Recognition Mode	WER	
Offline (PTT signal simulated)	3.1%	Male: 3.3% Female: 2.6%
Online (voice activity detection)	5.0%	Male: 5.5% Female: 3.7%

WER was evaluated for two different modes. Online recognition measures what recognition performance the S2T component achieved during the final validation trials in summer 2022. This means that these results contain a certain number of errors that are not induced by the S2T component but by voice activity detection (VAD), due to the missing PTT signal. In order to determine how large the influence of VAD is and what improvement can be expected by accessing the PTT signal, the offline recognition after summer 2022 was used to subsequently evaluate what the system would have recognized if the audio stream had been perfectly split by PTT. It can be seen that offline recognition again brings a significant improvement over online recognition, with a WER of 3.1% compared to 5.0%.

It is also interesting to observe that the average WER of female apron controllers (2.6% and 3.7%, respectively) was better than those of male apron controllers (3.3% and 5.5%, respectively). On the other hand, out of the total 14 apron controllers, only four were female. Performing unpaired *t*-tests with the 24 runs with female apron controllers versus the 62 runs of male controllers provides very statistically significant results, with a *p*-value of 0.02%.

The question of what WER is good enough for the intended purpose often arises. This question cannot be answered in a general way, because in the end, it is irrelevant how many words are recognized correctly. What is important is the ability of the system to extract the meaning behind the recognized words and finally to implement it appropriately in the application. Some errors on the word level can change the meaning of an utterance, while others have no influence at all. Therefore, it is not possible to define a general threshold for the WER, but a low WER allows conclusions on the quality of the implemented ABSR system.

5.1.2. Text-to-Concept Accuracy (Speech Understanding)

The performance of speech understanding is evaluated by comparing the commands automatically extracted by the system with the correct commands manually created and verified by human experts (gold annotations). The evaluation is based on three metrics: command recognition rate, command error rate, and command rejection rate. The command recognition rate is defined as the number of correctly recognized commands divided by the total number of commands actually given. A command is considered correctly recognized if and only if all elements of a command such as command type, callsign, value, unit, qualifier, condition, etc., as defined in the ontology, are correctly recognized. Command error rate is the percentage of incorrectly extracted commands divided by the total number of commands actually given. Command rejection rate is the percentage of actual commands given that were not extracted at all or were rejected by the system for some reason. Table 7 below shows the metrics defined above with an example. The example also illustrates that the sum of the recognition, error, and rejection rates can exceed 100%.

Table 7. Example for speech understanding metrics.

Actual Commands	Recognized Commands	Contribution to Metric
DLH695 TURN RIGHT	DLH695 TURN LEFT	⊖
DLH695 TAXI VIA N10 N	DLH695 TAXI VIA N10 N	⊕
	DLH695 TAXI TO V162	⊖
AUA1F PUSHBACK	AUA1F NO_CONCEPT	○
CCA644 NO_CONCEPT	CCA644 NO_CONCEPT	⊕
Recognition Rate (⊕) = 2/4 = 50%	Error Rate (⊖) = 2/4 = 50%	Rejection Rate (○) = 1/4 = 25%

In a similar manner to the extraction rates for commands, separately, the extraction rates for callsigns are determined. Again, there is a callsign recognition rate, error rate, and rejection rate. For each utterance, each callsign is considered only once, unless several different callsigns are extracted from the same utterance ("break break" utterances). Therefore, in the above example from Table 7, three callsigns are considered. Detailed information on the defined metrics can be found in [45]. Table 8 illustrates the performance of speech understanding based on the above-explained metrics, i.e., it contains the number of radio telephony utterances and commands as well as the recognition, error, and rejection rates for full commands and callsigns.

Table 8. Recognition (RecR), error (ErrR), and rejection (RejR) rate for commands (Cmds) and callsigns (Csgn) in [%] and absolute numbers of utterances (# Utterances) and commands (# Commands).

Recognition Mode	# Utterances	# Commands	Cmds [%]			Csgn [%]		
			RecR	ErrR	RejR	RecR	ErrR	RejR
Offline (PTT signal simulated)	5495	13,251	91.8	3.2	5.4	97.4	1.3	1.3
Online (voice activity detection)	5432	13,168	88.7	4.3	7.5	95.2	2.3	2.4
Offline (no callsign prediction used)			76.3	10.5	13.7	81.1	9.6	9.3
Delta to context			15.5	−7.3	−8.3	16.3	−8.3	−8.0

"Delta to context" is the difference of row 2 "Offline (PTT…)" and row 4 "(Offline no callsign…)".

Recognition rates of 91.8% and 88.7% are obtained when speech understanding is applied offline (simulated PTT) and online (VAD), respectively. The improvement in the speech understanding result for offline recognition comes from the better word-level recognition and the fact that the offline data does not include radio transmissions that were incorrectly split by VAD, allowing for a better interpretation of the content. Similarly, the recognition of aircraft callsigns in offline recognition is also better than in online recognition, with recognition rates of 97.4% and 95.2%, respectively. The last two rows of Table 8 show the influence of the predicted callsigns on the recognition performance of ABSR. With context information available, the recognition rate increases by 15.5% overall and 16.3% on the callsign level.

Table 9 shows the rates of offline recognition for different command types. The table lists only the most common or important command types that are relevant to the application.

Thus, speech understanding has error rates below 4% and recognition rates in the range of roughly 87% to 98%, depending on the command type with the exception of the GIVE_WAY command. The reason for the worse results of the GIVE_WAY command extraction, marked in red in Table 9, is its very complex nature, i.e., it can be given/uttered in many different ways, not all of which were modeled so far.

Table 9. Recognition (RecR), error (ErrR), and rejection (RejR) rate for specific command types in [%] and absolute number of commands (# Cmds) of that type.

Command Type	# Cmds	RecR	ErrR	RejR
TAXI VIA	2922	86.9	3.9	9.1
HOLD_SHORT	1837	89.3	0.8	9.9
TAXI TO	1406	89.0	1.1	9.9
CONTACT_FREQUENCY	1387	95.7	0.7	3.6
CONTINUE TAXI	1102	95.4	0.0	4.6
GIVE_WAY	728	69.6	10.2	20.3
CONTACT	672	98.4	0.3	1.3
PUSHBACK	663	92.3	1.2	6.5
TURN	359	89.2	3.9	6.9
HOLD_POSITION	223	93.4	0.0	6.6

Worst ErrR overall marked in red.

5.2. Interaction Count

To determine the amount of manual HMI interactions needed at the A-SMGCS with and without ABSR support, we recorded the HMI interactions at each position and per simulation run, counted them, and categorized them into 48 different task types, such as "edit route", "clear pushback", and "select label". Expectations were, of course, that the number of interactions would be significantly lower when ABSR was available. However, it was apparent from the pre-simulations that the controllers would not make all the required inputs without ABSR support, nor would they correct every error made by the ABSR system, because not performing an input has no direct consequences for the controller as long as the simulation pilot still follows the voice instructions, because the pilots control the simulation.

Therefore, the numbers of the simulation pilots are more meaningful, since here, any omitted input or correction leads to a delay or incorrect behavior in the simulator. On the other hand, not all interactions of the simulation pilots can be replaced by the ABSR input, because the pilot initiates the communication with the controller and needs information for this, which is only available when selecting the aircraft symbol via a mouse click.

Therefore, even for a perfect ABSR, the total number of interactions can never be zero and without ABSR, the number of interactions is higher for the simulation pilots, and the reduction in interactions for the simulation pilots is also not as extreme as for the controllers. Figure 10 shows the remaining portion of manual actions needed for the controllers and simulation pilots when being supported by ABSR for the most frequent interactions. A strong reduction of workload is apparent for both.

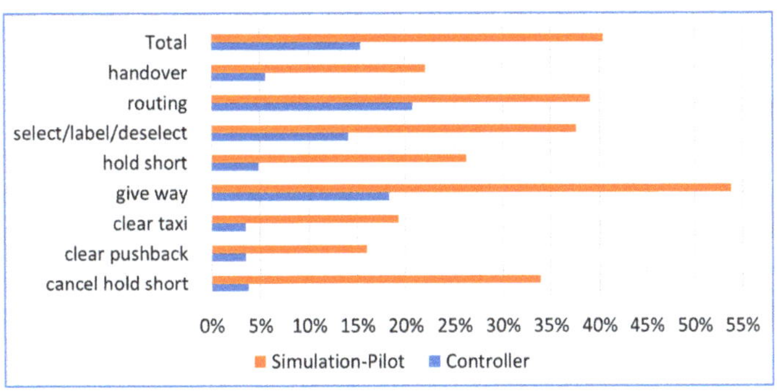

Figure 10. Portion of remaining manual interactions with the HMI (by type) of the controllers and simulation pilots with ABSR support compared to runs without ABSR support.

5.3. Workload, NASA TLX

The NASA Task Load Index (TLX) has been used for decades in different variants to assess perceived workload in six different aspects [46]. We used a simple unweighted questionnaire procedure in which a mark between 0 (very low) and 20 (very high) is to be entered for each aspect. "The Task", here, means the task of the apron control in the simulator operating the A-SMGCS.

This questionnaire was completed by the users at the controllers' working positions after each simulation run. The scores were aggregated by position, OD, and by use of the ABSR (or not). The six questions are as follows:

- Mental Demand: How mentally demanding was the task?
- Physical Demand: How physically demanding was the task?
- Temporal Demand: How hurried or rushed was the pace of the task?
- Performance: How successful were you in accomplishing what you were asked to do?
- Effort: How hard did you have to work to accomplish your level of performance?
- Frustration: How insecure, discouraged, irritated, stressed, and annoyed were you?

The question about the subjects' own performance (see list above) requires a reversal of the scale values: "how successful?" suggests a high value if the subjects were satisfied with their performance. This is also how the controllers expressed themselves in the feedback rounds, but this was not consistently evident in the questionnaires; in some cases, conspicuously low values were given here, although the values for the other aspects were also very low. Unfortunately, we have to assume that not all controllers understood this question correctly or answered it as expected, and we therefore excluded the performance aspect from the evaluation.

Table 10 below shows the average values by working position and overall, separately for the baseline runs (without ABSR) and the solution runs (with ABSR). Columns "α" show the statistical significance of a *t*-test.

Table 10. NASA TLX questionnaire results of the controllers on perceived workload.

	West			Center			East			All		
Workload	**Base**	**Sol**	α	**Base**	**Sol**	α	**Base**	**Sol**	α	**Base**	**Sol**	α
Mental Demand [MD]	7.9	6.4	14.5%	14.4	12.4	1.1%	14.7	13.8	10.3%	12.3	10.8	1.3%
Physical Demand [PD]	7.6	3.3	0.4%	9.4	5.6	0.3%	10.6	7.9	2.5%	9.2	5.6	3×10^{-4}
Temporal Demand [TD]	8.1	5.9	4.4%	12.8	10.7	2.4%	14.1	13.1	6.2%	11.7	9.9	0.3%
Effort [EF]	8.6	5.5	2.1%	12.6	10.5	1.3%	14.5	12.4	1.6%	11.9	9.5	1×10^{-3}
Frustration [FR]	4.0	3.4	23.9%	6.8	3.5	1.0%	7.2	5.0	6.1%	6.0	4.0	0.2%
ALL	7.2	4.9	2.5%	11.2	8.5	0.2%	12.2	10.4	1.4%	10.2	8.0	6×10^{-4}

Minimal α values, shaded in green for $0\% \leq \alpha < 5\%$, in light green for $5\% \leq \alpha < 10\%$, and in yellow for the rest ($|\alpha| \geq 10\%$). As the numbers show no evidence that results with ASRU support are worse, no further color coding is needed.

In general, the workload on the working position "West" was significantly lower than on the other two, while it was estimated to be slightly higher on East than on Center. At OD25, the workload is slightly higher at West and at Center, while the workload for East is estimated to be highest at OD07 (not shown in the table). On average, the workload is slightly higher for OD07, and only with regard to the time aspect is OD25 experienced as somewhat more stressful. Thus, although the working position makes a big difference, the OD does not have a noticeable effect on the average values for all positions.

Especially for physical demand, the value even decreases by almost four points. Significance tests (paired *t*-tests) on the data prove that the differences between with and without ABSR are not random. On the West position, the results for mental demand, and frustration does not decrease in a statistically significant way. The overall results with respect to workload reduction were very, very statistically significant. We obtained an alpha (*p*-value) of 0.06%, i.e., if we would have repeated the experiments with all the 15 participants 1000 times again, only in six cases could we expect that the workload without ABSR support is less than that with ABSR support.

5.4. Evaluation of Stroop Tests as Secondary Task

The results of the secondary task point in the expected direction: at the Center and West positions, subjects were able to perform significantly more tasks correctly in parallel with their work when the ABSR support was active, see Figure 11.

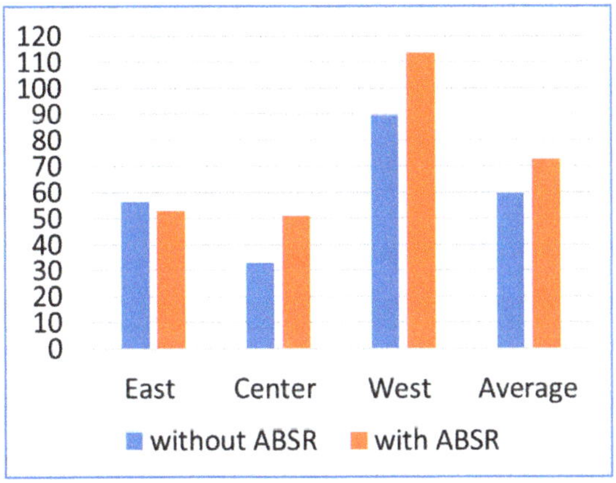

Figure 11. Average number of correct Stroop tests per working position with and without ABSR.

From a statistical point of view, the variance at the East working position was too high to make a reliable statement for this position. At Center and West, the figures support the hypothesis, but strictly speaking, this statement is still outside usually required statistical significance due to the relatively small total number of experiments. The qualitative observations made during the simulation runs support our decision to use the application and indicate that the secondary task used here is an objective measure of the cognitive capacity available:

- "If the Stroop tasks are done while R/T [radio telephony] must be done, selecting the correct button takes longer."
- "More complicated routes increase the error rate [in Stroop tasks]."

5.5. Situational Awareness, Shape-SASHA

The questionnaire Situational Awareness for Shape (SASHA) [47] was used to assess the controllers' situational awareness during the simulation runs. The test persons marked their assessment of the aspects on a Likert scale between 0 "never" and 6 "always". The negatively formulated statements (marked with an "*" below) have been inverted for later evaluation, i.e., the averages are calculated as 6 minus the raw average value. The statements in detail were as follows:

- In the previous working period(s), …
    - I was ahead of the traffic.
    - I started to focus on a single problem or a specific area of the sector.*
    - There was the risk of forgetting something important […].*
    - I was able to plan and to organize my work as I wanted.
    - I was surprised by an event I did not expect […].*
    - I had to search for an item of information.*

The subjects filled out this questionnaire after each simulation run. After evaluation, the situational awareness of the controllers is generally found to be good with and without ABSR (see Table 11). The average values over all simulation runs, positions, and aspects are above 4. The most important message is that with ABSR support, situational awareness increases on average over all aspects and at each position.

Table 11. Results of SASHA questionnaire with and without ABSR support.

Situational Awareness	without ABSR (W/C/E)	with ABSR (W/C/E)
ahead of traffic	4.4 (5.2/4.0/4.0)	4.7 (5.4/4.2/4.6)
focus single problem	4.3 (4.9/3.9/4.2)	4.3 (4.2/4.3/4.5)
risk of forgetting	3.9 (4.8/3.5/3.4)	4.4 (4.9/4.1/4.3)
able to plan	4.1 (4.7/3.9/3.6)	4.6 (4.9/4.6/4.3)
surprised by event	4.5 (5.1/4.5/4.0)	4.9 (5.4/4.7/4.7)
search information	3.9 (4.1/3.8/3.9)	4.4 (4.5/4.7/4.0)
ALL	4.2 (4.8/3.9/3.8)	4.6 (4.9/4.4/4.4)

Further analysis of the values not shown in the above table reveals some differences in situational awareness (SA) related to the working environment. The working position area has an influence on SA, i.e., at the West position, the value was 4.8, while East (4.1) and Center (4.2) have lower values. The OD makes a very small difference, with 4.3 for OD07 and 4.4 for OD25, respectively. However, SA is significantly lower (one whole scale point) at the Center position for OD25 and at the East position for OD07, regardless of the use of ABSR. These two working positions gain half a point with ABSR support, but this is not as clearly reflected at the West position. These results fit the NASA TLX results: where workload is lower, situational awareness is higher.

5.6. *Confidence in Automation, Shape-SATI*

To elicit trust in the automatic entry of commands into the controllers' or simulation pilots' HMI, we used the SHAPE Automation Trust Index questionnaire, SATI. We had each participant complete this questionnaire once at the end of the simulation day, with the request that they focus on the effects of the ABSR system. This allowed us to evaluate 15 questionnaires from the controllers and 6 from the pilots. The items that could be answered on a 7-point Likert scale from 0 "never" to 6 "always" in detail were as follows:

- In the previous working period(s), I felt that . . .
    - The system was <u>useful</u>.
    - The system was <u>reliable</u>.
    - The system worked <u>accurately</u>.
    - The system was <u>understandable</u>.
    - The system worked <u>robustly</u> (in difficult situations, with invalid inputs, etc.).
    - I was <u>confident</u> when working with the system.

A distinction whether the ABSR system or other automation features of the not completely familiar system triggered trust or distrust could probably not be made completely by the test persons. Hence, the results must be considered under this restriction. The overall impression turned out to be very positive, as shown in Table 12.

The controllers consider the system almost always useful. Regarding accuracy and robustness, the confidence is lowest but still high (>4). The simulation pilots are slightly more skeptical, but overall trust in the system is well above average.

Table 12. Results of SATI questionnaire.

Automation Trust	Controllers (15)	Simulation Pilots (6)
useful	5.1	4.6
reliable	4.5	4.2
accurate	4.3	4.2
understandable	4.9	5.3
robust	4.2	4.2
confident	4.8	4.3
ALL	4.6	4.5

5.7. Results with Respect to Safety

5.7.1. Software Failure Modes, Effects, and Criticality Analysis (SFMECA)

The risk analysis based on the SFMECA had the result that no error case would lead to an increased or unacceptable risk, so that no classification into good and bad recognitions is needed, as mentioned in Section 2.3, although our implementation of speech understanding provides this information. This result was not expected. However, it can be explained as follows: The apron control is not responsible for the runways, i.e., areas are excluded where wrong decisions have particularly severe effects and where the possibility of detecting errors quickly is reduced (due to higher speeds). There was also no indirect risk of causing distractions through false alarms and thus endangering situational awareness, since there were no automatic alarm functions available in the project (with which the controllers were familiar). Based on the safety analysis, it was therefore possible to make the decision to implement all commands directly in the A-SMGCS without exceptions (those that are not identified as nonsensical based on plausibility checks).

For the most critical command "Handover", it was decided to always offer an undo function. A mistakenly executed handover of a flight to another working position would cause all subsequent commands to be discarded: the aircraft would be assigned to another working position and therefore be unavailable for incoming commands at the actual working position. However, the error is very easy to detect with the implemented visualization, and we offered a one-click solution to undo it.

In addition to considering AI ABSR errors, this project also discussed and considered the issues of safety when introducing automation. In addition to the direct effects of automation errors, increased automation can affect safety in the following ways:

- Indirect impact due to automation errors (too many disruptive errors, either due to a lack of recognition or incorrect recognition);
- Lack of visibility of the automation result (a loss of "situational awareness");
- Lack of flexibility (no possibility of correction or override by the user and therefore a loss of control);
- Overconfidence/complacency.

The approaches in this project for addressing these risks were the following:

1. Achieve sufficient recognition rates and sufficiently low recognition error rates to prevent potential overload from occurring in the first place.
2. Make the results visible enough for users to retain situational awareness at all times.
3. Allow human operators to make corrections to automation errors in order to remain in control.
4. Assessments of risk by overconfidence through safety considerations: what can happen if automation errors are not corrected?

This was validated in multiple ways: (1) indirectly, by selecting particularly challenging simulation scenarios that go beyond the usual in terms of traffic density, by evaluating the required number of interactions with the user interface, and by measuring cognitive load using secondary tasks; and (2) directly through test subjects filling out standardized questionnaires on situational awareness and trust in automation.

5.7.2. Feedback from the Test Subjects on Safety

From simulations at the beginning of the project, in which significantly more errors happened, and significantly fewer commands were available for automatic execution, the following feedback was obtained:

- *Since the speech recognizer still makes mistakes and you have to check if everything is correct whenever you are spoken to, you are less free in your timing. One also expects that, e.g., the callsign is highlighted. If that doesn't happen, you're wondering why it didn't work.*

The feedback on safety became gradually less negative as the project progressed, and the number of errors decreased. At the beginning, there were definitely impairments of a smooth workflow, because the controllers had to wait for the implementation or were inclined to always check the correctness. When most commands worked and the error rate dropped significantly, there were no more comments suggesting a negative impact or reduced safety. This confirms the work on the safety of the overall validation system and the analysis from the safety assessment.

After the simulation runs, subjects were always questioned about safety. The following responses (translated analogously by the authors) are representative of the sentiments:

- *You could always see if there were errors or not.*
- *The delay is fine. You can already talk to the next pilot or you get the indication during the readback. That's sufficient.*
- *The errors were very few. They couldn't put us in critical situations.*
- *Here, the aircraft are controlled very directly because the simulator directly implements commands [with voice recognition enabled] [including errors]. A pilot would not do that. That's why it [emerging situations] would be less critical in real life.*
- *If something takes too long, you leave it out.—If the pilot executes it correctly, it's okay.—If incorrectly detected, the worst thing that can happen is false alarms.*

5.7.3. Summary of All Feedback Collected

Feedback was consistently positive toward the solution with ABSR support. The controllers were mostly surprised that the system worked so well, even though it is still in a research stage. It was emphasized that it made no qualitative difference to the ABSR system (1) whether the controller spoke quickly or slowly, or (2) whether the controller strictly adhered to International Civil Aviation Organization (ICAO) phraseology [16] in his/her speech utterances or deviated from it to a greater or lesser extent, caused by increased traffic density and high radio frequency use. The system was very well received because it did not require controllers to change. The controllers could simply speak as they were accustomed and still the correct action occurred in most cases. The controllers said that it could leave more time to keep an eye on traffic instead of staring at the display.

It was also noted that with ABSR support, the controllers sometimes instruct different taxi routes than when they have to input the route manually: If a route is pre-selected by the system, then it is easier to follow it than to change it manually. But if the controllers can simply use speech to change the route instead of having to enter it manually, then they are more likely to change the route, e.g., to shorten the aircraft's taxiing time.

The controllers as well as the simulation pilots indicated that the workload decreases significantly with ABSR support. The best feedback was for the working position West. Here, almost everything was correctly recognized for everybody. Recognition was also good for the East and Center positions, but there were also minor misrecognitions.

There were hardly any critical voices. Rather, there were suggestions on how to make it even better, e.g., that the recognition should be faster and could be better, so that there would be even fewer false recognitions. The command types "Hold Abeam" and "Pushback Abeam" were, e.g., not implemented within the resources of the STARFiSH project. Over the days, the feedback from the controllers involved was qualitatively repetitive, and so it became apparent that the different controllers had the same good experiences with the system.

5.8. Results with Respect to Validation Hypotheses

In Section 4.2.1, we formulated the Hypotheses H1 to H11. The results with respect to validation of the hypotheses and falsification were presented in the above sections. This subsection summarizes the results with respect to each hypothesis.

5.8.1. Hypotheses with Respect to "Number of Manual Inputs"

The results with respect to these hypotheses are presented in Section 5.2 in Figure 10. The number of inputs from the simulation pilots (dependent variable, *DV-Input-H-P-less_input*) is reduced by a factor of 2.5, and the number of manual inputs of the apron controllers (dependent variable, *DV-Input-H-C-less_input*) is even reduced by a factor of more than 6. Therefore, we mark the following two hypotheses as validated.

H1. *(H-C-less_input): Automatic documentation (conditions JC and CP) reduces the total number of manual inputs to guide taxiing traffic at the controllers' working position compared to full manual input (conditions NO and JP).* **Validated**

H2. *(H-P-less_input): Automatic command recognition for the simulation pilots (conditions JP and CP) reduces the total number of manual inputs to guide the taxiing traffic of the simulation pilots compared to full manual input (conditions NO and JC).* **Validated**

5.8.2. Hypothesis with Respect to "Free Cognitive Resources of Apron Controller"

The results with respect to this hypothesis are presented in Section 5.4 in Figure 11. The number of correct Stroop tasks increased for the West and Center positions and did not decrease for the East position. Therefore, we mark the following hypothesis as partially validated.

H3. *(H-C-more_cog_res): Automatic documentation (conditions JC and CP) increases the controller's free cognitive resources compared to full manual input (conditions JP and NO).* **Partially Validated**

5.8.3. Hypothesis with Respect to "Apron Controller Workload Reduction"

The results with respect to this hypothesis are presented in Section 5.3 in Table 10. The workload reduced by 2.2 scale units on the 20-unit NASA TLX scale on average. Therefore, we mark the following hypothesis as validated.

H4. *(H-C-less_workload): Automatic documentation (conditions JC and CP) reduces the workload of the controller compared to full manual input (conditions JP and NO).* **Validated**

5.8.4. Hypothesis with Respect to "Apron Controller's Situational Awareness"

The results with respect to this hypothesis are presented in Section 5.5 in Table 11. The situational awareness over all three positions and over both operating directions increased from 4.2 to 4.6 (maximum value of 6.0). The lowest effect was measured for the West position with an increase of 0.1 unit points. However, situational awareness was already high without ABSR support at this position (4.8). Therefore, we mark the following hypothesis as validated.

H5. *(H-C-sit_aw_ok): Automatic documentation (conditions JC and CP) does not limit the controller's situational awareness compared to full manual input (conditions JP and NO).* **Validated**

5.8.5. Hypotheses with Respect to "Apron Controller's Confidence"

The results with respect to this hypothesis are presented in Section 5.6 in Table 12. The average value for the apron controllers was 4.6 and that for simulation pilots was 4.5. These values are above the average of 3.0. In addition, the lowest individual value for both

(4.2) is far beyond the average of 3.0. Therefore, we mark the following hypotheses both as validated.

H6. *(H-C-conf): Controller confidence in command entry automation (conditions JC and CP) is above average.* **Validated**

H7. *(H-P-conf): Simulation pilot's confidence in command entry automation (conditions JP and CP) is above average.* **Validated**

5.8.6. Hypotheses with Respect to "Automatic Speech Understanding for Complete Commands"

The results with respect to this hypothesis are presented in Section 5.1.2 in Table 8. Assuming the availability of push-to-talk, we measured an average command recognition rate of 91.2%, which is fully above the threshold of 90%. We obtained 3.2% as the command recognition error rate, which is also better than the threshold of 5%. Therefore, we mark the following hypotheses both as validated.

H8. *(H-E-CmdRR): The command extraction rate (JC, JP, and CP) in the apron environment is comparable to the quality already achieved in the approach domain by ABSR (command extraction rate for simulation-relevant commands >90%).* **Validated**

H9. *(H-E-CmdER): The command extraction error rate (conditions JC, JP, and CP) in the apron environment is comparable to the quality already achieved in the approach domain by ABSR (command extraction error rate for simulation-relevant commands <5%).* **Validated**

The results with respect to callsign recognition are also presented in Table 8. The callsign recognition rate of 97.4% is better than the threshold of 97%, and the callsign recognition error rate of 1.3% is also better than the threshold of 2%. Therefore, we mark the following hypotheses both as validated.

H10. *(H-E-CsgRR): The callsign extraction rate (conditions JC, JP, and CP) in the apron environment is comparable to the quality already achieved in the approach domain by ABSR (>97%).* Validated

H11. *(H-E-CsgER): The callsign extraction error rate (conditions JC, JP, and CP) in the apron environment is comparable to the quality already achieved in the approach domain by ABSR (callsign extraction error rate <2%).* Validated

6. Discussion

The STARFiSH project was of course subject to some restrictions that determined what was possible to research in the given time and budget. This section, therefore, discusses possibilities for improvements that could be addressed in the future and highlights some aspects that proved to be useful within this project.

The SFMECA (see Sections 2.3 and 5.7) produced no RPNs that mandated mitigation actions. This was due to the environment in which the project was executed: Areas of responsibility for the apron control did not include runways and no automatic alerting functions were implemented at the baseline A-SMGCS. For an environment without these limitations, the SFMECA is expected to produce different results and additional challenges for usability and safety. In addition, while the SFMECA itself was chosen as a proven instrument, there are suggestions for amendments to the methodology which could be used in order to address its specific shortcomings [48].

The results in Sections 5.1.1 and 5.1.2 show that the use of voice activity detection significantly degrades the overall performance. The push-to-talk signal should therefore be used whenever possible. Especially in an operational scenario, voice activity detection should not be considered as an alternative, since the push-to-talk signal is in use anyway, and technical access should not be an issue. Nevertheless, in non-operational scenarios where, for technical reasons, the push-to-talk signal might not be available, more modern approaches to voice activity detection based on neural network architectures could be exploited [49].

In Section 3.2.3, the rule-based algorithm for speech understanding was mentioned. This approach, of course, offers a very precise control about what is extracted and how the extraction itself takes place. The disadvantage of this method is, on the other hand, that every adaptation has to be programmed manually, which can create a lot of effort. Future projects could ease the adaptation process by fine-tuning pre-trained language models such as BERT [50], which could then recognize the different elements of the ontology [51].

The iterative approach taken for the development of the whole system and also for the training and improvement of the speech recognition and understanding modules proved to be very useful throughout this project. The different prototypes made it possible to involve the apron controller (end-users) already at an early stage of development and to incorporate their feedback in future prototypes. That not only improved the system in itself but also made the controllers involved and interested in the system and in what can be achieved with such a technology. The iterative improvement of the speech recognition parts was also useful with respect to the transcription and annotation process of the recorded data. As the recognition performance of these components became better over the iterations, the manual work to correct and verify transcriptions and annotations could be reduced.

One of the next steps should be to move the developed system from the simulation into an operational environment to see how big the difference to real world operations is and what obstacles have to be overcome. A first step could be to run the system in shadow mode so that it does not interfere with the operating systems, but operational experts could monitor how the system would react.

7. Conclusions

The STARFiSH project was the first to implement a speech recognition and understanding system for a complex apron environment at Frankfurt Airport. DLR's ABSR system was successfully coupled with the commercial A-SMGCS system from ATRiCS, i.e., a previously prototypical technology from a scientific environment was integrated into a commercial system that is available on the market. The solution was iteratively improved and finally tested in validation trials with 14 different apron controllers in 29 simulation runs in the tower simulator of Fraport. A total of 43 h of validation data (radar, audio, HMI inputs, etc.) were recorded and subsequently analyzed.

A main objective of the STARFiSH project was to prepare the usage of an artificial intelligence powered speech recognition and understanding system in the safety-critical environment of the ops room at a European major hub airport. The formal method SFMECA (=Software Failure Modes, Effects, and Criticality Analysis) for risk assessment and subsequent identification of mitigation measures was applied with the very encouraging result that no error case would lead to an increased or unacceptable risk. At the same time, it could be shown that such an AI-equipped application can be operated safely in aviation and, moreover, does not have a negative impact on the controllers' situational awareness.

When supported by ABSR, the controllers made more than six times fewer manual entries into the A-SMGCS. This already includes the correction of wrong or missing recognitions from the speech recognition and understanding support. A recognition rate of 91.8% on the command level was observed, i.e., the callsign, the command type, the command values, e.g., taxi routes, and the command conditions were correctly extracted in 91.8% of the cases.

Author Contributions: Conceptualization, M.K. and H.H.; methodology, H.H.; software, M.K., H.H., O.O., S.S. (Shruthi Shetty) and H.E.; validation, M.K., H.W., M.M. and S.S. (Susanne Schacht); formal analysis, H.H.; investigation, H.H. and M.K.; resources, H.W. and S.S. (Susanne Schacht); data curation, H.W., M.K., H.H. and O.O.; writing—original draft preparation, M.K., O.O. and H.H.; writing—review and editing, O.O., H.W., M.M., H.E., S.S. (Shruthi Shetty) and S.S. (Susanne Schacht); visualization, H.H., M.K., H.E., O.O., M.M. and S.S. (Susanne Schacht); supervision, M.K.; project administration, M.K.; funding acquisition, M.K. All authors have read and agreed to the published version of the manuscript.

Funding: The project STARFiSH was funded by the German Federal Ministry of Education and Research, under support code 01IS20017C.

Data Availability Statement: Not applicable.

Acknowledgments: We would like to thank the Fraport controllers for their participation in this study.

Conflicts of Interest: The authors declare no conflict of interest. The funding sponsors had no role in the design of this study, in the collection, analyses, or interpretation of the data, in the writing of the manuscript, and in the decision to publish the results.

Appendix A

The following dependent variables of the final validation trials are considered. The respective results of a dependent variable are each compared between the different operational conditions within a scenario.

Appendix A.1. DV-Input: Number of Manual Inputs for Control by Controllers/Simulation Pilots

The manual inputs are counted. Since the inputs are identifiable by type, certain types are highlighted, if necessary, should it be found that some types occur particularly frequently or infrequently. The total count is compared between simulation runs with or without ABSR support.

These dependent variables are used to validate/falsify the following hypotheses:

- DV-Input-H-C-less_input (ABSR for the controllers reduces the number of manual inputs).
- DV-Input-H-P-less_input (ABSR for the simulation pilots reduces the number of manual inputs).

Appendix A.2. DV-Cog-Res: Measurement of Cognitive Resources by Secondary Task

The cognitive resources are measured by means of a secondary task, i.e., a task the test subject (controller or simulation pilot) performs during a scenario in parallel to the main task. This is a secondary task that the subject is only allowed to perform when no mental resources are needed for the main task. The secondary task consists of performing a repeated Stroop test in a web application, see Section 4.4. The number of correctly mastered tests in a given time period is a measure of free cognitive resources. For this purpose, the responses per item are categorized as correct/wrong, and the number per time is plotted as a histogram and compared for simulation runs with and without ABSR support, respectively. These dependent variables are used to validate/falsify the following hypothesis:

- DV-Cog-Res-H-C-more_cog_res (more free cognitive resources of the controller due to ABSR).

Appendix A.3. DV-Workload Scoring by NASA TLX

This dependent variable is used to validate/falsify the following hypothesis:

- DV-Workload-H-C-less_workload (less controller workload due to ABSR).

Appendix A.4. DV-Sit-Aw Scoring According to SHAPE-SASHA

This dependent variable is used to validate/falsify the following hypothesis:

- DV-Sit-Aw-H-C-sit_aw_ok (situational awareness of the controller).

Appendix A.5. DV-Trust: Scoring According to SHAPE-SATI

These dependent variables are used to validate/falsify the following hypotheses:

- DV-Trust-H-C-conf (automation trust of the controller).
- DV-Trust-H-P-conf (automation trust of the simulation pilot).

Appendix A.6. DV-CmdRR: Command Extraction Rate

This dependent variable is used to validate/falsify the following hypothesis:

- DV-CmdRR-H-E-CmdRR (comparable command extraction rate as in the approach environment).

Appendix A.7. DV-CmdER: Command Extraction Error Rate

This dependent variable is used to validate/falsify the following hypothesis:

- DV-CmdER-H-E-CmdRR (comparable command extraction error rate as in the approach environment).

Appendix A.8. DV-CsgRR: Callsign Extraction Rate

This dependent variable is used to validate/falsify the following hypothesis:

- DV-CsgRR-H-E-CsgRR (comparable callsign extraction rate as in the approach environment).

Appendix A.9. DV-CsgER: Callsign Extraction Error Rate

This dependent variable is used to validate/falsify the following hypothesis:

- DV-CsgER-H-E-CsgER (comparable callsign extraction error rate as in the approach environment).

Appendix B

The task of the module "Concept Interpretation" is to transfer only those commands into the assistance system that are plausible and fit into the current traffic context. In the following, we describe the steps already mentioned in Section 3.2.5 in more detail.

Appendix B.1. Preprocessing

The modules described in Sections 3.2.2 and 3.2.3 generate data telegrams for the respective assigned working position. These data telegrams contain, among other things, the extracted ATC concepts with the semantics according to the ontology for the annotation of ATC utterances. An example of the logical content of a data telegram is presented in Box A1.

Box A1. Logical content example of a data telegram which has to be preprocessed for the A-SMGCS.

```
Sender: MC East
Callsign: DLH4YE
Command: GREETING
Command: TAXI (TO) V106
Command: TAXI (VIA) L
```

The interpretation of such a data telegram requires several checking steps, depending on the commands contained, before one or more inputs can be safely made to the A-SMGCS. Basically, this step checks whether the working position assigned to the sender is authorized to make entries for the callsign or whether another working position is responsible for the aircraft of this callsign.

Appendix B.2. Highlighting the Aircraft Symbol on the Basis of the Recognized Callsign

If the basic check of this preprocessing is successful, the corresponding aircraft symbol is highlighted at the assigned working position to inform the human operator for which flight a command has been recognized.

Appendix B.3. Checking and Interpretation

Depending on the command type, the following checks and computation steps are performed depending on the characteristics of the command received.

Appendix B.3.1. Triggering Multiple Actions Based on a Single Command

Some commands require that multiple actions are triggered by the same command in the correct order. For example, a TAXI command should trigger a TAXI clearance in the system and create a taxi route. It may also be necessary to modify an existing route and cancel stop instructions in certain situations.

Appendix B.3.2. Discarding Commands Incompatible with the Traffic Situation

It may happen that a command is received that does not make sense in the current traffic situation, e.g., a TAXI command destination for an inbound flight that contains a runway as the destination of the route. If possible, such cases are detected by checking a set of rules. The command is then ignored, and a message is displayed to the human operator. The cause of an incompatible command cannot be determined here. It could be an error of the controller, or it could originate in the ABSR system.

Appendix B.3.3. Correctly Interpret Context-Dependent Commands

Some commands must be interpreted differently depending on the flight plan data or other circumstances identified by the A-SMGCS. For example, if a special routing procedure is set for a flight in the system depending on certain conditions in the database, the system must assign a different route than in the normal case. The same utterance by the controller, therefore, leads to different results in the system depending on the data situation.

Appendix B.3.4. Completing Incomplete Commands from Current Traffic Situation

There are commands that do not contain all the necessary information to be able to implement them directly. For example, the command "GIVE_WAY … A320 RIGHT" needs further analysis, assuming that there is more than one A320 aircraft moving at the airport. The transcription part "from the right" can be ambiguous, thus it needs to be determined algorithmically which aircraft is probably correct from the controller's and pilot's perspective. In such cases, configuration tables, algorithms, state machines, and rules stored in the code are used to generate the correct command appropriate to the traffic situation.

Appendix B.3.5. Conversion of Commands

Once a command has been established by the previous checks, it can be implemented. This means that the command is entered into the A-SMGCS, i.e., the internal system state is changed to reflect the command. For the controller, this results in visual feedback on the working position. For example, the callsign that the controller addressed is highlighted near the corresponding aircraft symbol on the ground situation display. A change in the route is illustrated by colored lines, and changes in clearances are indicated on the label of the corresponding flight.

Appendix B.3.6. Dealing with Detected Errors

If a command does not pass one of the plausibility checks, an error message is displayed. This gives the human operator the option to check the situation and either ignore it or correct it.

Appendix B.3.7. Undetected Errors and Identification of Error Sources

It is not possible to identify for each command whether it is operationally correct. It is also not possible to determine whether a detected error originates from the ABSR system or was made by the controller. The controller, as the user of the system, must therefore observe the output of the A-SMGCS and anticipate, detect, and correct error situations not detected by the system. In the training of the controllers, this behavior is trained specifically and repeatedly, since errors of the human actors involved (e.g., the pilots) and the electronic systems must always be expected. It is always necessary to make a trade-off between the recognition rate and the recognition error rate. For example, a 0% error rate can be achieved simply by discarding every recognized command.

References

1. Kleinert, M.; Shetty, S.; Helmke, H.; Ohneiser, O.; Wiese, H.; Maier, M.; Schacht, S.; Nigmatulina, I.; Sarfjoo, S.S.; Motlicek, P. Apron Controller Support by Integration of Automatic Speech Recognition with an Advanced Surface Movement Guidance and Control System. In Proceedings of the 12th SESAR Innovation Days, Budapest, Hungary, 5–8 December 2022.
2. International Civil Aviation Organization (ICAO). *Advanced Surface Movement Control and Guidance Systems (ASMGCS) Manual, Doc 9830 AN/452*, 1st ed.; International Civil Aviation Organization (ICAO): Montréal, QC, Canada, 2004.
3. Helmke, H.; Ohneiser, O.; Mühlhausen, T.; Wies, M. Reducing Controller Workload with Automatic Speech Recognition. In Proceedings of the 35th Digital Avionics Systems Conference (DASC), Sacramento, CA, USA, 25–29 September 2016.
4. Helmke, H.; Ohneiser, O.; Buxbaum, J.; Kern, C. Increasing ATM Efficiency with Assistant Based Speech Recognition. In Proceedings of the 12th USA/Europe Air Traffic Management Research and Development Seminar (ATM2017), Seattle, WA, USA, 26–30 June 2017.
5. European Commission. *L 36/10*, Commission Implementing Regulation (EU) 2021/116 of 1 February 2021 on the Establishment of the Common Project One Supporting the Implementation of the European Air Traffic Management Master Plan Provided for in Regulation (EC) No 550/2004 of the European Parliament and of the Council, Amending Commission Implementing Regulation (EU) No 409/2013 and Repealing Commission Implementing Regulation (EU) No 716/2014. Official Journal of the European Union: Luxembourg, 1 February 2021.
6. Helmke, H.; Rataj, J.; Mühlhausen, T.; Ohneiser, O.; Ehr, H.; Kleinert, M.; Oualil, Y.; Schulder, M. Assistant-Based Speech Recognition for ATM Applications. In Proceedings of the 11th USA/Europe Air Traffic Management Research and Development Seminar (ATM2015), Lisbon, Portugal, 23–26 June 2015.
7. Davis, K.H.; Biddulph, R.; Balashek, S. Automatic recognition of spoken digits. *J. Acoust. Soc. Am.* **1952**, *24*, 637–642. [CrossRef]
8. Juang, B.H.; Rabiner, L.R. Automatic speech recognition–a brief history of the technology development. *Ga. Inst. Technol. Atlanta Rutgers Univ. Univ. Calif. St. Barbar.* **2005**, *1*, 67.
9. Connolly, D.W. *Voice Data Entry in Air Traffic Control*; Report N93-72621; National Aviation Facilities Experimental Center: Atlantic City, NJ, USA, 1977.
10. Hamel, C.; Kotick, D.; Layton, M. *Microcomputer System Integration for Air Control Training*; Special Report SR89-01; Naval Training Systems Center: Orlando, FL, USA, 1989.
11. FAA. *National Aviation Research Plan (NARP)*; FAA: Washington, DC, USA, 2012.
12. Updegrove, J.A.; Jafer, S. Optimization of Air Traffic Control Training at the Federal Aviation Administration Academy. *Aerospace* **2017**, *4*, 50. [CrossRef]
13. Schäfer, D. Context-Sensitive Speech Recognition in the Air Traffic Control Simulation. Eurocontrol EEC Note No. 02/2001. Ph.D. Thesis, University of Armed Forces, Munich, Germany, 2001.
14. Tarakan, R.; Baldwin, K.; Rozen, R. An automated simulation pilot capability to support advanced air traffic controller training. In Proceedings of the 26th Congress of the International Council of the Aeronautical Sciences, Anchorage, Alaska, 14–19 September 2008.
15. Ciupka, S. Siris big sister captures DFS (original German title: Siris große Schwester erobert die DFS). *Transmission* **2012**, 1.
16. *Doc 4444 ATM/501*; ATM (Air Traffic Management): Procedures for Air Navigation Services. International Civil Aviation Organization (ICAO): Montréal, QC, Canada, 2007.
17. Cordero, J.M.; Dorado, M.; de Pablo, J.M. Automated speech recognition in ATC environment. In Proceedings of the 2nd International Conference on Application and Theory of Automation in Command and Control Systems (ATACCS'12), London, UK, 29–31 May 2012; IRIT Press: Toulouse, France, 2012; pp. 46–53.
18. Cordero, J.M.; Rodríguez, N.; de Pablo, J.M.; Dorado, M. Automated Speech Recognition in Controller Communications applied to Workload Measurement. In Proceedings of the 3rd SESAR Innovation Days, Stockholm, Sweden, 26–28 November 2013.
19. Nguyen, V.N.; Holone, H. N-best list re-ranking using syntactic score: A solution for improving speech recognition accuracy in Air Traffic Control. In Proceedings of the 2016 16th International Conference on Control, Automation and Systems (ICCAS), Gyeongju, Republic of Korea, 16–19 October 2016; pp. 1309–1314.
20. Nguyen, V.N.; Holone, H. N-best list re-ranking using syntactic relatedness and syntactic score: An approach for improving speech recognition accuracy in Air Traffic Control. In Proceedings of the 2016 16th International Conference on Control, Automation and Systems (ICCAS 2016), Gyeongju, Republic of Korea, 16–19 October 2016; pp. 1315–1319.
21. Helmke, H.; Kleinert, M.; Shetty, S.; Ohneiser, O.; Ehr, H.; Arilíusson, H.; Simiganoschi, T.S.; Prasad, A.; Motlicek, P.; Veselý, K.; et al. Readback Error Detection by Automatic Speech Recognition to Increase ATM Safety. In Proceedings of the 14th USA/Europe Air Traffic Management Research and Development Seminar (ATM2021), Virtual, 20–24 September 2021.
22. Ohneiser, O.; Helmke, H.; Shetty, S.; Kleinert, M.; Ehr, H.; Murauskas, Š.; Pagirys, T.; Balogh, G.; Tønnesen, A.; Kis-Pál, G.; et al. Understanding Tower Controller Communication for Support in Air Traffic Control Displays. In Proceedings of the 12th SESAR Innovation Days, Budapest, Hungary, 5–8 December 2022.

23. Helmke, H.; Kleinert, M.; Ahrenhold, N.; Ehr, H.; Mühlhausen, T.; Ohneiser, O.; Motlicek, P.; Prasad, A.; Zuluaga-Gomez, J. Automatic Speech Recognition and Understanding for Radar Label Maintenance Support Increases Safety and Reduces Air Traffic Controllers' Workload. In Proceedings of the 15th USA/Europe Air Traffic Management Research and Development Seminar (ATM2023), Savannah, GA, USA, 5–9 June 2023.
24. García, R.; Albarrán, J.; Fabio, A.; Celorrio, F.; de Oliveira, C.P.; Bárcena, C. Automatic Flight Callsign Identification on a Controller Working Position: Real-Time Simulation and Analysis of Operational Recordings. *Aerospace* **2023**, *10*, 433. [CrossRef]
25. Chen, S.; Kopald, H.D.; Elessawy, A.; Levonian, Z.; Tarakan, R.M. Speech inputs to surface safety logic systems. In Proceedings of the IEEE/AIAA 34th Digital Avionics Systems Conference (DASC), Prague, Czech Republic, 13–17 September 2015.
26. Chen, S.; Kopald, H.D.; Chong, R.; Wei, Y.; Levonian, Z. Read back error detection using automatic speech recognition. In Proceedings of the 12th USA/Europe Air Traffic Management Research and Development Seminar (ATM2017), Seattle, WA, USA, 26–30 June 2017.
27. Helmke, H.; Ondřej, K.; Shetty, S.; Arilíusson, H.; Simiganoschi, T.S.; Kleinert, M.; Ohneiser, O.; Ehr, H.; Zuluaga-Gomez, J.-P.; Smrz, P. Readback Error Detection by Automatic Speech Recognition and Understanding—Results of HAAWAII project for Isavia's Enroute Airspace. In Proceedings of the 12th SESAR Innovation Days, Budapest, Hungary, 5–8 December 2022.
28. Zuluaga-Gomez, J.-P.; Sarfjoo, S.S.; Prasad, A.; Nigmatulina, I.; Motlicek, P.; Ondřej, K.; Ohneiser, O.; Helmke, H. BERTRAFFIC: BERT-based joint Speaker Role and Speaker Change Detection for Air Traffic Control Communications. In Proceedings of the 2022 IEEE Spoken Language Workshop Technology Workshop (SLT 2022), Doha, Qatar, 9–12 January 2023.
29. Helmke, H.; Slotty, M.; Poiger, M.; Herrer, D.F.; Ohneiser, O.; Vink, N.; Cerna, A.; Hartikainen, P.; Josefsson, B.; Langr, D.; et al. Ontology for transcription of ATC speech commands of SESAR 2020 solution PJ.16-04. In Proceedings of the IEEE/AIAA 37th Digital Avionics Systems Conference (DASC), London, UK, 23–27 September 2018.
30. Kleinert, M.; Helmke, H.; Moos, S.; Hlousek, P.; Windisch, C.; Ohneiser, O.; Ehr, H.; Labreuil, A. Reducing Controller Workload by Automatic Speech Recognition Assisted Radar Label Maintenance. In Proceedings of the 9th SESAR Innovation Days, Athens, Greece, 2–6 December 2019.
31. Lin, Y. Spoken Instruction Understanding in Air Traffic Control: Challenge, Technique, and Application. *Aerospace* **2021**, *8*, 65. [CrossRef]
32. Ohneiser, O.; Helmke, H.; Kleinert, M.; Siol, G.; Ehr, H.; Hobein, S.; Predescu, A.-V.; Bauer, J. Tower Controller Command Prediction for Future Speech Recognition Applications. In Proceedings of the 9th SESAR Innovation Days, Athens, Greece, 2–5 December 2019.
33. Ohneiser, O.; Sarfjoo, S.; Helmke, H.; Shetty, S.; Motlicek, P.; Kleinert, M.; Ehr, H.; Murauskas, Š. Robust Command Recognition for Lithuanian Air Traffic Control Tower Utterances. In Proceedings of the InterSpeech 2021, Brno, Czech Republic, 30 August–3 September 2021.
34. Boehm, B. A Spiral Model of Software Development and Enhancement. *IEEE Comput.* **1988**, *21*, 61–72. [CrossRef]
35. Neufelder, A.M. *Effective Application of Software Failure Modes Effects Analysis*; Quanterion Solutions, Incorporated: New York, NY, USA, 2017.
36. Povey, D. Online Endpoint Recognition. 2013. Available online: https://github.com/kaldi-asr/kaldi/blob/master/src/online2/online-endpoint.h (accessed on 15 May 2023).
37. Povey, D.; Peddinti, V.; Galvez, D.; Ghahremani, P.; Manohar, P.; Na, X.; Wang, Y.; Khudanpur, S. Purely sequence-trained neural networks for ASR based on lattice-free MMI. *Interspeech* **2016**, *2016*, 2751–2755.
38. Povey, D.; Ghoshal, A.; Boulianne, G.; Burget, L.; Glembek, O.; Goel, N.; Hannemann, M.; Motlicek, P.; Qian, Y.; Schwarz, P.; et al. The Kaldi Speech Recognition Toolkit. In Proceedings of the IEEE 2011 Workshop on Automatic Speech Recognition and Understanding, IEEE Signal Processing Society, Waikoloa, Big Island, HI, USA, 11–15 December 2011.
39. Kleinert, M.; Helmke, H.; Shetty, S.; Ohneiser, O.; Ehr, H.; Prasad, A.; Motlicek, P.; Harfmann, J. Automated Interpretation of Air Traffic Control Communication: The Journey from Spoken Words to a Deeper Understanding of the Meaning. In Proceedings of the IEEE/AIAA 40th Digital Avionics Systems Conference (DASC), Virtual, 3–7 October 2021.
40. Nigmatulina, I.; Zuluaga-Gomez, J.; Prasad, A.; Sarfjoo, S.S.; Motlicek, P. A two-step approach to leverage contextual data: Speech recognition in air-traffic communications. In Proceedings of the ICASSP 2022—2022 IEEE International Conference on Acoustics, Speech and Signal Processing (ICASSP), Singapore, 23–27 May 2022.
41. Zuluaga-Gomez, J.; Nigmatulina, I.; Prasad, A.; Motlicek, P.; Vesely, K.; Kocour, M.; Szöke, I. Contextual Semi-Supervised Learning: An Approach to Leverage Air-Surveillance and Untranscribed ATC Data in ASR Systems. *Interspeech* **2021**, *2021*, 3296–3300.
42. Levenshtein, V.I. Binary codes capable of correcting deletions, insertions, and reversals. *Sov. Phys. Dokl.* **1966**, *10*, 707–710.
43. Stroop, J.R. Studies of interference in serial verbal reactions. *J. Exp. Psychol.* **1935**, *18*, 643–662. [CrossRef]
44. Maier, M. Workload-Gauge. Available online: https://github.com/MathiasMaier/workload-gauge (accessed on 15 May 2023).
45. Helmke, H.; Shetty, S.; Kleinert, M.; Ohneiser, O.; Prasad, A.; Motlicek, P.; Cerna, A.; Windisch, C. Measuring Speech Recognition Understanding Performance in Air Traffic Control Domain Beyond Word Error Rates. In Proceedings of the 11th SESAR Innovation Days, Virtual, 7–9 December 2021.
46. Hart, S.G. NASA-TASK LOAD INDEX (NASA-TLX); 20 years later. In Proceedings of the Human Factors and Ergonomics Society, San Francisco, CA, USA, 16–20 October 2006; Volume 50, pp. 904–908.

47. Dehn, D.M. Assessing the Impact of Automation on the Air Traffic Controller: The SHAPE Questionnaires. *Air Traffic Control Q.* **2008**, *16*, 127–146. [CrossRef]
48. Di Nardo, M.; Murino, T.; Osteria, G.; Santillo, L.C. A New Hybrid Dynamic FMECA with Decision-Making Methodology: A Case Study in an Agri-Food Company. *Appl. Syst. Innov.* **2022**, *5*, 45. [CrossRef]
49. Mihalache, S.; Burileanu, D. Using Voice Activity Detection and Deep Neural Networks with Hybrid Speech Feature Extraction for Deceptive Speech Detection. *Sensors* **2022**, *22*, 1228. [CrossRef] [PubMed]
50. Devlin, J.; Chang, M.; Lee, K.; Toutanova, K. BERT: Pre training of Deep Bidirectional Transformers for Language Understanding. *arXiv* **2018**, arXiv:1810.04805.
51. Zuluaga-Gomez, J.; Vesely, K.; Szöke, I.; Motlicek, P.; Kocour, M.; Rigault, M.; Choukri, K.; Prasad, A.; Sarfjoo, S.S.; Nigmatulina, I.; et al. ATCO2 corpus: A Large-Scale Dataset for Research on Automatic Speech Recognition and Natural Language Understanding of Air Traffic Control Communications. *arXiv* **2022**, arXiv:2211.04054.

aerospace

MDPI

Article

Assistant Based Speech Recognition Support for Air Traffic Controllers in a Multiple Remote Tower Environment

Oliver Ohneiser [1,*], Hartmut Helmke [1], Shruthi Shetty [1], Matthias Kleinert [1], Heiko Ehr [1], Sebastian Schier-Morgenthal [1], Saeed Sarfjoo [2], Petr Motlicek [2], Šarūnas Murauskas [3], Tomas Pagirys [3], Haris Usanovic [4], Mirta Meštrović [5] and Aneta Černá [6]

1 German Aerospace Center (DLR), Institute of Flight Guidance, Lilienthalplatz 7, 38108 Braunschweig, Germany; hartmut.helmke@dlr.de (H.H.); shruthi.shetty@dlr.de (S.S.); matthias.kleinert@dlr.de (M.K.); heiko.ehr@dlr.de (H.E.); sebastian.schier@dlr.de (S.S.-M.)
2 Idiap Research Institute, Centre du Parc, Rue Marconi 19, 1920 Martigny, Switzerland; saeed.sarfjoo@gmail.com (S.S.); petr.motlicek@idiap.ch (P.M.)
3 AB "Oro Navigacija" (ON), Air Navigation Service Provider of Lithuania, Balio Karvelio St. 25, 02184 Vilnius, Lithuania; murauskas.s@ans.lt (Š.M.)
4 Austro Control (ACG), Österreichische Gesellschaft für Zivilluftfahrt mbH, Air Navigation Service Provider of Austria, Schnirchgasse 17, 1030 Vienna, Austria
5 Croatia Control (CroControl), Air Navigation Service Provider of Croatia, Rudolfa Fizira 2, 10410 Velika Gorica, Croatia
6 Air Navigation Services of the Czech Republic (ANS CR), Navigační 787, 25261 Jeneč u Prahy, Czech Republic
* Correspondence: oliver.ohneiser@dlr.de; Tel.: +49-531-295-2566

Abstract: Assistant Based Speech Recognition (ABSR) systems for air traffic control radiotelephony communication have shown their potential to reduce air traffic controllers' (ATCos) workload. Related research activities mainly focused on utterances for approach and en-route traffic. This is one of the first investigations of how ABSR could support ATCos in a tower environment. Ten ATCos from Lithuania and Austria participated in a human-in-the-loop simulation to validate ABSR support within a prototypic multiple remote tower controller working position. The ABSR supports ATCos by (1) highlighting recognized callsigns, (2) inputting recognized commands from ATCo utterances in electronic flight strips, (3) offering correction of ABSR output, (4) automatically accepting ABSR output, and (5) feeding the digital air traffic control system. This paper assesses human factors such as workload, situation awareness, and usability when ATCos are supported by ABSR. Those assessments result from a system with a relevant command recognition rate of 82.9% and a callsign recognition rate of 94.2%. Workload reductions and usability improvement with p-values below 0.25 are obtained for the case when the ABSR system is compared to the baseline situation without ABSR support. This motivates the technology to be brought to a higher technology readiness level, which is also confirmed by subjective feedback from questionnaires and objective measurement of workload reduction based on a performed secondary task.

Keywords: air traffic controller; multiple remote tower; assistant-based speech recognition; automatic speech recognition and understanding; electronic flight strips

Citation: Ohneiser, O.; Helmke, H.; Shetty, S.; Kleinert, M.; Ehr, H.; Schier-Morgenthal, S.; Sarfjoo, S.; Motlicek, P.; Murauskas, Š.; Pagirys, T.; et al. Assistant Based Speech Recognition Support for Air Traffic Controllers in a Multiple Remote Tower Environment. *Aerospace* **2023**, *10*, 560. https://doi.org/10.3390/aerospace10060560

Academic Editor: Judith Rosenow

Received: 5 April 2023
Revised: 27 May 2023
Accepted: 7 June 2023
Published: 14 June 2023

1. Introduction

Speech recognition and speech understanding have found their way into use in daily life. While speech recognition has become quite robust with growing amounts of data, speech understanding remains a challenge given the complexity of verbal utterances' semantics. However, high accuracy in speech understanding is needed for human operators that supervise safety-critical processes, such as in aviation. Only then, users of speech recognition and understanding systems such as controllers will accept them and can benefit from their support, e.g., through workload reduction. Nowadays, tower controllers are burdened with manually maintaining flight strips, even if the content that needs to be entered in such flight strips is also communicated verbally in air traffic control radio telephony. This article presents one of the first prototypes of a speech recognition and understanding system to support ATCos in the tower environment in maintaining digital flight strips—in our case, even in a simulated multiple remote tower environment.

Our conducted validation study with ten air traffic controllers (1) quantifies any productivity enhancements in terms of mental workload, situation awareness, satisfaction, acceptance, trust, and usability through the advanced support functionalities in the digital system with automatic flight strip maintenance and highlighting features (independent variable); (2) quantifies the quality of speech-to-text and text-to-concept functionality; and (3) gathers feedback on the prototypes' functionality and visualization.

1.1. Related Work

1.1.1. Automatic Speech Recognition and Understanding in Air Traffic Management

During the last decades, a row of prototypes for speech recognition and understanding [1] in the air traffic management (ATM) domain has been developed. Early prototypes intended to support air traffic control (ATC) training and to reduce the number of required simulation pilots [2,3]. ATC events have been recognized from utterances to estimate controller workload [4,5]. The integration of contextual knowledge from an electronic assistant system for the speech recognition and understanding process [6] reduced recognition error rates [7]. These so-called assistant-based speech recognition (ABSR) systems initially focused on the approach environment [8]. For interoperability and comparability, rules for transcription (speech-to-text) and annotation (text-to-concepts)—so-called ontologies—have been defined and agreed upon between the major European ATM stakeholders [9]. Due to these rules, ATC utterances always comprise a callsign and at least one command that can consist of a type, unit, qualifier, and conditions. Later, ABSR systems were enhanced and enrolled on the en-route [10], apron [11,12], and tower environment [13]. This included the prediction and extraction of ATC commands [14]. Further research prototypes enhanced the ontologies, worked on speech recordings and radar data from real operations rooms, especially, but not limited to, recognizing callsigns [15–17], pre-filled aircraft radar labels that reduced the workload of ATCos [18,19], and implemented automatic readback error detection [10,20]. However, there was no validation of a sophisticated ABSR system's support for tower controllers, especially in a multiple remote tower setup using such a system in a high-fidelity laboratory environment.

1.1.2. Multiple Remote Air Traffic Control Tower and Human Operator Performance

The history of laboratory remote tower working positions started over two decades ago [21]. Recent research focused on human performance in multiple remote tower environments, i.e., where an ATCo is responsible for more than one remote airport at the same time. This started with analyzing eye-tracking data to characterize tower controllers' visual attention [22]. The research went on to investigate the changes in monitoring tasks and drafting multimodal interaction to support human operators at the controller working position (CWP) [23]. The latest research concentrated on workload assessment [24], operational feasibility and safety [25], as well as a supervisor position [26]. With fostering the technology maturity, questions regarding standardization with the European Organization for Civil Aviation Equipment (EUROCAE) and the European Union Aviation Safety Agency (EASA) guidelines have been developed [21]. Furthermore, the certification process for multiple remote tower operations has been sketched [27].

In the multiple remote tower environment, the human ATCo remains a central mean for the overall performance, with or without ABSR support. Related work on human performance assessment with standardized questionnaires is explained together with their results in the subsections of the result Section 3.

1.2. Structure of the Article

Section 2 describes the setup for the validation of ABSR support for ATCos and the conduction of this study. Section 3 presents the study results for the two aspects "Application of ABSR" and "ABSR in an ATM environment", i.e., results on speech recognition performance (Section 3.1) and speech understanding performance (Section 3.2) as well as on human factors such as mental workload, situation awareness, satisfaction, acceptance, trust, and usability (Sections 3.3–3.10), and ends with general feedback from ATCos (Section 3.11). Section 4 discusses the major study results for the fast readers who just quickly scanned Sections 2 and 3. For the very fast overview reader, Section 5 concludes and gives an outlook on future work. A list of abbreviations is provided before the Appendix. For more details and to follow some of the calculations, Appendix A lists results on speech-to-text performance, Appendix B lists results on text-to-concept performance, Appendix C lists the questionnaire statements of this study, and Appendix D details some validation setup views.

2. Materials and Methods

This section describes the hardware and software setup, as well as the methodology for the conduction of a human-in-the-loop simulation study to validate the benefits of an implemented ABSR prototype that was integrated with a prototypic electronic flight strip system for ATCos working within a simulated multiple remote tower environment. The technological validation exercise "006" was part of SESAR2020's wave 2 project PJ.05, "Digital Tower Technologies (DTT)" that received funding from the SESAR Joint Undertaking under the European Union's Horizon 2020 research and innovation program under grant agreement No 874470. More specifically, the exercise was conducted within solution 97, "HMI Interaction modes for Airport Tower," with its "Automatic Speech Recognition (ASR)" activity for "Improving controller productivity by ASR at the TWR CWP".

2.1. Hardware Setup of the Validation Study

Figure 1 shows the hardware setup of a prototypic CWP for a multiple remote tower environment in DLR's TowerLab [28]. Three horizontal rows of monitors (top of Figure 1) visualize the artificial outside view for the three configured airports. The airport layout is generic, but the three airports are named Vilnius, Kaunas, and Palanga.

Figure 1. Multiple remote tower environments with a row of monitors per each of the three airports under ATCo control, three radar screens, and the electronic flight strip system that is supported by the output of an assistant-based speech recognition system. The position for Vilnius is always top/left, Kaunas is middle, and Palanga is bottom/right.

The three monitors below on the desk (see Figure 1) depict the air traffic in the airport's vicinity. The touch display at the middle of the desk (see Figure 1) presents the electronic flight strips per airport per column. The ATCo wears a headset with speakers and a microphone that is triggered via a push-to-talk button at the headset's cable. The paper sheets on the left of the desk (see Figure 1) contained the airport layout, aircraft callsigns, and a legend for the symbols of the electronic flight strip system.

2.2. *Software Setup and Simulation Environment of the Validation Study*

All used software and displays are prototypic DLR developments. They consist of the most common elements that the usual controller working positions of European air navigation service providers offer. Thus, a wide range of ATCos from many different countries can use the systems of the validation study even if the details differ compared to their "usual" systems in daily-life operations. The aircraft and ground vehicle movements relevant to the tower and ground control were simultaneously simulated in three remote Lithuanian airports, i.e., Vilnius, Kaunas, and Palanga.

2.2.1. Outside View for Supervision of Movements on Ground and above the Airfield

The artificial outside view, such as out of a physical tower for those three airports, comprises the runway, taxiways, stands, and some environments, such as landscape and buildings, as shown in Figure 1. On the left and right side of each monitor row, there was a compass rose with additional information relevant to aircraft takeoff and landing (more details in Appendix D). If the validation condition "with ABSR support" was active, the ABSR output was also shown in the ATCo outside view.

2.2.2. Radar Displays to Monitor Air Traffic Close to the Airfield

A radar display for each of the three airports (see Figure 1 middle part) visualized the airspace structure with waypoints and the air traffic in the airport's vicinity. Each aircraft had a radar label displaying the aircraft callsign, weight category, current altitude, rate of descent/climb, speed, heading, and aircraft type. The biggest airport (Vilnius) also had a ground radar display showing the runway, taxiways, stands, and aircraft information, i.e., current and latest positions, aircraft callsign, relevant runway or stand, speed, and aircraft type, as well as a color indicating if the flight is an arrival or departure.

2.2.3. Electronic Flight Strip System (EFS)

The electronic flight strip system on the touch display consisted of one column per airport (see Figure 2). The column heads presented the airport's ICAO code, runways, automatic terminal information service (ATIS) letter, and radio frequency. Each of the three columns, in turn, comprised four different bays—air, runway, ground, and stand—in order to enable managing the flight progress in a procedural way.

Each flight strip (see zoomed white box in Figure 2) offered the option for hand written notes (pen symbol in upper left area), and showed aircraft callsign (BRU835), ICAO weight category (M), runway (34), stand (M1), estimated time of arrival/departure (08:39), aircraft type (A320), flight rules ("I" or "V" for instrument/visual flight rules), origin/destination airport (EDDK), standard instrument departure (such as BELED3D for aircraft GAF612 on the lower right blue flight strip), and squawk (3511).

The EFS for the ATCos further had a number of flight status icons on the right side (see Figure 2). The flight status icons depended on the flight intentions, i.e., blue departure flight strips/purple arrival flight strips, and on the progress, i.e., in which bay the flight strips currently are. Each flight status icon could be toggled, i.e., activated when a status change was initiated or deactivated, e.g., in case of activating by accident. The different flight status icons are shown in Figure 3. If they were activated through the tap of an electronic pen, they turned into a light green color in the electronic flight strip.

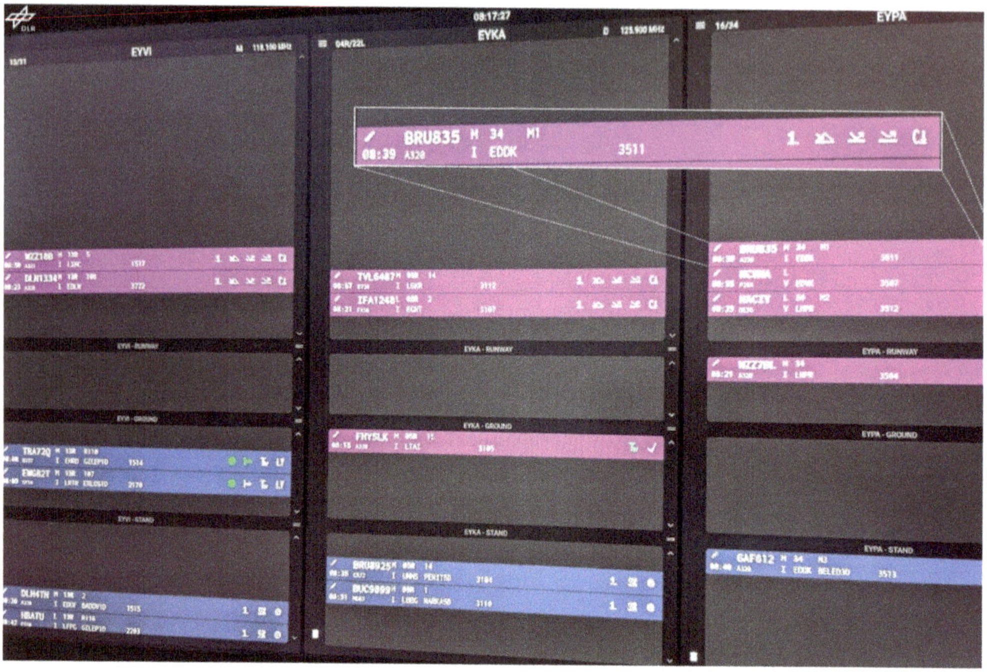

Figure 2. DLR's prototypic electronic flight strip system for aircraft at three remotely controlled airports (from left to right: Vilnius, Kaunas, Palanga).

Symbol	Name	Description
1	FIRST_CONTACT	First radio contact established
	START_UP	Aircraft has clearance for startup
	PUSHBACK_GIVEN	Aircraft has clearance for pushback
	TAXI_OUT	Aircraft has clearance to for taxi to runway
LT	LINE_UP	Aircraft has clearance to line up on the runway
CT	TAKEOFF_CLEARANCE	Aircraft has clearance for takeoff
	DEPARTING	Aircraft is flying away from airport
	EXIT_CTR	Aircraft is leaving control zone
	ENTER_CTR	Aircraft is entering control zone
	LANDING_CLEARANCE	Aircraft has clearance to land
	LANDED	Aircraft has landed
	TAXI_IN	Aircraft has clearance to taxi to apron
	TOUCH_AND_GO	Aircraft has clearance for touch and go landing
	LOW_APPROACH	Aircraft has clearance for low approach
√	CLOSED	Flightplan has been closed
SSR	SQUAWK_SET	Transponder code has been set (event, not a state the aircraft remains in)

Figure 3. Flight status icons of electronic flight strips available depending on the current flight status [29].

The electronic flight strips changed their bays with further progress of the flight status when arriving or departing, e.g., after setting the status "LINEUP," the flight strip moved from the ground bay to the runway bay.

2.2.4. Assistant-Based Speech Recognition and Understanding Prototype

The core development for the validation study was a prototypic system for speech recognition and understanding in a multiple remote tower environment. This ABSR system is based on a number of models based on deep neural networks trained by machine learning methods, respectively. The two main steps are (1) speech recognition, i.e., automatic speech-to-text transcription from tower controller audio input, and (2) speech understanding, i.e., automatic semantic text-to-concept annotations from the transcription input (see Figure 4). The speech recognition and understanding models were trained on in-domain and out-of-domain data, specifically 200 h from seven different datasets and 4.5 h (recorded in the later study environment) of manually transcribed speech data, as well as 400 h of untranscribed data from LiveATC (Homepage: https://www.liveatc.net/ (accessed on 4 April 2023)) [30]. Further references on the development of the speech recognition engine with artificial intelligence techniques can be found in [30].

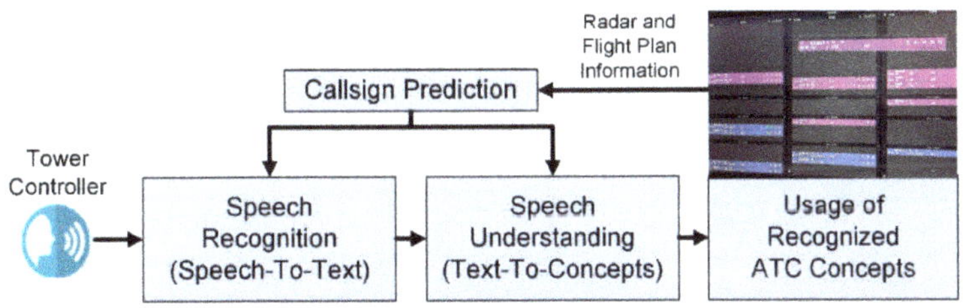

Figure 4. Components of assistant-based speech recognition (ABSR) in the multiple remote tower environment.

Both speech-to-text and text-to-concepts benefit from the use of contextual data, i.e., they consider radar data and flight plan data. The callsign prediction model is used to forecast aircraft callsigns for the next ATCo utterances, i.e., it predicts only those aircraft callsigns which are in the current area of responsibility of the ATCo. Those forecasted callsigns support the speech recognition engine in recognizing the correct word sequences and the speech understanding module in extracting the correct callsigns, especially in cases when not all words of the callsign are correctly recognized.

The command extraction model in the speech understanding module analyses the automatically transcribed ATCo utterances and extracts meaningful content, i.e., ATC concepts such as commands with callsigns, command types, values, units, etc., conform to the defined ontology. Two example transcriptions with their example annotations shall illustrate this:

- *wizz air two echo bravo good morning vilnius tower startup and pushback approved cleared to sofia* via *erlos one delta departure route seven thousand feet squawk two one seven seven QNH one zero one four*

  WZZ2EB GREETING
  WZZ2EB STATION VILNIUS_TOWER
  WZZ2EB STARTUP
  WZZ2EB PUSHBACK
  WZZ2EB CLEARED TO LBSF
  WZZ2EB CLEARED VIA ERLOS_1D
  WZZ2EB ALTITUDE 7000 ft
  WZZ2EB SQUAWK 2177
  WZZ2EB INFORMATION QNH 1014

- *hotel tango uniform when you are ready taxi to holding point runway three one correction one three right* via *[hes] golf vilnius*

  HBATU CORRECTION
  HBATU TAXI TO HP_13R WHEN READY
  HBATU TAXI VIA G C WHEN READY

The recognized ATC concepts, i.e., the annotations, are then used for highlighting purposes or supporting manual input in electronic ATC systems.

2.2.5. Visualization of ABSR Output on EFS and Outside View

The ABSR output was visible through different highlighting mechanisms in the electronic flight strips if the validation condition "with ABSR support" was active. If a callsign was recognized [31], the callsign was highlighted by displaying a rectangle in inverted colors for ten seconds at the callsign field of the flight strip (see "DLH4TN" in Figure 5). The callsign was highlighted immediately after being recognized and extracted even before the ATCo finished the utterance by releasing the push-to-talk button.

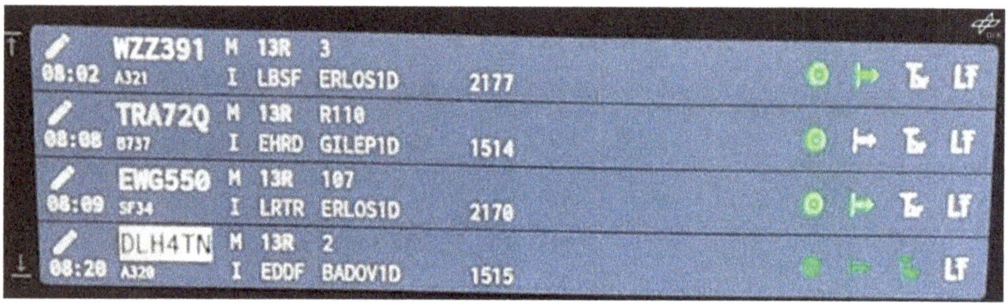

Figure 5. Prototypic electronic flight strips in the ground bay with a highlighted callsign as recognized from an ATCo utterance (DLH4TN), dark green automatically highlighted status icons for DLH4TN (STARTUP, PUSHBACK, TAXI), and five light green highlighted status icons of three other flights after being automatically accepted from the system or manually entered by the ATCo.

If one or more ATC concepts, such as commands and optionally command values, have been recognized, there was a dark green highlighting to support the ATCo in maintaining flight strips. This means the flight status icons on the right side of a flight strip or text values on the left side of a flight strip have been highlighted for ten seconds (see highlighted status icons for STARTUP, PUSHBACK, and TAXI of DLH4TN in Figure 5).

If the flight status icons in dark green mode remained unchanged by the ATCo for ten seconds, they were automatically accepted and turned into light green as with manual activation. In the case of a recognized HOLD_SHORT of runway command, the runway name was highlighted with color inversion for ten seconds as well.

2.3. *ATCo Tasks in the Different Validation Conditions*

Many of the tasks that ATCos needed to perform during the real-time human-in-the-loop validation study were identical under different validation conditions. Two conditions have been analyzed in the simulated multiple remote tower environment: baseline, i.e., without ABSR support and solution, i.e., with ABSR support. Section 2.3.1 describes the ATCo tasks in the baseline condition; Section 2.3.2 explains the changes induced for the ATCo when working in the solution condition.

2.3.1. ATCo Primary Tasks in Baseline Condition without ABSR Support

During the simulation runs, ATCos primarily needed to control the relevant traffic at three remote airports (tower and ground), with the above-described hardware and software setup consisting of an outside view, radar displays, and the electronic flight strip system.

Hence, they mainly gave ATC clearances, allowed for startup and pushback, instructed taxi, lineup/vacate and takeoff/landing/touch-and-go clearances for the single runway in use at each airport, as well as approved to enter/leave the control zone and to contact adjacent sectors. They also had to handle special situations on the ground with aircraft and ground vehicles being involved, such as a bird strike following a runway check and an emergency landing with the disembarkation of a sick passenger. The ATCos instructed all commands to the relevant traffic verbally in the English language via an emulated radio system.

Three simulation pilots (one for each airport) in another room communicated with the ATCo to run air and ground traffic with the support of a simulation pilot interface (see Appendix D). The ATCos were instructed to speak as usual at their working position. This also implies that some ATCos stick closer to the ICAO phraseology than others. The only continuous additional content for each ATCo utterance was the name of the station the ATCos are representing with the current utterance, i.e., "vilnius/kaunas/palanga tower," in order to fulfill safety requirements of the multiple remote tower concept.

The ATCos were asked to enter the semantic content of all utterances in terms of changed flight status into the electronic flight strip system with an electronic touch pen. Thus, they had to touch the flight status icon PUSHBACK in case they verbally instructed a pushback clearance or TAXI and the name of the taxiway if there were multiple options in case they issued to taxi via a certain taxiway (see Figure 6).

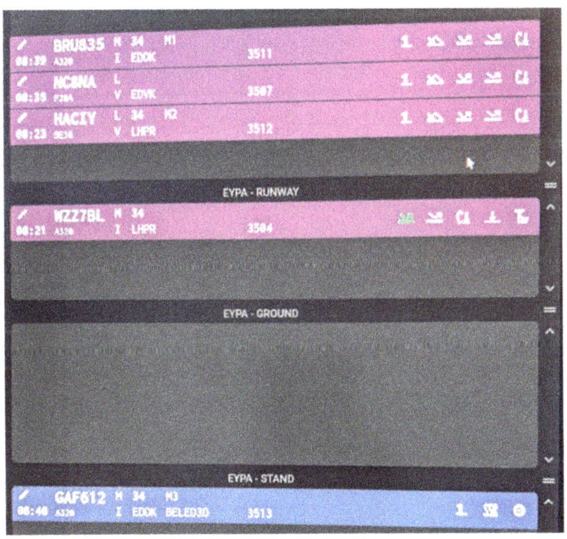

Figure 6. Prototypic electronic flight strips (blue departures; violet arrivals) in different bays (air, runway, ground, stand) with relevant information on the left (estimated time, callsign, aircraft type and weight category, flight rules, runway, destination airport, stand, departure route, squawk) and status icons on the right (e.g., CLEARED TOUCH_GO in green, ENTER_CTR, etc.).

The ontology defines 80 different command types as relevant for tower ATCos if they also include the role of ground control. All of these command types have been implemented within our command extraction algorithm.

The airport topologies were rather simple, i.e., the two smaller airports (Kaunas, Palanga) had just one taxiway each from the apron to the lineup. They vacated the single runway, and only the biggest airport (Vilnius) had two taxiway alternatives each for lining up and vacating the single runway. No runway change occurred during the simulation time. The weather conditions at all three airports remained visual meteorological conditions in the daytime throughout the simulation.

The relevant traffic in the two different one-hour simulation scenarios comprised twelve flights in Vilnius (plus two ground vehicles), six flights in Kaunas (plus one ground vehicle), and five flights in Palanga—at the latter airport, including training flights with multiple approaches—so 23 flights plus three ground vehicles (the ground vehicles make 11.5% of total relevant traffic) in total. For later evaluation, the results refer to all 26 traffic vehicles (flights plus ground vehicles) as ATC communication took place between ATCos and pilots or ground vehicle drivers, respectively. The callsigns and timing of appearance of the flights in these two scenarios were slightly different in order to reduce learning effects.

2.3.2. ATCo Tasks in Solution Condition with ABSR Support

In the solution scenario, ATCos had the same hardware setup as in the baseline scenario. The only difference was the support of the ABSR system. ATCos could majorly resign from using the electronic pen to maintain flight strips and benefit from automatic maintenance through the ABSR system, i.e., the ABSR output was used to highlight the flight status icons and callsigns in electronic flight strips automatically (see lower zoomed white box in Figure 7). The ATCos only needed to check the automatically highlighted output, i.e., representing issued commands and thus changes in the aircraft flight status, and correct if needed. A video about the simulation environment in the solution runs can be downloaded from https://www.youtube.com/watch?v=Y76kQmo_ANU&cbrd=1 (accessed on 4 April 2023). The ABSR output was only shown to the ATCos in solution scenarios. However, recording of verbal utterances, automatic transcription and automatic annotation was also performed in the background in baseline runs. The flow of using speech recognition and understanding output in the flight strips can be traced in Figure 7.

Figure 7. ATCo in front of electronic flight strip display with highlighted callsign and flight status icons, as well as outside view with transcription and annotation of ABSR output.

The complete transcription of words (first line) and the relevant annotation of commands in the agreed ontology format (second line) have been displayed in the outside view of the human-machine interface as shown in Figure 7 (zoomed white box on the upper area of the figure) if the validation condition "with ABSR support" was active.

2.4. Questionnaires and Further Tasks during and after Simulation Runs

Every five minutes, the ATCos were requested to rate their workload on a displayed graphical interface for an instantaneous self-assessment of workload (ISA) scale [32]. This interface offered values from 1 (low workload) to 5 (high workload) and appeared in the EFS system (see Figure 8).

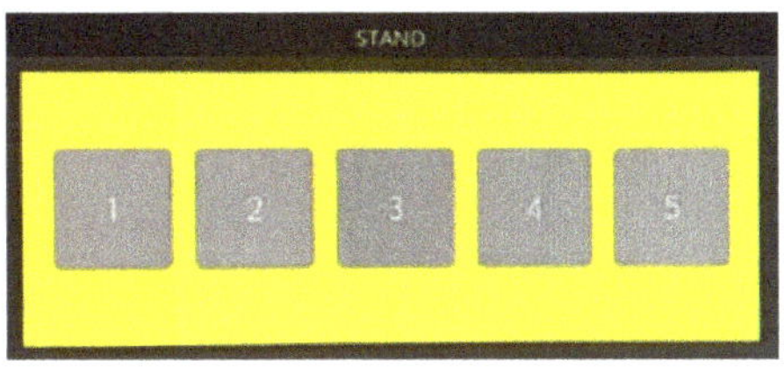

Figure 8. Instantaneous self-assessment of workload (ISA) scale to be responded to. "1" corresponds to "Under-utilized", "2" to "Relaxed", "3" to "Comfortable", "4" to "High Workload", and "5" to "Excessive Workload".

2.4.1. ATCo Secondary Tasks during Simulation Runs

Furthermore, the ATCos were asked to perform a secondary task next to their primary ATC task. After 10 and 40 min in the scenario, ATCos were requested to sort a deck of 48 cards and name one to four randomly missing cards (see Figure 9). This sorting of cards was repeated three times each or a maximum of 15 min (after 10 min) or 13 min (after 40 min), respectively. This secondary task is aimed to give a more objective impression about workload when comparing the time needed to sort and identify missing cards between baseline and solution scenarios. It is assumed that ATCos have more free cognitive capacity (less workload) if they can sort the cards quicker in one of the simulation conditions. The points in time (after 10 and 40 min) have been chosen as the ATCo workload should have been slightly increased due to the traffic situation at that time. The need to respond to ISA and to perform the card sorting remained identical in baseline and solution runs.

Figure 9. ATCo interrupts card sorting (secondary task) to check the outside view.

2.4.2. ATCo Post-Run Questionnaires after Simulation Runs

The post-run questionnaires needed to be filled by ATCos twice on each validation day, i.e., after each of the two simulation runs with the two different conditions. The well-established questionnaires cover the most important factors of air traffic controller work, such as situation awareness, workload, and trust [33] and are listed below:

- NASA-TLX (National Aeronautics and Space Administration Task Load Index) [34,35];
- Bedford Workload Scale [36];
- Three SHAPE questionnaires (Solutions for Human Automation Partnerships in European ATM) [37]:
    - AIM-s (Assessing the Impact on Mental Workload);
    - SASHA (Situation Awareness for SHAPE) ATCo;
    - SATI (SHAPE Automation Trust Index);
- CARS (Controller Acceptance Rating Scale) [38];
- SUS (System Usability Scale) [39,40].

2.4.3. Statistical Analysis Approach

When reporting the results of data that has been measured for baseline and solution runs, there will also be a statistical significance analysis, e.g., of all the above-mentioned questionnaires. Usually, there is a learning effect if ATCos perform multiple simulation runs in a row, i.e., they will perform better in the later runs, because they are used to the overall environment. Hence, better performance cannot simply be assigned to possibly different simulation run conditions such as baseline or solution. The sequence of baseline and solution runs is also an independent variable.

Therefore, two measures have been taken to compensate for the sequence effects as much as possible. First, the order of simulation runs alternate, i.e., half of ATCos start with a baseline run and end with a solution run and vice versa for the other half. The performance usually is, of course, better in the later runs, but the effect on baseline and solution runs should average out. Nevertheless, the standard deviations will be higher than they would be without sequence effects. Hence, secondly, the sequence effects will be compensated by considering the performance difference between the two runs. This sequence effect compensation technique (SECT) is described in more detail in [41]. An example shall illustrate the application of SECT. If any performance in all first runs of ATCos is 50 s and in all second runs 30 s, i.e., 20 s better, the performance difference is calculated as 50–30 = 20. Half of this difference (20/2), i.e., 10 s, is subtracted from each result of a first run and half of the difference is added to each result of a second run. Afterwards, the averages per run are the same. Furthermore, the averages of baseline and solution keep the same. We had exactly half of the ATCos having a baseline run and a solution run as the first run, respectively. However, the standard deviation will decrease, i.e., statistical significance will increase. This was already shown for earlier project result analyses such as of AcListant®-Strips when analyzing workload benefits [18].

Unpaired t-Tests can only reject hypotheses with some probability α. Therefore, the so-called null hypothesis H_0 is usually the opposite of the effect to be validated, e.g., "*ABSR support does not reduce workload as measured with a secondary task*". The test value T is calculated as the product of (1) the difference between the mean value of the performance measurement and μ_0, which is set to zero, and (2) the square root of the number of performance measurements, i.e., ten study subjects, divided by the standard deviation of the performance measurement. If the measurement values follow a Normal Gaussian distribution, the value T obeys a t distribution with n-1 degrees of freedom. Therefore, the resulting value T is compared with the value of the inverse t-distribution at the position $t_{n-1,1-\alpha}$ with n-1 degrees of freedom. If the calculated value T is bigger than the $t_{n-1,1-\alpha}$ threshold, we can reject the null hypothesis with probability α. As this falsifies the null hypotheses, we could assume that "ABSR support does reduce workload as measured with a secondary task." Additionally, the minimum α will be calculated, i.e., so that the value T threshold is still exceeded. These calculations will be performed on all single rated statements and answered questions, respectively, as well as for the group of statements/questions that belong together in a single questionnaire, e.g., the aggregating of the six items of NASA-TLX.

2.4.4. ATCo Post-Validation Overall Questionnaire

The post-validation questionnaire requested to be filled by ATCos only once after finishing all simulation runs, i.e., there is an overall rating on the ABSR prototype instead of a rating on baseline and solution each. It contained 28 statements to be rated regarding human performance, safety, operating methods, and technical feasibility. If answers on the post-validation questionnaire of the ten ATCos are reported in the following, the scale ranges from 1 (fully disagree) to 10 (fully agree), i.e., the scale mean is 5.5.

2.5. Validation Schedule and Participants

Each validation day with an ATCo began with organizational tasks such as the signature of informed consent, a briefing, and a demographics questionnaire. It was followed by 60 min training run with low to medium traffic (30 min each with baseline and solution condition, i.e., without ABSR and with ABSR support). Then, two simulation runs of 60 min each with baseline and solution conditions, respectively, and medium traffic were carried out. One run included a bird strike, and the other run included a sick passenger in an aircraft as special situations that the ATCos needed to handle and coordinate with ground vehicles. In order to average out the influence of the learning effect, baseline and solution scenarios have been alternated for ATCos throughout the validation campaign. After each run, the ATCos were requested to fill the mentioned questionnaires regarding workload, situation awareness, etc., as sketched in Section 2.4.2 and gave comments and answers in a debriefing. Finally, ATCos needed to fill out an overall tailor-made questionnaire (see Section 2.4.4) on the ABSR system after the last debriefing.

It has to be noted that the technical team of the validation campaign replaced a laptop and made a software update regarding the allowed central processing unit (CPU) load for the automatic speech recognition (ASR) engine after the eighth ATCo in the simulation campaign. However, no significant change in ABSR accuracy was noted due to this.

The validation campaign took place at DLR TowerLab in Braunschweig, Germany, from 14 February to 3 March 2022 (8:30 a.m. to 4:30 p.m.). This study was conducted with one ATCo per day for exactly ten days with five ATCos from Oro Navigacija (ON, Lithuania) and five ATCos from AustroControl (ACG, Austria). All participants were holders of an active tower ATCo license. The ten ATCos were not involved in the project in terms of participation in previous work sessions.

The nine male and one female ATCo had an arithmetic mean age of 31.9 years (standard deviation, SD: 5.5 years). The ATCos had 7.4 years of professional working experience as an ATCo (SD: 5.8 years), while ON ATCos were already longer on duty (9 years, SD: 7.3 years) compared to ACG ATCos (5.7 years, SD: 3.9 years).

3. Results

Each of the ten ATCos participated in a baseline run without ABSR support and a solution run with ABSR support, i.e., the data of twenty simulation runs with their succeeding post-run questionnaires as well as the final ten post-validation questionnaires' answers are analyzed in the following subsections. This section details:

(1) Objectively measured speech recognition performance;
(2) Objectively measured speech understanding performance;
(3) Perceived speech recognition and understanding performance;
(4) Operational and technical questions;
(5) Overall ratings on perceived workload, perceived situation awareness, satisfaction, acceptance, trust, and usability;

(6) Ratings per simulation run on perceived and more objectively measured workload, perceived situation awareness, satisfaction, acceptance, trust, and usability;
(7) General debriefing feedback.

The tailor-made statements of the questionnaires to be rated by ATCos described in the following contained the term ASR for brevity, even if automatic speech recognition and understanding was meant and experienced by the ATCos. Furthermore, the ABSR performance and the effect on subjective, as well as objective results are shown in more detail on a per-case basis by comparing ON and ACG ATCos for two reasons. First, the amount of training data differs by a factor of four between ON and ACG ATCos which influences the speech-to-text and text-to-concept performance. Second, the three controller working positions that (1) the Lithuanian ATCos are used to, (2) the Austrian ATCos are used to, and (3) is used as a prototypic environment in the simulation differ so that the familiarization with the system differs as well.

3.1. Results of Speech-to-Text Analysis

3.1.1. Audio Recordings with Transcriptions and Annotations

Verbal utterances of ATCos that were triggered with the push-to-talk button during twenty hours of simulation runs (radar data duration) have been recorded as wav-files. For each wav-file of the twenty simulation runs (baseline and solution) exists an automatic transcription and an automatic annotation. We recorded 2427 wav files with a net speech time of 4.5 h (i.e., when ATCos speak) during 20 h of radar simulation, i.e., the frequency load by ATCos was roughly 22%. The average duration per utterance was 6.6 s.

All wav-files have been manually transcribed and annotated ("gold") with DLR's Controller Command Logging Tool for Context Comparison (CoCoLoToCoCo, see Figure 10) to enable comparison and calculations about recognition and error rates on the word level and semantic level.

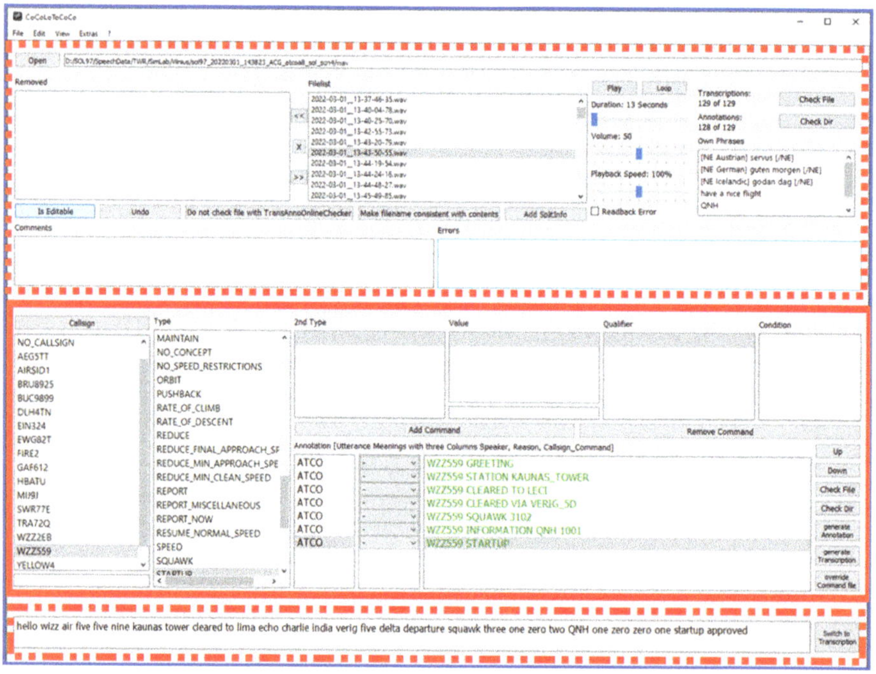

Figure 10. Software tool CoCoLoToCoCo to support transcription and annotation of ATC utterances.

The upper area of CoCoLoToCoCo (red dotted line) lists all audio files of a selected folder, has buttons and sliders to adjust the playback of the files, has a comment window and an error output window, as well as offers some further file-checking opportunities. The middle area (red solid line) shows the annotation view with a column per element of a controller command, the resulting annotation of an audio file in ontology format [9] (green font), and further buttons for rearranging and checking. The lower area (red point-dash line) visualizes the transcription of a selected audio file following defined transcription rules.

The gold transcriptions of the validation trials contain in total 37,238 words without words that are not fully uttered and thus contain a "*" such as *"lufthan*"* due to our transcription rules, i.e., each ATCo utterance contains roughly 15 words. Table A3 shows the top-25 1-grams, i.e., the uttered words with their absolute and relative frequency. The most often occurring words, "one" (6.43%) and "zero" (3.97%), are usually in the top three for other ATC communication corpora as well. However, the word at rank three, "tower" (3.96%), is specific for the multiple remote tower environment, in which the transmitting entity should always be named and, therefore, appears quite often. Normally, the digits from zero to nine fill the first ten ranks in ATC communication corpora.

Furthermore, the words *"runway," "to,"* and *"cleared"* appear in the top 12 as runway clearances and *"cleared to"* are often uttered. This latter result is confirmed by analyzing two real-life ATCo utterance corpora from Vilnius tower, as well as from Vienna tower, with roughly 7500 words in total each. This shows that the simulation setup and the challenges for the speech-to-text engine were quite realistic.

Table A4 lists the number of different words to reach a relevant portion of all uttered words, i.e., if speech-to-text performs well on the 100 most often occurring words, almost 90% of the total number of words are covered.

3.1.2. Speech-To-Text Performance

Some abbreviations that are used for analyzing purposes in the following and in the Appendices A and B are introduced:

- Onl = online (analysis as experienced by ATCos during simulation runs);
- Off = offline (analysis of audio files after the simulation runs);
- WER = Word Error Rate;
- Subs = Substitutions;
- Del = Deletions;
- Ins = Insertions;
- LevenDist = Levenshtein Distance [42] between automatic and gold transcription;

The speech-to-text accuracy is presented with details per each simulation run in the tables of Appendix A (see Tables A1 and A2). Table A1 visualizes the WER for offline recognition (Off) as evaluated after the end of the validation trials. It shows what results would be already achievable when the technical setup is improved to deliver the offline performance during the simulation runs. Table A2 visualizes the WER for online (i.e., real-time) recognition from the voice stream (Onl) as evaluated during the simulation runs, i.e., the WER are usually worse than for Off.

There were some technical problems with the ABSR setup: (1) the audio device continuously disconnected in one simulation run resulting in the loss of some data, and (2) there was partly CPU overload, especially for the first eight ATCos. The performance of the ASR engine was much worse in the online mode (as experienced by ATCos) than in the later offline analysis of recorded audio files. Worse speech-to-text performance, i.e., a higher WER being the sum of substitutions, insertions, and deletions regarding two-word sequences divided by the total number of correct words, of course also led to worse text-to-concepts performance. Some average and some specific results from these tables are analyzed deeper in the following.

The average WER for all twenty runs was 5.1% in Off mode. When just considering solution runs, the average WER even reached 4.4%, while baseline runs have an average

WER of 5.7%. When omitting the single run with audio device problems, the maximum WER was below 8% for all other 19 simulation runs in Off mode, i.e., the highest WER in that single run was 11.5%, and the lowest WER for any run was 1.3%. It needs to be admitted that the training data already contained a few speech samples from some ATCos that also participated in the final validation trials.

In Onl mode, the average WER was 13.6%, while the average WER for solution runs was 9.8% and for baseline runs 17.4% (see Table A2). There is a remarkable difference in the WER of ON ATCos (6.8%) compared to ACG ATCos (12.8%) in solution runs. This probably goes back to the amount of training data in the identical recording environment to the later validation trials, which was only 3.6 h for ON and even 0.9 h for ACG.

Four of twenty runs still achieved good performance with WER < 3%. However, three other runs that were affected by technical problems achieved a WER > 23%. Still, the Onl performance was sufficient in almost all solution runs to produce an acceptable text-to-concept quality. Nevertheless, the degradation of the speech-to-text performance is higher from offline mode to online mode than expected and offers room for improvement.

3.2. *Text-To-Concept Quality*

3.2.1. Description of Gold Annotation Data Set

All twenty simulation runs consist of 7560 commands (ALL), whereof 3701 are from baseline runs (BAS), and 3859 are from solution runs (SOL), respectively. Hence, there were 3.1 commands per ATCo utterance and 5.1 words per command if we assume that all words of an utterance are relevant to form a command.

However, it has to be noted that there are some word sequences annotated as commands that do neither influence the aircraft status nor include any request, report or traffic information from the ATCo side:

- First, the annotations GREETING (e.g., "hello"), FAREWELL (e.g., "bye"), and NO_CONCEPT (e.g., "thanks;" no relevant ATC command in the utterance) that are summing up to 9.8% of commands during this study. These command types can indicate that the ATCo workload might not be assumed as overwhelmingly high if they still have time for welcoming, saying goodbye, and thanking anybody.
- Second, the annotation CORRECTION and CALL_YOU_BACK (e.g., "standby") that sum up to 1% of the commands might indicate a higher workload as ATCos often correct themselves, are asking for repetition of the transmission or are telling to wait for further information. The annotation SAY_AGAIN, which also belongs to this command group, has not been used.
- Third, the annotation AFFIRM and one annotation of DISREGARD that sum up to 4.1% of the commands have ATC communication relevant content, even if they are no commands in a classical sense. The annotation NEGATIVE, that also belongs to this command group, has not been used.

Though, the above-listed annotations enable a workload analysis of human ATC operators that will be published in another paper. 15 of the 80 possible command types for tower ATCos as defined in the ontology, such as GO_AROUND and ABORT TAKEOFF, did not occur at all in the 7560 commands. This means 65 different command types have been used by the ten ATCos, e.g., PUSHBACK, TAXI TO, CLEARED TAKEOFF/LANDING, ENTER_CTR, etc. Table A5 lists the relative occurrence of all command types greater than 1%. The last type, "others", groups all command types that occurred between 0.33% and 1%, such as CONTACT, ENTER_CTR, LINEUP_BEHIND, CLIMB, and DIRECT_TO. In total, there are 36 different command types that appeared more than 25 times, i.e., more than 0.33%.

The most often used command type is—unsurprisingly—STATION, as ATCos were asked to utter it in each radio transmission. However, 1529 occurrences (20.2% of commands) in 2427 utterances mean that ATCos did not follow this multiple remote tower safety-related request in 37% of all utterances. This might not be critical if ATCos just uttered "bye," but in any case, it should be considered for the multiple remote tower concept. The (CONTINUE) TAXI TO/VIA commands sums up to 11.5% of commands. The INFORMATION WINDSPEED/DIRECTION even sum up to 15% of the commands as they were instructed for all takeoffs and landings/touch-and-gos. The exclusive runway clearances CLEARED TAKEOFF/LANDING/TOUCH_GO/VISUAL sum up to 6.8% of commands. The runway usage clearances LINEUP, LINEUP_BEHIND, VACATE (VIA), and BACKTRACK sum up to 4% of commands.

A total of 29 of those 65 used command types occurred a maximum of 25 times for all ATCos in total such as BACKTRACK, CLEARED VISUAL, HOLD_SHORT, JOIN_TRAFFIC _CIRCUIT, LEAVE_CTR VIA, and ORBIT. For the above considerations, we neglect that only 87% of all words that are available in the gold transcriptions have been used by the automatic command recognition algorithm to classify commands (see column "*Unknown Classified Rate*" in Tables A6, A8 and A10).

It needs to be mentioned that our prototype follows a more holistic approach than some very basic prototypes of other actors in the field of speech recognition and understanding [43]. Our command extraction algorithm does not only extract callsigns (DLH4TN), basic types (TAXI), and values, but more sophisticated command types of multiple parts (TAXI TO/VIA), units, qualifiers, conditions (WHEN READY), chain commands with multiple callsigns, tackles many types of corrections through the ATCo and even robustly recognizes elements of the ontology if there are minor and major (acceptable) deviations from ICAO phraseology [44] in the utterances. Furthermore, we support a bigger number of command types (from the agreed ontology) as defined by the different actors themselves. The execution time of the command extraction per utterance in offline mode on a standard laptop, i.e., on a complete transcription, has an arithmetic mean of 2 ms and a median of 1.2 ms with a minimum execution time below 0.1 ms and a maximum execution time below 40 ms independent of performing command extraction on gold, offline or online transcription files. In addition, our prototype is—to the best of our knowledge—the first to support multiple remote towers at the same time (not just one) and delivers recognition error rates on an acceptable level despite all the above-mentioned complex add-ons.

3.2.2. Description of Results of Automatically Extracted Commands on Different Versions of Speech-To-Text Transcriptions

The following three subsections present recognition and error rates on callsign and command level, as well as the portion of words from the utterances that have not been used for ATC concept extraction while referring to Appendix B. More details on the semantic level metrics can be found in [45]. The command extraction results will also be presented by comparing the different command type groups:

- "All;"
- "Relevant" if appearing more than 25 times in all 20 runs;
- "EFS" has a visible effect on the electronic flight strips;
- "Status" that changed the aircraft status in the electronic flight strips;
- "Outside" is just shown on the monitors for the outside view;
- "Hypo-EFS" could have been highlighted in the flight strips but have not been during the trials, such as recognizing the active runway in an utterance.

3.2.3. Speech Understanding Performance on Gold Transcriptions

In total, 65 different command types have been automatically extracted from the gold transcriptions, i.e., the same number as in gold annotations. Table A6 shows how well the ontology-conform automatic recognition of ATC commands is modeled. The command recognition rate is around 96% with an error rate below 2.5%; the rejection rate (not reported herein) causes a difference to 100% in the total sum of command rates. The callsign recognition rate even achieved 99.8% with an error rate of 0.2%. The command recognition rates in solution runs were 96.6% for ON and 95.4% for ACG.

A total of 18.3% of all problematic annotations (recognized commands) go back to the three ground vehicles in the scenario that make up 11.5% of all relevant traffic. Further,

7.3% of problematic annotations go back to the emergency aircraft, even if this makes up 3.8% of the flights.

18 of the 80 defined command types from the ontology had visible effects in the flight status icons of the electronic flight strips—hereinafter referred to as command type group *Status*. Three further commands had a visual effect on the textual data of the electronic flight strips. These 21 commands that influenced the appearance of the electronic flight strips are grouped in the command type group *EFS*. Three supported commands contained weather information from the *Outside* view (QNH, INFORMATION WINDDIRECTION and WINDSPEED); the values of four further supported commands could have been displayed in the relevant field of the electronic flight strip. However, this highlighting has not been fully implemented yet (command group *Hypo-EFS*), i.e., STATION, INFORMATION ATIS, INFORMATION ACTIVE_RWY, and HOLD_SHORT for all possible airfield elements such as taxiways. The command type group *Relevant* includes all commands that have been automatically extracted more than 25 times. Table A7 shows the command recognition performance on the above-mentioned command type groups, i.e., presenting command recognition rates of 96% and more.

3.2.4. Speech Understanding Performance on Offline Transcriptions

The command recognition results of Table A8 are based on the output of the speech recognition engine, i.e., the transcription from Off mode. The command recognition rate is above 91%, with an error rate below 5%. The callsign recognition rate achieved almost 98.5% with an error rate below 1%. The command recognition rate of command type group *EFS* is beyond 93%, as Table A9 shows. 16.2% of all problematic annotations go back to the three ground vehicles that comprise 11.5% of all relevant traffic.

3.2.5. Speech Understanding Performance on Online Transcriptions

Tables A10 and A11 present the command recognition results on transcriptions from Onl mode. The command recognition rates are roughly 10% worse than in Off mode. The command recognition rate for solution runs in which the ATCos saw the ABSR output was 82.9%, with an error rate of 6.6%. However, there is a huge difference in the command recognition rate for ON ATCos (88.0% based on WER of 6.8%) compared to ACG ATCos (77.7% based on WER of 12.8%). As the command recognition rates for ON and ACG ATCos were both close to 96% on gold transcriptions, the high WER resulting from the mentioned low amount of available training data was a major impact on the online command recognition next to some deviations of ATCos from ICAO phraseology. The online callsign recognition rate achieved 94.2% with an error rate of 2.4%. This again shows the influence of the high WER on the ATC concept extraction.

The following measurements, especially the questionnaire ratings of ATCos, are based on the Onl mode, as this performance was "experienced" by ATCos during simulation runs.

3.2.6. Subjectively Perceived Speech Recognition and Understanding Performance and Functionality (Post-Validation)

The post-validation questionnaire contained nine statements about technical feasibility with respect to the recognition and error rate of callsigns and commands as well as the ASR functionality:

1. The recognition rate and recognition error rates for callsigns by ASR were at an acceptable level. [CsgnRecRateOK];
2. The recognition rates and recognition error rates for commands by ASR were at an acceptable level. [CmdRecRateOK];
3. Overall, the level and quality of information provided by ASR were an acceptable level. [ASRQualInfOK];

The post-validation questionnaire contained four statements about the ASR interface:

4. The ASR tool interface (HMI) provides suitable access to relevant information in all situations. [ASRrelevInfo];
5. The ASR tool interface (HMI) does not display any non-essential information (clutter). [ASRessentInfo];
6. The ASR tool display is both comprehensible and acceptable. [ASRcomprehaccep];
7. The timeliness of the ASR tool display is within acceptable limits. [ASRtimeliness];
8. Automatic Speech Recognition (ASR) highlighting aircraft callsigns in the electronic flight strip display technically worked well. [Highl-Csgn];
9. Automatic Speech Recognition (ASR) highlighting aircraft callsigns in the electronic flight strip display supports recognizing which aircraft callsign has been (speech) recognized quickly. [Recog-Csgn].

The results are shown in Figure 11. ATCos rated the recognition of callsigns as almost perfect, with a mean value of around 9 on a scale from 1 to 10. The recognition rates of ATC commands were also perceived as good, with a mean value of around 7. The general quality level of information presentation from ASR was rated to be at an acceptable level with a mean value of slightly beyond 7. It has to be noted that the command recognition and overall ASR information displayed were rated much higher from ON than from ACG ATCos. This is most probably due to the underlying WER of 13% for ACG ATCos and 7% for ON ATCos, which is, however, still improvable to reach the 4% WER of offline analysis. Relevant information about the ABSR system can be assessed (mean value 7.4, but more than 1.5 points rated higher by ON than by ACG). The ASR tool seems to only present essential information with a mean value of 8.2 (again, ON rated almost 1.5 points higher than ACG). The ASR visualization is perceived as comprehensible with a mean value of 7.7 (again, ON rated almost 2 points higher than ACG). Finally, the output of the ABSR system was shown timely (mean value 7.5) due to the ATCo feedback.

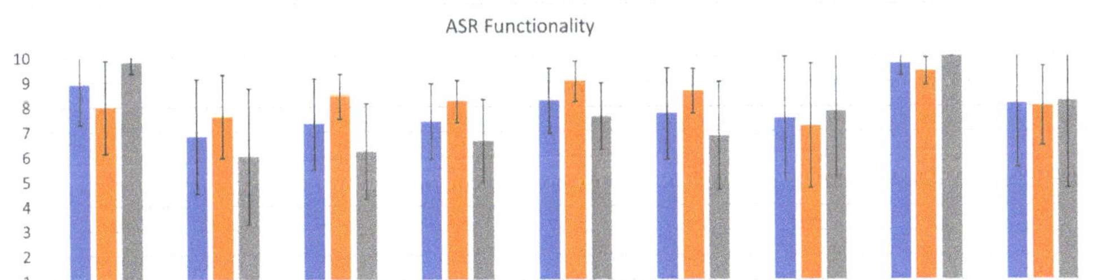

	CsgnRecRateOK	CmdRecRateOK	ASRQualInfOK	ASRrelevInfo	ASRessentInfo	ASRcompracc	ASRtimeliness	Highl-Csgn	Recog-Csgn
MEAN all	8.9	6.8	7.3	7.4	8.2	7.7	7.5	9.7	8.1
MEAN ON	8.0	7.6	8.4	8.2	9.0	8.6	7.2	9.4	8.0
MEAN ACG	9.8	6.0	6.2	6.6	7.6	6.8	7.8	10.0	8.2

Figure 11. Subjective ATCo ratings on ASR accuracy and functionality.

The highlighting of callsigns in the electronic flight strip display (*Highl-Csgn*) was perceived as working technically very well, with a mean of 9.7 on a 10-point scale and a low standard deviation of 0.5. The second statement *Recog-Csgn* rated with a mean value of 8.1, helped the ATCos to detect which aircraft callsign has been recognized by the ABSR system. This information is needed to decide whether the following recognized ATC commands are highlighted for the correct callsign. The interesting part of these answers is the comparison with the objective measurements, i.e., the online callsign recognition rates, which are 92.1% for Lithuanian ATCos and 91.3% for Austrian ATCos (see Table A10). The same applies to the callsign recognition error rates, which are 3.9% for ACG, and also much higher than the 2.4% for ON ATCos. We have no real answer for this discrepancy between subjective rating and objective measurement.

3.3. *Answers to Subjective Post-Validation Questionnaires*

3.3.1. Operational Use of ASR (Post-Validation)

The post-validation questionnaire contained five statements about the operational feasibility of the ASR system:

1. I can apply operating methods in an accurate, efficient, and timely manner with ASR. [AccOpMeth];
2. I think that operating methods are clearly identified and consistent in all operating conditions. [OpMethConsis];
3. Procedures and operating methods are acceptable when using the ASR tool. [ProcOKwASR];
4. There are no changes needed to current working methods/procedures to fully support the use of the ASR tool. [NoChgNeed];
5. The ASR tool would be operationally acceptable under either nominal or non-nominal conditions. [OpAccAllCond].

The results are shown in Figure 12. The operating methods with ASR seem to be accurate, efficient, timely, and consistent in different conditions, with mean values of 8 and 7.4, respectively. Procedures and operating methods seem to be fine, with a mean value of 8.5 and a standard deviation of only 1.0. There are some changes to current working methods needed to fully support the use of the ASR tool, as the mean value equals the scale mean value of 5.5. However, ON ATCos rated this statement with almost 7, while ACG ATCos rated it with slightly above 4 points. The ASR seems to be operationally acceptable under different conditions, most probably under the majority of nominal and a few non-nominal conditions, as the ATCo rating was just slightly beyond the scale mean value.

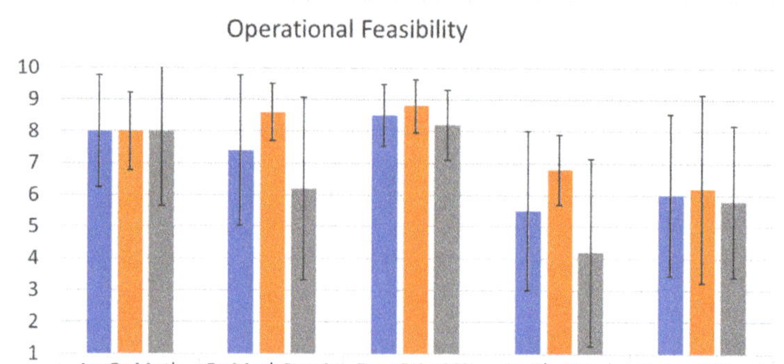

	AccOpMeth	OpMethConsis	ProcOKwASR	NoChgNeed	OpAccAllCond
MEAN all	8.0	7.4	8.5	5.5	6.0
MEAN ON	8.0	8.6	8.8	6.8	6.2
MEAN ACG	8.0	6.2	8.2	4.2	5.8

Figure 12. Subjective ATCo ratings on operational feasibility and operating methods.

3.3.2. Human Factors Questions (Post-Validation)

The post-validation questionnaire contained six statements on human factors:

1. I think that ASR supports me in maintaining my workload at an acceptable level. [ASRsupATCoWL];
2. I think that ASR supports me in maintaining an adequate level of situational awareness. [ASRsupATCoSAw];
3. My situational awareness is maintained at an acceptable level with Automated Speech Recognition (ASR). [ASRmaintSAw];

4. I see many safety-related issues to be solved regarding automatic speech recognition implementation. [ASRindSafeIssu];
5. I think that ASR did increase the potential for human errors. [ASRincrHumErr];
6. Overall, I was satisfied performing my task with ASR. [JobSatisf].

The results are shown in Figure 13.

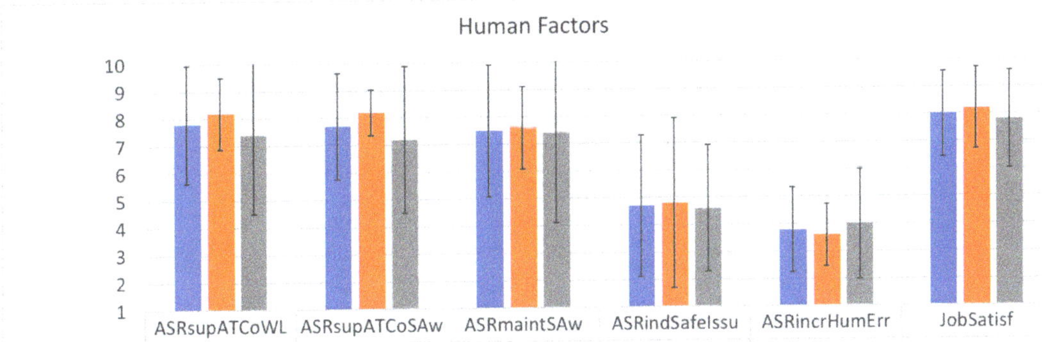

	ASRsupATCoWL	ASRsupATCoSAw	ASRmaintSAw	ASRindSafeIssu	ASRincrHumErr	JobSatisf
MEAN all	7.8	7.7	7.5	4.7	3.8	8.0
MEAN ON	8.2	8.2	7.6	4.8	3.6	8.2
MEAN ACG	7.4	7.2	7.4	4.6	4.0	7.8

Figure 13. Subjective ATCo ratings on human factors.

ASR seems to support maintaining situation awareness and workload of ATCos at an acceptable level with mean values of 7.5 and beyond on a 10-point scale. The *ASRsupATCoWL* statement was rated with 7.8 on a 10-point scale (90% of ATCos rated this item with 7 or above). The *ASRsupATCoSAw* statement was rated with 7.7 on a 10-point scale (90% of ATCos rated this item with 7 or above). The statement, if ASR induced safety issues or increased the potential for human errors, was rated with mean values below the scale mean of 5.5. ATCos rated their job satisfaction with using ASR high (mean value of 8 on the 10-point scale).

3.3.3. Acceptance (Post-Validation)

The post-validation questionnaire contained three statements about acceptance of and trust in the ASR system:

1. I think that the ASR system is adequately usable. [ASRadequse];
2. I would accept such an ASR system in my future tower CWP. [ASRacceptCWP];
3. My trust in the ASR system is at an acceptable level. [ASRtrust].

The results are shown in Figure 14. ATCos rated the adequate usage of ASR with a mean value of around 7. However, it has to be noted that it was rated much higher by ON than by ACG ATCos. All ATCos would accept such an ASR system in their future tower CWP with a mean value of 7.5. They trusted the ASR system with a mean value of around 7.

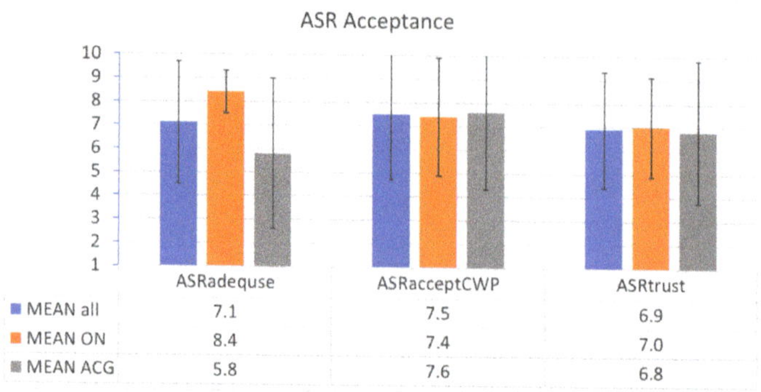

Figure 14. Subjective ATCo ratings on technical ASR acceptance.

3.4. *Answers to Subjective Post-Run Questionnaires*

3.4.1. Controller Acceptance Rating Scale (CARS) (Post-Run)

The post-run questionnaires contained the CARS statement to be rated on a scale from 1 to 10, with 10 being the best value, as listed in Appendix C.1. The results of the CARS questionnaire are shown in Figure 15. The acceptance was, on average, 0.6 points higher on the CARS scale for the baseline condition compared to the solution. The CARS questionnaire was filled out by each ATCo twice, once after the run with ABSR support and once after the run without ABSR support. Therefore, we are able to perform a paired *t*-test. After compensating sequence effects, the α was 0.1 to reject the inverse hypothesis that ABSR support reduces the controller acceptance due to CARS. The absolute value was 6.8 versus 6.2 (0.8 points higher for ON on average and 0.8 points lower for ACG on average).

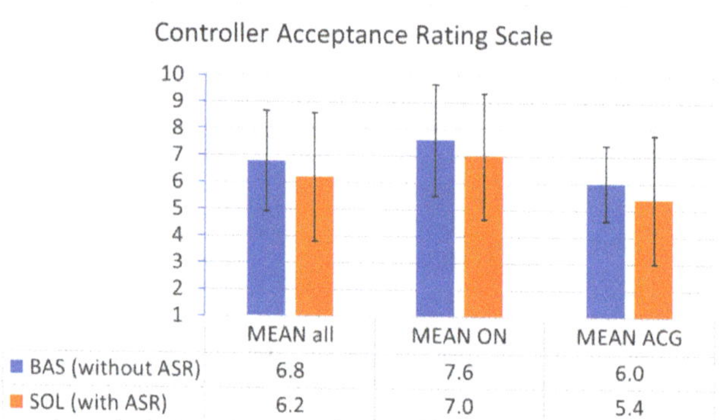

Figure 15. Subjective ATCo ratings on CARS.

3.4.2. Trust (SATI) (Post-Run)

The post-run questionnaires contained the six statements of SATI, as listed in Appendix C.2. The seven-item answer scale ranged from "Never, Seldom, Sometimes, Often, More Often, Very Often, and Always." To present the results in a bar diagram, "Never" is translated to 0%, "Seldom" to 1/6 %"… "Very Often" to "5/6 %" until "Always" to 100%. The results are shown in Figure 16.

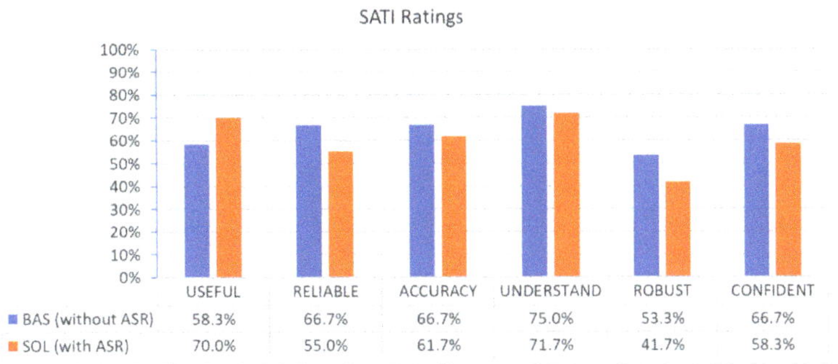

Figure 16. Subjective ATCo ratings on SATI questionnaire.

ABSR support reduced trust in automation due to SATI (α = 0.25). However, the usefulness of the system (*USEFUL* in Figure 16) was rated much better for SOL than for BAS (α = 0.05). The other five mean values are better for BAS than for the SOL condition. It is noteworthy that the four statements *RELIABLE*, *ACCURACY*, *UNDERSTAND*, and *ROBUST* from ON ATCos have better ratings for SOL than for BAS condition on average. The ambivalence of results will be discussed in Section 4.

3.4.3. Perceived Situational Awareness (SASHA ATCo) (Post-Run)

The post-run questionnaires contained the six statements of the SASHA ATCo, as listed in Appendix C.3. The seven-item answer scale ranged from "Never, Seldom, Sometimes, Often, More Often, Very Often, and Always." To present the results in a bar diagram, "Never" is translated to 0%, "Seldom" to 1/6 %"... "Very Often" to "5/6 %" until "Always" to 100%. The results are shown in Figure 17.

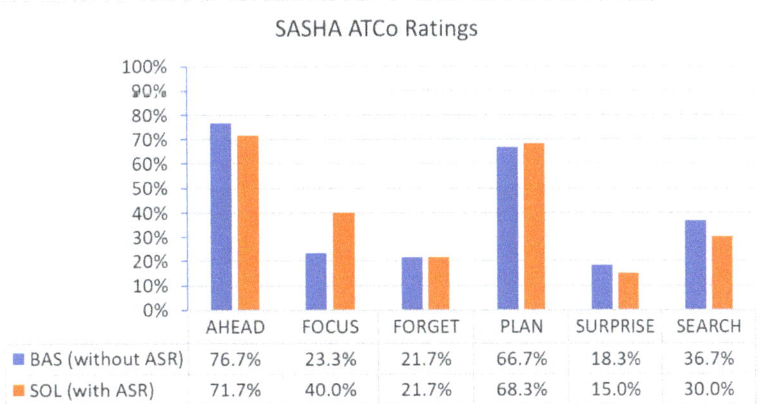

Figure 17. Subjective ATCo ratings on SASHA ATCo questionnaire.

ABSR support reduced the situation awareness of ATCos due to SASHA (α = 0.33). However, "searching for information" was less needed in the SOL condition (α = 0.15). The mean values of the first two items, *AHEAD* and *FOCUS*, are better for BAS than for SOL conditions. The mean values of the last four items, *FORGET*, *PLAN*, *SURPRISE*, and *SEARCH*, are equal or better for the SOL condition compared to the BAS condition without analyzing standard deviations, as differences in mean values are rather small.

3.5. Perceived Workload (High Workload Contribution) (Post-Run)

The post-run questionnaires contained a free-text question about high workload: "Which factors/events/conditions have contributed to potentially high workload?".

The structured answers and the number of ATCos noting this after each conducted simulation run (multiple notions in one questionnaire answer possible) were as follows:

- New/unknown airspace/airport layout (especially multiple remote towers): 15 times;
- New/unknown equipment/hardware/software/electronic flight strips: 7 times;
- Checking of ABSR output (only in solution condition): 4 times;
- Unexpected/unusual air traffic situations: 3 times;
- Other: Secondary task (2 times), tower view/runway perspective (2 times), slightly different phraseology to always name the calling tower (2 times), miscommunication, system errors.

Interpreting the above results, 15 of 20 ATCo answers stated that the unknown multiple remote tower environment with unknown airport layouts induced a higher workload. Furthermore, many ATCos remarked that the flight strip handling was difficult (as some details were different from "home"). This means that the majority of workload-increasing factors can be assigned to environmental aspects that should normally not be tested in the ABSR validation trials. The above-listed checking of ABSR output, as well as unexpected situations and some further aspects, seem to have been only a minor factor for the higher workload.

3.6. Perceived Workload (NASA-TLX and Bedford Workload Scale) (Post-Run)

The post-run questionnaires contained the six statements of NASA-TLX (National Aeronautics and Space Administration—Task Load Index) as listed in Appendix C.4 and the two statements of the Bedford workload scale to rate the average workload (AVG) and peak workload (PEAK) on a scale from 1 to 10 with 10 being the highest workload. In addition, the 15 pair-wise comparisons of workload contributing factors (as the other part of the weighted NASA-TLX questionnaire) were assessed with ATCos once.

The results of the weighted NASA-TLX and the Bedford workload scale are shown in Figure 18. Figure A1 in Appendix C shows the weight per each of the six dimensions for NASA-TLX, which is almost equally distributed except for more weight for mental workload than for physical workload. The overall weighted workload (OW) due to NASA-TLX was higher for the solution than for the baseline condition: 43.1 and 38.9 ($\alpha = 0.02$), respectively, with huge standard deviations around 17.5. However, the general difference between baseline and solution was only induced by the ON ATCo ratings, as the OW for ACG remained the same in baseline and solution.

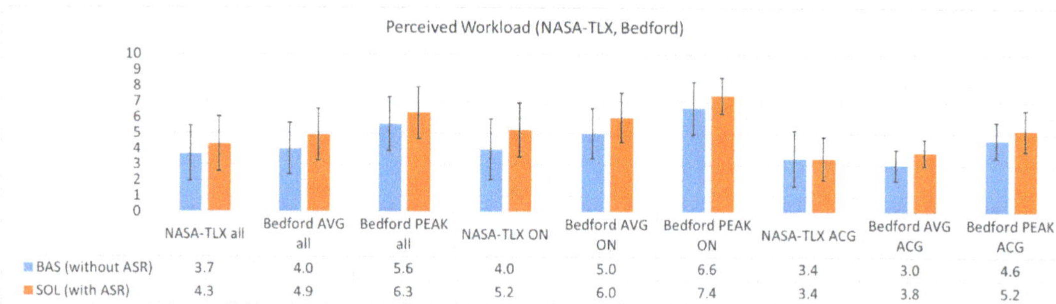

Figure 18. Subjective ATCo ratings on NASA-TLX (Weighted Overall Workload).

Furthermore, a clear learning effect during the validation day in terms of NASA-TLX OW can be seen. Those five ATCos who started with a baseline, rated the baseline (their first run) with an OW of 41.9; those five ATCos who started with a solution, rated the baseline (their second run) with an OW of 32. Those five ATCos who started with the solution, rated the solution (their first run) with an OW of 48.9; those five ATCos who started with baseline, rated the solution (their second run) with an OW of 37.2.

The average and peak Bedford workload were 0.9 and 0.7 points higher, respectively, in the solution condition with ABSR support compared to the baseline condition ($\alpha = 0.001$). The peak workload was roughly 1.5 points higher than the average workload. The workload level, in general, was roughly two points lower for ACG than for ON ATCos.

3.7. Perceived Workload through Automation Impact (AIM-s) (Post-Run)

The post-run questionnaires contained the sixteen statements of AIM-s as listed in Appendix C.5. The seven-item answer scale ranged from "None, Very Little, Little, Some, Much, Very Much, Extreme." To present the results in a bar diagram, "None" is translated to 0%, "Very Little" to 1/6 %"..."Very Much" to "5/6 %" until "Extreme" to 100%. The statements SHARE and TMN are not analyzed further as there were no team members during the simulation runs (fourteen statements remain). Figure 19 shows the results.

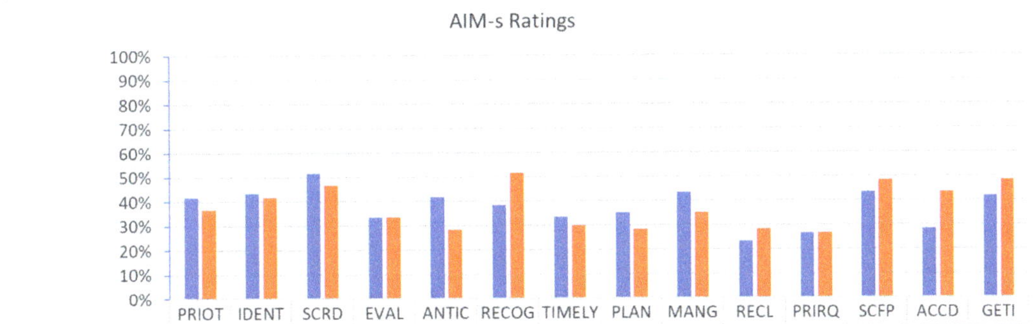

	PRIOT	IDENT	SCRD	EVAL	ANTIC	RECOG	TIMELY	PLAN	MANG	RECL	PRIRQ	SCFP	ACCD	GETI
BAS (without ASR)	41.7%	43.3%	51.7%	33.3%	41.7%	38.3%	33.3%	35.0%	43.3%	23.3%	26.7%	43.3%	28.3%	41.7%
SOL (with ASR)	36.7%	41.7%	46.7%	33.3%	28.3%	51.7%	30.0%	28.3%	35.0%	28.3%	26.7%	48.3%	43.3%	48.3%

Figure 19. Subjective ATCo ratings on AIM-s questionnaire.

After compensating sequence effects, the overall perceived workload due to AIM-s is not statistically better with or without ABSR support. We measured an α of 0.49, which is not better than throwing a coin. However, the anticipation of the future air traffic situation was much better for SOL than for BAS ($\alpha = 0.02$). Nine of the fourteen statements have been rated better on average (less) for the SOL condition than for the BAS condition. Only the five statements related to information *RECOG*, *RECL*, *SCFP*, *ACCD*, and *GETI* have been rated worse for SOL condition compared to BAS condition.

3.8. Perceived Workload (Instantaneous Self-Assessment of Workload (ISA)) (Within-Run)

During each simulation run, ATCos needed to rate their workload of the recent five minutes on a scale from 1 (bored) to 5 (almost overloaded). The results are shown in Figure 20. The average ISA workload was 0.1 points less, i.e., better, in solution condition with ASR support compared to baseline condition with $\alpha = 0.15$ (2.1 and 2.0 points, respectively).

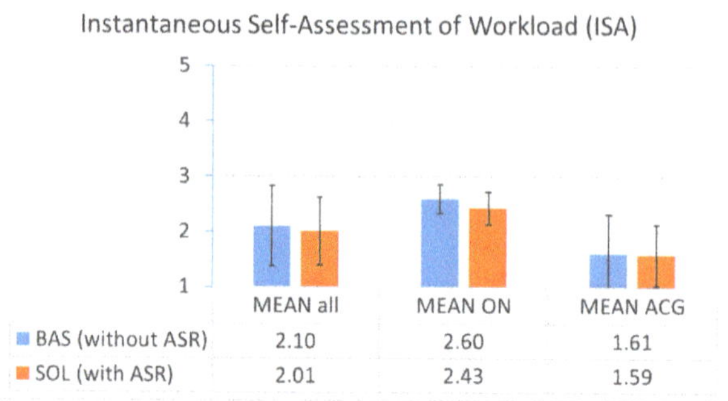

Figure 20. Subjective ATCo workload self-assessment (ISA).

The ISA of ON ATCos was on a higher level with 2.6 and 2.4, respectively, and had a much lower standard deviation of below 0.3. The ISA score of ACG ATCos was around 1.6, with a standard deviation more than twice as much as of ON ATCos.

3.9. Objectively Measured Workload with Secondary Task (Card Sorting) (Within-Run)

The ATCos always needed to make sure that their primary task of doing ATC remains safe and efficient. However, if they had time for a secondary task, i.e., free mental capacity, they should sort cards. This method has already been used in earlier ASR projects to generate a more objective measure of mental workload than just via self-ratings.

ATCos needed to sort 48 cards of a German Doppelkopf deck into six decks (Aces, Kings, Queens, Jacks, Tens, and Nines). In the beginning, all 48 cards are on one stack, with the picture side of the cards looking downwards. Each card needed to be turned around in a single move with just one hand to put it onto the correct of the six decks. After sorting, ATCos should name one to four randomly missing cards that the supervisor took out of the 48 cards deck prior to starting sorting. If there was an error in naming the missing cards, e.g., not all missing cards are named, ATCos must try again until all missing cards are named correctly. The time measurement in seconds started when the deck of 48 cards was put next to the electronic flight strip display. The time measurement ended when all missing cards were named correctly. Sorting cards were trained once in each of the thirty minutes training runs. Card sorting in the baseline and solution runs started after 10 min (for at least 15 min or at least three rounds) and again after 40 min (for at least 13 min or at least three rounds). Those time frames comprised higher traffic density to measure any difference in workload through ASR support.

The results are shown in Figure 21. ATCos finished their secondary task 8% slower in baseline runs when not being supported by ASR (395 s vs. 364 s with a standard deviation of 305 s and 262 s). This difference was 9% for ON and 7% for ACG ATCos. When compensating sequence effects with the SECT technique, ATCos were even 9% slower in baseline runs compared to solution runs. After compensating sequence effects, the α was 0.24 to reject the hypothesis that ABSR support does not reduce the workload of ATCos.

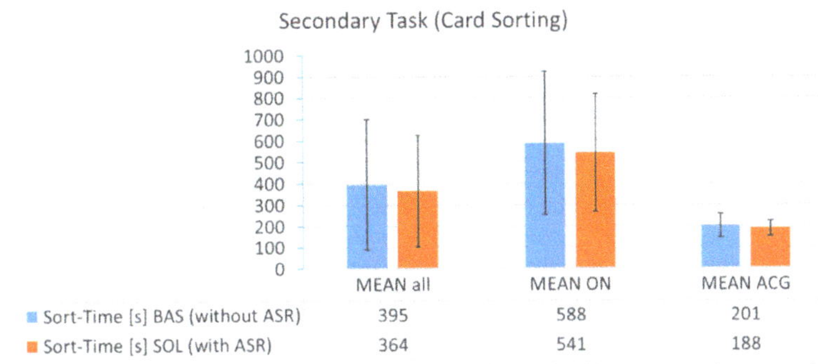

Figure 21. ATCo performance in the secondary task (card sorting).

When translating the timing results into workload, again, ON ATCos experienced a higher workload level (around 9 min sorting average) than ACG ATCos (around 3 min sorting average with more task repetitions than ON ATCos), but workload in solution condition seems to be lower than in baseline regarding the secondary task of card sorting. Additionally, the secondary task showed a great learning curve, i.e., ATCos were almost 19% slower in sorting the cards in their first simulation run compared to their second simulation run (baseline and solution alternated).

3.10. System Usability (Post-Run)

The post-run questionnaire contained the ten statements of the System Usability Scale (SUS), as listed in Appendix C.6. The results are shown in Figure 22 (one ATCo did not answer one of his ten statements both in baseline (without ASR) and solution (with ASR) condition. Therefore, the scale mean "3" ((5-1)/2) was chosen as a replacement to not heavily influence the overall result). ABSR support increases the system usability due to SUS ratings ($\alpha = 0.16$). There were three statements rated in the expected direction with an $\alpha < 0.075$, i.e., ATCos like to use the system, they do not deem it complex, and they hardly need support to use it.

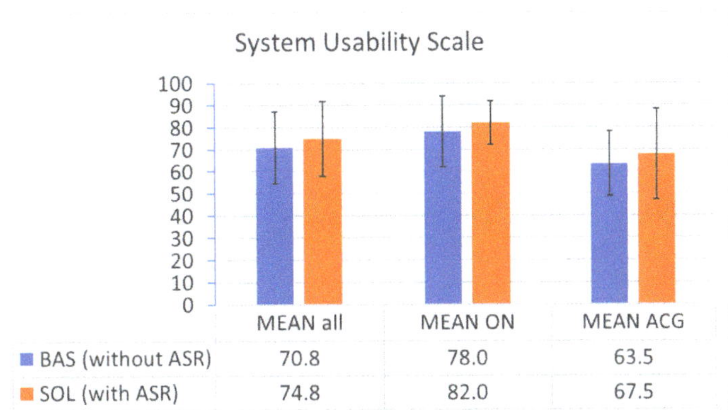

Figure 22. Subjective ATCo ratings on system usability.

Considering all ATCos, the SUS score was 4 percent absolute (5.7% relative) higher in the solution condition (SOL) with ABSR support compared to the baseline condition (BAS) without ABSR support. The difference of 4 percent remains when just analyzing the ON score or ACG score independently. However, the score itself is 14.5%, absolutely higher for ON than for ACG. This is probably due to the fact that ON really liked the electronic flight strip display (also in the baseline version), whereas ACG ATCos needed to adapt themselves more to the strip system due to the difference in their daily-life system.

3.11. Debriefing Feedback (Post-Validation)

The debriefing was conducted as a semi-structured interview with some pre-defined questions and some options for further thoughts and inputs. The feedback of ATCos is semantically reported per category in the following subsections—the most important feedback relevant for future usage of ABSR is listed after arrow symbol bullets. However, also the remaining feedback helps to improve future simulation planning, i.e., to know which aspects that are not the core part of the study do influence the subject's experience and study results. For example, the prototypic flight strip system induced a row of effects on how the ABSR output is perceived. The last question outlines further research or usage of ABSR systems.

3.11.1. Study Preparation and Conduction

- Briefing slides via e-mail two weeks before the trials and briefing at DLR was very good;
- All ATCos felt well-trained for the purpose of the validation after one hour of training;
- Simulation pilots performed well;
- Air traffic scenarios were rated to be fine for the study purpose;
- On the one hand, the baseline condition (manual work) was similar to everyday work, so performance might be better, therefore (2 ATCos);
- ➢ On the other hand, ASR in solution condition was good because it supported using a flight strip system that ATCos were not used to.

3.11.2. ABSR Functionality (also Related to Electronic Flight Strip Display)

- ➢ ABSR concept and implementation were found to be good by many ATCos;
- ➢ Checking ABSR output in the flight strip display slows some ATCos because, in the baseline mode, ATCos tick while speaking;
- ➢ Some ATCos judged the speed of ABSR output while speaking as sufficient; two ATCos wanted to have faster output;
- ➢ Non-standard situations should be covered well, i.e., better, by ASR;
- ➢ Speech understanding (annotation process) was good for covering errors in speech recognition (transcription process);
- ➢ Highlighting of callsigns and status icons (in green) and the 10s-highlighting mechanism in electronic flight strips were fine for all ATCos;
- ➢ When ASR worked fine, a tendency to over-rely on automatism existed;
- ➢ In case of non-recognition, a double effort to manually recognize the error and correct it compared to pen input (2 ATCos);
- ABSR output in outside view (complete transcription and annotation in solution condition) was just checked for curiosity by all ATCos.

3.11.3. Feedback to Colleagues Not having participated

When I am home in Lithuania/Austria, I tell my colleagues that working with DLR's speech recognition was:

- ➢ Interesting (said by all ON ATCos);
- ➢ Worked pretty well (2 ATCos);
- ➢ Positively surprising (even when speaking fast);
- ➢ Very good even if not being an early adaptor of new technologies and being very safety critical.

3.11.4. Usefulness of ASR

If you would use it tomorrow in your tower controller working position (not multiple remote towers), would ASR help?

- ➢ Yes, that would be great (3);
- ➢ Nothing to be changed to be used tomorrow (1);
- ➢ Great support is possible if some/many aspects are improved (4).

3.11.5. Used Phraseology in Baseline and Solution Runs

Did you think you have spoken differently in baseline and solution conditions?

- ➢ In baseline less carefully spoken because only simulation pilots needed to understand (3 ATCos);
- ➢ Spoken closer to phraseology in solution as being better supported (2 ATCos);
- ➢ Some stated that there was no difference in speaking;
- ➢ "ATCos automatically become more phraseology conform: That is one of the greatest advantages of such a technology."

3.11.6. Flight Strip System (More Related to 'Multiple Remote Tower" than the Core Study Purpose 'ABSR Support')

- Runway bay handling needs to be improved (sorting, highlighting, timing, etc.);
- Drag-and-drop functionality over the borders of flight strip bays for individual planning purposes was needed;
- Handling training flights (touch-and-go/low approach) that do not switch from an arrival flight strip to a departure flight strip were slightly difficult;
- Strip handling for aircraft crossing the control zone is difficult with status options;
- Visual flagging of strips (left/right) would be beneficial;
- Hide some non-frequent status icons;
- "Takeoff" status should include "lineup"-status (if not given explicitly);
- A combination of the selection of taxi status and taxiway would be easier;
- Suggestions for colors, e.g., ground vehicles, consistency with other systems;
- One ATCo loved the flight strip system; the majority of ATCos were ok with it;
- Many ATCos liked the fade-away functionality of flight strips;
- The portion of gazes at the three areas 'flight strip display,' 'outside view,' and 'radar view': too much on flight strips and too few on outside view where one can hardly identify small objects.

3.11.7. Further Applications/Ideas/Things to Be Changed?

- ➢ Callsign highlighting in flight strip display from pilot utterance would help to identify the communication partner;
- ➢ Speech log for pilot utterances (especially in emergency situations) anywhere on the controller screen;
- Connect ABSR output with:
    a. Radar information for automatic setting of landed/departed status;
    b. Lighting system to turn off stop bar lights in case of lineup clearance;
    c. Follow the greens for correct lighting;
    d. Airport phone conversation to automatically extract and include stand numbers given by the airport;
    e. Safety net functionality for dedicated aspects in case of good error rates, e.g., readback error detection;
    f. Transcription for incident analysis and searching for callsigns; other analysis on transcribed data;
    g. Great technology for on-the-job training.

4. Discussion on Major Study Results

The results on mental workload, situation awareness, satisfaction, acceptance, trust, and usability are ambivalent. The subjective post-run ratings on NASA-TLX, Bedford workload scale, and AIM-s, when interpreted as a whole, indicate a worse performance in solution runs with ABSR support compared to baseline runs without ABSR support.

However, the subjective post-validation rating on ABSR support for workload, the self-assessed workload ratings during the simulation runs by ISA, and the performance measurement of the objective secondary task indicate that ABSR support positively influences ATCo workload.

There might also be an influence through the usage of standardized and tailor-made questionnaires. The general low to medium workload level, as rated with roughly two on average on the five-point instantaneous self-assessment of workload scale, causes that it is hard to unambiguously measure a workload effect. Hence, the necessity for controller support functionalities might also be low in such a multiple remote tower environment.

The complexity of the task came with supervising three airports remotely at the same time with a working position the ATCos had not seen before. This could be the reason why especially the callsign highlighting was well-acknowledged by ATCos in order to reduce search times at the different displays. A workload reduction, especially in low workload conditions, is not always beneficial. Hence, it is also a success if the mental workload of ATCos is balanced at a medium level without peaks and boredom.

Similarly, the post-run rating on situation awareness (SASHA) indicates a negative influence, whereas the two rated post-validation statements on situation awareness at an acceptable level with ABSR support have answer values in the most positive scale third. Very similar effects were also seen for satisfaction, acceptance, and trust when comparing post-run ratings with overall post-validation answers.

The usability ratings (post-run and post-validation) seem to all indicate favor for ABSR support. The score of the system usability scale was four points better for the solution (with ABSR support) compared to the baseline (without ABSR support). A total of 80% of ATCos (with 8/10 or more points on the questionnaire scale) stated that they would accept such an ABSR system in their usual working position and that they could apply operating methods in a timely manner. Though, a row of adjustments were encouraged by ATCos, i.e., to make ABSR also reliable under non-nominal conditions where the pressure on ATCos is already high. The need for changes was rated very inhomogeneous by the different ATCos, i.e., some had already seen good support with the prototype's current technology readiness level, and others wanted to increase the number of covered situations and examples.

However, the comparison of a further objective measure with a subjective measurement again shows the ambivalence of some ATCo ratings: While ACG ATCos rated the perceived callsign recognition quality with 1.8 points higher than ON ATCos on a 10-point scale and the perceived command recognition quality with 1.6 points lower than ON ATCos such an effect cannot be seen in the online recognition rates where the callsign recognition rate and the command recognition rate in solution runs of ON ATCos was 2% and 10% (consistently both) better than of ACG ATCos, respectively.

Our study results based on text-to-concept analysis also revealed a potential safety issue for multiple remote towers: Even if ATCos were asked to utter the name of their current transmission station in each radio transmission, the station name, e.g., *vilnius tower*, was missing in every fifth utterance. This might confuse listening to cockpit crews being on or flying to one of the other two airports.

The subjective feedback through questionnaires etc., and the results from objective measurements at least are not consistent or even contradictory. This is a hint that ABSR's performance does not match with ATCos expectations. Objectively a word error rate of 10% with a command recognition rate of 80% might be sufficient to already have positive effects on workload. The ATCos are then, however, not trusting the system, which will be a showstopper. Objective improvements are not enough. ATCos also need to be convinced by their subjective feelings. Previous validation trials for Frankfurt airport to support apron controllers by ABSR to reduce workload for pre-filling electronic flight strips [12] and for Vienna approach controllers [41] indicate that a command recognition rate greater than 90% is needed.

5. Conclusions and Outlook

5.1. Conclusions on ABSR Study in Multiple Remote Tower Environments

Human-in-the-loop trials were conducted with five Austrian and five Lithuanian air traffic controllers (ATCos) to validate whether an assistant-based speech recognition (ABSR) system can support air traffic controllers in a multiple remote tower environment. In baseline runs, controllers needed to manually maintain electronic flight strips without ABSR support, whereas in solution runs, they were supported by ABSR through callsign highlighting and automatically inputting recognized commands from ATCo utterances into electronic flight strips.

This study recorded a huge amount of data with results analyses that are shared with other researchers by this article. The chosen "within-subject design" [46] assessed the dependent variables mental workload, situation awareness, satisfaction, acceptance, trust, and usability with the independent variable "availability of ABSR support". Further qualitative feedback was gathered on ABSR accuracy, technical functionality, and operating methods. Although a very small number of training data of 3.6 and 0.9 h, respectively, was available for the adaption of the ABSR models to Lithuanian and Austrian tower phraseology, some results show statistical significance and are in line with findings of earlier ABSR projects from an approach environment [8]. The text-to-concept accuracy of the speech understanding module performed well, i.e., correcting wrong word recognition by context information. A callsign recognition rate of 94.2% and a command recognition rate of 82.9% were achieved, although each 10th word was wrongly recognized due to the observed word error rate of 9.8%. Given an independent distribution of word errors and an average callsign length of five words, a word error rate of 10% would result in a callsign recognition rate of below 60%, i.e., $(1–0.1)^5$. For an average command length of six words, including values, qualifiers, and conditions plus the five words for the callsign, the expected command recognition rate would be below 35%, i.e., $(1–0.1)^{11}$. These theoretical values were outperformed by our speech understanding module (command recognition) by using context information.

The study results on human factors comprised subjective ratings on mental workload, situation awareness, satisfaction, acceptance, trust, and usability via standardized and tailor-made questionnaires, the self-assessed workload during simulation runs, and an objective method to assess workload based on a secondary task.

The analysis results on the dependent variables were ambivalent. The reasons are the small number of study subjects, the prototype of a non-operational user interface, and the low workload resulting from low to medium traffic in the multiple remote tower environment of the chosen airports. A positive influence on workload was found with the self-assessed workload ratings during the simulation runs and the performance in the secondary task as a more objective measurement during simulation runs. Future validation trials involving ATCos should focus more on objective or live measurements than on retrospective ratings.

Our study results with ATCos reporting on benefits and drawbacks raise detailed awareness and give recommendations on which aspects of automatic speech recognition and understanding for a multiple remote tower environment are already solved and which aspects require deeper research to go beyond the now achieved technology readiness level four.

The speech-to-text performance is a prerequisite to enable good text-to-concept performance. An error analysis after the validation trials revealed processor overload as a factor in decreasing our speech-to-text performance. When applying our command extraction on offline speech-to-text analysis results having a word error rate of 4.4%, we achieve a command recognition rate of 91.8% and a callsign recognition rate of 98.2%. The data analysis showed that ABSR support has a statistically significant positive effect on the usage of ICAO phraseology: The above-reported solution runs have higher command recognition rates than baseline runs because ATCos obtain better support if recognition rates are higher. If ATCos are sticking closer to ICAO phraseology just by the pure presence of an ABSR system, that will already be a safety feature. Some ATCos, i.e., the human operators that would use the operating system later on, highlighted that such an ABSR system would be a great support in their working position.

5.2. Outlook on Future Work

The amount of training data must be further increased, given representative samples. Furthermore, a large amount of data must be recorded from operations rooms (not from labs) because this is the operational environment. The European-wide agreed ontology for the annotation of ATC utterances was successfully used and enhanced in this study and should be further exploited or standardized. The continuous mutual enhancements of the ontology for en-route/oceanic, approach, tower, and apron traffic within the ASR projects HAAWAII (Highly Automated Air Traffic Controller Workstations with Artificial Intelligence Integration (HAAWAII), Homepage: https://www.haawaii.de (accessed on 4 April 2023)) (as the successor of MALORCA (Machine Learning of Speech Recognition Models for Controller Assistance (MALORCA), Homepage: https://www.malorca-project.de (accessed on 4 April 2023)), and STARFiSH (Safety and Artificial Intelligence Speech Recognition (STARFiSH), Homepage: https://www.dlr.de/fl/desktopdefault.aspx/tabid-1149/1737_read-74905/ (accessed on 4 April 2023)) tremendously build a base for interoperability of systems. Hence, following ASR activities can build on strong shoulders and reuse the achieved good results and methods of such ABSR projects.

For the specific case of electronic flight strips, eye tracking technology could be of further help to make sure that ATCos checked the ABSR output [47]. This technology could also be used to assess the time to recognize and correct an ABSR error (Times to correct ABSR errors in an ATM environment have been investigated in "Automatic Speech Recognition and Understanding for Radar Label Maintenance Support Increases Safety and Reduces Air Traffic Controllers' Workload" of Helmke et al. presented at the 15th USA/Europe Air Traffic Management Research and Development Seminar (ATM2023), Savannah, GA, USA, 5–9 June 2023). Furthermore, the support through callsign highlighting when recognized from pilot utterances should be investigated and potentially feed attention guidance systems at the controller working position. To summarize, the validation trials have shown the potential of using the output of an ABSR system in the multiple remote tower environment and revealed aspects to be considered when moving forward to higher technology readiness levels.

Author Contributions: Conceptualization, O.O.; Methodology, H.H., S.S. (Shruthi Shetty), M.K. and H.E.; Software, O.O., S.S. (Shruthi Shetty), M.K., H.E., S.S. (Saeed Sarfjoo) and P.M.; Validation, O.O., H.H. and S.S.-M.; Formal analysis, O.O. and M.M.; Investigation, O.O.; Resources, Š.M., T.P., H.U. and A.Č.; Data curation, O.O.; Writing—original draft, O.O.; Writing—review and editing, H.H., S.S. (Shruthi Shetty), M.K., H.E., S.S.-M., S.S. (Saeed Sarfjoo), P.M., Š.M., T.P., H.U., M.M. and A.Č.; Supervision, O.O. All authors have read and agreed to the published version of the manuscript.

Funding: This project has received funding from the SESAR Joint Undertaking under the European Union's Horizon 2020 research and innovation program under grant agreement No 874470.

Institutional Review Board Statement: Not applicable.

Informed Consent Statement: Informed consent was obtained from all subjects involved in the study.

Data Availability Statement: Not applicable.

Acknowledgments: We like to thank the ten air traffic controllers of ACG and ON who participated in the validation study in spring 2022 under a strict hygienic protocol due to COVID-19 as well as roughly one dozen of air traffic controllers from ON, ACG, and the Polish Air Navigation Services Agency (PANSA) who contributed to the recording of training data (surveillance data and speech utterances) in earlier multiple remote tower trials at DLR Braunschweig. Further, we thank the air traffic management simulation department and the simulation pilots at DLR's Institute of Flight Guidance as well as our colleague Lennard Nöhren for supporting software preparation of simulation environment and conduction of a row of multiple remote tower human-in-the-loop validation studies in the course of our automatic speech recognition activities.

Conflicts of Interest: The authors declare no conflict of interest. The founding sponsors had no role in the design of this study; in the collection, analysis, or interpretation of data; in the writing of the manuscript, and in the decision to publish the results.

Abbreviations

ABSR	Assistant Based Speech Recognition
ACG	Austro Control
AIM-s	Assessing the Impact on Mental Workload
ASR	Automatic Speech Recognition
ATC	Air Traffic Control
ATCo	Air Traffic Controller
ATIS	Automatic Terminal Information Service
ATM	Air Traffic Management
BAS	Baseline Runs
CARS	Controller Acceptance Rating Scale
CoCoLoToCoCo	Controller Command Logging Tool for Context Comparison
CPU	Central Processing Unit
CWP	Controller Working Position
Del	Deletions
DLR	German Aerospace Center
DTT	Digital Tower Technologies
EASA	European Union Aviation Safety Agency
EFS	Electronic Flight Strip System
EUROCAE	European Organization for Civil Aviation Equipment
HMI	Human Machine Interface
ICAO	International Civil Aviation Organization
Ins	Insertions
ISA	Instantaneous Self-Assessment
LevenDist	Levenshtein Distance
NASA-TLX	National Aeronautics and Space Administration Task Load Index
Off	Offline (analysis of audio files after the simulation runs)
ON	Oro Navigacija
Onl	Online (analysis as experienced by ATCos during simulation runs)
OW	Overall Weighted Workload
SASHA	Situation Awareness for SHAPE
SATI	SHAPE Automation Trust Index
SD	Standard Deviation
SECT	Sequence Effect Compensation Technique
SHAPE	Solutions for Human Automation Partnerships in European ATM
SOL	Solution Runs
Subs	Substitutions
SUS	System Usability Scale
TWR	Tower
WER	Word Error Rate

Appendix A. Speech-To-Text Accuracy

The following tables in this Appendix A show the speech recognition performance on the word level, i.e., the word error rates (WER). The first row must be read like this; 1,944 words were spoken. Ninety-seven errors occurred, i.e., 43 words were substituted by another word, 38 words were not recognized at all (deleted), and 16 words were inserted, i.e., not said, but a word was recognized. This results in a word error rate of 5.1% (97/1944).

Table A1. Speech-To-Text performance for offline recognition on audio files (Off).

Sample	# Words	LevenDist	# Subs	# Del	# Ins	% WER
MEAN all	1944	97	43	38	16	5.1
MEAN ON	1966	94	38	36	20	5.0
MEAN ACG	1921	99	48	39	13	5.1
MEAN w/o outlier run	1971	90	40	34	16	4.5
MEAN BAS all	1902	104	46	43	15	5.7
MEAN BAS ON	1891	100	41	43	16	5.7
MEAN BAS ACG	1913	109	51	44	14	5.7
MEAN BAS w/o outlier run	1961	98	44	39	15	5.0
MEAN SOL all	1985	89	40	32	17	4.4
MEAN SOL ON	2041	88	36	30	23	4.3
MEAN SOL ACG	1929	90	44	34	11	4.6
MEAN SOL w/o outlier run	1980	81	36	28	17	4.1

Rows are shaded, when containing all ATCos, i.e., both from ACG and ON.

Table A2. Speech-To-Text accuracy for real-time online recognition from voice stream (Onl).

Sample	# Words	LevenDist	# Subs	# Del	# Ins	% WER
MEAN all	1936	245	46	175	24	13.6
MEAN ON	1954	199	38	140	21	11.9
MEAN ACG	1918	290	54	209	27	15.3
MEAN w/o outlier run	1967	212	41	152	19	11.0
MEAN BAS all	1891	300	54	219	27	17.4
MEAN BAS ON	1871	261	42	196	23	17.1
MEAN BAS ACG	1911	339	66	241	32	17.8
MEAN BAS w/o outlier run	1959	254	50	181	23	13.2
MEAN SOL all	1980	189	38	131	21	9.8
MEAN SOL ON	2037	136	34	83	19	6.8
MEAN SOL ACG	1924	242	42	178	22	12.8
MEAN SOL w/o outlier run	1976	171	32	123	15	8.9

Rows are shaded, when containing all ATCos, i.e., both from ACG and ON.

The following two tables show the frequency of certain words appearing in the gold transcriptions and the number of unique words needed to reach a certain portion of all words in the gold transcriptions, respectively.

Table A3. 1-grams of gold transcriptions.

Rank	Word	Count	Portion
1	one	2393	6.43%
2	zero	1479	3.97%
3	tower	1473	3.96%
4	three	1356	3.64%
5	runway	1154	3.10%
6	five	1085	2.91%
7	seven	925	2.48%
8	two	923	2.48%
9	four	898	2.41%
10	to	888	2.38%
11	cleared	808	2.17%
12	right	795	2.13%
13	vilnius	747	2.01%
14	eight	721	1.94%
15	nine	720	1.93%

Table A3. *Cont.*

Rank	Word	Count	Portion
16	via	601	1.61%
17	air	571	1.53%
18	degrees	556	1.49%
19	and	539	1.45%
20	knots	531	1.43%
21	bravo	465	1.25%
22	wind	456	1.22%
23	alfa	409	1.10%
24	taxi	408	1.10%
25	kaunas	390	1.05%
	others	15,947	42.8%
1-505	SUM	37,238	100%

Table A4. The number of different words needed to reach a certain portion of all uttered words.

Count	Portion
61	80%
101	90%
145	95%
283	99%
505	100%

Appendix B. Text-To-Concept Accuracy

The following tables lists the relative frequency of supported air traffic control command types from the gold annotations.

Table A5. Percentage of used command types in gold annotations occurring more often than 1% (7560 commands in total).

Command Type	Portion of All Commands
STATION	20.2%
INFORMATION WINDSPEED	7.5%
INFORMATION WINDDIRECTION	7.5%
TAXI TO	6.4%
GREETING	5.6%
TAXI VIA	4.8%
AFFIRM	4.0%
INFORMATION QNH	3.3%
CLEARED VIA	2.9%
STARTUP	2.9%
CLEARED TO	2.9%
CLEARED TAKEOFF	2.8%
FAREWELL	2.8%
CLEARED LANDING	2.8%
SQUAWK	2.8%
LINEUP	2.4%
REPORT	1.5%
PUSHBACK	1.4%
INFORMATION ACTIVE_RWY	1.4%
NO_CONCEPT	1.4%
REPORT_MISCELLANEOUS	1.4%
VACATE VIA	1.2%
CLEARED TOUCH_GO	1.1%
others	8.9%

The following six tables present the speech understanding performance per study subject group and per command type group for gold, offline, and online transcriptions, respectively.

Table A6. Text-to-concept quality for gold transcriptions (assumed to be 100% correct).

Gold Transcription	Command Recognition Rate	Command Error Rate	Callsign Recognition Rate	Callsign Error Rate	Unknown Classified Rate	Amount of Data
all ATCos ALL	95.9%	2.4%	99.8%	0.2%	13.3%	100.0%
ON ATCos ALL	97.1%	1.5%	99.7%	0.2%	12.5%	49.9%
ACG ATCos ALL	94.8%	3.2%	99.9%	0.1%	14.2%	50.1%
ATCos ALL w/o outlier run	95.8%	2.5%	99.8%	0.2%	13.2%	91.8%
all ATCos BAS	95.9%	2.4%	99.7%	0.3%	13.8%	49.0%
ON ATCos BAS	97.6%	1.3%	99.7%	0.3%	13.0%	24.1%
ACG ATCos BAS	94.1%	3.5%	99.8%	0.2%	14.7%	24.8%
all ATCos SOL	96.0%	2.3%	99.8%	0.1%	12.8%	51.0%
ON ATCos SOL	96.6%	1.8%	99.7%	0.2%	12.0%	25.8%
ACG ATCos SOL	95.4%	2.9%	100.0%	0.0%	13.7%	25.3%

Table A7. Text-to-concept quality for gold transcriptions (assumed to be 100% correct) per command type groups.

Command Type Group	# Command Types	Command Recognition Rate
Relevant	34	97.3%
EFS	21	97.4%
Status	18	96.7%
Outside	3	96.0%
Hypo-EFS	4	99.2%

Table A8. Text-to-concept quality for Off transcriptions (current best word error rates of automatic speech-to-text with callsign boosting on audio files).

Offline	Command Recognition Rate	Command Error Rate	Callsign Recognition Rate	Callsign Error Rate	Unknown Classified Rate	Amount of Data
all ATCos ALL	91.4%	4.5%	98.4%	0.9%	14.0%	100.0%
ON ATCos ALL	92.7%	3.9%	98.6%	0.6%	12.8%	49.9%
ACG ATCos ALL	90.1%	5.1%	98.2%	1.2%	15.2%	50.1%
ATCos ALL w/o outlier run	91.7%	4.4%	98.7%	0.9%	13.9%	91.8%
all ATCos BAS	91.0%	4.6%	98.6%	0.8%	14.5%	49.0%
ON ATCos BAS	92.8%	3.6%	99.0%	0.3%	13.2%	24.1%
ACG ATCos BAS	89.3%	5.5%	98.1%	1.2%	15.8%	24.8%
all ATCos SOL	91.8%	4.5%	98.2%	1.1%	13.6%	51.0%
ON ATCos SOL	92.7%	4.1%	98.1%	0.9%	12.6%	25.8%
ACG ATCos SOL	90.9%	4.8%	98.3%	1.2%	14.6%	25.3%

Table A9. Text-to-concept quality for Off transcriptions (current best word error rates of automatic speech-to-text with callsign boosting on audio files) per command type groups.

Command Type Group	# Command Types	Command Recognition Rate
Relevant	31	92.4%
EFS	21	93.4%
Status	18	92.7%
Outside	3	90.5%
Hypo-EFS	4	96.3%

Table A10. Text-to-concept quality for Onl transcriptions (automatic speech-to-text with callsign boosting from continuous stream).

Online	Command Recognition Rate	Command Error Rate	Callsign Recognition Rate	Callsign Error Rate	Unknown Classified Rate	Amount of Data
all ATCos ALL	79.4%	7.0%	91.7%	3.1%	15.4%	100.0%
ON ATCos ALL	84.2%	5.5%	92.1%	2.4%	13.8%	49.9%
ACG ATCos ALL	74.6%	8.6%	91.3%	3.9%	17.0%	50.1%
ATCos ALL w/o outlier run	81.2%	6.6%	94.0%	2.5%	14.9%	91.8%
all ATCos BAS	75.7%	7.5%	89.1%	3.8%	16.2%	49.0%
ON ATCos BAS	80.1%	5.6%	88.9%	2.8%	14.6%	24.1%
ACG ATCos BAS	71.4%	9.3%	89.3%	4.8%	17.9%	24.8%
all ATCos SOL	82.9%	6.6%	94.2%	2.4%	14.5%	51.0%
ON ATCos SOL	88.0%	5.4%	95.2%	2.0%	13.2%	25.8%
ACG ATCos SOL	77.7%	7.9%	93.2%	2.9%	16.1%	25.3%

Table A11. Text-to-concept quality for Onl transcriptions (automatic speech-to-text with callsign boosting from continuous stream) per command type groups.

Command Type Group	# Command Types	Command Recognition Rate
Relevant	31	80.7%
EFS	21	79.2%
Status	18	80.0%
Outside	3	81.0%
Hypo-EFS	4	87.2%

Appendix C. Questions and Statements of Questionnaires

The following full-text questions and statements were contained within the listed post-run questionnaires:

Appendix C.1. Statement and Answer Scale from CARS

The color coding shows worse answers in red and good answers in green.

"Please read the descriptors and score your overall level of user acceptance experienced during the run. Please check the appropriate number."

▪ Improvement mandatory. Safe operation could not be maintained.
▪ Major Deficiencies. Safety not compromised, but system is barely controllable and only with extreme controller compensation.
▪ Major Deficiencies. Safety not compromised but system is marginally controllable. Considerable compensation is needed by the controller.
▪ Major Deficiencies. System is controllable. Some compensation is needed to maintain safe operations.
▪ Very Objectionable Deficiencies. Maintaining adequate performance requires extensive controller compensation.
▪ Moderately Objectionable Deficiencies. Considerable controller compensation to achieve adequate performance.
▪ Minor but Annoying Deficiencies. Desired performance requires moderate controller compensation.
▪ Mildly unpleasant Deficiencies. System is acceptable and minimal compensation is needed to meet desired performance.
▪ Negligible Deficiencies. System is acceptable and compensation is not a factor to achieve desired performance.
▪ Deficiencies are rare. System is acceptable and controller does not have to compensate to achieve desired performance.

Appendix C.2. Statements from SATI Questionnaire

1. In the previous working period, I felt that the system was useful. [USEFUL]
2. In the previous working period, I felt that the system was reliable. [RELIABLE]
3. In the previous working period, I felt that the system worked accurately. [ACCURACY]
4. In the previous working period, I felt that the system was understandable. [UNDERSTAND]
5. In the previous working period, I felt that the system worked robustly (in difficult situations, with invalid inputs, etc.). [ROBUST]
6. In the previous working period, I felt that I was confident when working with the system. [CONFIDENT]

Appendix C.3. Statements from SASHA Questionnaire

1. In the previous run, I was ahead of the traffic. [AHEAD]
2. In the previous run, I started to focus on a single problem or a specific aircraft. [FOCUS]
3. In the previous run, there was a risk of forgetting something important (such as inputting the spoken command values into the labels). [FORGET]
4. In the previous run I was able to plan and organize my work as wanted. [PLAN]
5. In the previous run I was surprised by an event I did not expect (such as an aircraft call). [SURPRISE]
6. In the previous run I had to search for an item of information. [SEARCH]

Appendix C.4. Questions from NASA-TLX Questionnaire

1. How mentally demanding was the task? [Mental Demand, MD]
2. How physically demanding was the task? [Physical Demand, PD]
3. How hurried or rushed was the pace of the task? [Temporal Demand, TD]
4. How successful were you in accomplishing what you were asked to do? [Operational Performance, OP]
5. How hard did you have to work to accomplish your level of performance? [Effort, EF]
6. How insecure, discouraged, irritated, stressed, and annoyed were you? [Frustration, FR]

Furthermore, the 15 pairwise comparisons of workload contributing factors have been analyzed. When looking at the subscores for all six NASA-TLX dimensions, half of them (three) were rated equal or better in SOL compared to BAS (PD, EF, FR), and the other half was rated vice versa (MD, TD, OP). In general, physical demand (PD, 3.3%) was rated as being a less important contributor to workload, and mental demand (MD, 23.3%) was the most important contributor to workload. The other four dimensions were rather equally important contributors to the overall workload (TD 22%, OP 18%, EF 16.7%, FR 16.7%). The horizontal axis in Figure A1 shows the weight; the area shows the contribution of this very dimension to the OW of BAS and SOL conditions, respectively.

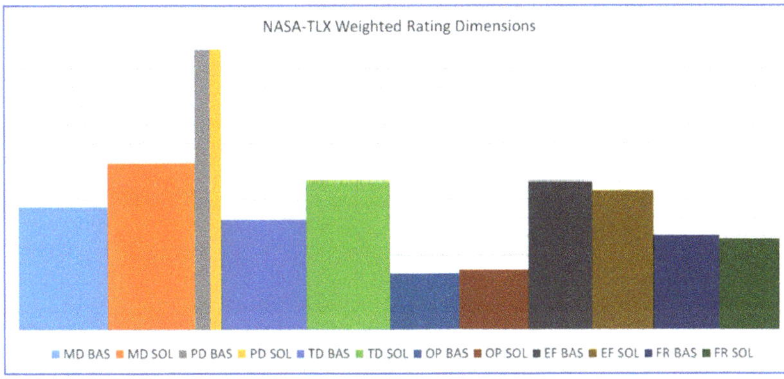

Figure A1. ATCo ratings on NASA-TLX (Weighted Workload Factors).

Appendix C.5. Questions from AIM-s Questionnaire

1. In the previous run, how much effort did it take to prioritize tasks? [PRIOT]
2. In the previous run, how much effort did it take to identify potential conflicts? [IDENT]
3. In the previous run, how much effort did it take to scan radar or any display? [SCRD]
4. In the previous run, how much effort did it take to evaluate conflict resolution options against the traffic situation and conditions? [EVAL]
5. In the previous run, how much effort did it take to anticipate the future traffic situation? [ANTIC]
6. In the previous run, how much effort did it take to recognize a mismatch of available data with the traffic picture? [RECOG]
7. In the previous run, how much effort did it take to issue timely commands? [TIMELY]
8. In the previous run, how much effort did it take to evaluate the consequences of a plan? [PLAN]
9. In the previous run, how much effort did it take to manage flight data information? [MANG]
10. In the previous run, how much effort did it take to share information with team members? [SHARE]
11. In the previous run, how much effort did it take to recall necessary information? [RECL]
12. In the previous run, how much effort did it take to anticipate team members' needs? [TMN]
13. In the previous run, how much effort did it take to prioritize requests? [PRIRQ]
14. In the previous run, how much effort did it take to scan flight progress data? [SCFP]
15. In the previous run, how much effort did it take to access relevant aircraft or flight information? [ACCD]
16. In the previous run, how much effort did it take to gather and interpret information? [GETI]

Appendix C.6. Statements from SUS Questionnaire

1. I think that I would like to use this system frequently.
2. I found the system unnecessarily complex.
3. I thought the system was easy to use.

4. I think that I would need the support of a technical person to be able to use this system.
5. I found the various functions in this system were well integrated.
6. I thought there was too much inconsistency in this system.
7. I would imagine that most people would learn to use this system very quickly.
8. I found the system very cumbersome to use.
9. I felt very confident using the system.
10. I needed to learn a lot of things before I could get going with this system.

Appendix D. Validation Setup Details

The left and right sides of the outside view areas presented current meteorological data as relevant for aircraft takeoff and landing (see Figure A2), i.e., wind speed in knots (here 10) and wind direction with an additional red arrow (here 070°) according to the runway orientation (grey rectangle), the active runway name (here 05), the airport International Civil Aviation Organization (ICAO) code (EYKA), the QNH (here 1001), the visibility conditions (here 9999, i.e., no visibility restrictions), and cloud information (in green circles).

Figure A2. Remote tower outside view with a small aircraft passing a parking aircraft on the apron and meteorological information in and around the compass rose on the right.

An adjacent laboratory room accommodated three simulation pilot workstations. Each workstation consisted of a monitor to visualize the simulation pilot interface (see Figure A3) for one of the three simulated airports, a keyboard, and a mouse.

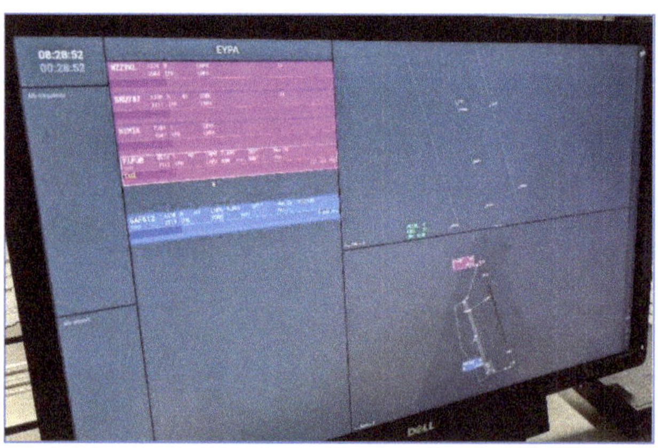

Figure A3. Simulation pilot interface for a simulated airport with time, pseudo flight strips for arrival and departure traffic, and radar views for airport surface and surrounding.

References

1. Lin, Y. Spoken Instruction Understanding in Air Traffic Control: Challenge, Technique, and Application. *Aerospace* **2021**, *8*, 65. [CrossRef]
2. Schäfer, D. Context-Sensitive Speech Recognition in the Air Traffic Control Simulation. Ph.D. Thesis, The University of Armed Forces, Munich, Germany, 2001.
3. Updegrove, J.A.; Jafer, S. Optimization of Air Traffic Control Training at the Federal Aviation Administration Academy. *Aerospace* **2017**, *4*, 50. [CrossRef]
4. Cordero, J.M.; Rodriguez, N.; de Pablo, J.M.; Dorado, M. Automated speech recognition in controller communications applied to workload measurement. In Proceedings of the 3rd SESAR Innovation Days, Stockholm, Sweden, 26–28 November 2013.
5. Cordero, J.M.; Dorado, M.; de Pablo, J.M. Automated speech recognition in ATC environment. In Proceedings of the 2nd International Conference on Application and Theory of Automation in Command and Control Systems, London, UK, 29–31 May 2012; pp. 46–53.
6. Kleinert, M.; Helmke, H.; Shetty, S.; Ohneiser, O.; Ehr, H.; Prasad, A.; Motlicek, P.; Harfmann, J. Automated Interpretation of Air Traffic Control Communication: The Journey from Spoken Words to a Deeper Understanding of the Meaning. In Proceedings of the IEEE/AIAA 40th Digital Avionics Systems Conference (DASC), Virtual, 3–7 October 2021. [CrossRef]
7. Helmke, H.; Rataj, J.; Mühlhausen, T.; Ohneiser, O.; Ehr, H.; Kleinert, M.; Oualil, Y.; Schulder, M. Assistant-Based Speech Recognition for ATM Applications. In Proceedings of the 11th USA/Europe Air Traffic Management Research and Development Seminar (ATM2015), Lisbon, Portugal, 23–26 June 2015.
8. Helmke, H.; Ohneiser, O.; Mühlhausen, T.; Wies, M. Reducing Controller Workload with Automatic Speech Recognition. In Proceedings of the 35th Digital Avionics Systems Conference (DASC), Sacramento, CA, USA, 25–29 September 2016.
9. Helmke, H.; Slotty, M.; Poiger, M.; Herrer, D.F.; Ohneiser, O.; Vink, N.; Cerna, A.; Hartikainen, P.; Josefsson, B.; Langr, D.; et al. Ontology for transcription of ATC speech commands of SESAR 2020 solution PJ.16-04. In Proceedings of the IEEE/AIAA 37th Digital Avionics Systems Conference (DASC), London, UK, 23–27 September 2018. [CrossRef]
10. Helmke, H.; Ondřej, K.; Shetty, S.; Arilíusson, H.; Simiganoschi, T.S.; Kleinert, M.; Ohneiser, O.; Ehr, H.; Zuluaga-Gomez, J.-P.; Smrz, P. Readback Error Detection by Automatic Speech Recognition and Understanding—Results of HAAWAII project for Isavia's Enroute Airspace. In Proceedings of the 12th SESAR Innovation Days, Budapest, Hungary, 5–8 December 2022.
11. Chen, S.; Kopald, H.D.; Elessawy, A.; Levonian, Z.; Tarakan, R.M. Speech inputs to surface safety logic systems. In Proceedings of the IEEE/AIAA 34th Digital Avionics Systems Conference (DASC), Prague, Czech Republic, 13–17 September 2015. [CrossRef]
12. Kleinert, M.; Shetty, S.; Helmke, H.; Ohneiser, O.; Wiese, H.; Maier, M.; Schacht, S.; Nigmatulina, I.; Sarfjoo, S.S.; Motlicek, P. Apron Controller Support by Integration of Automatic Speech Recognition with an Advanced Surface Movement Guidance and Control System. In Proceedings of the 12th SESAR Innovation Days, Budapest, Hungary, 5–8 December 2022.
13. Ohneiser, O.; Helmke, H.; Kleinert, M.; Siol, G.; Ehr, H.; Hobein, S.; Predescu, A.-V.; Bauer, J. Tower Controller Command Prediction for Future Speech Recognition Applications. In Proceedings of the 9th SESAR Innovation Days, Athens, Greece, 2–5 December 2019.
14. Ohneiser, O.; Helmke, H.; Shetty, S.; Kleinert, M.; Ehr, H.; Murauskas, Š.; Pagirys, T. Prediction and extraction of tower controller commands for speech recognition applications. *J. Air Transp. Manag.* **2021**, *95*, 102089. [CrossRef]
15. Badrinath, S.; Balakrishnan, H. Automatic Speech Recognition for Air Traffic Control Communications. *Transp. Res. Rec.* **2021**, *2676*, 798–810. [CrossRef]
16. Pellegrini, T.; Farinas, J.; Delpech, E.; Lancelot, F. The Airbus Air Traffic Control Speech Recognition 2018 Challenge: Towards ATC Automatic Transcription and Call Sign Detection. In Proceedings of the Interspeech, Graz, Austria, 15–19 September 2019. [CrossRef]
17. García, R.; Albarrán, J.; Fabio, A.; Celorrio, F.; Pinto de Oliveira, C.; Bárcena, C. Automatic Flight Callsign Identification on a Controller Working Position: Real-Time Simulation and Analysis of Operational Recordings. *Aerospace* **2023**, *10*, 433. [CrossRef]
18. Helmke, H.; Ohneiser, O.; Buxbaum, J.; Kern, C. Increasing ATM Efficiency with Assistant Based Speech Recognition. In Proceedings of the 12th USA/Europe Air Traffic Management Research and Development Seminar (ATM2017), Seattle, WA, USA, 26–30 June 2017.
19. Kleinert, M.; Helmke, H.; Moos, S.; Hlousek, P.; Windisch, C.; Ohneiser, O.; Ehr, H.; Labreuil, A. Reducing Controller Workload by Automatic Speech Recognition Assisted Radar Label Maintenance. In Proceedings of the 9th SESAR Innovation Days, Athens, Greece, 2–5 December 2019.
20. Chen, S.; Kopald, H.D.; Chong, R.; Wei, Y.; Levonian, Z. Read back error detection using automatic speech recognition. In Proceedings of the 12th USA/Europe Air Traffic Management Research and Development Seminar (ATM2017), Seattle, WA, USA, 26–30 June 2017.
21. Fürstenau, N.; Jakobi, J.; Papenfuss, A. Introduction: Basics, History, and Overview. In *Virtual and Remote Control Tower Research Topics in Aerospace*; Fürstenau, N., Ed.; Springer: Cham, Switzerland, 2022; pp. 3–22. [CrossRef]
22. Möhlenbrink, C.; Papenfuß, A. Eye-data metrics to characterize tower controllers' visual attention in a multiple remote tower exercise. In Proceedings of the 6th International Conference on Research in Air Transportation (ICRAT2014), Istanbul, Turkey, 26–30 May 2014.
23. Papenfuss, A.; Friedrich, M. Head Up Only—A design concept to enable multiple remote tower operations. In Proceedings of the IEEE/AIAA 35th Digital Avionics Systems Conference (DASC), Sacramento, CA, USA, 25–29 September 2016.

24. Fürstenau, N.; Papenfuss, A. Model Based Analysis of Subjective Mental Workload During Multiple Remote Tower Human-In-The-Loop Simulations. In *Virtual and Remote Control Tower. Research Topics in Aerospace*; Fürstenau, N., Ed.; Springer: Cham, Switzerland, 2022; pp. 293–342. [CrossRef]
25. Hamann, A.; Jakobi, J. Changing of the Guards: The Impact of Handover Procedures on Human Performance in Multiple Remote Tower Operations. In *Virtual and Remote Control Tower. Research Topics in Aerospace*; Fürstenau, N., Ed.; Springer: Cham, Switzerland, 2022; pp. 343–363. [CrossRef]
26. Friedrich, M.; Timmermann, F.; Jakobi, J. Active supervision in a Remote Tower Center: Rethinking of a new position in the ATC Domain. In Proceedings of the 19th International Conference on Engineering Psychology and Cognitive Ergonomics, EPCE 2022 as part of the 24th HCI International Conference, HCII 2022, Virtual, 26 June—1 July 2022; Springer: Cham, Switzerland, 2022; pp. 265–278. [CrossRef]
27. Li, W.-C.; Kearney, P.; Braithwaite, G. The Certification Processes of Multiple Remote Tower Operations for Single European Sky. In *Virtual and Remote Control Tower. Research Topics in Aerospace*; Fürstenau, N., Ed.; Springer: Cham, Switzerland, 2022; pp. 511–541. [CrossRef]
28. Schier, S.; Rambau, T.; Timmermann, F.; Metz, I.; Stelkens-Kobsch, T.H. Designing the Tower Control Research Environment of the Future. Deutscher Luft- und Raumfahrtkongress, DLRK2013. In Proceedings of the English: German Aerospace Congress, Stuttgart, Germany, 10–12 September 2013.
29. Schier, S.; Manske, P. *VisiTop II—Briefing-Unterlagen*; Section 4.2. DLR-internal report; DLR Institute of Flight Guidance: Braunschweig, Germany, 2015.
30. Ohneiser, O.; Sarfjoo, S.; Helmke, H.; Shetty, S.; Motlicek, P.; Kleinert, M.; Ehr, H.; Murauskas, Š. Robust Command Recognition for Lithuanian Air Traffic Control Tower Utterances. In Proceedings of the InterSpeech, Brno, Czech Republic, 30 August–3 September 2021. [CrossRef]
31. Shetty, S.; Helmke, H.; Kleinert, M.; Ohneiser, O. Early Callsign Highlighting using Automatic Speech Recognition to Reduce Air Traffic Controller Workload. In *Human Factors in Transportation, Proceedings of the International Conference on Applied Human Factors and Ergonomics (AHFE2022), New York, NY, USA, 24–28 July 2022*; Plant, K., Praetorius, G., Eds.; AHFE International: New York, NY, USA, 2022; Volume 60. [CrossRef]
32. Jordan, C.S.; Brennen, S.D. *Instantaneous Self-Assessment of Workload Technique (ISA)*; Defence Research Agency: Portsmouth, UK, 1992.
33. Bongo, M.F.; Seva, R.R. Evaluating the Performance-Shaping Factors of Air Traffic Controllers Using Fuzzy DEMATEL and Fuzzy BWM Approach. *Aerospace* **2023**, *10*, 252. [CrossRef]
34. Hart, S.G.; Staveland, L.E. Development of NASA-TLX (Task Load Index): Results of empirical and theoretical research. In *Human Mental Workload*; Hancock, P.A., Meshkati, N., Eds.; North Holland Press: Amsterdam, The Netherlands, 1988; p. 198. [CrossRef]
35. Hart, S.G. NASA-Task Load Index (NASA-TLX); 20 years later. In Proceedings of the Human Factors and Ergonomics Society, San Francisco, CA, USA, 16–20 October 2006; Volume 50, pp. 904–908. [CrossRef]
36. Roscoe, A.H. Assessing Pilot Workload in Flight. In Proceedings of the AGARD Conference Proceedings Flight Test Techniques, Lisbon, Portugal, 2–5 April 1984.
37. Dehn, D.M. Assessing the Impact of Automation on the Air Traffic Controller: The SHAPE Questionnaires. *Air Traffic Control Q.* **2008**, *16*, 127–146. [CrossRef]
38. Lee, K.K.; Kerns, K.; Bone, R.; Nickelson, M. The Development and Validation of the Controller Acceptance Rating Scale (CARS): Results of Empirical Research. In Proceedings of the 4th USA/Europe Air Traffic Management R&D Seminar, Santa Fe, NM, USA, 3–7 December 2001.
39. Brooke, J. SUS—A quick and dirty usability scale. In *Usability Evaluation in Industry*; Jordan, P.W., Thomas, B., McClelland, I.L., Weerdmeester, B.A., Eds.; Taylor and Francis: London, UK, 1996; pp. 189–194.
40. Bangor, A.; Kortum, P.T.; Miller, J.T. An empirical evaluation of the system usability scale. *Intl. J. Hum.-Comput. Interact.* **2008**, *24*, 574–594. [CrossRef]
41. Helmke, H.; Kleinert, M.; Ahrenhold, N.; Ehr, H.; Mühlhausen, T.; Ohneiser, O.; Motlicek, P.; Prasad, A.; Zuluaga-Gomez, J. Automatic Speech Recognition and Understanding for Radar Label Maintenance Support Increases Safety and Reduces Air Traffic Controllers' Workload. In Proceedings of the 15th USA/Europe Air Traffic Management Research and Development Seminar (ATM2023), Savannah, GA, USA, 5–9 June 2023.
42. Levenshtein, V.I. Binary codes capable of correcting deletions, insertions and reversals. *Sov. Phys. Dokl.* **1966**, *10*, 707–710.
43. Ohneiser, O.; Helmke, H.; Shetty, S.; Kleinert, M.; Ehr, H.; Murauskas, Š.; Pagirys, T.; Balogh, G.; Tønnesen, A.; Kis-Pál, G.; et al. Understanding Tower Controller Communication for Support in Air Traffic Control Displays. In Proceedings of the 12th SESAR Innovation Days, Budapest, Hungary, 5–8 December 2022.
44. ICAO. *ATM (Air Traffic Management): Procedures for Air Navigation Services*; DOC 4444 ATM/501; International Civil Aviation Organization (ICAO): Montréal, QC, Canada, 2007.
45. Helmke, H.; Shetty, S.; Kleinert, M.; Ohneiser, O.; Ehr, H.; Prasad, A.; Motlicek, P.; Cerna, A.; Windisch, C. Measuring Speech Recognition and Understanding Performance in Air Traffic Control Domain Beyond Word Error Rates. In Proceedings of the 11th SESAR Innovation Days, Virtual, 7–9 December 2021.

46. Charness, G.; Gneezy, U.; Kuhn, M.A. Experimental methods: Between-subject and within-subject design. *J. Econ. Behav. Organ.* **2012**, *81*, 1–8. [CrossRef]
47. Ohneiser, O.; Adamala, J.; Salomea, I.-T. Integrating Eye- and Mouse-Tracking with Assistant Based Speech Recognition for Interaction at Controller Working Positions. *Aerospace* **2021**, *8*, 245. [CrossRef]

aerospace

MDPI

Article

A Virtual Simulation-Pilot Agent for Training of Air Traffic Controllers

Juan Zuluaga-Gomez [1,2,*], Amrutha Prasad [1,3], Iuliia Nigmatulina [1,4], Petr Motlicek [1,4] and Matthias Kleinert [5]

1 Speech & Audio Processing Group, Idiap Research Institute, 1920 Martigny, Switzerland; aprasad@idiap.ch (A.P.); iuliia.nigmatulina@idiap.ch (I.N.)
2 LIDIAP, Ecole Polytechnique Fédérale de Lausanne, 1015 Lausanne, Switzerland
3 Faculty of Information Technology, Brno University of Technology, 60190 Brno, Czech Republic
4 Institute of Computational Linguistics, University of Zurich, 8006 Zurich, Switzerland
5 Institute of Flight Guidance, German Aerospace Center (DLR), 38108 Braunschweig, Germany; matthias.kleinert@dlr.de
* Correspondence: juan-pablo.zuluaga@idiap.ch

Abstract: In this paper we propose a novel virtual simulation-pilot engine for speeding up air traffic controller (ATCo) training by integrating different state-of-the-art artificial intelligence (AI)-based tools. The virtual simulation-pilot engine receives spoken communications from ATCo trainees, and it performs automatic speech recognition and understanding. Thus, it goes beyond only transcribing the communication and can also understand its meaning. The output is subsequently sent to a response generator system, which resembles the spoken read-back that pilots give to the ATCo trainees. The overall pipeline is composed of the following submodules: (i) an automatic speech recognition (ASR) system that transforms audio into a sequence of words; (ii) a high-level air traffic control (ATC)-related entity parser that understands the transcribed voice communication; and (iii) a text-to-speech submodule that generates a spoken utterance that resembles a pilot based on the situation of the dialogue. Our system employs state-of-the-art AI-based tools such as Wav2Vec 2.0, Conformer, BERT and Tacotron models. To the best of our knowledge, this is the first work fully based on open-source ATC resources and AI tools. In addition, we develop a robust and modular system with optional submodules that can enhance the system's performance by incorporating real-time surveillance data, metadata related to exercises (such as sectors or runways), or even a deliberate read-back error to train ATCo trainees to identify them. Our ASR system can reach as low as 5.5% and 15.9% absolute word error rates (WER) on high- and low-quality ATC audio. We also demonstrate that adding surveillance data into the ASR can yield a callsign detection accuracy of more than 96%.

Keywords: air traffic controller training; simulation-pilot agent; BERT; automatic speech recognition and understanding; speech synthesis

Citation: Zuluaga-Gomez, J.; Prasad, A.; Nigmatulina, I.; Motlicek, P.; Kleinert, M. A Virtual Simulation-Pilot Agent for Training of Air Traffic Controllers. *Aerospace* **2023**, *10*, 490. https://doi.org/10.3390/aerospace10050490

Academic Editor: Judith Rosenow

Received: 13 April 2023
Revised: 5 May 2023
Accepted: 9 May 2023
Published: 22 May 2023

1. Introduction

The exponential advances in artificial intelligence (AI) and machine learning (ML) have opened the door of automation to many applications. Examples are automatic speech recognition (ASR) [1] applied to personal assistants (e.g., SIRI® or Amazon's ALEXA®) and natural language processing (NLP) and understating [2] for different tasks such as sentiment analysis [3] and user intent detection [4]. Even though these advances are remarkable, many applications have lagged behind due to their critical matter, imperative near-to-perfect performance or simply because the users or administrators only trust the already existing legacy systems. One clear example is air traffic control (ATC) communications.

In ATC communications, ATCos are required to issue verbal commands to pilots in order to keep control and safety of a given area of airspace, although there are different means of communication, such as controller–pilot data link communications (CPDLC).

CPDLC is a two-way data link system by which controllers can transmit non-urgent strategic messages to an aircraft as an alternative to voice communications. These messages are displayed on a flight deck visual display.

Research targeted at understanding spoken ATC communications in the military domain can be traced back to the 1970s [5], late 1980s [6], and 1990s [7]. Recent projects are aiming at integrating AI-based tools into ATC processes by developing robust acoustic-based AI systems for transcribing dialogues. For instance, MALORCA [8,9], HAAWAAI [10] and ATCO2 [11,12]. These latest projects have shown mature-enough ASR and NLP systems that demonstrate potential for deployment in real-life operation control rooms. Other fields of work are voice activity detection (VAD), diarization [13] and ASR [14–16]. In addition, a few researchers have gone further by developing techniques to understand the ATCo–pilot dialogues [9,11,17]. However, previous works are mostly disentangled from each other. Some researchers only focus on ASR [18,19], while a few prior studies have integrated natural language understanding into their ASR pipelines [14,20].

Another key application that has seen growth in interest is the ATCo training framework. Training ATCos usually involves a human simulation-pilot. The simulation-pilot responds to or issues a request to the ATCo trainee in order to simulate an ATC communication with standard phraseology [21]. It is a human-intensive task, where a specialized workforce is needed during ATCo training and the overall cost is usually high. An example is the EUROCONTROL's ESCAPE lite simulator https://www.eurocontrol.int/simulator/escape, accessed on 12 May 2023) which still requires a human simulation-pilot. In a standard training scenario, the default simulation-pilots (humans) are required to execute the steps given by ATCo trainees, as in the case of real pilots (directly introduced to the simulator). The pilots, on the other hand, update the training simulator, so that the ATCos can see whether the pilots are following the desired orders. Therefore, this simulation is very close to a real ATCo–pilot communication. One well-known tool for ATCo training is Eurocontrol's ESCAPE simulator. It is an air traffic management (ATM) real-time simulation platform that supports: (i) airspace design for en-route and terminal maneuvering areas; (ii) the evaluation of new operational concepts and ATCo tools; (iii) pre-operational validation trials; and most importantly, (ii) the training of ATCos [22]. In this paper, we develop a virtual simulation-pilot engine that understands ATCo trainees' commands and possibly can replace current simulators based on human simulation-pilots. In practice, the proposed virtual simulation-pilot can handle simple ATC communications, e.g., the first phase of the ATCo trainee's training. Thus, humans are still required for more complex scenarios. Analogous efforts of developing a virtual simulation-pilot agent (or parts of it) have been covered in [23,24].

In this paper, we continue our previous work presented at SESAR Innovation Days 2022 [25]. There, a simple yet efficient 'proof-of-concept' virtual simulation-pilot was introduced. This paper formalizes the system with additional ATM-related modules. It also demonstrates that open-source AI-based models are a good fit for the ATC domain. Figure 1 contrasts the proposed pipeline (left side) and the current (default) human-based simulation-pilot (right side) approaches for ATCo training.

Main contributions Our work proposes a novel virtual simulation-pilot system based on fine-tuning several open-source AI models with ATC data. Our mains contributions are:

- Could human simulation pilots be replaced (or aided) by an autonomous AI-based system? This paper presents an end-to-end pipeline that utilizes a virtual simulation-pilot capable of replacing human simulation-pilots. Implementing this pipeline can speed up the training process of ATCos while decreasing the overall training costs.
- Is the proposed virtual simulation-pilot engine flexible enough to handle multiple ATC scenarios? The virtual simulation-pilot system is modular, allowing for a wide range of domain-specific contextual data to be incorporated, such as real-time air surveillance data, runway numbers, or sectors from the given training exercise. This flexibility boosts the system performance, while making its adaptation easier to various simulation scenarios, including different airports.

- Are open-source AI-based tools enough to develop a virtual simulation-pilot system? Our pipeline is built entirely on open-source and state-of-the-art pre-trained AI models that have been fine-tuned on the ATC domain. The Wav2Vec 2.0 and XLSR [26,27] models are used for ASR, BERT [28] is employed for natural language understanding (NLU), and FastSpeech2 [29] is used for the text-to-speech (TTS) module. To the best of our knowledge, this is the first study that utilizes open-source ATC resources exclusively [11,30–32].
- Which scenarios can a virtual simulation-pilot handle? The virtual simulation-pilot engine is highly versatile and can be customized to suit any potential use case. For example, the system can employ either a male or a female voice or simulate very high-frequency noise to mimic real-life ATCo–pilot dialogues. Additionally, new rules for NLP and ATC understanding can be integrated based on the target application, such as approach or tower control.

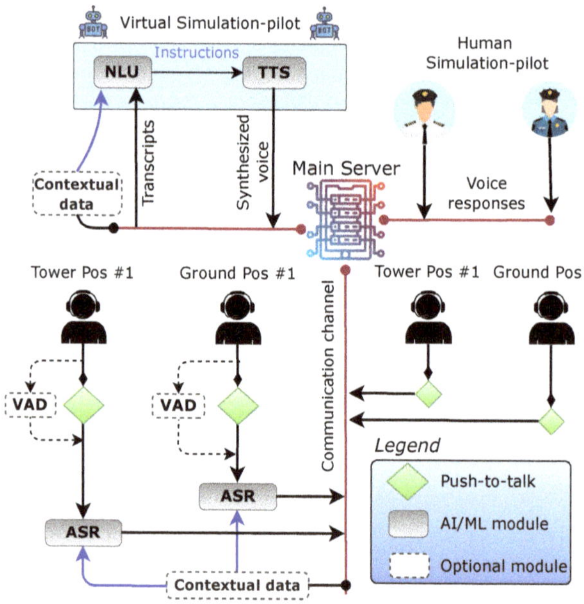

Figure 1. Virtual simulation-pilot pipeline for ATCo training. A traditional ATCo training setup is depicted on the right side, while our proposed virtual simulation-pilot is on the left side. The pipeline starts with an ATCo trainee issuing a communication and its capture after the end of the push-to-talk (PTT) signal, or voice-activity detection if not available. Then, the ASR and ***high-level entity parser*** (NLP) modules transcribe and extract the ATC-related entities from the voice communication. The output is later rephrased with simulation-pilot grammar. The speech synthesizer uses the generated text to create a WAV file containing the spoken textual prompt. In the end, a response is generated by the virtual simulation-pilot that matches the desired read-back.

The authors believe this research is a game changer in the ATM community due to two aspects. First, a novel modular system that can be adjusted to specific scenarios, e.g., aerodrome control or area control center. Second, it is demonstrated that open-source models such as XLSR [27] (for ASR) or BERT [28] (for NLP and ATC understanding) can be successfully adapted to the ATC scenario. In practice, the proposed virtual simulation-pilot engine could become the starting point to develop more inclusive and mature systems aimed at ATCo training.

The rest of the paper is organized as follows. Section 2 describes the virtual simulation-pilot system, covering the fundamental background for each of the base (Section 2.1) and optional modules (Section 2.2) of the system. Section 3 describes the databases used. Then, Section 4 covers the experimental setup followed for adapting the virtual simulation-pilot and the results for each module of the system. Finally, brief future research directions are provided in Section 5 and the paper is concluded in Section 7.

2. Virtual Simulation-Pilot System

The virtual simulation-pilot system manages the most commonly used commands in ATC. It is particularly well-suited for the early stages of ATCo training. Its modular design allows the addition of more advanced rules and grammar to enhance the system's robustness. Our goal is to enhance the foundational knowledge and skills of ATCo trainees. Furthermore, the system can be customized to specific conditions or training scenarios, such as when the spoken language has a heavy accent (e.g., the case of foreign English) or when the ATCo trainee is practicing different positions.

In general, ATC communications play a critical role in ensuring the safe and efficient operation of aircraft. These communications are primarily led by ATCos, who are responsible for issuing commands and instructions to pilots in real-time. The training process of ATCos involves three stages: (i) initial, (ii) operational, and (iii) continuation training. The volume and complexity of these communications can vary greatly depending on factors such as the airspace conditions and seasonal fluctuations, with ATCos often facing increased workloads during peak travel seasons [25]. As such, ATCo trainees must be prepared to handle high-stress and complex airspace situations through a combination of intensive training and simulation exercises with human simulation-pilots [33]. In addition to mastering the technical aspects of air traffic control, ATCo trainees must also develop strong communication skills, as they are responsible for ensuring clear and precise communication with pilots at all times.

Due to the crucial aspect of ATC, efforts have been made to develop simulation interfaces for their training [33–35]. Previous works includes optimization of the training process [36], post-evaluation of each training scenario [37,38], and virtual simulation-pilot implementation, for example, a deep learning (DL)-based implementation [39]. In [24], the authors use sequence-to-sequence DL models to map from spoken ATC communications to high-level ATC entities. They use the well-known Transformer architecture [40]. Transformer is the base of the recent, well-known encoder–decoder models for ASR (Wav2Vec 2.0 [26]) and NLP (BERT [28]). The subsections address in more detail each module that is a part of the virtual simulation-pilot system.

2.1. Base Modules

The proposed virtual simulation-pilot system (see Figure 1) is built with a set of base modules, and possibly, optional submodules. The most simple version of the engine contains only the base modules.

2.1.1. Automatic Speech Recognition

Automatic speech recognition (ASR) or speech-to-text systems convert speech to text. An ASR system uses an acoustic model (AM) and a language model (LM). The AM represents the relationship between a speech signal and phonemes/linguistic units that make up speech and is trained using speech recordings along with their corresponding text. The LM provides a probability distribution over a sequence of words, provides context to distinguish between words and phrases that sound similar and is trained using a large corpus of text data. A decoding graph is built as a weighted finite state transducer (WFST) [41–43] using the AM and LM that generates text output given an observation sequence. Standard ASR systems rely on a lexicon, LM and AM, as stated above. Currently, there are two main ASR paradigms, where different strategies, architectures and procedures are employed for blending all these modules in one system. The first is hybrid-based ASR, while the second is a more recent approach, termed end-to-end ASR. A comparison of both is shown in Figure 2.

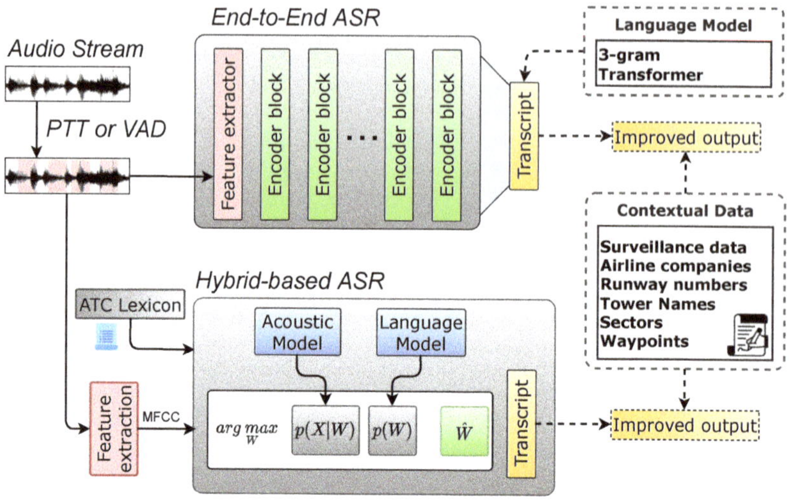

Figure 2. Traditional hybrid-based and more recent end-to-end automatic speech recognition systems. These systems take an ATCo voice communication as input and then produce transcripts as output. The dotted blocks refer to modules that are optional. For instance, surveillance data or other types of data (e.g., sector or waypoints) can be added to increase the overall performance of the system.

Hybrid-Based Automatic Speech Recognition. ASR with hybrid systems is based on hidden Markov models (HMM) and deep neural networks (DNN) [44]. DNNs are an effective module for estimating the posterior probability of a given set of possible outputs (e.g., phone-state or tri-phone-state probability estimator in ASR systems). These posterior probabilities can be seen as pseudo-likelihoods or "scale likelihoods", which can be interfaced with HMM modules. HMMs provide a structure for mapping a temporal sequence of acoustic features, X, e.g., Mel-frequency cepstral coefficients (MFCCs), into a sequence of states [45]. Hybrid systems remain one of the best approaches for building ASR engines based on lattice-free maximum mutual information (LF-MMI) [46]. Currently, HMM-DNN-based ASR is the state-of-the-art system for ASR in ATC domain [15].

Recent work in ASR has targeted different areas in ATC. For instance, a benchmark for ASR on ATC communications databases is established in [47]. Leveraging non-transcribed ATC audio data using semi-supervised learning has been covered in [48,49] and using self-supervised learning for ATC in [18]. The previous work related to the large-scale automatic collection of ATC audio data from different airports worldwide is covered in [15,50]. Additionally, innovative research aimed at improving callsign recognition by integrating surveillance data into the pipeline is covered by [10,12]. ASR systems are also employed for more high-level tasks such as pilot report extractions from very-high frequency (VHF) communications [51]. Finally, multilingual ASR has also been covered in ATC applications in [19].

The main components of a hybrid system are a pronunciation lexicon, LM and AM. One key advantage of a hybrid system versus other ASR techniques is that the text data (e.g., words, dictionary) and pronunciation of new words are collected and added beforehand, hoping to match the target domain of the recognizer. Standard hybrid-based ASR approaches still rely on word-based lexicons, i.e., missing or out-of-vocabulary words from the lexicon cannot be hypothesized by the ASR decoder. The system is composed of an explicit acoustic and language model. A visual example of hybrid-based ASR systems is in the bottom panel of Figure 2. Most of these systems can be trained with toolkits such as Kaldi [52] or Pkwrap [53].

End-to-End Automatic Speech Recognition. End-to-end (E2E) systems are based on a different paradigm compared to hybrid-based ASR. E2E-ASR aims at directly transcribing speech to text without requiring alignments between acoustic frames (i.e., input features) and output characters/words, which is a necessary separate component in standard hybrid-based systems. Unlike the hybrid approach, E2E models are learning a direct mapping between acoustic frames and model label units (characters, subwords or words) in one step toward the final objective of interest.

Recent work on encoder–decoder ASR can be categorized into two main approaches: connectionist temporal classification (CTC) [54] and attention-based encoder–decoder systems [55]. First, CTC uses intermediate label representation, allowing repetitions of labels and occurrences of 'blank output', which labels an output with 'no label'. Second, attention-based encoder–decoder or only-encoder models directly learn a mapping from the input acoustic frames to character sequences. For each time step, the model emits a character unit conditioned on the inputs and the history of the produced outputs. The important lines of work for E2E-ASR can be categorized as self-supervised learning [56–58] for speech representation, covering bidirectional models [26,59] and autoregressive models [60,61].

Moreover, recent innovative research on E2E-ASR for the ATC domain is covered in [62]. Here, the authors follow the practice of fine-tuning a Wav2Vec 2.0 model [26] with public and private ATC databases. This system reaches on-par performances with hybrid-based ASR models, demonstrating that this new paradigm for ASR development also performs well in the ATC domain. In E2E-ASR, the system encodes directly an acoustic and language model, and it produces transcripts in an E2E manner. A visual example of an only-encoder E2E-ASR system is in the top panel of Figure 2.

2.1.2. Natural Language Understanding

Natural language understanding (NLU) is a field of NLP that aims at reading comprehension. In the field of ATC, NLU is related to intent detection and slot filling. The slot-filling task is akin to named entity recognition (NER). In intent detection, the commands from the communication are extracted, while slot filling refers to the values of these commands and callsigns. Furthermore, throughout the paper, the system that extracts the high level ATC-related knowledge from the ASR outputs is called a ***high-level entity parser*** system. The NER-based understanding of ATC communications has been previously studied in [11,23,24], while our earlier work [25] includes the integration of named-entity recognition (NER) into the virtual simulation-pilot framework.

The ***high-level entity parser*** system is responsible for identifying, categorizing and extracting crucial keywords and phrases from ATC communications. In NLP, these keywords are classified into pre-defined categories such as parts of speech tags, locations, organizations or individuals' names. In the context of ATC, the key entities include callsigns, commands and values (which includes units, e.g., flight level). For instance, consider the following transcribed communication (taken from Figure 3):

ASR transcript: ryanair nine two bravo quebec turn right heading zero nine zero,

would be parsed to high-level ATC entity format:

Output: <callsign> ryanair nine two bravo quebec </callsign> <command> turn right heading </command> <value> zero nine zero </value>.

The previous output is then used for further processing tasks, e.g., generating a simulation-pilot-like response, metadata logging and reporting, or simply to help ATCos in their daily tasks. Thus, NLU is mostly focused on NER [63]. Initially, NER relied on the manual crafting of dictionaries and ontologies, which led to complexity and human error when scaling to more entities or adapting to a different domain [64]. The advancement of ML-based methods for text processing, including NER, has been introduced by [65]. The previous work [66] continued to advance NER techniques. A ***high-level entity parser*** system (such as ours) can be implemented by fine-tuning a pre-trained LM for the NER task. Currently, state-of-the-art NER models utilize pre-trained LMs such as BERT [28], RoBERTa [67] or DeBERTa [68]. For the proposed virtual simulation-pilot, we use a fine-tuned BERT on ATC text data.

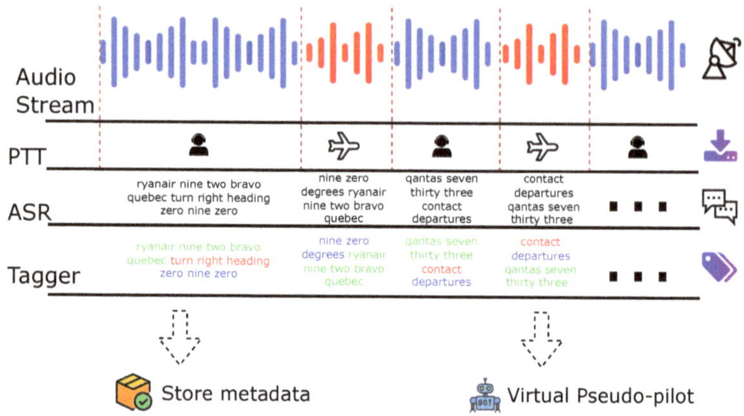

Figure 3. Detailed outputs of the main ML-based submodules of our proposed simulation-pilot system. It includes pre-processing from the input audio stream, speaker role detection by push-to-talk (PTT) signal, transcript generation and callsign/command/value extraction with the high-level entity with ASR and NER modules, respectively. All the data are later aggregated, packaged and sent to the response generator and TTS module. Note that these data can also be logged into a database for control and recording. Figure adapted from our previous work [25].

2.1.3. Response Generator

The response generator (RG) is a crucial component of the simulation-pilot agent. It processes the output from the ***high-level entity parser*** system, which includes the callsign, commands and values uttered by the ATCo, and then later generates a spoken response. The response is then delivered in the form of a WAV file, which is played through the headphones of the ATCo trainee. Additionally, the response, along with its metadata, can be stored for future reference and evaluation. The RG system is designed to generate responses that are grammatically consistent with what a standard simulation-pilot (or pilot) would say in response to the initial commands issued by the ATCo. The RG system comprises three submodules: (i) grammar conversion, (ii) a word fixer (e.g., ATCo-to-pilot phrase fixer), and (iii) text-to-speech, also known as a speech synthesizer. A visual representation of the RG system split by submodules is in Figure 4.

Grammar Conversion Submodule. A component designed to generate the response of the virtual simulation-pilot. First, the output of the ***high-level entity parser*** module (discussed in Section 2.1.2) is input to the grammar conversion submodule. At this stage, the communication knowledge has already been extracted, including the callsign, commands and their values. This is followed by a grammar-adjustment process, where the order of the high-level entities is rearranged. For example, we take into account the common practice of pilots mentioning the callsign at the end of the utterance while ATCos mention it at the beginning of the ATC communication. Thus, our goal is to align the grammar used by the simulation-pilot with the communication style used by the ATCo. See the first left panel in Figure 4.

Word Fixer Submodule. This is a crucial component of the virtual simulation-pilot system that ensures that the output from the response generator aligns with the standard ICAO phraseology. This is achieved by modifying the commands to match the desired response based on the input communication from the ATCo. The submodule applies specific mapping rules, such as converting *descend* $\rightarrow$ *descending* or *turn* $\rightarrow$ *heading*, to make the generated reply as close to standard phraseology as possible. Similar efforts have been covered in a recent study [39] where the authors propose a *copy mechanism* that copies the key entities from the ATCo communication into the desired response of the virtual simulation-pilot, e.g., *maintain* $\rightarrow$ *maintaining*. In real-life ATC communication, however, the wording of ATCos and pilots slightly differs. Currently, our *word fixer* submodule contains a list of 18 commands but can be easily updated by adding additional mapping rules to a `rules.txt` file. This allows the system to adapt to different environments, such as aerodrome control, departure/approach control or area control center. The main conversion rules used by the word fixer submodule are listed in Table 1. The ability to modify and adapt the word fixer submodule makes it a versatile tool for training ATCos to recognize and respond to standard ICAO phraseology. See the central panel in Figure 4.

Text-to-Speech Submodule. Speech synthesis, also referred to as text-to-speech (TTS), is a multidisciplinary field that combines various areas of research such as linguistics, speech signal processing and acoustics. The primary objective of TTS is to convert text into an intelligible speech signal. Over the years, numerous approaches have been developed to achieve this goal, including formant-based parametric synthesis [69], waveform concatenation [70] and statistical parametric speech synthesis [71]. In recent times, the advent of deep learning has revolutionized the field of TTS. Models such as Tacotron [72] and Tacotron2 [73] are end-to-end generative TTS systems that can synthesize speech directly from text input (e.g., characters or words). Most recently, FastSpeech2 [29] has gained widespread recognition in the TTS community due to its simplicity and efficient non-autoregressive manner of operation. Finally, TTS is a complex field that draws on a variety of areas of research and has made significant strides recently, especially with the advent of deep learning. For a more in-depth understanding of the technical aspects of TTS engines, readers are redirected to [74] and novel diffusion-based TTS systems in [75]. The TTS system for ATC is depicted in the right panel in Figure 4.

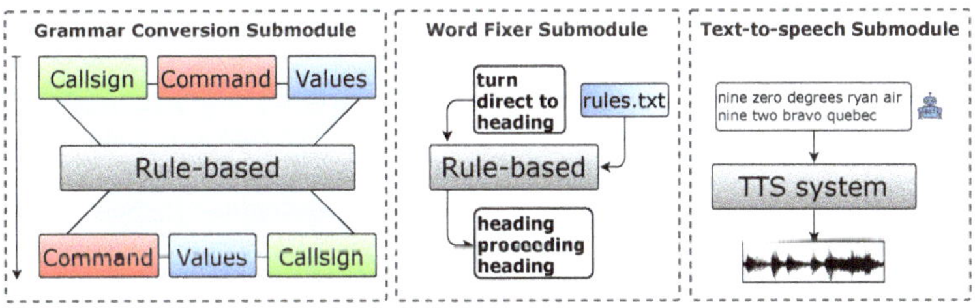

Figure 4. Detailed submodules of the response generator.

Table 1. Word-fixing rules. The rules are used to convert ATCos input communications into a virtual simulation-pilot response.

Word Fixer Submodule—Rules.txt	
Horizontal commands	**Handover commands**
continue heading → continuing altitude	contact tower → contact tower
heading → heading	station radar → station radar
turn → heading	squawk → squawk
turn by → heading	squawking → squawk
direct to → proceeding direct	contact frequency → NONE
Level commands	**Speed commands**
maintain altitude → maintaining altitude	reduce → reducing
maintain altitude → maintain	maintain speed → maintaining
descend → descending	reduce speed → reduce speed
climb → climbing	speed → NONE

2.2. *Optional Modules*

In contrast to the base modules, covered in Section 2.1, the optional modules are blocks that can be integrated into the virtual simulation-pilot to enhance or add new capabilities.

An example is the PTT (push-to-talk) signal. In some cases a PTT signal is not available; thus, voice activity detection can be integrated. Below, each of the proposed optional modules is covered in more detail.

2.2.1. Voice Activity Detection

Voice activity detection (VAD) is an essential component in standard speech-processing systems to determine which portions of an audio signal correspond to speech and which are non-speech, i.e., background noise or silence. VAD can be used for offline purpose decoding, as well as for online streaming recognition. The offline VAD is used to split a lengthy audio into shorter segments that can then be used for training or evaluating ASR or NLU systems. The online VAD is particularly crucial for ATC ASR when the PTT signal is not available. An example of an online VAD is the WebRTC (developed by Google https://webrtc.org/, accessed on 12 May 2023). In ATC communications, VAD is used to filter out the background noise and keep only the speech segments that carry the ATCo's (or pilot's) voice messages. One of the challenges for VAD in ATC communications is the presence of a high level of background noise. The noise comes from various sources, e.g., the engines of aircraft, wind or even other ATCos. ATC communications can easily have signal-to-noise (SNR) ratios lower than 15 dB. If VAD is not applied (and there is not a PTT signal available), the ASR system may degrade the accuracy of speech transcription, which may result in incorrect responses from the virtual simulation-pilot agent.

VAD has been explored before in the framework of ATC [76]. A general overview of recent VAD architecture and research directions is covered in [77]. Some other researchers have targeted how to personalize VAD systems [78] and how this module plays its role in the framework of diarization [79]. There are several techniques used for VAD, ranging from traditional feature-based models to hidden Markov models to Gaussian mixture-based models [80]. On the other hand, machine-learning-based models have proven to be more accurate and robust, particularly deep neural network-based methods. These techniques can learn complex relationships between the audio signal and speech and can be trained on large annotated datasets. For instance, convolutional and deep-neural-network-based VAD has received much interest [76]. VAD can be used in various stages of the ATC communication pipeline. VAD can be applied at the front-end of the ASR system to pre-process the audio signal and reduce the processing time of the ASR system. Figures 1 and 2 depict where a VAD module can be integrated into the virtual simulation-pilot agent.

2.2.2. Contextual Biasing with Surveillance Data

In order to enhance the accuracy of an ASR system's predictions, it is possible to use additional context information along with speech input. In the ATC field, radar data can serve as context information, providing a list of unique identifiers for aircraft in the airspace called "callsigns". By utilizing these radar data, the ASR system can prioritize the recognition of these registered callsigns, increasing the likelihood of correct identification. Callsigns are typically a combination of letters, digits and an airline name, which are translated into speech as a sequence of words. The lattice, or prediction graph, can be adjusted during decoding by weighting the target word sequences using the finite state transducer (FST) operation of composition [12]. This process, called lattice rescoring, has been found to improve the recognition accuracy, particularly for callsigns. Multiple experiments using ATC data have demonstrated the effectiveness of this method, especially in improving the accuracy of callsign recognition. The results of contextual biasing are presented and discussed below in Section 4.1.

Re-ranking module based on Levenshtein distance. The ***high-level entity parser*** system for NER (see Section 2.1.2) allows us to extract the callsign from a given transcript or ASR 1-best hypotheses. Recognition of this entity is crucial where a single error produced by the ASR system affects the whole entity (normally composed of three to eight words). Additionally, speakers regularly shorten callsigns in the conversation, making it impossible for an ASR system to generate the full entity (e.g., *'three nine two papa'* instead of *'austrian* *three nine two papa'*, *'six lima yankee'* instead of *'hansa six lima yankee'*). One way to overcome this issue is to re-rank the entities extracted by the ***high-level entity parser*** system with the surveillance data. The output of this system is a list of tags that match words or sequences of words in an input utterance. As our only available source of contextual knowledge is callsigns registered at a certain time and location, we extract callsigns with the ***high-level entity parser*** system and discard other entities. Correspondingly, each utterance has a list of callsigns expanded into word sequences. As input, the re-ranking module takes (i) a callsign extracted by the ***high-level entity parser*** system and (ii) an expanded list of callsigns. The re-ranking module compares a given n-gram sequence against a list of possible n-grams and finds the closest match from the list of surveillance data based on the *weighted Levenshtein distance*. In order to use contextual knowledge, it is necessary to know which words in an utterance correspond to a desired entity (i.e., a callsign), which is why it is necessary to add into the pipeline the ***high-level entity parser*** system. We skip the re-ranking in case the output is a 'NO_CALLSIGN' flag (no callsign recognized).

2.2.3. Read-Back Error-Insertion Module

The approach of using the virtual simulation-pilot system can be adapted to meet various communication requirements in ATC training. This includes creating a desirable read-back error (RBE), which is a plausible scenario in ATC, where a pilot or ATCo misreads or misunderstands a message [81]. By incorporating this scenario in ATCos' training, they can develop the critical skills for spotting these errors. This is a fundamental aspect of ensuring the safety and efficiency of ATM [82]. The ability to simulate (by inserting a desired error) and practice these scenarios through the use of the virtual simulation-pilot system offers a valuable tool for ATCo training and can help to improve the overall performance of ATC. An example could look like: `ATCo: turn right` → `Pilot (RBE): turning left`.

The structure of the generated RBE could depend on the status of the exercise, for instance, whether the ATCo trainee is in the aerodrome control or approach/departure control position. These positions should, in the end, change the behavior of this optional module. The proposed, optional RBE insertion module is depicted in Figure 5.

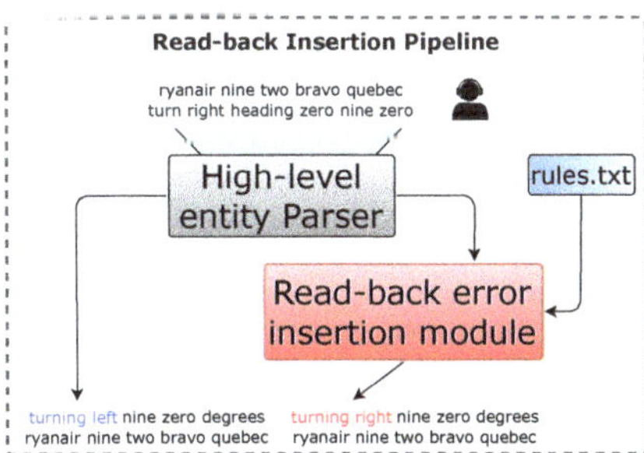

Figure 5. Read-back insertion module. At first, an input transcribed communication is sent to the ***high-level entity parser*** in order to extract the knowledge, i.e., callsign, commands and values. Later, with a defined probability, a desired read-back error can be inserted

3. Datasets

This section describes the public databases used for training and evaluating the different modules of our virtual simulation-pilot system. In addition, Table 2 summarizes well-known private and public ATC-related databases [83]. Our goal is to conduct a thorough and comprehensive study on the use of virtual simulation-pilots in ATC. To ensure the reproducibility of our research, we use either only open-source databases or a combination of these and private databases during the training phase of our base models. The exception is the TTS module. The TTS system is a pre-trained out-of-the-box module downloaded from HuggingFace (the TTS is part of the response generator). Despite this, we aim at demonstrating the potential of the virtual simulation pilot in a more realistic setting. Hence, the system is also evaluated on highly challenging private databases that the authors have access to. These databases cover real-life ATC communications, which might contain high levels of background or cockpit noise. The results achieved in this work can provide a better idea of the performance of our approach in practical applications, while also highlighting its strengths and weaknesses in a real-world scenario. In any case, our focus remains on ensuring that our research is thoroughly documented and that it can be easily replicated by other researchers in the ATC and ATM field.

3.1. Public Databases

LDC-ATCC corpus: The air traffic control corpus (ATCC) (available for download in: https://catalog.ldc.upenn.edu/LDC94S14A, accessed on 12 May 2023) consists of recorded speech initially designed for research on ASR. Here, the metadata is also used for NLU research, e.g., speaker role detection. The audio data contain voice communication traffic between various ATCos and pilots. The audio files are sampled at 8 kHz, 16-bit linear, representing continuous monitoring without squelch or silence elimination. Each file has a single frequency over one to two hours of audio. The corpus contains gold annotations and metadata. The metadata cover voice activity segmentation details, speaker role information (who is talking) and callsigns in ICAO format. In addition, the corpus consists of approximately 25 h of ATCo and pilot transmissions (after VAD).

UWB-ATCC corpus: The UWB-ATCC corpus (released by the University of West Bohemia, see: https://lindat.mff.cuni.cz/repository/xmlui/handle/11858/00-097C-0000-0001-CCA1-0, accessed on 12 May 2023) is a free and public resource for research on ATC. It contains recordings of communication between ATCos and pilots. The speech is manually transcribed and labeled with the speaker information, i.e., pilot/controller. The total amount of speech after removing silences is 13 h. The audio data are mono-channel sampled at 8 kHz and 16-bit PCM.

ATCO2 corpus: The dataset was built for the development and evaluation of ASR and NLP technologies for English ATC communications. The dataset consists of English coming from several airports worldwide (e.g., LKTB, LKPR, LZIB, LSGS, LSZH, LSZB, YSSY). We used this corpus twofold. First, we employed up to 2500 h of audio data of the official pseudo-annotated set (see more information in [11]) for training our ASR systems; this training set is labeled *ATCO2-PL set corpus*. It is worth mentioning that the transcripts of the ATCO2 training corpus were automatically generated by an ASR system. Despite this, recent work has shown its potential to develop robust ASR systems for ATC from scratch, e.g., [11]. Second, for completeness, we use the two official partitions of the ATCO2 test set, namely, the *ATCO2 test set 1h corpus* and the *ATCO2 test set 4h corpus*, as evaluation sets. The first corpus contains 1.1 h of open-source transcribed annotations, and it can be accessed for free at: https://www.atco2.org/data (accessed on 12 May 2023). The latter contains $\sim$3 h of extra annotated data, and the full corpus is available for purchase through ELDA at: http://catalog.elra.info/en-us/repository/browse/ELRA-S0484 (accessed on 12 May 2023). The recordings are mono-channel sampled at 16 kHz and 16-bit PCM.

ATCOSIM corpus: This is a free public database for research on ATC communications. It comprises 10 h of speech data recorded during real-time ATC simulations using a close-talk headset microphone. The utterances are in the English language and pronounced by ten non-native speakers. The speakers are split by gender. Even though we do not pursue this direction, the ATCOSIM corpus can be used for the development or adaptation of already-existing TTS systems to ATC with voices from different genders, i.e., males or females. ATCOSIM also includes orthographic transcriptions and additional information about the speakers and recording sessions [32]. This dataset can be accessed for free at: https://www.spsc.tugraz.at/databases-and-tools (accessed on 12 May 2023).

Table 2. Air traffic control communications-related databases. * abbreviation in IETF format. † research directions that can be explored based on the annotations provided by the dataset. †† ASR and TTS are interchangeable; the same annotations of each recording can be used to fine-tune or train a TTS module. § denotes datasets that contain annotations on the callsign or/and command level. SpkID: Speaker role identification.

Characteristics			Research Topics †			Other	
Database	**Accents ***	**Hrs**	**ASR/TTS ††**	**SpkID**	**NLU §**	**License**	**Ref.**
			Private databases				
MALORCA	cs, de	14	✓	✓	✓	✗	[48]
AIRBUS	fr	100	✓	-	✓	✗	[83]
HAAWAII	is, en-GB	47	✓	✓	✓	✗	[62]
Internal	several	44	✓	✗	✗	✗	-
			Public databases				
ATCOSIM	de, fr, de-CH	10.7	✓	✗	✗	✓	[32]
UWB-ATCC	cs	13.2	✓	✓	✗	✓	[31]
LDC-ATCC	en-US	26.2	✓	✓	✓	✓	[30]
HIWIRE	fr, it, es, el	28.7	✓	✗	✗	✓	[84]
ATCO2	several	5285	✓	✓	✓	✓	[11]

3.2. *Private Databases*

MALORCA corpus: MAchine Learning Of speech Recognition models for Controller Assistance: http://www.malorca-project.de/wp/ (accessed on 12 May 2023). This dataset is based on a research project that focuses to propose a general, cheap and effective solution to develop and automate speech recognition for controllers using the speech data and contextual information. The data collected are mainly from the Prague and Vienna airports, which is around 14 h. The data are split into training and testing sets with a split amount of 10 h and 4 h (2 h from each airport), respectively.

HAAWAII corpus: Highly Advanced Air Traffic Controller Workstation with Artificial Intelligence Integration: https://www.haawaii.de (accessed on 12 May 2023): This dataset is based on an exploratory research project that aims to research and develop a reliable and adaptable solution to automatically transcribe voice commands issued by both ATCos and pilots. The controller and pilot conversations are obtained from two air navigation service providers (ANSPs): (i) NATS for London approach and (ii) ISAVIA for Icelandic en route. The total amount of manually transcribed data available is around 47 h (partitioned into 43 h for training and 4 h for testing). The 4 h test set is taken—2 h each—from both London and Iceland airports. Similar to another corpus, the audio files are sampled at 8 kHz and 16-bit PCM. This corpus is only used as an out-of-domain dataset; thus, we only report the results on the ASR level.

Internal data: In addition to the above-mentioned datasets, we have data from some industrial research projects that amount to a total duration of 44 h of speech recordings of ATCos and pilots along with their manual transcripts.

4. Experimental Setup and Results

In this section, we present the experimental results for some modules described in Section 2. These modules are trained with the datasets from Section 3. Note that not all datasets are used during the training and testing phases.

4.1. *Automatic Speech Recognition*

This subsection list the results related to ASR, previously covered in Section 2.1.1. We analyze (i) the three proposed ASR architectures, (ii) the training datasets used during the training phase, and (iii) the experimental setup and results obtained on different public and private test sets.

4.1.1. Architectures

The results of ASR are split in two. First, we evaluate hybrid-based ASR models, which are the default in current ATC-ASR research [8,9]. Second, we train ASR models with state-of-the-art end-to-end architectures, e.g., Transformer-based [27,40] and Conformer-based [85]. The experimental setup and results analysis (below) for each proposed model refers to the results from Table 3.

Hybrid-based ASR: For the hybrid-based ASR experiments, we use conventional biphone convolutional neural network (CNN) [86] + TDNN-F [46]-based acoustic models trained with the Kaldi [52] toolkit (i.e., nnet3 model architecture). The AMs are trained with the LF-MMI training framework, considered to produce a state-of-the-art performance for hybrid ASR. In all the experiments, 3-fold speed perturbation with MFCCs and i-vector features is used. The LM is trained as a statistical 3-gram model using manual transcripts. Previous work related to ATC with this architecture is in [11,15].

XLSR-KALDI ASR: In [87], the authors propose to use the LF-MMI criterion (similar to hybrid-based ASR) for the supervised adaptation of the self-supervised pre-trained XLSR model [27]. They also show that this approach outperforms the models trained with only the supervised data. Following that technique, we use the XLSR [27] model pre-trained with a dataset as large as 50 k h of speech data, and later we fine-tune it with the supervised ATC data using the LF-MMI criterion. Further details about the architecture and experimental setup for pre-training the XLSR model can be found in the original paper [27]. The results for this model are in the row tagged as '`XLSR-KALDI`' in Table 3.

End-to-End ASR: We use the SpeechBrain [88] toolkit to train a Conformer [85] ASR model with ATC audio data. The Conformer model is composed of 12 encoder layers and an additional 4 decoder layers (transformer-based [40]). We reuse the `Conformer-small` recipe from LibriSpeech [89] and adapt it to the ATC domain. See the recipe at: https://github.com/speechbrain/speechbrain/tree/develop/recipes/LibriSpeech/ASR/transformer (accessed on 12 May 2023). The dimension of the encoder and decoder model is set to $d_{model} = 144$ with $d_{ffn} = d_{model} * 4$. This accounts for a total of 11M parameters. We use dropout [90] with a probability of $dp = 0.1$ for the attention and hidden layers, while Gaussian error linear units (GELU) is used as the activation function [91]. We use the Adam [92] optimizer with an initial learning rate of $\gamma = 1\text{e}{-3}$. We also use the default dynamic batching, which speeds up the training. During training, we combine the per-frame conformer decoder output and CTC probabilities [93]. The CTC loss [94] is weighted by $\alpha = 0.3$. During inference and evaluation, the beam size is set to 66 with a CTC weight of $ctc_w = 0.4$.

4.1.2. Training and Test Data

Training data configuration: To see the effectiveness of using automatically transcribed data, as well as comparing the performance on the in-domain VS out-of-domain sets, we train both the hybrid (CNN-TDNNF) and E2E (Conformer) models twice. First, we employ a mix between public and private supervised (recordings with gold annotations) ATC resources, which comprises around 190 h. We tag these models as ***scenario (a)—only supervised data***. Second, we use a subset of 500 h of pseudo-annotated recordings (a seed ASR system is used to transcribe the ATC communications from different airports) from the open-source ATCO2-PL set corpus (see introductory paper [11]). We tag this model as ***scenario (b)—only ATCO2-PL 500 h data***. The results referencing both scenarios are in Tables 3 and 4.

Test data configuration: Six different test sets are used for ASR evaluation, as shown in Table 3. The first four test sets (highlighted in orange) are private and the last two test sets (highlighted in blue) are open data. Each two consecutive test sets are taken from one project: (i) NATS and ISAVIA are part of the HAAWAII corpus, (ii) Prague and Vienna are part of the MALORCA corpus, and (iii) ATCO2-1h and ATCO2-4h are from the ATCO2 project. Each dataset, along with the test split, is described in Section 3. We aimed at evaluating how the model's architecture, training paradigm (hybrid-based and E2E-ASR) and training data directly affect the performance of ASR.

4.1.3. Evaluation Metric

The preeminent technique for evaluating the efficacy of an ASR system is the word error rate (WER). This metric entails a meticulous comparison between the transcription of an utterance and the word sequence hypothesized by the ASR model. The WER is determined by computing the aggregate of three types of errors, specifically, substitutions (S), insertions (I) and deletions (D), over the total count of words within the transcription. Should the reference transcript comprise N words, the WER can be computed using Equation (1), outlined below.

$$WER = \frac{I + D + S}{N} \times 100. \quad (1)$$

We evaluate all the models from Tables 3 and 4 with WERs. For the boosting experiments (see Section 2.2.2 and Table 4) we additionally use *EntWER*, which evaluates WER only on the callsign word sequence, and *ACC*, which evaluates the accuracy of the system in capturing the target callsign in ICAO format.

4.1.4. Speech Recognition Results

The results of all the compared ASR models are in Table 3.

CNN-TDNNF model: This is our default architecture, as it has already been shown to work properly on the ATC domain. It has also been used largely in prior ATC work, such as ATCO2, HAAWAII and MALORCA (see Section 1). In our experiments, we trained this model for both *scenario (a)* and *scenario (b)*. Not surprisingly, we noted that the WERs are heavily impacted by the training and testing data. If we compare CNN-TDNNF scenario (a) VS scenario (b), we see a systematic drop in performance for NATS (7.5% WER → 26.7% WER) and ISAVIA (12.4% WER → 34.1% WER). However, in Prague and Vienna, which are still out-of-domain for scenario (b), less degradation in WERs is seen: 6.6% WER → 11.7% WER for Prague and 6.3% WER → 11.8% WER for Vienna.

XLSR-KALDI model: As mentioned earlier, we fine-tune the XLSR model (pre-trained based on wav2vec 2.0 [26]) with the supervised data from *scenario (a)*. We do this as a proof of concept. The results show that the performance is consistent over all the private test sets compared to the CNN-TDNNF model trained with the same data. Though the model has not seen the noisy ATCO2 data during fine-tuning, since this model is pre-trained with large amounts of data, the WER on the ATCO2 test sets significantly improves compared to the CNN-TDNNF model. We see an absolute improvement of 9.4% (27.4% → 18%) and 10.9% (36.6% → 25.7%) for the ATCO2-1h and ATCO2-4h test sets, respectively.

Conformer model: We evaluate Conformer [85], an encoder–decoder Transformer-based [40] model. With the Conformer architecture, we again train two models: on supervised data (*scenario (a)*) and on the 500 h ATCO2-PL set (*scenario (b)*). Both are tagged as *CONFORMER* in Table 3. Likewise, for CNN-TDNNF models, the first four test sets are deemed in-domain for the baseline model, whereas the last two test sets (ATCO2-1h and ATCO2-4h test sets) are considered out-of-domain. Conversely, the second model is optimized for the out-of-domain test sets, while the first four are considered out-of-domain. Our goal is to demonstrate the effectiveness of the *ATCO2-PL* dataset as an optimal resource for training models when only limited in-domain data are available. The second model demonstrates competitive performance when tested on close-mic speech datasets such as Prague and Vienna, which exclusively use the ATCo recordings. Yet, the model's performance deteriorates on more complex datasets, such as NATS and ISAVIA, which include pilot speech. We also note significant improvements on the ATCO2-1h and ATCO2-4h test sets when training with the ATCO2-PL dataset. Scenario (b) exhibits a 62% and 48% relative

WER reduction compared to scenario (a) on ATCO2-1h and ATCO2-4h, respectively. In contrast, the first model performed poorly on both: 41.8 and 46.2% WER, respectively. A critical consideration arises when examining the performance of the Conformer and CNN-TDNNF models under the same training scenario, scenario (b). Notably, the Conformer model outperforms the CNN-TDNNF model across all datasets, except for the Vienna test set. This leads us to hypothesize that the Conformer architecture shows greater proficiency when being trained over extensive datasets when compared to the CNN-TDNNF model in this particular scenario.

Table 3. WER for various public and private test sets with different ASR engines. The top results per block are **highlighted in bold**. The best result per test set is marked with an underline. ‡ datasets from HAAWAII corpus and †datasets from MALORCA project [8].

Model	Test Sets					
	NATS ‡	ISAVIA ‡	Prague †	Vienna †	ATCO2-1h	ATCO2-4h
	scenario (a)—only supervised data					
CNN-TDNNF	7.5	12.4	6.6	6.3	27.4	36.6
XLSR-KALDI	**7.1**	**12.0**	6.7	**5.5**	**18.0**	**25.7**
CONFORMER	9.5	13.7	**5.7**	7.0	41.8	46.2
	scenario (b)—only ATCO2-PL 500 h data					
CNN-TDNNF	26.7	34.1	11.7	**11.8**	19.1	25.1
CONFORMER	**21.6**	**32.5**	**7.6**	12.5	**15.9**	**24.0**

Callsign boosting with surveillance data (≈contextual biasing): The contextual biasing approach is introduced in Section 2.2.2. Table 4 demonstrates the effect of callsign boosting on the NATS test set (part of HAAWAII). The results of two ASR models are compared. Both models have the same architecture (Kaldi CNN-TDNNf) but are trained on different data. The first scenario (a), as in the experiments above, is trained on a combination of open-source and private annotated ATC databases that includes in-domain data (NATS); the second scenario (b) is trained on the 500 hours of automatically transcribed data collected and prepared for the ATCO2 project, which is out-of-domain data. As expected, the in-domain model performs better on the NATS dataset. At the same time, for both models, we can see a considerable improvement when contextual biasing is applied. The best results are achieved when only a ground-truth callsign is boosted, i.e., 86.7% → 96.1% ACC for scenario (a) and 39.9% → 70.0% ACC for scenario (b). As in real life, we usually do not have the ground-truth information, the improvement we can realistically obtain with radar data is shown in the line tagged as **N-grams**. The effectiveness of biasing also depends on the number of callsigns used to build the biasing FST, as the more false callsigns are boosted the noisier the final rescoring is. According to previous findings, the ideal size of the biasing FST for improving performance depends on the data, but typically, the performance begins to decline when there are more than 1000 contextual entities [95]. In our data, we have an average of 200 contextual entities per spoken phrase. For the n-gram-boosting experiments, we achieved a relative improvement in callsign WERs of 51.2% and 34% for callsign recognition with models (a) and (b), and 9.5% and 12.4% for the entire utterance, respectively (see Table 4).

Table 4. Results for boosting on NATS test set corpus (HAAWAII). We ablate two models: scenario (a), a general ATC model trained only on supervised data, and scenario (b), a model trained on the ATCO2-PL 500 h set. Results are obtained with offline CPU decoding. ¶ word error rates only on the sequence of words that compose the callsign in the utterance.

Boosting	General ATC Model			ATCO2 Model-500h		
	WER	EntWER ¶	ACC	WER	EntWER ¶	ACC
	scenario (a)—only supervised data			scenario (b)—only ATCO2-PL 500 h data		
Baseline	7.4	4.1	86.7	26.7	30.0	39.9
Unigrams	7.4	3.6	88.0	25.6	24.1	46.2
N-grams	6.7	2.0	93.3	23.4	19.8	61.3
GT boosted	6.4	1.3	96.1	22.0	16.2	70.0

4.2. High-Level Entity Parser

A NER system is trained to parse text into high-level entities relevant to ATC communications. The NER module (or tagger) is depicted in Figure 3. First, a BERT [28] model is downloaded from HuggingFace [96,97] which is then fine-tuned on the NER task with 3k sentences (∼3 h of speech) using the *ATCO2 test set corpus* (the pre-trained version of `BERT-base-uncased` with 110 million parameters is used, see at: https://huggingface.co/bert-base-uncased, accessed on 12 May 2023). In this corpus, each word has a tag that corresponds to either callsign, command, values or UNK (everything else). The final layer of the BERT model is replaced by a linear layer with a dimension of 8 (this setup follows the class structures from Section 3.3 of the paper: [13], i.e., two outputs for each class). As only 3k sentences are used, a 5-fold cross-validation is conducted to avoid overfitting. Further details about experimentation are covered in [10]. We redirect the reader to the public and open-source GitHub repository of the ATCO2 corpus (ATCO2 GitHub repository: https://github.com/idiap/atco2-corpus, accessed on 12 May 2023).

Experimental setup: we fine-tune each model on an NVIDIA GeForce RTX 3090 for 10k steps. During experimentation, we use the same learning rate of $\gamma = 5\text{e}{-5}$, with a linear learning rate scheduler. The dropout [90] is set to $dp = 0.1$ for the attention and hidden layers, while GELU is used as an activation function [91]. We also employ gradient norm clipping [98]. We fine-tune each model with an effective batch size of 32 for 50 epochs with the AdamW optimizer [99] ($\beta_1 = 0.9, \beta_2 = 0.999, \epsilon = 1\text{e}{-8}$).

Evaluation metric: we evaluate our NER system with the F-score. The F-score or F-measure is a statistical measure utilized in binary classification analysis to evaluate a test's accuracy. The F1-score, defined in Equation (4), represents the harmonic mean of precision and recall. Precision, as described in Equation (2), is the ratio of true positive (TP) results to all positive results (including false positives (FP)), while recall, as defined in Equation (3), is the ratio of TP to all samples that should have been identified as positive (including false negatives (FN)):

$$Precision = \frac{TP}{TP + FP} \tag{2}$$

$$Recall = \frac{TP}{TP + FN} \tag{3}$$

$$F1 = \frac{2 * Precision * Recall}{Precision + Recall} = \frac{2 * TP}{2 * TP + FP + FN} \tag{4}$$

Results: the ***high-level entity parser*** system is evaluated on the only available public resource, the ATCO2-4h test set, which contains word-level tags, i.e., callsign, command and values. The results for precision, recall and F1-score over each of the proposed classes are listed in Table 5. Our BERT-based system achieves a high level of performance in callsign detection, with an F1-score of 97.5%. However, the command and values classes show an average worse performance, with F1-scores of 82.0% and 87.2%, respectively. Notably, the command class presents the greatest challenge due to its inherent complexity when compared to values and callsigns. Values are predominantly composed of keywords, such as "flight level" followed by cardinal numbers such as "two", "three hundred" or "one thousand". These characteristics make them easy for a system to recognize. Similarly, callsigns are highly structured, consisting of an airline designator accompanied by numbers and letters spoken in the radiotelephony alphabet [21]. Given their importance in communication, as in callsign highlighting [11] or read-back error detection [81], additional validation is necessary for real-life scenarios or when working with proprietary/private data.

Although our BERT-based system achieves a high performance in callsign recognition, there is still room for improvement. One potential method for enhancing the performance is to incorporate real-time surveillance data into the system [10], which was introduced in Section 2.2.2 and Table 4.

4.3. Response Generator

The response generator is composed of three submodules (see Section 2.1.3). The **grammar conversion** and **word fixer** submodules are based on hard-coded rules. Thus, quantitative results are not provided.

4.4. Text to Speech

Text-to-speech (TTS) is a technology that facilitates the conversion of written text into spoken language. When employed in ATC, TTS can be integrated with virtual simulation-pilot systems to train ATCos. In this study, a state-of-the-art non-autoregressive speech synthesizer, the FastSpeech2 model [29], is utilized. A pre-trained TTS model is downloaded from the HuggingFace hub [96]. Access to the model can be obtained at https://huggingface.co/facebook/fastspeech2-en-ljspeech, accessed on 12 May 2023. FastSpeech2 is used in inference mode with the sentence produced by the grammar conversion submodule, and subsequently, the word fixer submodule, serving as the prompt of the virtual simulation-pilot. Other models, such as Tacotron [72] or Tacotron2 [73] (free access to the Tacotron2 model can be found at https://huggingface.co/speechbrain/tts-tacotron2-ljspeech, accessed on 12 May 2023), can be fine-tuned and implemented to handle ATC data.

System analysis: In our experiments, we discovered that the model is capable of handling complex word sequences, such as those commonly encountered in ATC, including read-backs from virtual simulation-pilots that contain multiple commands and values. However, we did not conduct any qualitative analysis of the TTS-produced voice or speech, leaving this as a future area of exploration. We also did not investigate the possibility of fine-tuning the TTS module with ATC audio data, as our main focus was on developing a simple and effective virtual simulation-pilot system using pre-existing open-source models.

Future lines of work: Although we did not pursue this area, it is indeed possible to fine-tune the TTS module using in-domain ATC data. In Table 2, we provide a list of both public and private databases that could be utilized for this purpose. Generally, the same annotations used for ASR can also be applied to fine-tune a TTS system. However, there are two different approaches that could be explored simultaneously or sequentially. First, by considering the speaker role (ATCo or pilot), the TTS module could be biased to produce speech that is more appropriate for different roles, such as noisy speech from pilots. Second, if datasets are available that provide information on gender and accents, such as the ATCOSIM dataset [32], TTS models with different accents and gender could be developed.

Table 5. F1-score (@F1), precision (@P) and recall (@R) metrics for callsign, command and value classes of the ***high-level entity parser*** system. Results are averaged over a 5-fold cross-validation scheme on the *ATCO2-4h corpus* in order to mitigate overfitting. We run fine-tuning five times with different training seeds (2222/3333/4444/5555/6666).

Model	Callsign			Command			Values		
	@P	**@R**	**@F1**	**@P**	**@R**	**@F1**	**@P**	**@R**	**@F1**
`bert-base-uncased`	97.1	97.8	97.5	80.4	83.6	82.0	86.3	88.1	87.2

5. Limitations and Future Work

In our investigation of ASR systems, we have explored the potential of hybrid-based and E2E ASR systems, which can be further enhanced by incorporating data relevant to the specific exercise undertaken by ATCo trainees, such as runway numbers or waypoint lists. Moving forward, we suggest that research should continue to explore E2E training techniques for ASR, as well as methods for integrating contextual data into these E2E systems.

The repetition generator currently in use employs a simple grammar converter and a pre-trained TTS system. However, we believe that additional efforts could be made to enhance the system's ability to convey more complex ATC communications to virtual simulation-pilots. In particular, the TTS system could be fine-tuned to produce female or male voices, as well as modify key features such as the speech rate, noise artifacts or cues to synthesize voices in a stressful situation. Additionally, a quantitative metric for evaluating the TTS system could be integrated to further enhance its efficacy. We also list some optional modules (see Section 2.2) that can be further explored, e.g., the read-back insertion module or voice activity detection.

Similarly, there is scope for the development of multimodal and multitask systems. Such systems would be fed with real-time ATC communications and contextual data simultaneously, later generating transcripts and high-level entities as the output. Such systems could be considered a dual ASR and high-level entity parser. Finally, the legal and ethical challenges of using ATC audio data are another important field that needs to be further explored in future work. We redirect the reader to the **legal and privacy aspects for collection of ATC recordings** section in [11].

6. How to Develop Your Own Virtual Simulation-Pilot

If you would like to replicate this work with in-domain data, i.e., for a specific scenario or airport, you can follow the steps below:

1. Start by defining the set of rules and grammar to use for the annotation protocol. You can follow the cheat-sheet from the ATCO2 project [11]. See https://www.spokendata.com/atc and https://www.atco2.org/, accessed on 12 May 2023. In addition, one can use previous ontologies developed for ATC [17].
2. For training or adapting the virtual simulation-pilot engine, you need three sets of annotations: (i) gold annotations of the ATCo–pilot communications for ASR adaptation; (ii) high-level entity annotations (callsign, command and values) to perform NLU for ATC; and (iii) a set of rules to convert ATCo commands into pilots read-backs, e.g., "descend to" → "descending to".
3. Gather and annotate at least 1 hour of ATCo speech and 1k samples for training your ***high-level entity parser*** system. If the reader is interested in obtaining a general idea of how much data are needed for reaching a desired WER or F1-score, see [62] for ASR and [11] for ATC-NLU.
4. Fine-tune a strong pre-trained ASR model, e.g., Wav2Vec 2.0 or XLSR [26,27] with the ATC audio recordings. For instance, if the performance is not sufficient, you can use open-source corpora (see Table 2) to increase the amount of annotated samples (see [11,30–32]). We recommend acquiring the ATCO2-PL dataset [11], which has

proven to be a good starting point when no in-domain data are available. This is related to ASR and NLU for ATC.

5. Fine-tune a strong pre-trained NLP model, e.g., BERT [28] or RoBERTa [67], with the NLP tags. If the performance is not sufficient, one can follow several text-based data-augmentation techniques. For example, it is possible to replace the callsign in one training sample with different ones from a predefined callsign list. In that way, one can generate many more valuable training samples. It is also possible to use more annotations during fine-tuning, e.g., see the `ATCO2-4h` corpus [11].
6. Lastly, in case you need to adapt the TTS module to pilot speech, you could adapt the FastSpeech2 [29] system. Then, you need to invert the annotations used for ASR, i.e., using the transcripts as input and the ATCo or pilot recordings as targets. This step is not strictly necessary, as already-available pre-trained modules possess a good quality.

7. Conclusions

In this paper, we have presented a novel virtual simulation-pilot system designed for ATCo training. Our system utilizes cutting-edge open-source ASR, NLP and TTS systems. To the best of our knowledge, this is the first such system that relies on open-source ATC resources. The virtual simulation-pilot system is developed for ATCo training purposes; thus, this work represents an important contribution to the field of aviation training.

Our system employs a multi-stage approach, including ASR transcription, a ***high-level entity parser*** system and a repetition-generator module to provide pilot-like responses to ATC communications. By utilizing open-source AI models and public databases, we have developed a simple and efficient system that can be easily replicated and adapted for different training scenarios. For instance, we tested our ASR system on different well-known ATC-related projects, i.e., HAAWAII, MALORCA and ATCO2. We reached as low as 5.5% WER on high-quality data (MALORCA, ATCo speech in operations room) and 15.9% WER on low-quality ATC audio such as the test sets from the ATCO2 project (noise levels below 15 dB).

Going forward, there is significant potential for further improvements and expansions to the proposed system. Incorporating contextual data, such as runway numbers or waypoint lists, could enhance the accuracy and effectiveness of the ASR and high-level entity parser modules. In this work, we evaluated the introduction of real-time surveillance data, which proved to further improve the system's performance in recognizing and responding to ATC communications. For instance, our boosting technique brings a 9% absolute amelioration in callsign-detection accuracy levels (86.7% $\rightarrow$ 96.1%) for the NATS test set. It is also important to recall that additional efforts could be made to fine-tune the TTS system for the improved synthesis of male or female voices, as well as modifying the speech rate, noise artifacts and other features.

The proposed ASR system can reach as low as 5.5% and 15.9% word error rates (WERs) on high- and low-quality ATC audio (Vienna and ATCO2-test-set-1h, respectively). It is also proven that adding surveillance data to the ASR can yield a callsign detection accuracy of more than 96%. Overall, this work represents a promising first step towards developing advanced virtual simulation-pilot systems for ATCo training, and it is expected that future work will continue to explore this research direction.

Author Contributions: Conceptualization, J.Z.-G., A.P., I.N. and P.M.; Data curation, J.Z.-G., A.P. and I.N.; Formal analysis, J.Z.-G., I.N. and M.K.; Funding acquisition, P.M.; Investigation, J.Z.-G., A.P. and I.N.; Methodology, J.Z.-G., A.P., I.N., P.M. and M.K.; Project administration, P.M.; Resources, J.Z.-G. and A.P.; Software, J.Z.-G., A.P., I.N. and P.M.; Supervision, P.M.; Validation, J.Z.-G. and I.N.; Visualization, J.Z.-G.; Writing—original draft, J.Z.-G., I.N. and M.K.; Writing—review and editing, J.Z.-G., P.M. and M.K. All authors have read and agreed to the published version of the manuscript.

Funding: Ideas presented in paper are partially related to needs on simulator pilot operator services described by Eurocontrol in 2020: call for tenders No.: 20-220319-A, title: Simulator Pilot Operator in support to Real-Time Simulations at: https://www.eurocontrol.int/sites/default/files/2020-08/20-220319-a.pdf, accessed on 12 May 2023. This work was also partly supported by SESAR Joint Undertaking under Grant Agreement No. 884287—HAAWAII (**H**ighly **A**utomated **A**ir traffic controller **W**orkstations with **A**rtificial **I**ntelligence **I**ntegration).

Data Availability Statement: Private and public databases are used in this paper. They are covered in detail in Table 2.

Conflicts of Interest: The authors declare no conflict of interest.

Abbreviations

The following abbreviations are used in this manuscript:

ASR	Automatic Speech Recognition
NLP	Natural Language Processing
ATCo	Air Traffic Controller
ATC	Air Traffic Control
CPDLC	Controller-Pilot Data Link Communications
AI	Artificial Inteligencce
WER	Word Error Rate
ML	Machine Learning
VAD	Voice Activity Detection
ATM	Air Traffic Management
TTS	Text-To-Speech
NLU	Natural Language Understanding
PTT	Push-To-Talk
LM	Language Model
AM	Acoustic Model
WFST	Weighted Finite State Transducer
FST	Finite State Transducer
HMM	Hidden Markov Models
DNN	Deep Neural Networks
MFCCs	Mel-frequency Cepstral Coefficients
LF MMI	Lattice Free Maximum Mutual Information
VHF	Very-High Frequency
E2E	End-To-End
CTC	Connectionist Temporal Classification
NER	Named Entity Recognition
RG	Response Generator
ICAO	International Civil Aviation Organization
SNR	Signal-To-Noise
dB	Decibel
RBE	Read-back Error
ATCC	Air Traffic Control Corpus
ELDA	European Language Resources Association
ANSPs	Air Navigation Service Providers
CNN	Convolutional Neural Network
GELU	Gaussian Error Linear Units
Conformer	Convolution-augmented Transformer

References

1. Nassif, A.B.; Shahin, I.; Attili, I.; Azzeh, M.; Shaalan, K. Speech recognition using deep neural networks: A systematic review. *IEEE Access* **2019**, *7*, 19143–19165. [CrossRef]
2. Otter, D.W.; Medina, J.R.; Kalita, J.K. A survey of the usages of deep learning for natural language processing. *IEEE Trans. Neural Netw. Learn. Syst.* **2020**, *32*, 604–624. [CrossRef] [PubMed]
3. Zhang, L.; Wang, S.; Liu, B. Deep learning for sentiment analysis: A survey. *Wiley Interdiscip. Rev. Data Min. Knowl. Discov.* **2018**, *8*, e1253. [CrossRef]

4. Lugosch, L.; Ravanelli, M.; Ignoto, P.; Tomar, V.S.; Bengio, Y. Speech Model Pre-Training for End-to-End Spoken Language Understanding. In Proceedings of the Interspeech 2019, 20th Annual Conference of the International Speech Communication Association, Graz, Austria, 15–19 September 2019; pp. 814–818. [CrossRef]
5. Beek, B.; Neuberg, E.; Hodge, D. An assessment of the technology of automatic speech recognition for military applications. *IEEE Trans. Acoust. Speech Signal Process.* **1977**, *25*, 310–322. [CrossRef]
6. Hamel, C.J.; Kotick, D.; Layton, M. *Microcomputer System Integration for Air Control Training*; Technical Report; Naval Training Systems Center: Orlando, FL, USA, 1989.
7. Matrouf, K.; Gauvain, J.; Neel, F.; Mariani, J. Adapting probability-transitions in DP matching processing for an oral task-oriented dialogue. In Proceedings of the International Conference on Acoustics, Speech, and Signal Processing (ICASSP), Albuquerque, NM, USA, 3–6 April 1990; pp. 569–572.
8. Helmke, H.; Ohneiser, O.; Mühlhausen, T.; Wies, M. Reducing controller workload with automatic speech recognition. In Proceedings of the 2016 IEEE/AIAA 35th Digital Avionics Systems Conference (DASC), Sacramento, CA, USA, 25–29 September 2016; pp. 1–10.
9. Helmke, H.; Ohneiser, O.; Buxbaum, J.; Kern, C. Increasing ATM efficiency with assistant based speech recognition. In Proceedings of the 13th USA/Europe Air Traffic Management Research and Development Seminar, Seattle, WA, USA, 27–30 June 2017.
10. Nigmatulina, I.; Zuluaga-Gomez, J.; Prasad, A.; Sarfjoo, S.S.; Motlicek, P. A two-step approach to leverage contextual data: Speech recognition in air-traffic communications. In Proceedings of the ICASSP, Singapore, 23–27 May 2022.
11. Zuluaga-Gomez, J.; Veselý, K.; Szöke, I.; Motlicek, P.; Kocour, M.; Rigault, M.; Choukri, K.; Prasad, A.; Sarfjoo, S.S.; Nigmatulina, I.; et al. ATCO2 corpus: A Large-Scale Dataset for Research on Automatic Speech Recognition and Natural Language Understanding of Air Traffic Control Communications. *arXiv* **2022**, arXiv:2211.04054.
12. Kocour, M.; Veselý, K.; Blatt, A.; Zuluaga-Gomez, J.; Szöke, I.; Cernocký, J.; Klakow, D.; Motlícek, P. Boosting of Contextual Information in ASR for Air-Traffic Call-Sign Recognition. In Proceedings of the Interspeech 2021, 22nd Annual Conference of the International Speech Communication Association, Brno, Czechia, 30 August–3 September 2021; pp. 3301–3305. [CrossRef]
13. Zuluaga-Gomez, J.; Sarfjoo, S.S.; Prasad, A.; Nigmatulina, I.; Motlicek, P.; Ondre, K.; Ohneiser, O.; Helmke, H. BERTraffic: BERT-based Joint Speaker Role and Speaker Change Detection for Air Traffic Control Communications. In Proceedings of the 2022 IEEE Spoken Language Technology Workshop (SLT), Doha, Qatar, 9–12 January 2023.
14. Lin, Y.; Li, Q.; Yang, B.; Yan, Z.; Tan, H.; Chen, Z. Improving speech recognition models with small samples for air traffic control systems. *Neurocomputing* **2021**, *445*, 287–297. [CrossRef]
15. Zuluaga-Gomez, J.; Veselý, K.; Blatt, A.; Motlicek, P.; Klakow, D.; Tart, A.; Szöke, I.; Prasad, A.; Sarfjoo, S.; Kolčárek, P.; et al. Automatic call sign detection: Matching air surveillance data with air traffic spoken communications. *Proceedings* **2020**, *59*, 14.
16. Fan, P.; Guo, D.; Lin, Y.; Yang, B.; Zhang, J. Speech recognition for air traffic control via feature learning and end-to-end training. *arXiv* **2021**, arXiv:2111.02654.
17. Helmke, H.; Slotty, M.; Poiger, M.; Herrer, D.F.; Ohneiser, O.; Vink, N.; Cerna, A.; Hartikainen, P.; Josefsson, B.; Langr, D.; et al. Ontology for transcription of ATC speech commands of SESAR 2020 solution PJ. 16-04. In Proceedings of the IEEE/AIAA 37th Digital Avionics Systems Conference (DASC), London, UK, 23–27 September 2018; pp. 1–10.
18. Guo, D.; Zhang, Z.; Yang, B.; Zhang, J.; Lin, Y. Boosting Low-Resource Speech Recognition in Air Traffic Communication via Pretrained Feature Aggregation and Multi-Task Learning. In *IEEE Transactions on Circuits and Systems II: Express Briefs*; IEEE: Piscataway, NJ, USA, 2023.
19. Fan, P.; Guo, D.; Zhang, J.; Yang, B.; Lin, Y. Enhancing multilingual speech recognition in air traffic control by sentence-level language identification. *arXiv* **2023**, arXiv:2305.00170 .
20. Guo, D.; Zhang, Z.; Fan, P.; Zhang, J.; Yang, B. A context-aware language model to improve the speech recognition in air traffic control. *Aerospace* **2021**, *8*, 348. [CrossRef]
21. International Civil Aviation Organization. *ICAO Phraseology Reference Guide*; ICAO: Montreal, QC, Canada, 2020.
22. Bouchal, A.; Had, P.; Bouchaudon, P. The Design and Implementation of Upgraded ESCAPE Light ATC Simulator Platform at the CTU in Prague. In Proceedings of the 2022 New Trends in Civil Aviation (NTCA), Prague, Czech Republic, 26–27 October 2022; pp. 103–108.
23. Lin, Y. Spoken instruction understanding in air traffic control: Challenge, technique, and application. *Aerospace* **2021**, *8*, 65. [CrossRef]
24. Lin, Y.; Wu, Y.; Guo, D.; Zhang, P.; Yin, C.; Yang, B.; Zhang, J. A deep learning framework of autonomous pilot agent for air traffic controller training. *IEEE Trans. Hum.-Mach. Syst.* **2021**, *51*, 442–450. [CrossRef]
25. Prasad, A.; Zuluaga-Gomez, J.; Motlicek, P.; Sarfjoo, S.; Nigmatulina, I.; Vesely, K. Speech and Natural Language Processing Technologies for Pseudo-Pilot Simulator. In Proceedings of the 12th SESAR Innovation Days, Budapest, Hungary, 5–8 December 2022.
26. Baevski, A.; Zhou, Y.; Mohamed, A.; Auli, M. wav2vec 2.0: A Framework for Self-Supervised Learning of Speech Representations. In Proceedings of the Advances in Neural Information Processing Systems 33: Annual Conference on Neural Information Processing Systems 2020, NeurIPS 2020, Virtual, 6–12 December 2020.
27. Conneau, A.; Baevski, A.; Collobert, R.; Mohamed, A; Auli, M. Unsupervised cross-lingual representation learning for speech recognition. *arXiv* **2021**, arXiv:2006.13979.

28. Devlin, J.; Chang, M.W.; Lee, K.; Toutanova, K. BERT: Pre-training of Deep Bidirectional Transformers for Language Understanding. In *Proceedings of the 2019 Conference of the North American Chapter of the Association for Computational Linguistics: Human Language Technologies, Volume 1 (Long and Short Papers)*; Association for Computational Linguistics: Minneapolis, MN, USA, 2019; pp. 4171–4186. [CrossRef]
29. Ren, Y.; Hu, C.; Tan, X.; Qin, T.; Zhao, S.; Zhao, Z.; Liu, T. FastSpeech 2: Fast and High-Quality End-to-End Text to Speech. In Proceedings of the 9th International Conference on Learning Representations, ICLR 2021, Virtual Event, Austria, 3–7 May 2021.
30. Godfrey, J. *The Air Traffic Control Corpus (ATC0)—LDC94S14A*; Linguistic Data Consortium: Philadelphia, PA, USA, 1994. .
31. Šmídl, L.; Švec, J.; Tihelka, D.; Matoušek, J.; Romportl, J.; Ircing, P. Air traffic control communication (ATCC) speech corpora and their use for ASR and TTS development. *Lang. Resour. Eval.* **2019**, *53*, 449–464. [CrossRef]
32. Hofbauer, K.; Petrik, S.; Hering, H. The ATCOSIM Corpus of Non-Prompted Clean Air Traffic Control Speech. In *Proceedings of the Sixth International Conference on Language Resources and Evaluation (LREC'08)*; European Language Resources Association (ELRA): Marrakech, Morocco, 2008.
33. Pavlinović, M.; Juričić, B.; Antulov-Fantulin, B. Air traffic controllers' practical part of basic training on computer based simulation device. In Proceedings of the International Convention on Information and Communication Technology, Electronics and Microelectronics (MIPRO), Opatija, Croatia, 22–26 May 2017; pp. 920–925.
34. Juričić, B.; Varešak, I.; Božić, D. The role of the simulation devices in air traffic controller training. In Proceedings of the International Symposium on Electronics in Traffic, ISEP 2011 Proceedings, Berlin, Germany, 26–28 September 2011.
35. Chhaya, B.; Jafer, S.; Coyne, W.B.; Thigpen, N.C.; Durak, U. Enhancing scenario-centric air traffic control training. In Proceedings of the 2018 AIAA modeling and Simulation Technologies Conference, Kissimmee, FL, USA, 8–12 January 2018; p. 1399.
36. Updegrove, J.A.; Jafer, S. Optimization of air traffic control training at the Federal Aviation Administration Academy. *Aerospace* **2017**, *4*, 50. [CrossRef]
37. Eide, A.W.; Ødegård, S.S.; Karahasanović, A. A post-simulation assessment tool for training of air traffic controllers. In *Human Interface and the Management of Information. Information and Knowledge Design and Evaluation*; Springer: Cham, Switzerland, 2014; pp. 34–43.
38. Némethová, H.; Bálint, J.; Vagner, J. The education and training methodology of the air traffic controllers in training. In Proceedings of the International Conference on Emerging eLearning Technologies and Applications (ICETA), Starý Smokovec, Slovakia, 21–22 November 2019; pp. 556–563.
39. Zhang, J.; Zhang, P.; Guo, D.; Zhou, Y.; Wu, Y.; Yang, B.; Lin, Y. Automatic repetition instruction generation for air traffic control training using multi-task learning with an improved copy network. *Knowl.-Based Syst.* **2022**, *241*, 108232. [CrossRef]
40. Vaswani, A.; Shazeer, N.; Parmar, N.; Uszkoreit, J.; Jones, L.; Gomez, A.N.; Kaiser, L.; Polosukhin, I. Attention is All you Need. In Proceedings of the Advances in Neural Information Processing Systems 30: Annual Conference on Neural Information Processing Systems 2017, Long Beach, CA, USA, 4–9 December 2017; pp. 5998–6008.
41. Mohri, M.; Pereira, F.; Riley, M. Weighted finite-state transducers in speech recognition. *Comput. Speech Lang.* **2002**, *16*, 69–88. [CrossRef]
42. Mohri, M.; Pereira, F.; Riley, M. Speech recognition with weighted finite-state transducers. In *Springer Handbook of Speech Processing*; Springer: Berlin/Heidelberg, Germany, 2008; pp. 559–584.
43. Riley, M.; Allauzen, C.; Jansche, M. OpenFst: An Open-Source, Weighted Finite-State Transducer Library and its Applications to Speech and Language. In *Proceedings of Human Language Technologies: The 2009 Annual Conference of the North American Chapter of the Association for Computational Linguistics, Companion Volume: Tutorial Abstracts*; Association for Computational Linguistics: Boulder, CO, USA, 2009; pp. 9–10.
44. Veselý, K.; Ghoshal, A.; Burget, L.; Povey, D. Sequence-discriminative training of deep neural networks. *Interspeech* **2013**, *2013*, 2345–2349.
45. Bourlard, H.A.; Morgan, N. *Connectionist Speech Recognition: A Hybrid Approach*; Springer Science & Business Media: Berlin/Heidelberg, Germany, 1993; Volume 247.
46. Povey, D.; Cheng, G.; Wang, Y.; Li, K.; Xu, H.; Yarmohammadi, M.; Khudanpur, S. Semi-Orthogonal Low-Rank Matrix Factorization for Deep Neural Networks. In Proceedings of the Interspeech 2018, 19th Annual Conference of the International Speech Communication Association, Hyderabad, India, 2–6 September 2018; pp. 3743–3747. [CrossRef]
47. Zuluaga-Gomez, J.; Motlicek, P.; Zhan, Q.; Veselý, K.; Braun, R. Automatic Speech Recognition Benchmark for Air-Traffic Communications. *Proc. Interspeech* **2020**, 2297–2301. [CrossRef]
48. Srinivasamurthy, A.; Motlícek, P.; Himawan, I.; Szaszák, G.; Oualil, Y.; Helmke, H. Semi-Supervised Learning with Semantic Knowledge Extraction for Improved Speech Recognition in Air Traffic Control. In Proceedings of the Interspeech 2017, 18th Annual Conference of the International Speech Communication Association, Stockholm, Sweden, 20–24 August 2017; pp. 2406–2410.
49. Zuluaga-Gomez, J.; Nigmatulina, I.; Prasad, A.; Motlicek, P.; Veselý, K.; Kocour, M.; Szöke, I. Contextual Semi-Supervised Learning: An Approach to Leverage Air-Surveillance and Untranscribed ATC Data in ASR Systems. *Proc. Interspeech* **2021**, 3296–3300. [CrossRef]
50. Kocour, M.; Veselý, K.; Szöke, I.; Kesiraju, S.; Zuluaga-Gomez, J.; Blatt, A.; Prasad, A.; Nigmatulina, I.; Motlíček, P.; Klakow, D.; et al. Automatic processing pipeline for collecting and annotating air-traffic voice communication data. *Eng. Proc.* **2021**, *13*, 8.
51. Chen, S.; Kopald, H.; Avjian, B.; Fronzak, M. Automatic Pilot Report Extraction from Radio Communications. In Proceedings of the 2022 IEEE/AIAA 41st Digital Avionics Systems Conference (DASC), Portsmouth, VA, USA, 18–22 September 2022; pp. 1–8.

52. Povey, D.; Ghoshal, A.; Boulianne, G.; Burget, L.; Glembek, O.; Goel, N.; Hannemann, M.; Motlicek, P.; Qian, Y.; Schwarz, P.; et al. The Kaldi speech recognition toolkit. In Proceedings of the IEEE Workshop on Automatic Speech Recognition and Understanding, Waikoloa, HI, USA, 11–15 December 2011.
53. Madikeri, S.; Tong, S.; Zuluaga-Gomez, J.; Vyas, A.; Motlicek, P.; Bourlard, H. Pkwrap: A pytorch package for lf-mmi training of acoustic models. *arXiv* **2020**, arXiv:2010.03466.
54. Graves, A.; Jaitly, N. Towards End-To-End Speech Recognition with Recurrent Neural Networks. In Proceedings of the 31th International Conference on Machine Learning, ICML 2014, Beijing, China, 21–26 June 2014; Volume 32, pp. 1764–1772.
55. Chorowski, J.; Bahdanau, D.; Serdyuk, D.; Cho, K.; Bengio, Y. Attention-Based Models for Speech Recognition. In Proceedings of the Advances in Neural Information Processing Systems 28: Annual Conference on Neural Information Processing Systems 2015, Montreal, QC, Canada, 7–12 December 2015; pp. 577–585.
56. Schneider, S.; Baevski, A.; Collobert, R.; Auli, M. wav2vec: Unsupervised Pre-Training for Speech Recognition. In Proceedings of the Interspeech 2019, 20th Annual Conference of the International Speech Communication Association, Graz, Austria, 15–19 September 2019; pp. 3465–3469. [CrossRef]
57. Baevski, A.; Mohamed, A. Effectiveness of Self-Supervised Pre-Training for ASR. In Proceedings of the 2020 IEEE International Conference on Acoustics, Speech and Signal Processing, ICASSP 2020, Barcelona, Spain, 4–8 May 2020; pp. 7694–7698. [CrossRef]
58. Zhang, Z.Q.; Song, Y.; Wu, M.H.; Fang, X.; Dai, L.R. Xlst: Cross-lingual self-training to learn multilingual representation for low resource speech recognition. *arXiv* **2021**, arXiv:2103.08207.
59. Chen, S.; Wang, C.; Chen, Z.; Wu, Y.; Liu, S.; Chen, Z.; Li, J.; Kanda, N.; Yoshioka, T.; Xiao, X.; et al. Wavlm: Large-scale self-supervised pre-training for full stack speech processing. *arXiv* **2021**, arXiv:2110.13900.
60. Oord, A.v.d.; Li, Y.; Vinyals, O. Representation learning with contrastive predictive coding. *arXiv* **2018**, arXiv:1807.03748.
61. Baevski, A.; Schneider, S.; Auli, M. vq-wav2vec: Self-Supervised Learning of Discrete Speech Representations. In Proceedings of the 8th International Conference on Learning Representations, ICLR 2020, Addis Ababa, Ethiopia, 26–30 April 2020.
62. Zuluaga-Gomez, J.; Prasad, A.; Nigmatulina, I.; Sarfjoo, S.; Motlicek, P.; Kleinert, M.; Helmke, H.; Ohneiser, O.; Zhan, Q. How Does Pre-trained Wav2Vec2.0 Perform on Domain Shifted ASR? An Extensive Benchmark on Air Traffic Control Communications. In Proceedings of the IEEE Spoken Language Technology Workshop (SLT), Doha, Qatar, 9–12 January 2023.
63. Yadav, V.; Bethard, S. A Survey on Recent Advances in Named Entity Recognition from Deep Learning models. In Proceedings of the 27th International Conference on Computational Linguistics, Santa Fe, NM, USA, 20–26 August 2018; Association for Computational Linguistics: Santa Fe, NM, USA, 2018; pp. 2145–2158.
64. Grishman, R.; Sundheim, B. Message Understanding Conference- 6: A Brief History. In Proceedings of the COLING 1996 Volume 1: The 16th International Conference on Computational Linguistics, Copenhagen, Denmark, 5–9 August 1996 .
65. Collobert, R.; Weston, J.; Bottou, L.; Karlen, M.; Kavukcuoglu, K.; Kuksa, P. Natural language processing (almost) from scratch. *J. Mach. Learn. Res.* **2011**, *12*, 2493–2537.
66. Piskorski, J.; Pivovarova, L.; Šnajder, J.; Steinberger, J.; Yangarber, R. The First Cross-Lingual Challenge on Recognition, Normalization, and Matching of Named Entities in Slavic Languages. In *Proceedings of the 6th Workshop on Balto-Slavic Natural Language Processing*; Association for Computational Linguistics: Valencia, Spain, 2017; pp. 76–85. [CrossRef]
67. Liu, Y.; Ott, M.; Goyal, N.; Du, J.; Joshi, M.; Chen, D.; Levy, O.; Lewis, M.; Zettlemoyer, L.; Stoyanov, V. Roberta: A robustly optimized bert pretraining approach. *arXiv* **2019**, arXiv:1907.11692.
68. He, P.; Liu, X.; Gao, J.; Chen, W. Deberta: Decoding-Enhanced Bert with Disentangled Attention. In Proceedings of the 9th International Conference on Learning Representations, ICLR 2021, Virtual Event, 3–7 May 2021.
69. Klatt, D.H. Review of text-to-speech conversion for English. *J. Acoust. Soc. Am.* **1987**, *82*, 737–793. [CrossRef]
70. Murray, I.R.; Arnott, J.L.; Rohwer, E.A. Emotional stress in synthetic speech: Progress and future directions. *Speech Commun.* **1996**, *20*, 85–91. [CrossRef]
71. Tokuda, K.; Nankaku, Y.; Toda, T.; Zen, H.; Yamagishi, J.; Oura, K. Speech synthesis based on hidden Markov models. *Proc. IEEE* **2013**, *101*, 1234–1252. [CrossRef]
72. Wang, Y.; Skerry-Ryan, R.J.; Stanton, D.; Wu, Y.; Weiss, R.J.; Jaitly, N.; Yang, Z.; Xiao, Y.; Chen, Z.; Bengio, S.; et al. Tacotron: Towards End-to-End Speech Synthesis. In Proceedings of the Interspeech 2017, 18th Annual Conference of the International Speech Communication Association, Stockholm, Sweden, 20–24 August 2017; pp. 4006–4010.
73. Shen, J.; Pang, R.; Weiss, R.J.; Schuster, M.; Jaitly, N.; Yang, Z.; Chen, Z.; Zhang, Y.; Wang, Y.; Ryan, R.; et al. Natural TTS Synthesis by Conditioning Wavenet on MEL Spectrogram Predictions. In Proceedings of the 2018 IEEE International Conference on Acoustics, Speech and Signal Processing, ICASSP 2018, Calgary, AB, Canada, 15–20 April 2018; pp. 4779–4783. [CrossRef]
74. Kaur, N.; Singh, P. Conventional and contemporary approaches used in text to speech synthesis: A review. *Artif. Intell. Rev.* **2022**, 1–44. [CrossRef]
75. Jeong, M.; Kim, H.; Cheon, S.J.; Choi, B.J.; Kim, N.S. Diff-TTS: A Denoising Diffusion Model for Text-to-Speech. In Proceedings of the Interspeech 2021, 22nd Annual Conference of the International Speech Communication Association, Brno, Czechia, 30 August–3 September 2021; pp. 3605–3609. [CrossRef]
76. Sarfjoo, S.S.; Madikeri, S.R.; Motlícek, P. Speech Activity Detection Based on Multilingual Speech Recognition System. In Proceedings of the Interspeech 2021, 22nd Annual Conference of the International Speech Communication Association, Brno, Czechia, 30 August–3 September 2021; pp. 4369–4373. [CrossRef]

77. Ariav, I.; Cohen, I. An end-to-end multimodal voice activity detection using wavenet encoder and residual networks. *IEEE J. Sel. Top. Signal Process.* **2019**, *13*, 265–274. [CrossRef]
78. Ding, S.; Wang, Q.; Chang, S.y.; Wan, L.; Moreno, I.L. Personal VAD: Speaker-conditioned voice activity detection. *arXiv* **2019**, arXiv:1908.04284.
79. Medennikov, I.; Korenevsky, M.; Prisyach, T.; Khokhlov, Y.Y.; Korenevskaya, M.; Sorokin, I.; Timofeeva, T.; Mitrofanov, A.; Andrusenko, A.; Podluzhny, I.; et al. Target-Speaker Voice Activity Detection: A Novel Approach for Multi-Speaker Diarization in a Dinner Party Scenario. In Proceedings of the Interspeech 2020, 21st Annual Conference of the International Speech Communication Association, Virtual Event, Shanghai, China, 25–29 October 2020; pp. 274–278. [CrossRef]
80. Ng, T.; Zhang, B.; Nguyen, L.; Matsoukas, S.; Zhou, X.; Mesgarani, N.; Veselý, K.; Matějka, P. Developing a speech activity detection system for the DARPA RATS program. In Proceedings of the Thirteenth Annual Conference of the International Speech Communication Association, Portland, OR, USA, 9–13 September 2012.
81. Helmke, H.; Ondřej, K.; Shetty, S.; Arilíusson, H.; Simiganoschi, T.; Kleinert, M.; Ohneiser, O.; Ehr, H.; Zuluaga-Gomez, J.; Smrz, P. Readback Error Detection by Automatic Speech Recognition and Understanding—Results of HAAWAII Project for Isavia's Enroute Airspace. In Proceedings of the 12th SESAR Innovation Days, Budapest , Hungary, 5–8 December 2022.
82. Cordero, J.M.; Dorado, M.; de Pablo, J.M. Automated speech recognition in ATC environment. In Proceedings of the 2nd International Conference on Application and Theory of Automation in Command and Control Systems, London, UK, 29–31 May 2012; pp. 46–53.
83. Delpech, E.; Laignelet, M.; Pimm, C.; Raynal, C.; Trzos, M.; Arnold, A.; Pronto, D. A Real-life, French-accented Corpus of Air Traffic Control Communications. In *Proceedings of the Eleventh International Conference on Language Resources and Evaluation (LREC 2018)*; European Language Resources Association (ELRA): Miyazaki, Japan, 2018.
84. Segura, J.; Ehrette, T.; Potamianos, A.; Fohr, D.; Illina, I.; Breton, P.; Clot, V.; Gemello, R.; Matassoni, M.; Maragos, P. The HIWIRE Database, a Noisy and Non-Native English Speech Corpus for Cockpit Communication. 2007. Available online: http://www.hiwire.org (accessed on 12 May 2023).
85. Gulati, A.; Qin, J.; Chiu, C.; Parmar, N.; Zhang, Y.; Yu, J.; Han, W.; Wang, S.; Zhang, Z.; Wu, Y.; et al. Conformer: Convolution-augmented Transformer for Speech Recognition. In Proceedings of the Interspeech 2020, 21st Annual Conference of the International Speech Communication Association, Virtual Event, Shanghai, China, 25–29 October 2020; pp. 5036–5040. [CrossRef]
86. LeCun, Y.; Bengio, Y. Convolutional networks for images, speech, and time series. *Handb. Brain Theory Neural Netw.* **1995**, *3361*, 1995.
87. Vyas, A.; Madikeri, S.; Bourlard, H. Lattice-Free Mmi Adaptation of Self-Supervised Pretrained Acoustic Models. In Proceedings of the ICASSP 2021—2021 IEEE International Conference on Acoustics, Speech and Signal Processing (ICASSP), Toronto, ON, Canada, 6–11 June 2021; pp. 6219–6223. [CrossRef]
88. Ravanelli, M.; Parcollet, T.; Plantinga, P.; Rouhe, A.; Cornell, S.; Lugosch, L.; Subakan, C.; Dawalatabad, N.; Heba, A.; Zhong, J.; et al. SpeechBrain: A general-purpose speech toolkit. *arXiv* **2021**, arXiv:2106.04624.
89. Panayotov, V.; Chen, G.; Povey, D.; Khudanpur, S. Librispeech: An ASR corpus based on public domain audio books. In Proceedings of the 2015 IEEE International Conference on Acoustics, Speech and Signal Processing, ICASSP 2015, South Brisbane, Australia, 19–24 April 2015; pp. 5206–5210. [CrossRef]
90. Srivastava, N.; Hinton, G.; Krizhevsky, A.; Sutskever, I.; Salakhutdinov, R. Dropout: A simple way to prevent neural networks from overfitting. *J. Mach. Learn. Res.* **2014**, *15*, 1929–1958.
91. Hendrycks, D.; Gimpel, K. Gaussian error linear units (gelus). *arXiv* **2016**, arXiv:1606.08415.
92. Kingma, D.P.; Ba, J. Adam: A Method for Stochastic Optimization. In Proceedings of the 3rd International Conference on Learning Representations, ICLR 2015, San Diego, CA, USA, 7–9 May 2015.
93. Karita, S.; Chen, N.; Hayashi, T.; Hori, T.; Inaguma, H.; Jiang, Z.; Someki, M.; Soplin, N.E.Y.; Yamamoto, R.; Wang, X.; et al. A comparative study on transformer vs rnn in speech applications. In Proceedings of the 2019 IEEE Automatic Speech Recognition and Understanding Workshop (ASRU), Singapore, 14–18 December 2019; pp. 449–456.
94. Graves, A.; Graves, A. Connectionist temporal classification. In *Supervised Sequence Labelling with Recurrent Neural Networks*; Springer: Berlin/Heidelberg, Switzerland, 2012; pp. 61–93.
95. Chen, Z.; Jain, M.; Wang, Y.; Seltzer, M.L.; Fuegen, C. End-to-end Contextual Speech Recognition Using Class Language Models and a Token Passing Decoder. In Proceedings of the IEEE International Conference on Acoustics, Speech and Signal Processing, ICASSP 2019, Brighton, UK, 12–17 May; pp. 6186–6190. [CrossRef]
96. Wolf, T.; Debut, L.; Sanh, V.; Chaumond, J.; Delangue, C.; Moi, A.; Cistac, P.; Rault, T.; Louf, R.; Funtowicz, M.; et al. Transformers: State-of-the-Art Natural Language Processing. In Proceedings of the 2020 Conference on Empirical Methods in Natural Language Processing: System Demonstrations, Online, 16–20 November 2020; pp. 38–45. [CrossRef]
97. Lhoest, Q.; Villanova del Moral, A.; Jernite, Y.; Thakur, A.; von Platen, P.; Patil, S.; Chaumond, J.; Drame, M.; Plu, J.; Tunstall, L.; et al. Datasets: A Community Library for Natural Language Processing. In Proceedings of the 2021 Conference on Empirical Methods in Natural Language Processing: System Demonstrations, Online, 7–11 November 2021; pp. 175–184. [CrossRef]

98. Pascanu, R.; Mikolov, T.; Bengio, Y. On the difficulty of training recurrent neural networks. In Proceedings of the 30th International Conference on Machine Learning, ICML 2013, Atlanta, GA, USA, 16–21 June 2013; Volume 28, pp. 1310–1318.
99. Loshchilov, I.; Hutter, F. Decoupled Weight Decay Regularization. In Proceedings of the 7th International Conference on Learning Representations, ICLR 2019, New Orleans, LA, USA, 6–9 May 2019.

aerospace MDPI

Article

An Automatic Speaker Clustering Pipeline for the Air Traffic Communication Domain

Driss Khalil [1,*], Amrutha Prasad [1,2], Petr Motlicek [1,2], Juan Zuluaga-Gomez [1,3], Iuliia Nigmatulina [1,4], Srikanth Madikeri [1] and Christof Schuepbach [5]

1 Idiap Research Institute, 9120 Martigny, Switzerland; amrutha.prasad@idiap.ch (A.P.); petr.motlicek@idiap.ch (P.M.); juan-pablo.zuluaga@idiap.ch (J.Z.-G.); iuliia.nigmatulina@idiap.ch (I.N.); srikanth.madikeri@idiap.ch (S.M.)
2 Faculty of Information Technology, Brno University of Technology, 60190 Brno, Czech Republic
3 LIDIAP, Ecole Polytechnique Fédérale de Lausanne, 1015 Lausanne, Switzerland
4 Institute of Computational Linguistics, University of Zurich, 8006 Zurich, Switzerland
5 Armasuisse Science and Technology, 3602 Thun, Switzerland; christof.schuepbach@armasuisse.ch
* Correspondence: dkhalil@idiap.ch

Abstract: In air traffic management (ATM), voice communications are critical for ensuring the safe and efficient operation of aircraft. The pertinent voice communications—air traffic controller (ATCo) and pilot—are usually transmitted in a single channel, which poses a challenge when developing automatic systems for air traffic management. Speaker clustering is one of the challenges when applying speech processing algorithms to identify and group the same speaker among different speakers. We propose a pipeline that deploys (i) speech activity detection (SAD) to identify speech segments, (ii) an automatic speech recognition system to generate the text for audio segments, (iii) text-based speaker role classification to detect the role of the speaker—ATCo or pilot in our case—and (iv) unsupervised speaker clustering to create a cluster of each individual pilot speaker from the obtained speech utterances. The speech segments obtained by SAD are input into an automatic speech recognition (ASR) engine to generate the automatic English transcripts. The speaker role classification system takes the transcript as input and uses it to determine whether the speech was from the ATCo or the pilot. As the main goal of this project is to group the speakers in pilot communication, only pilot data acquired from the classification system is employed. We present a method for separating the speech parts of pilots into different clusters based on the speaker's voice using agglomerative hierarchical clustering (AHC). The performance of the speaker role classification and speaker clustering is evaluated on two publicly available datasets: the ATCO2 corpus and the Linguistic Data Consortium Air Traffic Control Corpus (LDC-ATCC). Since the pilots' real identities are unknown, the ground truth is generated based on logical hypotheses regarding the creation of each dataset, timing information, and the information extracted from associated callsigns. In the case of speaker clustering, the proposed algorithm achieves an accuracy of 70% on the LDC-ATCC dataset and 50% on the more noisy ATCO2 dataset.

Keywords: speaker clustering; speaker role detection

Citation: Khalil, D.; Prasad, A.; Motlicek, P.; Zuluaga-Gomez, J.; Nigmatulina, I.; Madikeri, S.; Schuepbach, C. An Automatic Speaker Clustering Pipeline for the Air Traffic Communication Domain. *Aerospace* **2023**, *10*, 876. https://doi.org/10.3390/aerospace10100876

Academic Editor: Joost Ellerbroek

Received: 30 April 2023
Revised: 25 September 2023
Accepted: 26 September 2023
Published: 10 October 2023

1. Introduction

Air traffic control (ATC) communication ensures safe and efficient flight operations [1]. The rise of new artificial intelligence/machine learning technologies provides opportunities for a fundamental change in automation and it becomes a central enabler for future air traffic management concepts. Machine learning technologies are typically data-driven and require a large amount of data for training and development. In the case of voice communication, these data are available through Air Traffic Navigation Service Providers (ANSPs). However, obtaining such data through ANSPs is a legally very complex task, as it typically requires access to the operational control rooms of the ANSPs. A cheap and easy alternative (if allowed by national data privacy laws) is the use of data collected by various initiatives worldwide, such as LiveATC1 (https://www.liveatc.net, accessed on 29 April 2023) in the U.S. and ATCO2 (https://www.atco2.org, accessed on 29 April 2023) in Europe, which collect and store freely available voice communications from Very-High-Frequency (VHF) radio channels. In the case of ATCO2, a large set of volunteers collect the voice as well as the supporting contextual data using relatively cheap VHF radio receivers, and the data are then collected through a centralized server. This approach can easily deliver thousands of unlabeled transmissions. Although such data are typically noisier [2], it has been shown that they can be valuable for training machine learning technologies, including the ATC domain [3].

The average length of each utterance in the collected data is around 3.3 s. However, this presents a unique challenge in the ATC domain, where rapid exchanges between pilots and air traffic controllers occur in communication scenarios, and where utterances are often brief. Accurately identifying speaker roles and clustering them can therefore be challenging, especially when multiple speakers communicate simultaneously on the same channel. The task becomes further complicated due to variations in speech patterns, accents, and communication styles. These challenges underscore the need for advanced machine learning techniques that can handle noisy, short-duration audio data while accurately distinguishing between different speakers and their roles in the communication process.

Besides collecting free data for ATM-oriented machine learning technologies, there are also other use cases that are of serious interest to governmental agencies such as pre-screening the VHF radio channels and detecting their potential abuse by anonymous private persons. This use case is a principal motivator for this paper. VHF radio channels carry the utterances of both pilots and ATCos as one single-channel recording (i.e., a huge wave file that is not segmented). Even if the segmentation algorithm is applied (i.e., typically, voice/speech activity detection can separate communication into short chunks), there is no other information about whether the utterance comes from the ground (ATCo) or the cockpit (pilot).

This paper focuses on clustering speakers appearing either in the same VHF radio channel or across many channels over a given period of time. This is a principal question of security officers when dealing with the abuse of non-encrypted radio communications. The solution given in this paper is tested by analyzing ATCo–pilot communication captured by ATCO2 data. More specifically, our paper is partially built on the concept of separating ATCos and pilots, as investigated in [4]. As ATCos typically appear in the same VHF radio channel over a relatively long period (up to several hours, depending on the length of their shift), the appearance of a new pilot in the analyzed VHF radio channel is very probable. This paper, therefore, focuses on clustering pilot audio recordings to emulate the reality of automatically clustering random speakers in VHF radio channels.

Recent advances in machine learning, particularly deep neural networks (DNNs), have shown promise in addressing these issues by modeling the complex relationships between acoustic features and speakers' identities. DNNs can be trained on large amounts of speech data and can learn to extract high-level features from the speech audio, which can be used for speaker clustering [5]. Despite these advances, the task of speaker clustering in ATC communications remains challenging due to the presence of multiple speakers in the same channel, in addition to the lack of ground-truth information. This absence of labeled data poses significant problems in developing such systems. Nevertheless, the development of such automatic pipelines presents great value in both operational and forensic contexts.

The proposed pipeline presented in this paper comprises four stages: speech segment separation, automatic speech recognition (ASR), speaker role classification, and speaker clustering. The first stage separates the speech segments from a single channel, followed by ASR to transcribe into English. The transcribed text is then fed into a speaker classification model, which detects the pilot segments. The speaker role classification is used only to filter the pilot audio required for speaker clustering. Finally, a speaker clustering method separates and groups the pilot speaker from the audio segments. The proposed pipeline aims to improve the accuracy of speaker clustering in the ATC domain and facilitate effective communication between controllers and pilots.

In air traffic control (ATC) communication analysis, most similar works tend to address only a portion of the complete pipeline outlined in this research. Notably, speaker clustering, a critical part of this pipeline, has received limited attention due to the absence of reliable ground truths on publicly available datasets. It is in this context that the importance of our proposed integrated framework becomes evident. By including speech activity detection, automatic speech recognition, text-based speaker role classification, and unsupervised speaker clustering, this pipeline offers a comprehensive solution. This approach not only addresses the limitations of existing methods but also has the potential to significantly improve the analysis of ATC communication. It bridges the crucial gap between individual components, enabling a deeper understanding of speaker roles and ultimately enhancing safety and efficiency in air traffic control. The rest of this paper is organized as follows. In Section 2, we present the different steps of the pipeline through a discussion of related works, as well as the method used in each step in our work. In Section 3, we describe the different datasets used for training and evaluation in each of the two main components of our pipeline. In Section 4, we present the experiments, the method of evaluation, and the results of each experiment. Finally, we conclude the paper in Section 5 and discuss potential future directions for research in this area.

2. Automatic Speaker Clustering Pipeline

In the ATC domain, the communication of the pilots is of particular interest compared to that of ATC controllers (ATCos) because pilots are responsible for executing flight plans and maneuvering the aircraft, making their communication critical for ensuring safe and efficient flights. Therefore, it is vital to separate pilots' communications from those of ATCos to train automatic systems for each group. Separating the communications of individual pilots is essential for post-flight analysis, incident investigations, and pilot training tasks. The proposed method starts by extracting the speech segments using the SAD system and then using ASR to transcribe those extracted segments. The transcripts obtained are used as input to classify the pilot's speech segments, which are used in the final step as input to the speaker clustering model, as shown in Figure 1. The following subsections describe each step of this pipeline.

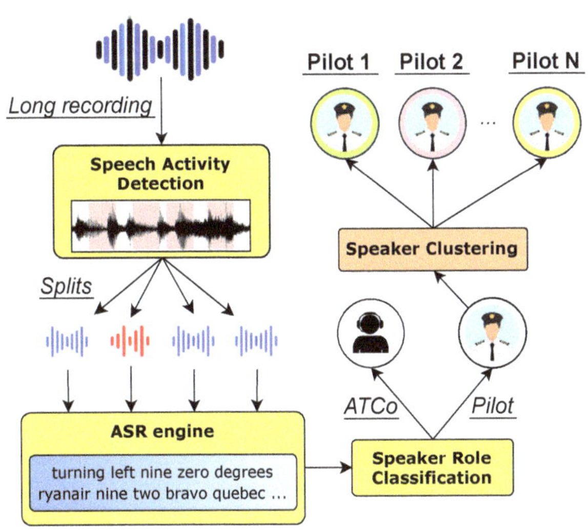

Figure 1. Overview of the automatic speaker clustering pipeline.

2.1. Speech Activity Detection

Speech activity detection (SAD) is a crucial process in speech processing that involves identifying speech segments within an audio utterance. This system splits the audio based on long-silence regions to generate a subset of audio files without silence. It plays a vital role in many speech-based applications such as automatic speech recognition (ASR), speaker recognition, and speaker diarization. Researchers are actively working on developing a SAD system that can accurately operate in noisy environments. The approach is based on [6], which leverages multilingual ASR to improve speech activity detection. The acoustic model (AM) was trained using a lattice-free maximum mutual information loss to extract contextual information from acoustic frames. Multilingual training enhances robustness to noise and language variability. The proposed multilingual acoustic model was trained on 18 languages from the BABEL datasets (https://catalog.ldc.upenn.edu/byyear, accessed on 29 April 2023), including LDC2018S07, LDC2018S13, LDC2018S02, LDC2017S03, LDC2017S22, LDC2017S08, LDC2017S05, LDC2017S13, LDC2017S01, LDC2017S19, LDC2016S06, LDC2016S08, LDC2016 S02, LDC2016S12, LDC2016S09, LDC2016S13, and LDC2016S10. The primary objective of using this dataset was to develop a SAD system that can operate accurately in noisy environments and is robust to language variability. Within each language-dependent part of the acoustic model, speech and non-speech acoustic frames were mapped to a different set of output context-dependent phones or posteriors. For each language, the index of the maximum output posterior was used as a frame-level speech/non-speech decision function. Conventional logistic regression [7] and majority voting were employed to combine decisions from different languages.

2.2. Automatic Speech Recognition (ASR)

Automatic speech recognition (ASR) is a sub-field of speech processing that involves converting speech to text, typically in one language. Hence, this is also termed speech-to-text. A typical ASR system employs an AM and a language model (LM) for converting a speech signal to text. The former is trained on speech recordings with corresponding (ideally manually corrected) text, also referred to as transcripts. The AM represents the relationship between a speech signal and phonemes or other linguistic units that make up the speech. The latter is trained on a large corpus of text data. A probability distribution over sequences of words usually represents the LM. The LM provides context to distinguish between words and phrases that sound similar. Using the knowledge of the AM and LM, a decoding graph is usually built as a weighted Finite State Transducer (FST) [8–10], which generates text output given an observation sequence.

To build a robust speech recognition engine, the artificial intelligence behind it has to be adept at handling challenges such as different acoustic conditions, background noise, model size, and performance. Development in natural language processing and neural network technology has improved speech and voice technology. Past research projects in ATM have provided a platform to develop and improve ASR systems for ATCos and pilots. In [11,12], the authors developed ASR for ATCos to help increase their efficiency and reduce workloads. The authors of [1] provided a benchmark on ASR for different ATC databases. An approach for leveraging non-transcribed audio data to improve ASR was investigated in [13]. A semi-supervised learning approach for enhancing ASR in the ATM domain was employed in [11,12,14]. In [15–17], the authors aimed to improve the recognition of the callsigns in ASR by integrating surveillance data. Finally, the authors of [18] investigated the effect of fine-tuning large pre-trained models, trained using a Transformer architecture, for application in the ATC domain.

This work presents ASR systems that employ two approaches: (i) a hybrid system and (ii) end-to-end training. The hybrid system for ASR uses deep neural network (DNN)-based AMs trained with the lattice-free maximum mutual information (LF-MMI) [19] criterion and n-gram models for the LM. Current state-of-the-art systems use a Transformer architecture, which uses unsupervised [20] or self-supervised [21] learning for speech representations.

2.3. Speaker Role Classification

The task of sequence labeling (SL) assigns labels to words that share a specific role and meaning within the grammatical structure of a sentence. In [22], these groups of words/sentences had similar grammatical properties, and the work focused on two subtasks of SL: named entity recognition (NER) [23,24] and sequence classification (SC) [22,25]. Early work on NER and SC was based on handcrafted ontology, dictionaries, and lexicons, which made them prone to human error. Nowadays, deep learning-based systems are cataloged as state-of-the-art on NER [24] and SC. These models are primarily based on convolutional neural networks [26], recurrent neural networks [22], and Transformers [27].

ATC communications are a rich source of information and follow explicit grammar and ontology. Additionally, ATC communications are built on a well-defined lexicon and dictionary that speakers' errors can sometimes disrupt. One example is the order in which the named entities (e.g., callsign) are uttered in the communication. ATCos utter the callsign (lufthansa seven eight two) at the beginning, whereas pilots invariably do so at the end:

ATCo: *"lufthansa seven eight two descend flight level seven zero"* and,
PILOT: *"descend flight level seven zero lufthansa seven eight two"*.

Following the pros and cons described in Section 2.3, we demonstrate that state-of-the-art NER and SC can be leveraged to automatically identify speaker roles. For instance, one can apply NER to identify ATC-related named entities such as *callsigns*, *command types*, or *units*. Similarly, the structure and type of these 'entities' used in a given communication can be leveraged to identify speaker roles. Our previous research on identifying speaker roles [4] mainly focused on a grammar-based bag-of-words system that was capable of performing speaker role identification with precision/recall values of 0.82/0.81 for ATCos and 0.84/0.85 for pilots, respectively. Also, in [28–30], we explored speaker change detection for ATC text. In [31], the authors mentioned that manually annotating pilot recordings was more challenging than annotating ATCo recordings due to their quality, speech rate, speaker accent, etc. Another reason is that the audio of ATCos is obtained directly from the source, whereas the pilot audio is recorded through the radio receiver. This is one of the reasons why speech processing systems (ASR, diarization, and speaker role identification) perform considerably worse for pilots' recordings compared to ATCos' recordings.

2.4. Speaker Clustering

Over the past few years, there has been growing interest in applying speech processing techniques to the air traffic control (ATC) domain. Specifically, researchers have explored various methods for automatically analyzing and classifying speech in ATC conversations. Although speaker clustering is an essential task in the ATC domain, only a few research studies have focused on it due to the need for ground truths for speaker identity. However, speaker clustering is essential for improving safety and efficiency, especially for pilots, by accurately tracking and managing communication flow, identifying instances of miscommunication and errors, and enabling timely interventions and corrective actions. In [32], the author proposed a method based on graph neural networks (GNNs) to enhance clustering procedures in speaker diarization. The approach aims to purify the similarity matrix used in spectral clustering and assumes a sequence of speaker embeddings that the GNN processes. The GNN outputs a distance metric between the reference and estimated affinity matrices and is trained using a combination of a histogram loss and nuclear norm. Another approach for speaker diarization was proposed in [33], using deep neural networks to learn representations and scoring functions for speaker diarization without relying on i-vector clustering. The proposed method aims to reduce the computational cost and improve the efficiency of speaker diarization in the presence of multiple speakers.

As described above, the purpose of speaker clustering is to classify segmented speech into clusters so that each group only contains speech from one specific speaker. Our approach is based on the methodology described in [34], in which speech segments were preprocessed using the Kaldi FBank features with 40 dimensions, a 16k Hz sampling rate, and 40 filter-bank channels. These features were used as input to the RESNET34 neural network, which processed them using 2-dimensional CNN layers to generate fixed-size embeddings for each speaker. To train the model, we used 500,000 utterances by thousands of speakers from the publicly available VOXCeleb 2 dataset. We applied Probabilistic Linear Discriminant Analysis (PLDA) to the embeddings, which were trained on the VOXCeleb 2 data. The x-vector features generated by the neural network were centered using the training data mean, and Linear Discriminant Analysis (LDA) was applied to further improve the system's performance. For speaker clustering, we used the unweighted pair group method with arithmetic mean (UPGMA), which is a variant of agglomerative hierarchical clustering (AHC). The method consists of grouping similar objects or data points based on their pairwise distances. The algorithm follows the following steps:

1. Calculate the distance matrix D:

$$D_{ij} = \begin{cases} 0, & i = j \\ d_{ij}, & i \neq j, \end{cases} \tag{1}$$

where d_{ij} is the distance between objects (i, j).

2. Calculate the minimum distance pair (i^*, j^*) in the distance matrix D:

$$(i^*, j^*) = \arg\min_{i,j} D_{ij} \tag{2}$$

3. Calculate the new cluster k by averaging the distances between i^* and all objects in the cluster containing i^* and j^*:

$$k = \frac{1}{|C_i| + |C_j|} \sum_{l \in C_i \cup C_j} d_{il} \tag{3}$$

where C_i and C_j are the clusters containing objects i* and j*, and k represents the distance value associated with the newly formed cluster.

4. Update the distance matrix D by removing rows and columns i^* and j^* and adding a new row and column for the newly formed cluster k:

$$D_{ik} = D_{ki} = \frac{d_{ik} + d_{jk}}{2}, \quad \forall i \neq i, j \tag{4}$$

$$D_{kj} = D_{jk} = \infty \tag{5}$$

$$D_{kl} = D_{lk} = \frac{|C_i| d_{il} + |C_j| d_{jl}}{|C_i| + |C_j|}, \quad \forall l \neq i, j, k \tag{6}$$

In this step, k is an index representing the newly formed cluster, whereas i' and j' represent the indices of the objects selected for merging.

5. Repeat steps 2–4 until all objects are in a single cluster or the process is stopped based on a fixed threshold.

In our case, we obtain a pairwise log-likelihood ratio scores matrix using our PLDA model. We represent the distance between clusters by subtracting this matrix from zero, which is then fed into our clustering algorithm. The output generated by the clustering algorithm groups the audio files into clusters, with files that are potentially spoken by the same speaker being assigned to the same cluster.

3. Datasets

In this section, we describe the datasets used to train and evaluate our speaker role classification and speaker clustering components. For training, we employed different datasets for each of the two components. However, for evaluation, we utilized identical testing datasets to evaluate the performance of both components.

3.1. Training

The following sub-section describes the datasets used in the training of speaker role classification and speaker clustering.

3.1.1. Speaker Role Classification

LDC-ATCC corpus: The Air Traffic Control Corpus (LDC-ATCC: https://catalog.ldc.upenn.edu/LDC94S14A, accessed on 29 April 2023). (ATCC) consists of recorded speech for use in ATC research in the area of ASR and NLP. The audio data contains voice communication traffic between various ATCos and pilots. The audio files are sampled at 8 kHz, 16-bit linear, representing continuous monitoring without squelch or silence elimination. Each file captures a single radio frequency channel over one to two hours of audio. The corpus contains gold annotations and metadata (metadata covers voice activity segmentation details, speaker role information (who is talking), and callsigns in ICAO format). The corpus consists of approximately 25 h of ATCo and pilot transmissions (after SAD).

UWB-ATCC corpus: This corpus (released by the University of West Bohemia: https://lindat.mff.cuni.cz/repository/xmlui/handle/11858/00-097C-0000-0001-CCA1-0, accessed on 29 April 2023) is a free and public resource for research on ATC. The communication between ATCos and pilots is manually transcribed and labeled with the speaker information, i.e., pilot/controller. The total duration of speech after removing silences is 13 h. The audio data are single-channel and sampled at 8kHz and 16-bit PCM.

3.1.2. Speaker Clustering

VoxCeleb 1 [35]: This is a dataset comprising 100,000 utterances by 1251 celebrities, which were extracted from videos uploaded to YouTube. The dataset ensures a gender balance, with 55% of the speakers being male, and it features a wide range of speakers of different ages, accents, and ethnicities. The dataset includes speech audio taken in different environments. The diversity of the speakers and environments makes the dataset valuable for training and evaluating different systems for speaker and speech recognition.

VoxCeleb 2 [36]: This is the second version of the dataset, and it builds upon the success of its predecessor, VoxCeleb 1. Like VoxCeleb 1, it contains a large number of utterances by celebrities extracted from YouTube videos, and it features a diverse set of speakers in terms of age, ethnicity, and accent. However, VoxCeleb 2 has more than 1,000,000 utterances by 6000 celebrities, and it has no overlap with the identities of VoxCeleb 1. As a result, it provides an even larger and more diverse dataset for training and evaluating speech processing systems.

Librispeech [37]: This is a corpus of approximately 1000 h of read English speech sampled at 16 kHz from the LibriVox project. The LibriVox project is responsible for the creation of approximately 8000 public domain audio books, the majority of which are in English. Most of the recordings are based on texts from Project Gutenberg2, also in the public domain.

3.2. Evaluation

We employed the same testing datasets for evaluating the performance of automatic speech recognition, speaker role classification, and speaker clustering components.

LDC-ATCC dataset: The test set of LDC-ATCC was used for evaluation. This set consists of 2961 utterances, featuring dialogues between ATCos and pilots, and uses the same speech audio data described in the previous section.

ATCO2 corpus: This dataset was built for the development and evaluation of ASR and NLP technologies for English ATC communications from several airports worldwide (e.g., LKTB, LKPR, LZIB, LSGS, LSZH, LSZB, and YSSY). We used the *ATCO2 test set corpus*, which comprises ∼4 h of annotated data. The corpus is available for purchase through ELDA (http://catalog.elra.info/en-us/repository/browse/ELRA-S0484, accessed on 29 April 2023). The recordings are mono-channel sampled at 16 kHz and 16-bit PCM [38].

4. Experiments and Results

In this section, we present the experimental setup and the results obtained for the different modules of the automatic speaker clustering pipeline described in Section 2. The results include ASR, speaker role classification, speaker clustering modules, as well as the overall performance of the pipeline when all modules are combined. The results are discussed in detail in the following subsections.

4.1. Automatic Speech Recognition

As mentioned in Section 2.2, we adopted two approaches for training an ASR engine: (i) a hybrid-based approach and (ii) an end-to-end training approach. The automatic transcripts were generated using automatic speech recognition systems trained with ∼190 h of annotated ATC data.

Baseline: The main blocks of the hybrid ASR system are the acoustic model (AM) and language model (LM). In our experiments, conventional biphone convolutional neural network (CNN) [26] + TDNN-F [39]-based acoustic models trained with the Kaldi [40] toolkit (i.e., nnet3 model architecture) were used. AMs were trained with the LF-MMI training framework, which is considered to achieve state-of-the-art performance for hybrid ASR. Threefold speed perturbation with MFCC features was used, and i-vectors were used for speaker representation. The 3-gram LM was trained on all the manual transcripts available in the ATC datasets.

XLSR-KALDI: As mentioned earlier, the self-supervised learning approaches using the wav2vec framework facilitated the state-of-the-art performance in ASR. These models were pre-trained with 50k h of speech data. One such model is the XLSR [41], which can then be fine-tuned to ATC data. The authors of [42] proposed to use the LF-MMI criterion (similar to hybrid-based ASR) for the supervised adaptation of the self-supervised pretrained XLSR model [41]. We employed this technique to fine-tune the pre-trained model on our annotated ATC data.

The performance of our ASR system is presented in Table 1 using the Word Error Rate (WER) metric. The system that achieved the lowest WER on the test data was used as the input for the speaker role classification system.

Table 1. WER (%) of our ASR used in our experiments for speaker role classification evaluated on the ATCO2 and LDC-ATCC test sets described in Section 3.2.

Model	ATCO2	LDC-ATCC
Baseline	36.6	**13.5**
XLSR-KALDI	**25.7**	18.7

4.2. Speaker Role Classification

A BERT-based speaker role identification module was implemented that allowed us to attribute a speaker role (i.e., ATCo or pilot) to a given ATC communication. We fetched a BERT (BERT-base-uncased model: 110 M parameters) model [27] from Huggingface [43,44]. We then used ground-truth speaker labels to fine-tune the model on the sequence classification task with the data defined in Section 3.1.1.

Fine-tuning: the BERT model was fine-tuned for 3k steps (∼5 epochs), with a 500-step warm-up phase. The learning rate was increased linearly up to 5×10^{-5} during warm-up, and then it decayed linearly. We fine-tuned each model using the Adam optimizer, a batch size of 32, and a gradient accumulation of 2. After the training, we simply performed inference on either the manual transcripts or automatic transcripts generated through ASR.

Results: Table 2 shows the performance of the data-driven model trained for speaker role classification. The performance is shown for the test sets—ATCO2 and LDC ATCC—trained with all combinations of the training data sets mentioned in Section 3.1.1. We also report the F1 score of the system when (i) manual transcripts and (ii) automatic transcripts are used for classification.

Table 2. Averaged F1 score [0–1] for speaker role classification using different training (column 1) and test sets. All the experiments used the same model (BERT-base-uncased) and the same hyperparameters. We report the mean of five runs with different seeds (the standard deviation was less than 0.01 for all cases, thus we omit it). **Bold** refers to the best performance in each column. Metrics reported on ground-truth transcripts and automatic transcripts generated using the speech recognition system.

Model	ATCO2	LDC-ATCC
Manual Transcripts		
LDC-ATCC	0.83	**0.94**
UWB	0.85	0.87
LDC-ATCC + UWB	**0.87**	0.93
Automatic Transcripts		
LDC-ATCC	0.5	0.9
UWB	0.51	0.8
LDC-ATCC + UWB	0.53	0.9

4.3. Speaker Clustering

For all the experiments in this study, hypotheses concerning the ground truth of pilot identities were generated based on information about the creation of the datasets. Two datasets, ATCO2 and LDC-ATCC, were used to evaluate the performance of our model. In ATCO2, the ground truth was generated using both the callsign and flight date as the pilot identity information. In LDC-ATCC, only the callsign was used as the ground truth for the pilot identity. To determine the optimal threshold for hierarchical clustering, we randomly selected a representative subset of the LDC-ATCC training set consisting of three files per callsign from 259 different callsigns. We fine-tuned the threshold on this selected set as a whole, extracting the value that resulted in the highest accuracy, as shown in Figure 2. The resulting threshold was then used for evaluation on both datasets.

Upon evaluating the test set using this ground-truth generation approach, we observed a total of 929 distinct speakers in the ATCO2 dataset. The ATCO2 dataset covers a span of 7 months from October 2020 to May 2021. Additionally, in the LDC-ATCC dataset, we identified 189 distinct speakers. This indicates the number of unique speakers identified within each respective dataset. The output generated by the clustering algorithm represents the different clusters.

To evaluate the accuracy of our system, we proposed the following evaluation approach: Using the ground truth, we assigned to each cluster the speaker that was assigned to it the most. The utterances that were not assigned to that specific cluster but had this speaker as their label are considered errors. The idea was to map each speaker with one cluster, while all the remaining clusters would be considered errors. Using the same approach, when evaluating the performance of the entire pipeline, we added a constraint to our evaluation method. The pipeline first extracts the speech segments of the pilots. All utterances that are incorrectly classified as belonging to a pilot are also considered errors in our speaker clustering accuracy.

We conducted experiments using two datasets, LDC-ATCC and ATCO2, and utilized the speaker role classification (SRC) method to extract the speech segments of pilots from the datasets. The number of speech utterances for pilots was initially 1350 for ATCO2 and 1446 for LDC-ATCC using the SRC ground truth. However, 243 and 281 utterances were, respectively, removed from ATCO2 and LDC-ATCC datasets due to their short duration (less than 1 s). We further applied the SRC on the manual transcripts of the same dataset, resulting in 1455 utterances for ATCO2 and 1563 utterances for LDC-ATCC. However, 322 and 300 of these utterances were excluded due to their short duration. Lastly, we used SRC on the ASR transcripts of the datasets, resulting in 1705 and 1563 speech utterances for ATCO2 and LDC-ATCC, respectively. However, 389 and 288 of the ASR transcripts were excluded due to their short duration. These excluded segments were not used in the speaker clustering part of the experiment. The details of the pipeline's output are summarized in Table 3, whereas the performance of our model across all experiments is summarized in Table 4.

In our experiments, we found that the level of noise in the data had a significant impact on the accuracy of the speaker clustering pipeline. We observed that the speaker clustering model performed better on the LDC-ATCC dataset, which contained less noise, compared to the noisier ATCO2 dataset. After analyzing the results, we concluded that the difference in the accuracy of all pipelines was mainly due to the performance of the automatic speech recognition (ASR) and speaker role classification (SRC) components of the pipeline, which exhibited lower performance on the noisier ATCO2 data. However, on the LDC-ATCC dataset, we observed that both ASR and SRC exhibited better performance, resulting in a smaller decrease in accuracy. In addition, we found that the difference in performance between clustering alone and the complete pipeline was 8% on the LDC-ATCC dataset and 16% on the ATCO2 dataset. These findings suggest that more research is necessary to improve the performance of the ASR and SRC components, especially in datasets with higher levels of noise like ATCO2 to achieve optimal results with the speaker clustering pipeline.

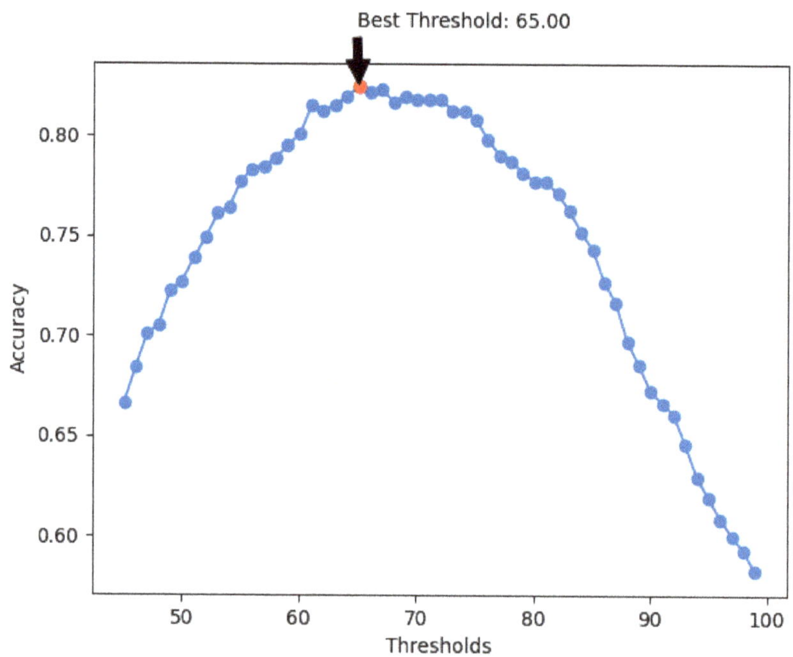

Figure 2. Accuracy vs. thresholds plot used to fine-tune the threshold on a representative subset of the LDC-ATCC dataset for the speaker clustering algorithm. The x-axis shows the threshold values and the y-axis shows the corresponding accuracy values. The red circle indicates the best threshold value (65) with a maximum accuracy of 82%.

Table 3. Automatic speaker clustering performance details for ATCO2 and LDC-ATCC datasets. **Number of segments classified as ATCo**: The segments that were classified as ATCo instead of pilot using the SRC model. **Number of Correct segments**: The segments that were assigned the same label as the ground truth after the evaluation mapping. **Number of Incorrect segments**: The segments that were assigned a different label from the ground truth after the evaluation mapping.

Experiment	Number of Segments Classified as ATCo	Number of Correct Segments	Number of Incorrect Segments
		ATCO2	
SRC Ground Truth	-	739	368
Manual transcript	118	663	470
ASR transcript	337	661	655
		LDC-ATCC	
SRC Ground Truth	-	908	257
Manual transcript	111	897	366
ASR transcript	128	896	379

Table 4. Accuracy (%) of the speaker clustering on ATCO2 and LDC-ATCC datasets.

Dataset	SRC Ground Truth	Manual Transcript	ASR Transcript
ATCO2	66%	58%	50%
LDC-ATCC	78%	71%	70%

5. Discussion and Conclusions

In conclusion, the presented pipeline offers a viable solution to the speaker clustering problem in ATC communication. By using a combination of speech activity detection, automatic speech recognition, text-based speaker role classification, and unsupervised speaker clustering, the pipeline can accurately identify and group speech segments from the same pilot among different speakers. The reported accuracies of 70% and 50% on the LDC-ATCC and ATCO2 datasets, respectively, signify the pipeline's proficiency in identifying pilot speakers within the ATC domain. It is important to note that these accuracies reflect the overall performance of the entire pipeline. There is an observed variation in accuracy when dealing with datasets of different noise levels, such as the LDC-ATCC and ATCO2 datasets, which show a notable deviation of approximately 20%. This discrepancy can be attributed to the effect of noise appearing in VHF data. Specifically, when noise levels increase, it not only challenges the initial component (SAD) by making it harder to accurately identify speech segments but also significantly impacts the ASR component, leading to transcription errors. These inaccuracies propagate through the pipeline and affect the performance of all the remaining components. Consequently, the decrease in speaker clustering accuracy from 70% to 50% on the LDC-ATCC and ATCO2 datasets illustrates the sensitivity of the entire pipeline to noise interference. Nevertheless, when considering the speaker clustering step alone and utilizing the speaker role classification as ground truth, even higher accuracy rates of 78% and 66% can be achieved on the same LDC-ATCC and ATCO2 datasets. This technology has the potential to improve ATC safety, facilitating post-flight analysis and incident investigation. As such, further research in this area is warranted to refine and improve these automated methods for speaker clustering in ATC communication.

Potential future work could focus on enhancing the performance of speaker clustering models with noisy data such as the ATCO2 dataset. We aim to adapt the embedding used for the speaker clustering model on ATC data to improve its performance with such types of noisy data. Another approach is to investigate some speech processing methods to reduce noise and improve the quality of the input data. We also plan to incorporate language identification (LID) as prior information for the speaker clustering in our proposed pipeline. This could potentially improve the accuracy of the clustering by providing additional information about the language and dialect being spoken. Another approach could be to expand the pipeline to support a variety of languages and accents, which would make it more suitable for use in actual ATM systems. By making these modifications, we believe that we can enhance the performance of the speaker clustering model and make it more appropriate for use in real-world scenarios.

Author Contributions: Writing—original draft, D.K. and A.P.; Writing—review & editing, P.M., J.Z.-G., I.N., S.M. and C.S.; Supervision, P.M. All authors have read and agreed to the published version of the manuscript.

Funding: This work was supported by Armasuisse Science and Technology.

Data Availability Statement: Private and public databases are used in this paper. They are covered in detail in Section 3.

Conflicts of Interest: The authors declare no conflict of interest.

References

1. Zuluaga-Gomez, J.; Motlicek, P.; Zhan, Q.; Veselý, K.; Braun, R. Automatic Speech Recognition Benchmark for Air-Traffic Communications. In *Proceedings of the Interspeech*; ISCA: Singapore, 2020; pp. 2297–2301. [CrossRef]
2. Szöke, I.; Kesiraju, S.; Novotný, O.; Kocour, M.; Veselý, K.; Černocký, J. Detecting English Speech in the Air Traffic Control Voice Communication. In *Proceedings of the Interspeech*; ISCA: Singapore, 2021; pp. 3286–3290. [CrossRef]
3. Zuluaga-Gomez, J.; Veselý, K.; Blatt, A.; Motlicek, P.; Klakow, D.; Tart, A.; Szöke, I.; Prasad, A.; Sarfjoo, S.; Kolčárek, P.; et al. Automatic call sign detection: Matching air surveillance data with air traffic spoken communications. *Proc. Multidiscip. Digit. Publ. Inst.* **2020**, *59*, 14.
4. Prasad, A.; Juan, Z.G.; Motlicek, P.; Sarfjoo, S.S.; Iuliia, N.; Ohneiser, O.; Helmke, H. Grammar Based Speaker Role Identification for Air Traffic Control Speech Recognition. *arXiv* **2022**, arXiv:2108.12175.
5. Lukic, Y.X.; Vogt, C.; Dürr, O.; Stadelmann, T. Learning embeddings for speaker clustering based on voice equality. In Proceedings of the 2017 IEEE 27th International Workshop on Machine Learning for Signal Processing (MLSP), Tokyo, Japan, 25–28 September 2017; pp. 1–6. [CrossRef]
6. Sarfjoo, S.S.; Madikeri, S.; Motlicek, P. Speech Activity Detection Based on Multilingual Speech Recognition System. *arXiv* **2020**, arXiv:2010.12277.
7. Kleinbaum, D.G.; Dietz, K.; Gail, M.; Klein, M.; Klein, M. *Logistic Regression*; Springer: Berlin/Heidelberg, Germany, 2002.
8. Mohri, M.; Pereira, F.; Riley, M. Weighted finite-state transducers in speech recognition. *Comput. Speech Lang.* **2002**, *16*, 69–88. [CrossRef]
9. Mohri, M.; Pereira, F.; Riley, M. Speech recognition with weighted finite-state transducers. In *Springer Handbook of Speech Processing*; Springer: Berlin/Heidelberg, Germany, 2008; pp. 559–584.
10. Riley, M.; Allauzen, C.; Jansche, M. OpenFst: An Open-Source, Weighted Finite-State Transducer Library and its Applications to Speech and Language. In Proceedings of the Human Language Technologies: The 2009 Annual Conference of the North American Chapter of the Association for Computational Linguistics, Companion Volume: Tutorial Abstracts, Boulder, Colorado, 31 May–5 June 2009; Association for Computational Linguistics: Boulder, CO, USA, 2009; pp. 9–10.
11. Srinivasamurthy, A.; Motlicek, P.; Himawan, I.; Szaszak, G.; Oualil, Y.; Helmke, H. Semi-supervised learning with semantic knowledge extraction for improved speech recognition in air traffic control. In Proceedings of the 18th Annual Conference of the International Speech Communication Association, Stockholm, Sweden, 20–24 August 2017.
12. Kleinert, M.; Helmke, H.; Siol, G.; Ehr, H.; Cerna, A.; Kern, C.; Klakow, D.; Motlicek, P.; Oualil, Y.; Singh, M.; et al. Semi-supervised adaptation of assistant based speech recognition models for different approach areas. In Proceedings of the 37th Digital Avionics Systems Conference (DASC), London, UK, 23–27 September 2018; IEEE: Piscataway, NJ, USA, 2018, pp. 1–10.
13. Khonglah, B.; Madikeri, S.; Dey, S.; Bourlard, H.; Motlicek, P.; Billa, J. Incremental semi-supervised learning for multi-genre speech recognition. In Proceedings of the ICASSP 2020-2020 IEEE International Conference on Acoustics, Speech and Signal Processing (ICASSP), Barcelona, Spain, 4–8 May 2020; IEEE: Piscataway, NJ, USA, 2020, pp. 7419–7423.
14. Zuluaga-Gomez, J.; Nigmatulina, I.; Prasad, A.; Motlicek, P.; Veselý, K.; Kocour, M.; Szöke, I. Contextual Semi-Supervised Learning: An Approach to Leverage Air-Surveillance and Untranscribed ATC Data in ASR Systems. In *Proceedings of the Interspeech*; ISCA: Singapore, 2021; pp. 3296–3300. [CrossRef]
15. Kocour, M.; Veselý, K.; Blatt, A.; Gomez, J.Z.; Szöke, I.; Cernocky, J.; Klakow, D.; Motlicek, P. Boosting of Contextual Information in ASR for Air-Traffic Call-Sign Recognition. In *Proceedings of the Interspeech*; ISCA: Singapore, 2021; pp. 3301–3305. [CrossRef]
16. Nigmatulina, I.; Braun, R.; Zuluaga-Gomez, J.; Motlicek, P. Improving callsign recognition with air-surveillance data in air-traffic communication. *arXiv* **2021**, arXiv:2108.12156.
17. Nigmatulina, I.; Zuluaga-Gomez, J.; Prasad, A.; Sarfjoo, S.S.; Motlicek, P. A two-step approach to leverage contextual data: Speech recognition in air-traffic communications. In *Proceedings of the ICASSP*; ISCA: Singapore, 2022.

18. Zuluaga-Gomez, J.; Prasad, A.; Nigmatulina, I.; Sarfjoo, S.; Motlicek, P.; Kleinert, M.; Helmke, H.; Ohneiser, O.; Zhan, Q. How Does Pre-trained Wav2Vec2.0 Perform on Domain Shifted ASR? An Extensive Benchmark on Air Traffic Control Communications. *arXiv* **2023**, arXiv:2203.16822.
19. Povey, D.; Peddinti, V.; Galvez, D.; Ghahremani, P.; Manohar, V.; Na, X.; Wang, Y.; Khudanpur, S. Purely sequence-trained neural networks for ASR based on lattice-free MMI. In *Proceedings of the Interspeech*; ISCA: Singapore, 2016; pp. 2751–2755.
20. Schneider, S.; Baevski, A.; Collobert, R.; Auli, M. wav2vec: Unsupervised Pre-Training for Speech Recognition. In Proceedings of the Interspeech 2019, Graz, Austria, 15–19 September 2019; ISCA: Singapore, 2019; pp. 3465–3469. [CrossRef]
21. Baevski, A.; Zhou, Y.; Mohamed, A.; Auli, M. wav2vec 2.0: A Framework for Self-Supervised Learning of Speech Representations. In Proceedings of the Advances in Neural Information Processing Systems, Virtual, 6–12 December 2020.
22. He, Z.; Wang, Z.; Wei, W.; Feng, S.; Mao, X.; Jiang, S. A Survey on Recent Advances in Sequence Labeling from Deep Learning Models. *arXiv* **2020**, arXiv:2011.06727.
23. Grishman, R.; Sundheim, B. Message Understanding Conference-6: A Brief History. In Proceedings of the COLING 1996 Volume 1: The 16th International Conference on Computational Linguistics, Copenhagen, Denmark, 5–9 August 1996.
24. Yadav, V.; Bethard, S. A Survey on Recent Advances in Named Entity Recognition from Deep Learning models. In Proceedings of the 27th International Conference on Computational Linguistics, Santa Fe, NM, USA, 20–26 August 2018; Association for Computational Linguistics: Santa Fe, NM, USA, 2018; pp. 2145–2158.
25. Zhou, C.; Cule, B.; Goethals, B. Pattern based sequence classification. *IEEE Trans. Knowl. Data Eng.* **2015**, *28*, 1285–1298. [CrossRef]
26. LeCun, Y.; Bengio, Y. Convolutional networks for images, speech, and time series. *Handb. Brain Theory Neural Netw.* **1995**, *3361*, 1995.
27. Devlin, J.; Chang, M.W.; Lee, K.; Toutanova, K. BERT: Pre-training of Deep Bidirectional Transformers for Language Understanding. *arXiv* **2018**, arXiv:1810.04805.
28. Zuluaga-Gomez, J.; Sarfjoo, S.S.; Prasad, A.; Nigmatulina, I.; Motlicek, P.; Ondre, K.; Ohneiser, O.; Helmke, H. BERTraffic: BERT-based Joint Speaker Role and Speaker Change Detection for Air Traffic Control Communications. In Proceedings of the IEEE Spoken Language Technology Workshop (SLT), Doha, Qatar, 9–12 January 2023.
29. Prasad, A.; Zuluaga-Gomez, J.; Motlicek, P.; Sarfjoo, S.; Nigmatulina, I.; Veselý, K. Speech and Natural Language Processing Technologies for Pseudo-Pilot Simulator. *arXiv* **2022**, arXiv:2212.07164.
30. Zuluaga-Gomez, J.; Prasad, A.; Nigmatulina, I.; Motlicek, P.; Kleinert, M. A Virtual Simulation-Pilot Agent for Training of Air Traffic Controllers. *Aerospace* **2023**, *10*, 490. [CrossRef]
31. Pellegrini, T.; Farinas, J.; Delpech, E.; Lancelot, F. The Airbus Air Traffic Control speech recognition 2018 challenge: Towards ATC automatic transcription and call sign detection. *arXiv* **2018**, arXiv:1810.12614.
32. Wang, J.; Xiao, X.; Wu, J.; Ramamurthy, R.; Rudzicz, F.; Brudno, M. Speaker diarization with session-level speaker embedding refinement using graph neural networks. *arXiv* **2020**, arXiv.2005.11371.
33. Garcia-Romero, D.; Snyder, D.; Sell, G.; Povey, D.; McCree, A. Speaker diarization using deep neural network embeddings. In Proceedings of the 2017 IEEE International Conference on Acoustics, Speech and Signal Processing (ICASSP), New Orleans, LA, USA, 5–9 March 2017; pp. 4930–4934. [CrossRef]
34. Zeinali, H.; Wang, S.; Silnova, A.; Matějka, P.; Plchot, O. BUT System Description to VoxCeleb Speaker Recognition Challenge *arXiv* **2019**, arXiv.1910.12592.
35. Nagrani, A.; Chung, J.S.; Zisserman, A. VoxCeleb: A Large-Scale Speaker Identification Dataset. *arXiv* **2017**, arXiv:1706.08612.
36. Chung, J.S.; Nagrani, A.; Zisserman, A. VoxCeleb2: Deep Speaker Recognition. *arXiv* **2018**, arXiv:1806.05622.
37. Panayotov, V.; Chen, G.; Povey, D.; Khudanpur, S. Librispeech: An asr corpus based on public domain audio books. In Proceedings of the 2015 IEEE International Conference on Acoustics, Speech and Signal Processing (ICASSP), South Brisbane, QLD, Australia, 19–24 April 2015; IEEE: Piscataway, NJ, USA, 2015; pp. 5206–5210.
38. Zuluaga-Gomez, J.; Veselý, K.; Szöke, I.; Motlicek, P.; Kocour, M.; Rigault, M.; Choukri, K.; Prasad, A.; Sarfjoo, S.S.; Nigmatulina, I.; et al. ATCO2 corpus: A Large-Scale Dataset for Research on Automatic Speech Recognition and Natural Language Understanding of Air Traffic Control Communications. *arXiv* **2022**, arXiv:2211.04054.
39. Povey, D.; Cheng, G.; Wang, Y.; Li, K.; Xu, H.; Yarmohammadi, M.; Khudanpur, S. Semi-Orthogonal Low-Rank Matrix Factorization for Deep Neural Networks. In *Proceedings of the Interspeech*; ISCA: Singapore, 2018; pp. 3743–3747.
40. Povey, D.; Ghoshal, A.; Boulianne, G.; Burget, L.; Glembek, O.; Goel, N.; Hannemann, M.; Motlicek, P.; Qian, Y.; Schwarz, P.; et al. The Kaldi speech recognition toolkit. In Proceedings of the IEEE Workshop on Automatic Speech Recognition and Understanding, Waikoloa, HI, USA, 11–15 December 2011; IEEE Signal Processing Society: Piscataway, NJ, USA, 2011.
41. Conneau, A.; Baevski, A.; Collobert, R.; Mohamed, A.; Auli, M. Unsupervised cross-lingual representation learning for speech recognition. *arXiv* **2020**, arXiv:2006.13979.
42. Vyas, A.; Madikeri, S.; Bourlard, H. Lattice-Free Mmi Adaptation of Self-Supervised Pretrained Acoustic Models. In Proceedings of the ICASSP 2021-2021 IEEE International Conference on Acoustics, Speech and Signal Processing (ICASSP), Toronto, ON, Canada, 6–11 June 2021; pp. 6219–6223. [CrossRef]

43. Wolf, T.; Debut, L.; Sanh, V.; Chaumond, J.; Delangue, C.; Moi, A.; Cistac, P.; Rault, T.; Louf, R.; Funtowicz, M.; et al. Transformers: State-of-the-Art Natural Language Processing. In Proceedings of the 2020 Conference on Empirical Methods in Natural Language Processing: System Demonstrations, Online, 16–20 November 2020; pp. 38–45.
44. Lhoest, Q.; del Moral, A.V.; Jernite, Y.; Thakur, A.; von Platen, P.; Patil, S.; Chaumond, J.; Drame, M.; Plu, J.; Tunstall, L.; et al. Datasets: A Community Library for Natural Language Processing. In Proceedings of the 2021 Conference on Empirical Methods in Natural Language Processing: System Demonstrations, Online, 7–11 November 2021; pp. 175–184.

aerospace

MDPI

Article

Lessons Learned in Transcribing 5000 h of Air Traffic Control Communications for Robust Automatic Speech Understanding

Juan Zuluaga-Gomez [1,2,*], Iuliia Nigmatulina [1,3], Amrutha Prasad [1,4], Petr Motlicek [1,4,*], Driss Khalil [1], Srikanth Madikeri [1], Allan Tart [5], Igor Szoke [4], Vincent Lenders [6], Mickael Rigault [7] and Khalid Choukri [7]

1 Speech & Audio Processing Group, Idiap Research Institute, 1920 Martigny, Switzerland; iuliia.nigmatulina@idiap.ch (I.N.)
2 LIDIAP, Ecole Polytechnique Fédérale de Lausanne, 1015 Lausanne, Switzerland
3 Institute of Computational Linguistics, University of Zurich, 8050 Zurich, Switzerland
4 Faculty of Information Technology, Brno University of Technology, 60190 Brno, Czech Republic; szoke@replaywell.com
5 OpenSky Network, 3400 Burgdorf, Switzerland
6 Cyber-Defence Campus, Armasuisse, 3602 Thun, Switzerland
7 Evaluations and Language Resources Distribution Agency (ELDA), 75013 Paris, France; mickael@elda.org (M.R.)
* Correspondence: juan-pablo.zuluaga@idiap.ch (J.Z.-G.); petr.motlicek@idiap.ch (P.M.)

Abstract: Voice communication between air traffic controllers (ATCos) and pilots is critical for ensuring safe and efficient air traffic control (ATC). The handling of these voice communications requires high levels of awareness from ATCos and can be tedious and error-prone. Recent attempts aim at integrating artificial intelligence (AI) into ATC communications in order to lessen ATCos's workload. However, the development of data-driven AI systems for understanding of spoken ATC communications demands large-scale annotated datasets, which are currently lacking in the field. This paper explores the lessons learned from the ATCO2 project, which aimed to develop an unique platform to collect, preprocess, and transcribe large amounts of ATC audio data from airspace in real time. This paper reviews (i) robust automatic speech recognition (ASR), (ii) natural language processing, (iii) English language identification, and (iv) contextual ASR biasing with surveillance data. The pipeline developed during the ATCO2 project, along with the open-sourcing of its data, encourages research in the ATC field, while the full corpus can be purchased through ELDA. ATCO2 corpora is suitable for developing ASR systems when little or near to no ATC audio transcribed data are available. For instance, the proposed ASR system trained with ATCO2 reaches as low as 17.9% WER on public ATC datasets which is 6.6% absolute WER better than with "out-of-domain" but gold transcriptions. Finally, the release of 5000 h of ASR transcribed speech—covering more than 10 airports worldwide—is a step forward towards more robust automatic speech understanding systems for ATC communications.

Keywords: air traffic control communications; automatic speech recognition and understanding; OpenSky Network; callsign recognition; ADS-B data

check for updates

Citation: Zuluaga-Gomez, J.; Nigmatulina, I.; Prasad, A.; Motlicek, P.; Khalil, D.; Madikeri, S.; Tart, A.; Szoke, I.; Lenders, V.; Rigault, M.; et al. Lessons Learned in Transcribing 5000 h of Air Traffic Control Communications for Robust Automatic Speech Understanding. *Aerospace* **2023**, *10*, 898. https://doi.org/10.3390/aerospace10100898

Academic Editor: Joost Ellerbroek

Received: 5 September 2023
Revised: 10 October 2023
Accepted: 11 October 2023
Published: 20 October 2023

1. Introduction

There has been a growing interest in the development of automatic speech recognition (ASR) and understanding systems for air traffic control (ATC) due to their potential to enhance the safety and efficiency of the aviation industry. The application of ASR and understanding technologies in ATC has resulted in the creation of advanced proof-of-concept engines that can assist air traffic controllers (ATCos) in their daily tasks. These systems are designed to analyze spoken ATC communications and convert them into machine-readable texts, allowing for faster and more accurate processing. Previous works such as MALORCA [1], HAAWAII [2] or SESAR2020's Solution 97.2 [3] have shown mature enough methods to reduce ATCos' workload while increasing safeness, e.g., see [4,5]. The authors concluded that integrating novel ASR-based tools can reduce the total amount of time that ATCos expend on entering and confirming the clearances in their workstations by 20% absolute points. As a result, ASR and understanding technologies are becoming more advanced and capable of handling the complexities of ATC communications, leading to improved safety and efficiency in the aviation industry. The paragraphs below summarize three current challenges—while working with ATC voice communications—addressed by this paper:

(1) Previous works on ASR to analyze air traffic communication is built for a specific domain, e.g., one airport or en-route/approach scenarios. The process of adapting machine learning models to different airports or control areas requires new in-domain data, which remain challenging to collect and annotate. For instance, ATC audio data collected from one airport, e.g., `airport` X, in general, do not transfer well to `airport` Y.

(2) ATC data collection and their transcription [6] are expensive and time-consuming tasks. The data collection phase includes data capture and preprocessing; this task can be automated. The data transcription phase aims at producing the word-by-word transcript of the given ATC utterance; this task is carried out by hand by humans. It becomes expensive as several man hours are needed to transcribe one hour of ATC speech without silence. For some solutions, such as those targeting small airports, this cost may be prohibitive. This raises the question of what is the most efficient manner to collect and process large-scale ATC audio data.

(3) In addition, audio data from ATC communication are considerably noisier with regard to standard ASR corpora when they are captured via very-high-frequency (VHF) receivers. In some cases, SNR (signal-to-noise) levels may range from 5 to 20 dB. Thus, it becomes challenging to develop an ASR system and later use its outputs for downstream tasks, e.g., natural language processing (NLP), due to the high word error rates (WER). In contrast, higher SNR ATC data sourced from operation rooms and characterized by close-mic recordings and substantially reduced noise can be obtained from air navigation service providers (ANSPs), albeit being limited to private use in most cases.

In this paper, we answer these questions by extending our previous work on the ATCO2 project and its resulting corpora [7]; see detailed information in Appendix A. The ATCO2 project aimed to reduce the human effort required to collect, preprocess, and transcribe ATC voice communications by employing state-of-the-art ASR and NLP systems [8]. ATCO2 releases the largest corpus of ATC voice communications to date, consisting of more than 5000 h of automatically transcribed audio data and their correspondent surveillance data [9]. In addition, four hours of human-transcribed data (i.e., gold transcriptions) were also released, where we quantified that the transcription process can be significantly accelerated by providing the annotators with automatically transcribed data (i.e., output from an in-domain ASR system), rather than requiring them to produce transcriptions from scratch. According to [7], the real-time factor (RTF; time needed to generate the gold word-by-word transcription of the ATC audio with regard to its duration) for transcribing the data can be reduced from 50 to 20. An overview of the composition of ATCO2 corpora is given in Figure 1.

This paper covers several aspects and lessons learned (see Section 6) related to the data collection and the transcription pipeline, including its primary actors. Also, it covers the main AI-based systems that can be developed with the ATCO2 corpora, and we set baselines on ASR and understanding.

The rest of the paper is organized as follows. Section 2 covers related work on automatic speech recognition and understanding for ASR. We describe the ATCO2 system, data collection pipeline, and the main contributions of this paper in Section 3. In Section 4, we cover technical details about the data collection platform (front end and back end) and how the community of volunteers interacts with them. In Section 5, we cover the main technologies that can be developed with ATCO2 corpora. We conclude the paper and discuss the main lessons learned in Section 6.

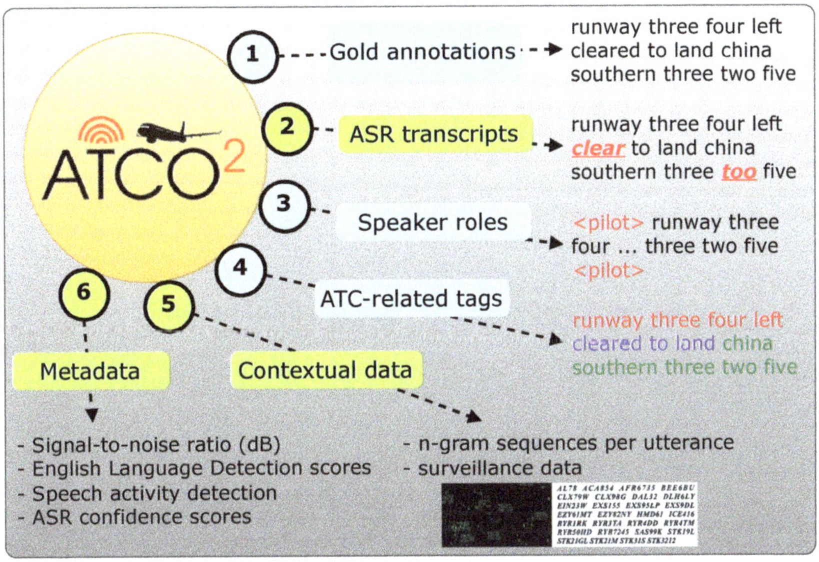

Figure 1. ATCO2 corpora. Blue circles denote transcriptions only available for ATCO2 test set corpus. Green circles denote transcriptions and metadata available for both ATCO2 test set and ATCO2 transcribed corpus sets. Taken from previous work in [7].

2. Early Work on Automatic Speech Recognition and Understanding in ATC

Recent work in ASR and understanding of ATC communications has been documented for trainee's ATCo training by AENA (Aeropuertos Españoles y Navegación Aérea) [10] and MITRE corporation [11], also including workload estimation with ASR systems [12]. In recent years, more research-oriented work has focused on pure ASR. For example, ref. [13] established the first benchmark on ASR for different ATC communications-focused databases. Furthermore, there has been a significant effort to integrate novel semisupervised learning algorithms for boosting the ASR performance with surveillance data such as [14]. This supports the idea of the growing interest in research in ASR and understanding towards ATC, with mature proof-of-concept engines that can assist ATCos in their daily tasks. Our previous work related to the large-scale automatic collection of ATC audio data from different airports worldwide was in [9]. Additionally, recent work targeted to improve callsign recognition by integrating surveillance data into the pipeline has been explored in [15] or, for instance, automating pilots report extraction with ASR tools [16].

Another line of work has been directed at open-sourcing ATC-related databases: for US-based communications [17], in Czechia [18], and [19] for several accents in English. Recently, there was an Airbus-led challenge [20] for ATC communications, with French-accented recordings from France [21]. Private databases such as VOCALISE [22] and ENAC [23] have also targeted ATC communications. For a general overview of ATC-related databases, we redirect the reader to Table 1 in ref. [7], and for the databases released by the ATCO2 project, to Table 1 in this paper.

Table 1. Air traffic control communications corpora released by ATCO2 project. † Full database after silence removal. †† Speaker accents depend on the airport's location and on the airline origin (e.g., Air France in Australia may contain French-accented audio); accents of pilots are not known at any time of the communication due to privacy regulations.

Database	Details	Licensed	Accents	Hours †	Ref.
	Released corpora by ATCO2 project				
ATCO2 corpora	Data from different airports and countries: public corpora catalog.elra.info/en-us/repository/browse/ELRA-S0484/ (accessed on 10 October 2023)		Several ††		[7]
ATCO2-test-set	Real life data for ASR and NLP research.	✓	Several ††	4	[7]
ATCO2-T set	ASR-based transcribed dataset. Real data for research in ASR and NLU.	✓	Several ††	5281	[7,9]
	Free access databases released by ATCO2 project				
ATCO2-test-set-1h	'ASR dataset': public 1 h sample, a subset of *ATCO2-test-set*. https://www.atco2.org/data (accessed on 10 October 2023)	✓	Several ††	1	[9]
ATCO2-ELD set	'ELD dataset': public dataset for English language detection. https://www.atco2.org/data (accessed on 10 October 2023)	✓	Several ††	26.5	[24]

3. ATCO2 Corpora

It is well known that AI-based tools need large amounts of reliably transcribed data during their training process. For instance, ASR or NLP tools for ATC could work better if we had large-scale data. The ATCO2 corpora was designed to target this data scarcity issue by solving four big challenges:

(1) Current corpora related to air traffic control are primarily focused on automatic speech recognition. However, for an AI engine to be successfully deployed in the control room, it must not only accurately transcribe ATC communication but also understand it. This includes the ability to detect speaker roles (SRD) as well as extract and parse callsigns and commands. The ATCO2 corpora provides a comprehensive solution to this challenge by including detailed tags for SRD and callsign and command extraction. This, in turn, will improve the accuracy and efficiency of AI-based systems in ATC operations.

(2) Out-of-domain ASR and NLP-based corpora transfer poorly to the ATC domain. ATC communication follows an unique grammatical structure and employs a specific set of the vocabulary defined by ICAO [25], making it a niche application. This poses a significant limitation to the use of out-of-domain corpora (previous studies [13] have shown that employing non-ATC related corpora such as LibriSpeech [26], CommonVoice [27] or SWITCHBOARD [28], does not match the acoustics of ATC communication, and therefore does not contribute substantially to ASR training). As such, the ATCO2 project collected and publicly released a large amount of ATC-specific data to aid in the development of ASR and understanding engines for ATC.

(3) The research community working on ATC is hindered by a severe lack of openly available annotated data. To address this issue, the ATCO2 project has released a vast corpus of over 5000 h of automatically transcribed data (i.e., *ATCO2-T set*), as well as 4 h of manually annotated data (i.e., *ATCO2-test-set-4h*). It is worth noting from Table 1, that the transcriptions generated by the automatic tools have been proven to be robust, with WERs as low as 9%. These errors are achieved when training an ASR engine with ATCO2 corpora only. See the prior results for the `Malorca-Vienna-test` set coming from the MALORCA project in [7].

(4) There is no standardized metric to evaluate quality of nontranscribed data prior to their transcription process. Currently, when a new corpus for ASR is in its collection and labeling phase, few filtering stages are performed to ensure high-quality audio data selection. In contrast, in Section 3.3, and specifically in Equation (1), ATCO2 uncovers the quality estimation that helped to select the best audio files for gold transcription generation by humans.

3.1. ATCO2 System and Generalities

The ATCO2 system is described in Figure 1. During the collection of the ATCO2 corpora, we followed several preprocessing steps in order to normalize the generated transcriptions. Here, we aim at minimizing errors produced by phonetic dissimilarities, e.g., "descent to two thousand" and "descend two two thousand". We performed several text normalization steps in order to unify the gold and automatic transcriptions following ICAO rules [25] and well-known ontologies for ATC communications [5]. A summary of the transcription protocol is depicted in Figure 2. Additionally, we direct the reader to a more detailed overview on text normalization and lexicon for transcript generation in Section 3 of ref. [7]. Furthermore, the ATCO2 corpora are composed of *ATCO2-T set* corpus and *ATCO2-test-set* corpus, described below:

- First, the ***ATCO2-T set corpus*** is the first ever release of a large-scale dataset targeted to ATC communications. We recorded, preprocessed, and automatically transcribed ~5281 h of ATC speech from ten different airports (see Table 2). To the best of the authors' knowledge, this is the largest and richest dataset in the area of ATC ever created that is accessible for research and commercial use. Further information and details are available in [7].
- Second, ***ATCO2-test-set-4h corpus*** was built for the evaluation and development of automatic speech recognition and understanding systems for English ATC communications. This dataset was annotated by humans. There are two partitions of the dataset, as stated in Table 1. The *ATCO2-test-set-1h corpus* is a ~1 h long open-sourced corpus, and it can be accessed for free at https://www.atco2.org/data (accessed on 10 October 2023). The *ATCO2-test-set-4h corpus* contains *ATCO2-test-set-1h corpus* and adds to it ~3 more hours of manually annotated data. The full corpus is available for purchase through ELDA at http://catalog.elra.info/en-us/repository/browse/ELRA-S0484 (accessed on 10 October 2023).

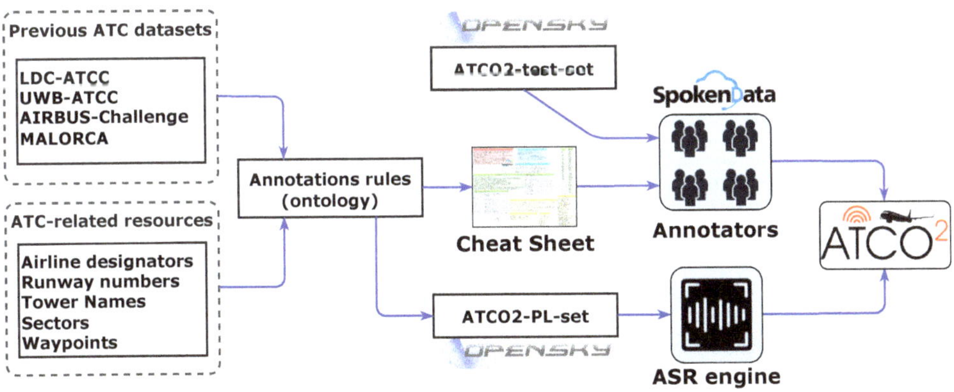

Figure 2. Transcription protocol. ATCO2 corpora follow a rigorous transcription protocol based on previous ATC-related corpora and resources. Additionally, a cheat sheet for ATC transcript generation was developed during the project. The cheat sheet is available in Appendix E.

Table 2. Total accumulated duration. (in hours) of speech after voice activity detection per airport in *ATCO2-T set*. † English language detection (ELD) (0–1) score. This score shows how confident our ELD system is in detecting whether there is only English spoken inside the ATC communication. Note that the first word for each name denotes the ICAO airport identifier.

ICAO (airport identifier)—City										
EETN—Tallinn	EPLB—Lublin	LKPR—Prague	LKTB—Brno	LSGS—Sion	LSZB—Bern	LSZH—Zurich	LZIB—Bratislava	YBBN—Brisbane	YSSY—Sydney	others—others
English Data (language score ≥0.5) †										
131	<1	1762	888	330	699	921	24	170	77	<1
Non-English Data (language score <0.5) †										
2	<1	187	611	83	55	49	26	10	3	<1

3.2. Data Collection Pipeline

The processing pipeline is implemented as a Python script that follows a configuration file → `worker.py`. The configuration file allows us to modify the logic and flow of the data in the pipeline on-the-fly. It allows parallelism, forking, and conditions. In principle, `worker.py` consists of global definitions (constants), blocks (local definitions), and links (an acyclic oriented graph) between blocks. The processing pipeline is given in Figure 3. For instance, we address earlier implementations of each technology from the previous work [9], e.g., segmentation and diarization, ASR, or named entity recognition (NER). All the technologies and tools are encapsulated in `BASH scripts` with an unified interface.

The first row of blocks from Figure 3 refers to segmentation and demodulation. Initially, an antenna and a recording device jointly capture the radio signal, which is divided into segments containing portions where the transmission was "active", and the silent parts are not recorded (push-to-talk is used in ATC voice communication). This functionality is part of the RTLSDR-Airband audio recording software, from which we dump the raw I/Q signal. Second, we convert this complex I/Q radio signal into a waveform signal by a software-defined radio CSDR. The first part is performed in the recording device, while the second is performed at the OpenSky Network (OSN). The OSN is a nonprofit community-based receiver network which has been continuously collecting air traffic surveillance data since 2013. Unlike other networks, OpenSky keeps the complete unfiltered raw data and makes them accessible to academic and institutional researchers).

Next, we perform "signal-to-noise ratio (SNR) filtering" (second row); the purpose is to remove the recordings that are too noisy. In bad recording conditions, we can end up in a situation in which the voice is not intelligible. The following step is "diarization" (third row). In the automatically segmented data, some recordings contain more than one speaker. This is a problem because we intend to automatically transcribe speaker turns of single speakers. And, for subsequent NLP/SLU tasks, it is important to separate the speaker turns as well. The diarization solves this by splitting the audio into segments with single speakers and assigning them speaker labels. In the ASR step, we simply convert "speech-to-text". This is performed by our ASR system that we build with tools from the Kaldi toolkit [29]. The outputs from this step are transcripts, which inevitably contain some errors. To improve the accuracy of the transcripts, we use callsign lists from surveillance as contextual information. The callsign lists come from the air traffic monitoring databases of OpenSky Network. Further details can be found in Section 5.2.2 and [15].

Next, the transcripts are used as input for the English language detection (ELD) system. The purpose is to be able to discard non-English audio data. The typical state-of-the-art language identification system is based on acoustic modeling and uses audio as input. For the ATC speech, we do not need to "identify" the non-English languages, so we developed a "lexical English detection system" which uses transcripts and confidence scores produced by ASR as its inputs (see previous work at Interspeech in 2021 [24]). For ATC speech, this worked better than the "traditional" acoustic language identification method. The last automatic operation is "post-processing by NLP". Currently, the pipeline performs a callsign-code extraction step. It returns the callsign in ICAO format, like "DLH77RM", belonging to an aircraft. Finally, some processed data go through "human correction", and some data are kept with automatic labels. The former case produced *ATCO2-test-set-4h corpus*, while the latter, *ATCO2-T set corpus*. A more detailed description of the data collection flow and data transcription is given in Appendices C and D.

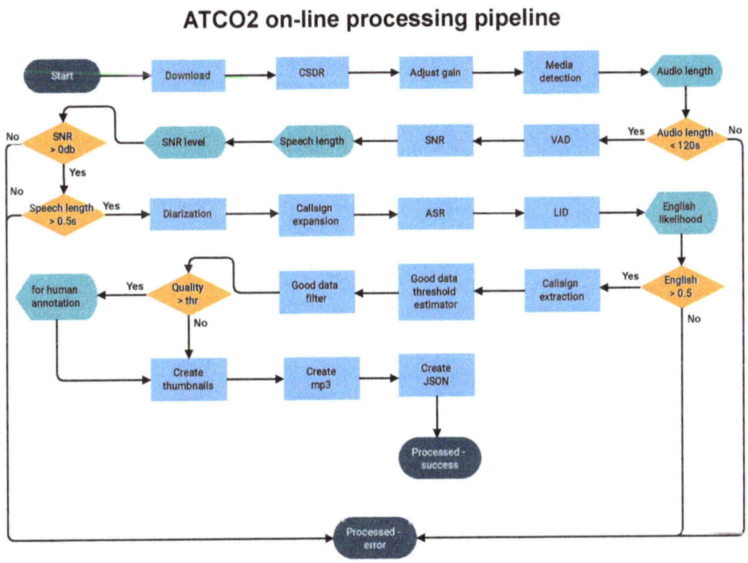

Figure 3. ATCO2 workflow for processing data collected by a community of feeders. Initially, the data are sent and stored on OSN servers. The audio data go through several modules to filter out recordings with a high level of noise and too-long or too-short segments. Blue rectangles are processes. The cyan arrow blocks are internal callback events, where the pipeline informs the master node about progress and sends intermediate results. The orange rhombuses are conditions, where intermediate results are taken into account (e.g., an SNR level), i.e., whether to continue (clean audio) or stop processing. A final internal callback is run when the pipeline finishes. It triggers the API to call the OSN server with the particular callback, for instance, the processing has finalized as *OK* or *ERROR*.

3.3. *Quality Estimation for Data Transcription*

As mentioned at the end of Section 3.2, the captured, processed, and automatically transcribed data (see Figure 3) can be annotated by humans. This in turn would generate "gold transcriptions" that we use to evaluate the proposed ASR and NLP systems. The *ATCO2-test-set-4h corpus* went through all these steps. As the data are continuously being recorded by OSN, we need to select the most intelligible and clean data. We developed a score that ranks the recordings depending on their quality. This score integrates seven metrics that assess the quality of each recording present in *ATCO2 corpora*. For instance, we used Equation (1) to measure, rank, and select the ATC communications with the highest quality. Later, these recordings were shortlisted for human transcription (see Section 4.2). The data annotators generated the ground truth transcripts and tags that are part of the *ATCO2-test-set-4h corpus*. The ranking score is given as follows:

$$Score = \log(avg_{SNR} + e) + \log(num_{spk} + e) + \log\left(\frac{speech_{len}}{audio_{len}} + e\right) + \\ ELD_{score} \times 3 \ + avg_{WordConf} \times 3 + \log(wrd_{cnt} + e), \quad (1)$$

where

- avg_{SNR}—provides average SNR of speech in range <0, 40>. SNR needs to be as high as possible;
- num_{spk}—provides the number of speakers in the audio in the range of <1, 10>. The more speakers detected in audio, the better;
- $speech_{len}$—provides the amount of speech in seconds;
- $audio_{len}$—provides the overall audio length. More speech detected in audio is better;
- ELD_{score} provides "probability" of audio being English in the range <0.0, 1.0>. The higher the ELD score, the better;
- $avg_{WordConf}$—provides average confidence of the speech recognizer <0.0, 1.0>. We want data where the recognizer is confident. Higher is better;
- wrd_{cnt}—provides the number of words spoken in the range of <0, ~150>. The more words, the better.

A breakdown of the outputs of these steps for a single day is given in Figure 4. For instance, ~0.6 h of data are selected for gold transcriptions from an initial 26 h pool of audio data. We believe this is a robust quality scoring method because it gathers information from different systems, e.g., ASR, SNR, and ELD estimation. A day-to-day estimation of the output of each of these steps is available on the SpokenData website: https://www.spokendata.com/atc (accessed on 10 October 2023).

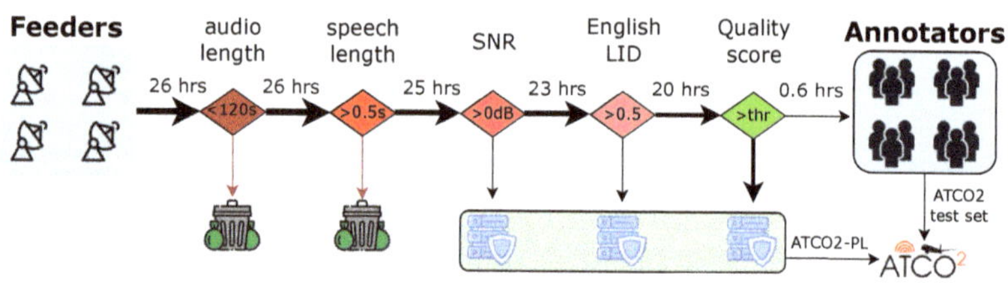

Figure 4. Breakdown of data flow yield from raw data (recordings from data feeders) w.r.t the human-annotated transcripts throughout the pipeline. This is a one-day snapshot from 9 February 2022.

3.4. Runtime Characteristics

We also measured the running time for individual components of our processing pipeline. In Table 3, we list the relative time spent by each module, such as ASR and speaker diarization; both accounting for 65% of the overall processing time. This disparity compared to other modules is due to the fact that both are AI-powered modules, which, in principle, needs more processing time. Other important parts are preprocessing, voice activity detection (VAD) segmentation, and ELD. Audio data preprocessing involves obtaining data, demodulation by software radio, segmental gain control, detecting media format, and plotting waveform. A key metric is the real-time factor of the whole pipeline. The real-time factor is the ratio of "processing time" over "length of the audio". Our processing pipeline has a real-time factor of 4.47. In other words, the processing is computationally demanding. For an average five-second-long recording, the processing time is 22 s. The actual running times of each component for the "average" five-second-long recording are shown in Table 3.

Table 3. Processing time per component in the transcription pipeline. The values in the second column are for an average 5.016 s long recording. The average was computed over 10,334 recordings (14.4 h), recorded on 4 December 2021.

Processing Step	Time [s]	Percentage [%]
Preprocessing	2.5	11.6
VAD segmentation	2.4	11.1
SNR estimation	0.6	3.0
Diarization	7.1	32.6
Callsign expansion	0.5	2.1
Speech-to-text (ASR)	7.0	32.1
English detection	1.3	5.9
Callsign extraction	0.1	0.4
Post-processing	0.2	1.2
Total time	21.6	100.0

4. Collection Platform and Community of Volunteers

In this section, we summarize the data collection and distribution. In addition, a short description of the roles involved in data processing is provided. We also cover some high-level statistics about the collected data. First, data are captured and fed into the OpenSky Network by the volunteers who operate their own receiver equipment (see Figure 5). These individuals are often aviation enthusiasts with previous operational experience, or people with an interest in aviation technology, e.g., conducting domain-related research. But anyone with little to no background in aviation or technology can become a feeder. To become a feeder, one must have an internet connection and access to a VHF receiver. An affordable low-complexity setup is covered in the ATCO2 corpora paper [7] and the guide for setting it up is provided https://ui.atc.opensky-network.org/set-up (accessed on 10 October 2023). It is important to recall that in some countries, it is prohibited by law to record air traffic management (ATM) data. Readers interested in the legal aspect are directed to the legal and privacy aspects for collection of ATC recordings section in [7].

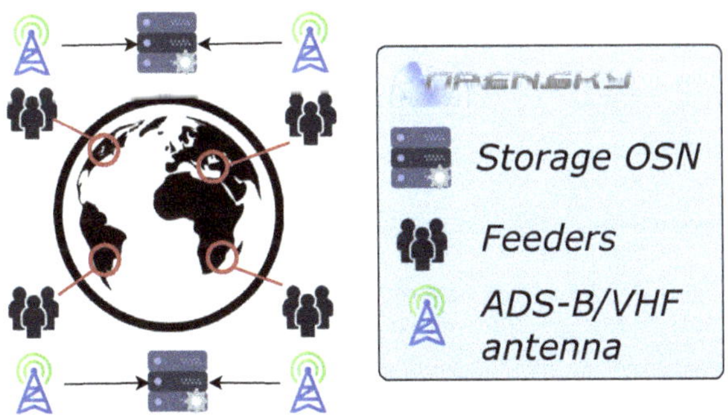

Figure 5. Data feeders pipeline. The data users have set up a VHF receiver and feed data to OSN servers.

4.1. The Platform

The high-level architecture is given in Figure 6. As one can observe, the platform has been divided into three distinct groups: (i) feeder equipment, (ii) back-end, and (iii) front-end. The architecture was decided during the design phase of ATCO2, with the main objective to achieve scalability of the entire system. That means keeping the complexity relatively low within all the groups, which allows it to

- Support a similar number of users to the current OpenSky Automatic Dependent Surveillance–Broadcast system (ADS-B);
- Keep the feeder equipment simple and affordable;
- Provide data to different types of users in a simple and intuitive way;

- Interface external services (e.g., voice annotation) in a simple and intuitive way;
- Keep maintenance and error handling as simple as possible.

A better overview of the OSN platform is also listed in Appendix B. As mentioned above, the platform has been divided into three parts. Below, we describe each of these platform's groups.

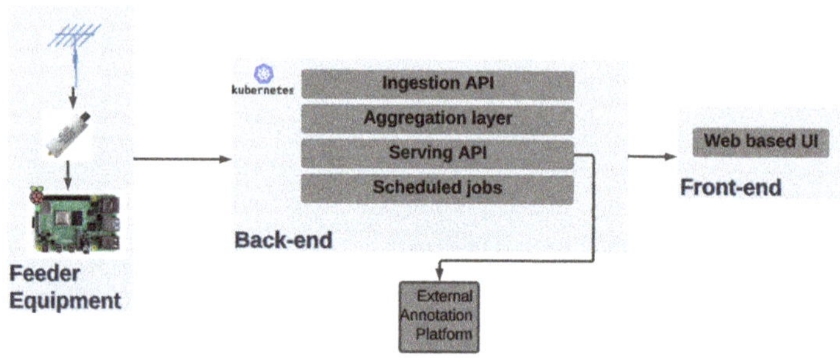

Figure 6. The high-level architecture of the data collection platform.

Feeder equipment: the main task of the feeder equipment is to capture the conversation between the pilot and the ATCo and feed the data, together with some relevant metadata, to the back-end. For the recording part, we recommend using RTLSDR-Airband together with RTLSDR dongle. RTLSDR is a set of tools that enables USB dongles based on the Realtek RTL2832U chipset to be used as cheap software defined radios, given that the chip allows transferring raw I/Q samples from the tuner straight to the host device (see further documentation in https://osmocom.org/projects/rtl-sdr/wiki/Rtl-sdr; accessed on 10 October 2023). The latter is an affordable and widely used combination within the aviation enthusiast community for this exact purpose—to capture and stream ATC voice.

The feeder software is responsible for transmitting the recordings from the receiver to the remote server. It is a rather simple piece of software that monitors the output directory of the RTLSDR-Airband and transfers any new data it finds to the back-end using a gRPC (gRPC; remote procedure calls) connection. The fact that the feeder software only looks for specific types of data from the output folder suggests that the feeder is free to choose any other software for capturing and storing the voice data. Care must be taken to assure that the output is suitable for the feeder software. A simple, step-by-step guide is provided to simplify the setup process. It can be found at https://ui.atc.opensky-network.org/set-up (accessed on 10 October 2023).

Back-end: the main tasks for the back-end are (i) to store recordings, transcripts, and any other relevant metadata, and (ii) to provide interfaces for external users. The external users in this are data feeders, transcription service providers, data users, or any other parties contributing to the dataset or making use of it. The back-end is deployed on Kubernetes, an open-source container orchestration system. As one can observe from Figure 6, there are several processing layers involved. These layers are as follows:

- Ingestion API: receives recording segments and metadata and queues them for processing in Kafka/S3 compatible object storage;
- Aggregation layer: converts raw data to flac audio, stores metadata, and triggers transcription using Kafka Streams, S3, and Serving API;
- Serving API: provides external interfaces to consume metadata, store, and consume transcript and statistics;
- Scheduled jobs: run processes that are not part of the streaming process like statistics aggregation and data housekeeping.

Interfacing the back-end is performed using API, which is well documented in https://api.atc.opensky-network.org/q/swagger-ui (accessed on 10 October 2023). In order to access the back-end and make use of the available APIs, one needs to register on https://auth.

opensky-network.org/auth/ (accessed on 10 October 2023), contact OpenSky Network (mailto: `contact@opensky-network.org`), and give a short description of what one needs the access for.

Front-end: the front-end is a web-page (https://ui.atc.opensky-network.org/; accessed on 10 October 2023) and it provides access to public stats, links to documentation, e.g., API documentation, and external web pages, e.g., SpokenData transcription service. In addition, this is a place for an user to set up their receivers, see some statistics about the receiver performance, and so on.

Statistics: since the public opening of the service (5 March 2023), the ATCO2 project has recorded speech from 24 different airports in 14 different countries. In Figures 7 and 8, names of countries and airports, together with corresponding recording lengths, are shown. Please note that only the airports/areas with the length of regrinding $\geq$ 1 h are included. This also applies for the ATCO2 corpora released in ELDA.

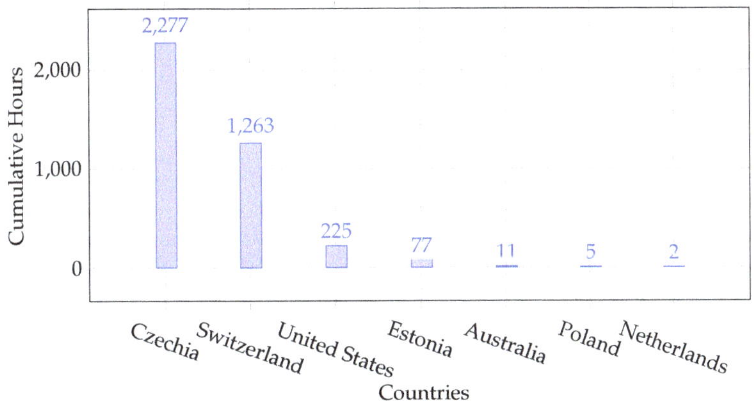

Figure 7. Length of recordings per country from the beginning of the service until 5 March 2023. Countries where the length of recordings is longer than 1 h are given. Note that some countries (e.g., United States) were not part of the official release of the ATCO2 corpora (see Table 2). Still, they are currently being collected in the OSN Platform.

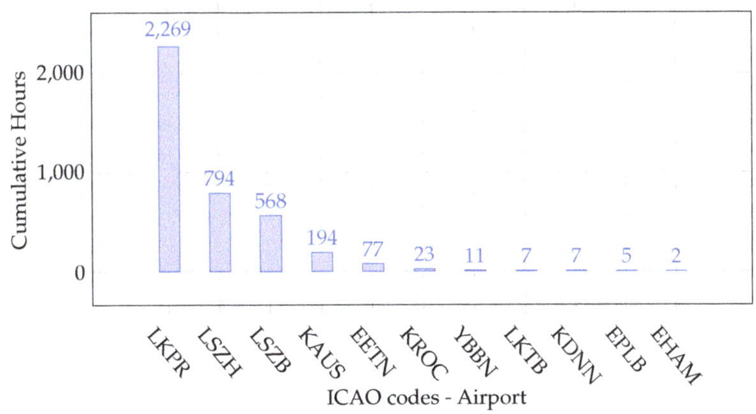

Figure 8. Length of recordings per airport from the beginning of the service until 5 March 2023. Airports where the length of recordings is longer than 1 h are given. LKPR: Vaclav Havel Airport Prague; LZSH: Zurich Airport; LSZB: Bern Airport; KAUS: Austin-Bergstrom International Airport; KROC: Frederick Douglass Greater Rochester International Airport; YBBN: Brisbane Airport; LKTB: Brno Airport; KDNN: Dalton Municipal Airport; EPLB: Lublin Airport; EHAM: Amsterdam Airport Schiphol.

4.2. Data Annotators

Apart from the data feeder, there is another type of volunteers who have contributed to the project and will continue to contribute in the future. These are called "Annotators". The data annotators are volunteers who write down the transcripts of the ATC voice communications, including assigning speakers and annotating named entities, i.e., callsigns and commands. For the ATCO2 project, we relied on both volunteers and paid transcribers. Our data processing pipeline (as seen in Figure 3) generates transcripts and NLP tags for each communication. By generating transcriptions with AI tools, we are able to speed up the overall transcription process (if you are interested in becoming an annotator, please create an account on the SpokenData transcription platform: http://www.spokendata.com/atco2; accessed on 10 October 2023). The amount of human transcribed data is the package of a four-hour test set, i.e., *ATCO2-test-set-4h corpus*. The data annotators are the final actors involved in the transcription step, as shown in Figure 4.

5. Technologies

In this section, we cover the main tools developed with the ATCO2 corpora. We also list some potential topics that can be explored with it. Moreover, note that the ATCO2 corpora are not limited to the fields covered in this paper e.g., ASR or NLP, but also can be used for text-to-speech (TTS), which is somehow opposite to ASR. We expect the community will build on top of ATCO to foster and advance speech and text-based technologies for ATC.

5.1. Automatic Speech Recognition

One of the principal components of the ATCO2 project is the strong ASR system, used in order to provide high-quality automatic transcriptions for the collected ATC data. An ASR system is trained to predict the best text translation for the input acoustic signal. Formally speaking, ASR aims to find the best probability candidate output sequence of words from a set of all possible word combinations (or sentences) in a language given a noisy acoustic observation sequence. End-to-end ASR models learn a direct mapping of speech S, to the output text W:

$$\hat{W} = \underset{W \in \mathcal{V}^*}{\operatorname{argmax}}\, p(W|S).$$

The hybrid (conventional) ASR systems combine three separately trained models: acoustic model (AM), pronunciation model, and language model (LM). The model calculates the conditional probability $p(W|S)$, where W is a sequence of words ($W = w_1, \ldots, w_n$), S is a sequence of input feature vectors representing the acoustic observations ($S = s_1, \ldots, s_t$), and $\mathcal{V}$ is the vocabulary of all possible words [30,31] or subwords [32], as shown in Equation (4).

$$\begin{aligned} \hat{W} &= \underset{W \in \mathcal{V}^*}{\operatorname{argmax}}\, p(W|S) && (2) \\ &= \underset{W \in \mathcal{V}^*}{\operatorname{argmax}}\, p(S|W)p(W) && (3) \\ &= \underset{W \in \mathcal{V}^*}{\operatorname{argmax}} \sum_{P} p(S|P)p(P|W)p(W), && (4) \end{aligned}$$

where $p(S|P)$ is an AM, $p(P|W)$ is a pronunciation model, and $p(W)$ is an LM; we use $\mathcal{V}^*$ to represent the collection of all word sequences formed by words in $\mathcal{V}$. One of the advantages of conventional pipeline models is a more transparent optimization of an objective function [33]. Moreover, the LM is trained with unpaired text data and can be easily adapted to a specific domain. This gives conventional models more flexibility and makes them convenient for use in industrial projects, such as ATC.

5.1.1. Training Data Configuration

To measure the effectiveness of using automatically transcribed data (ATCO2-T set) versus using fully supervised gold transcriptions, we defined three training scenarios.

- **Scenario (a) only supervised data**: we employ a mix of public and private supervised ATC databases (recordings with gold transcriptions). It comprises ∼190 h of audio data (or 573 h after speed perturbation);
- **Scenario (b) only ATCO2-T 500 h dataset**: we use only a subset of 500 h from the ATCO2-T corpus (see introductory paper [7]);
- **Scenario (c) only ATCO2-T 2500 h dataset**: same as **scenario (b)**, but instead of only using 500 h subset, we use five times more, i.e., a 2500 h subset. This subset is only used to train a hybrid-based ASR model (CNN-TDNNF; convolutional neural network and time-delay neural network) to test the boosting experiments in Section 5.2.2.

5.1.2. Test Data Configuration

Two ATCO2 test sets are used for ASR evaluation, as shown in Table 4: ATCO2-test-set-1h (in short ATCO2-1h) and ATCO2-test-set-4h (in short ATCO2-4h). The same test sets are used for boosting experiments presented in Section 5.2.2.

Table 4. WER results for the public ATCO2 test sets with CNN-TDNNF models trained on different data; this includes from scenario (a) to scenario (c).

Model	Test Sets	
CNN-TDNNF	**ATCO2-1h**	**ATCO2-4h**
Scenario (a) only supervised 573 h dataset	24.5	32.5
Scenario (b) only ATCO2-T 500 h dataset	18.1	25.1
Scenario (c) only ATCO2-T 2500 h dataset	17.9	24.9

5.2. Conventional ASR

To obtain automatic transcriptions of the best possible quality for ATCO2 corpora audio, we use a strong hybrid model trained on ATC data only. We train a hybrid-based model for each of the scenarios described above. For **scenario (a)**, an AM was built to include all available 190 h datasets, speech augmentation accounting for 573 h of data. The model dictionary consists of 30,832 words coming from diverse sources. This includes (i) a list of airline designators for callsigns taken from Wikipedia: https://en.wikipedia.org/wiki/List_of_airline_codes (accessed on 10 October 2023); (ii) all five-letter waypoint names in Europe retrieved from the Traffic project, see https://pypi.org/project/traffic/ (accessed on 10 October 2023); (iii) additional words, such as countries, cities, airport names, airplane models and brands, and some ATC acronyms. For training the acoustic model, we use the Kaldi toolkit [29]. The system follows the standard Kaldi recipe, which uses MFCC and i-vectors features [34] with time-delay neural networks (TDNN) [35,36]. The standard chain training is based on lattice-free maximum mutual information (LF-MMI [37], which includes threefold speed perturbation and one-third frame subsampling). The acoustic model is a CNN-TDNNF [38], which comprises a convolutional network and a factorized-TDNN. The LM is 3G trained on the same data as the acoustic model with additional text data coming from additional public resources such as airlines names, airports, ICAO alphabet, and way-points in Europe.

Results and analysis: the results are presented in Table 4. We compared three models trained with the same conventional CNN-TDNNF architecture but on different data: scenarios (a), (b), and (c) (see Section 5.1.1). The model (a) in Table 4 is trained on the "out-of-domain" for ATCO2 but supervised data. The models (b) and (c) are trained on the "in-domain" ATCO2 data and the difference is only in the size of the training set: 500 h vs. 2500 h. We can see that training on completely unsupervised data yields good performance in comparison to (a). Increasing the size of unsupervised data from 500 h to 2500 h, however, does not bring too much improvement: the WER goes from 18.1% to 17.9% and from 25.1% to 24.9% only for ATCO2-1h and ATCO2-4h, respectively.

Our main hypothesis is that ATCO2 test sets contain higher levels of noise compared to the audio data present in (a), i.e., mainly clean data from ATCos. Moreover, ATCO2 test sets also contain speech from pilots collected via VHF receivers, which in turn degrades the

SNR levels, i.e., reduced audio quality. Hence, when the system is trained on "clean data", i.e., scenario (a) and later tested on ATCO2, it creates a large train–test set mismatch. Yet, when we use ATCO2 training data, scenario (b) or scenario (c), this mismatch is reduced, and therefore we obtain substantially better results.

5.2.1. End-to-End ASR

Differently from hybrid-based ASR, there exists another paradigm for performing ASR [39], named end-to-end (E2E) ASR [40]. Here, we aim at directly transcribing speech to text without requiring alignments between input features and output words or characters (i.e., standard procedure in hybrid-based ASR); see Equation (4). Recent work on encoder–decoder ASR has shown that this step can be removed [41]. E2E can be divided into connectionist temporal classification (CTC)-based [42], attention-based encoder–decoder modeling [43], or hybrid [44]. Previous work based on self-supervised learning [45] for ASR includes Wav2Vec2.0 [46], vq-Wav2Vec [47], and, most recently, WavLM [48] and multilingual XLS-R [49] models. E2E ASR aims at reducing the expert knowledge needed. This makes the overall ASR development simpler; thus, it could have a significant impact on ATC [50]. This work focuses on data novelty (including their collection and preparation) rather than investigating (i) different E2E architectures for ASR, e.g., Conformer [51], HyperConformer [52], Conmer [53], or BranchFormer [54]; or (ii) toolkits for E2E ASR such as SpeechBrain [55], ESPnet [56], NeMo [57], or WeNet toolkits [58]. Therefore, we leave these lines of research for future work.

5.2.2. Callsign Boosting

To further improve the prediction made by an ASR system, along with speech input, one can use other information available from context. For the ATC domain, such context information may be the data received from radar. Every moment, radar registers aircraft that are currently in the airspace, listing unique identifiers of those aircraft, i.e., "callsigns". With the radar data, we know exactly what callsigns are especially likely to appear in the conversation. This knowledge allows us to bias the system outputs towards these registered callsigns and to increase the probability that they are recognized correctly. A callsign is typically a sequence of an ICAO airline identifier, letters, and digits, which in speech turns into a sequence of words. In ASR, the target sequences of words can be boosted during decoding with WFST (weighted finite state transducer) by adjusting the weights in the prediction graphs, called "lattices". The rescoring technique with WFST was proposed earlier and applied for biasing towards use's play lists [59], contact names [60], and named entities [61]. Recently, a similar biasing approach has proved to be useful in improving callsign recognition [9]. The rescoring of lattices is performed with the finite state transducer (FST) operation of composition between lattices produced by an ASR system and an FST created with the target transcript and discount weights (Equation (5)):

$$biased\_Lattices = Lattices \circ biasing\_FST \quad (5)$$

Biasing the lattice toward the context callsigns usually allows us to considerably improve their recognition in the final outputs (Table 5). The results of different experiments on the ATC data proved that applying the lattice rescoring method on top of ASR predictions leads to higher accuracy of automatic transcriptions, first of all, callsigns [14]. Therefore, lattice rescoring was used for all transcriptions of the ATCO2 data.

Table 5. Results for boosting experiment on ATCO2 corpora. Results are listed for the CNN-TDNNF model trained with either all supervised data or 500 h or 2500 h of ATCO2 corpora. The top results per block are **highlighted in bold**. The best result per column is marked with underline. † 1h public test set. ‡ 4h full test set. Results are obtained with offline CPU decoding. ¶ word error rates only on the sequence of words that compose the callsign in the utterance. CallWER: callsign word error rate; ACC: accuracy.

Boosting	ATCO2-test-set-1h †			ATCO2-test-set-4h ‡		
	WER	CallWER ¶	ACC	WER	CallWER ¶	ACC
scenario (a) only supervised dataset						
Baseline	24.5	26.9	61.3	32.5	36.7	42.4
Unigrams	24.4	25.5	63.2	33.1	35.0	45.8
N-grams	23.8	23.8	66.4	31.3	33.7	47.9
GT boosted	**22.9**	**19.1**	**75.2**	**29.7**	**29.1**	**58.5**
scenario (b) only ATCO2-T 500 h dataset						
Baseline	18.1	16.2	71.2	25.1	24.8	62.6
Unigrams	19.1	14.6	74.2	26.0	22.8	65.6
N-grams	17.5	13.5	75.3	24.3	21.4	66.6
GT boosted	**16.3**	**6.9**	**88.9**	**22.5**	**13.0**	**82.9**
scenario (c) only ATCO2-T 2500 h dataset						
Baseline	17.9	16.7	70.5	24.9	24.2	62.0
Unigrams	18.3	14.4	73.8	25.6	22.0	65.9
N-grams	17.3	14.2	74.3	24.3	21.1	66.5
GT boosted	**15.9**	**6.5**	**89.4**	**22.2**	**12.5**	**83.9**

Results and analysis: in Table 5, we report the results for the out-of-domain (ATC supervised) and in-domain (ATCO2-500 h/2500 h) ATC models. Both acoustic models are trained with CNN-TDNNF architecture following the standard Kaldi recipe, as described in Section 5.2. The results are reported with three metrics: WER (word error rate), Call-WER (WER calculated on the sequence of n-grams that correspond to callsigns only), and ACC (accuracy).

To rescore a decoding lattice according to the current context, we perform the following steps: (1) we receive all the callsigns registered by the radar at the current timestamp in the ICAO format; (2) we expand the ICAO callsigns to word sequences to include all possible callsign variations, i.e., ways this callsign can be spoken; (3) we use the expanded callsigns to bias the decoding lattice towards the current context. See our previous work [15] for more details on callsign verbalization.

Biasing multiple callsigns registered by the radar, compared to biasing only a ground truth (GT) callsign, can be used in a real-life scenario and with real-time ASR. To allow it, a new contextual FST with expanded callsigns is generated on the fly every time when new data come from radar. The results of biasing a GT callsign are given in Table 5 to illustrate the oracle performance of the biasing method. Overall, decoding with n-grams biasing always helps to achieve better performance, especially for callsigns, with a relative improvement of 15.0% and 12.8% for callsign recognition and of 3.4% and 2.4% for the entire utterance on ATCO2-test-set-1h and ATCO2-test-set-4h test sets, respectively.

The size of biasing FST depends on the number of callsigns and their variations we want to boost. Too many callsigns may decrease the effectiveness of the biasing method, as the more nontrue callsigns are boosted, the less the correct sequence is prominent. The previous results show that the optimal size of biasing FST highly depends on the data, but generally, the performance begins to degrade when the number of biased word sequences exceeds 1000 [62]. For our experiments, we have, on average, 214 biased callsigns variations per utterance in the ATCO2-test-set-4h and 140 biased callsigns variations per utterance in the ATCO2-test-set-1h corpus.

5.3. Natural Language Understanding of Air Traffic Control Communications

Natural language understanding (NLU) is a subfield of NLP that focuses on the ability to understand and interpret human language. NLU involves the development of algorithms and models that can extract meaning and intent from text and/or spoken communication. NLU involves several subtasks, including (i) named entity recognition [63], which aims at identifying entities in text, such as people, places, and organizations [64]; (ii) part-of-speech tagging (POS), identifying the grammatical role of each word in a sentence [65], similar to sequence classification (see Section 5.3.2); (iii) sentiment analysis, identifying the emotional tone of a piece of text [66]; (iv) relationship extraction, identifying the relationships between entities in text [67]; (v) question answering, understanding, and answering natural language questions [68]. The following subsections cover each of the proposed NLU submodules that can be developed with ATCO2 corpora, like the ones presented in Figure 9.

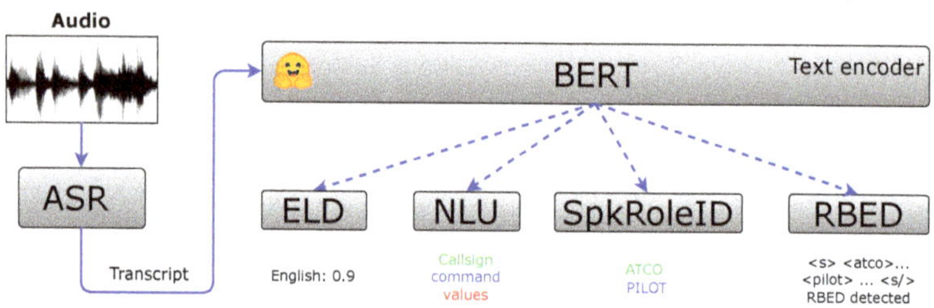

Figure 9. Main automatic speech recognition and understanding tasks that can be achieved with the ATCO2 corpora. ELD: English language detection; NLU: natural language understanding, e.g., callsign highlighting; SPKRoleID: speaker role identification; RBED: read-back error detection.

5.3.1. Named Entity Recognition for Air Traffic Control Communications

In ATC communications, NLU can be used to automatically analyze and interpret the meaning of spoken messages between pilots and ATCos, which can aid ATCos in downstream tasks, such as assisting in identifying emergency situations and other critical events. NLU can help to extract important information, such as flight numbers, callsigns, or airport codes, which in turn can aid ATCos to manage traffic more efficiently.

Overall, the use of NLU in ATC helps improve communication accuracy and efficiency, aids in reduction of ATCos' workload by prefilling aircraft radar labels, and provides valuable data for analysis and decision making. In this work, one of the main tasks is to understand and extract high-level information within ATC communication. Therefore, we develop an NER system tasked to extract this information, as depicted in Figure 10a. For instance, consider the following transcribed communication (taken from Figure 1):

ASR transcript: runway three four left cleared to land china southern three two five,

would be converted to high-level entity format with the NER system to:

Output: <value> runway three four left </value> <command> cleared to land </command> <callsign> china southern three two five </callsign> .

In this work, we developed two systems based on transformers [69] to extract and tag this information from ATC communications, i.e., a pretrained BERT [70] model and RoBERTa [71] model.

Experimental setup: we fine-tune a pretrained BERT and RoBERTa model on the NER task, as shown in Figure 9). We employed the pretrained version of `BERT-base-uncased` [70] with 110M parameters, URL: https://huggingface.co/bert-base-uncased (accessed on 10 October 2023). Also, the pretrained version of `RoBERTa-base` [71] is composed of 123M parameters, URL: https://huggingface.co/roberta-base (accessed on 10 October 2023). We download the pretrained models from HuggingFace [72,73]. For training, we use the full ATCO2-test-set-4h, which contains ~3k sentences. In this dataset, each word is annotated together with a predefined class, as follows: callsign, command, values, and

UNK (everything else). In order to fine-tune the model, we append a layer on top of the BERT model by using a feedforward network with a dimension of 8 (we define two outputs per class, see the class structures in Section 3.3 of ref. [8] and in [15]). Due to the lack of gold transcriptions, we perform a fivefold cross-validation scheme to avoid overfitting. The reader interested in developing their own NER system for ATC is redirected to the open-source GitHub repository of the ATCO2 corpora (GitHub repository: https://github.com/idiap/atco2-corpus; accessed on 10 October 2023). We fine-tune each model on an NVIDIA GeForce RTX 3090 for ~10 k steps. During experimentation, we use a linear learning rate scheduler with an initial learning rate of $\gamma = 5 \times 10^{-5}$, dropout [74] of $dp = 0.1$, and GELU (Gaussian error linear unit) activation function [75]. We also employ gradient norm clipping [76] for regularization and AdamW as optimizer [77]. Each model during the cross-validation scheme uses an effective batch size of 32.

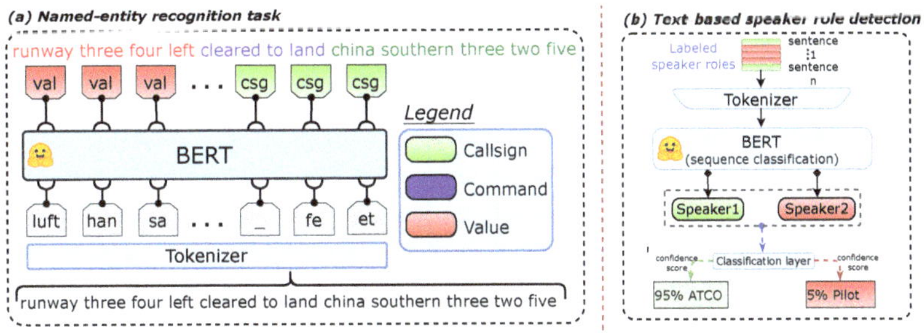

Figure 10. (**a**) Named entity recognition and (**b**) speaker role detection module based on sequence classification (SC). Both systems are based on fine-tuning a pretrained BERT [70] model on ATC data. The NER systems recognize callsign, command, and values, while the SC assigns a speaker role to the input sequence. Taken from [7].

Evaluation metric: we evaluate both BERT and RoBERTa NER systems with a binary classification metric named, F-score. Particularly, the F1-score, defined in Equation (8), represents the harmonic mean of precision and recall. Recall, as defined in Equation (7), is the ratio of TP to all samples that should have been identified as positive (including false negatives (FN)). Precision, as described in Equation (6), is the ratio of true positive (TP) results to all positive results (including false positives (FP)):

$$Precision = \frac{TP}{TP + FP} \tag{6}$$

$$Recall = \frac{TP}{TP + FN} \tag{7}$$

$$F1 = \frac{2 * Precision * Recall}{Precision + Recall} = \frac{2 * TP}{2 * TP + FP + FN} \tag{8}$$

Results and analysis: the NER system's performance is evaluated on the ATCO2-test-set-4h corpus using a fivefold cross-validation scheme, with five fine-tuning runs using different training seeds. Table 6 presents the performance metrics for callsign, command, and values classes of two transformer-based [69] models, namely, BERT-base and RoBERTa-base. Although both models achieve similar F1-scores, we provide analysis for the BERT-based NER system, which achieves an F1-score of over 97% for the callsign class, while the command and values classes lag behind with F1-scores of 81.9% and 87.1%, respectively. We hypothesize that the command class contains higher complexity when compared to the other two classes, values and callsigns. Values are mostly composed of defined keywords (e.g., flight level) followed by cardinal numbers (e.g., "one hundred"), while callsigns follow a well-defined structure of airline designators and a set of numbers or letters spoken in ICAO format [25]. These characteristics make it easier for the NER system to correctly detect them.

One potential method for increasing the performance of the NER system for the command and values classes is to incorporate plausible commands and values in real time, depending on the situation of the surveillance data. This can be achieved using the boosting technique, as described in Section 5.2.2. Although the results with boosting callsigns are reported in Table 5, further investigation is needed to assess the impact of boosting on the command and values classes.

Table 6. Different performance metrics for callsign, command, and values classes of the NER system. Results are averaged over a fivefold cross-validation scheme on *ATCO2-test-set-4h corpus* in order to mitigate overfitting. We run five-times fine-tuning with different training seeds (2222/3333/4444/5555/6666). Results are reported on two transformer-based models. @P, @R, and @F1 refer to precision, recall, and F1-score, respectively.

Model	Callsign			Command			Values		
	@P	@R	@F1	@P	@R	@F1	@P	@R	@F1
Bert-base	97.1	97.8	97.5	80.4	83.6	82.0	86.3	88.1	87.2
RoBERTa-base	97.1	97.7	97.5	80.2	83.7	81.9	85.6	88.6	87.1

5.3.2. Text-Based Speaker Role Detection

Sequence classification (SC) is a type of machine learning (ML) task that involves assigning a label or a category to a sequence of data points [78,79]. The data points in the sequence can be of various types, such as text, audio, or numerical data, and the label assigned to the sequence can also be of different types, such as binary (e.g., positive or negative sentiment [66]) or multiclass. Sequence classification can also be used to automatically classify ATC communication sequences into various categories. This technique can be applied to both audio and text data, making it a versatile tool to provide a high-level understanding of the communication at hand.

In scenarios where only a monaural communication channel exists, it can be challenging to recognize the identity of the speaker. Hence, it is especially important to distinguish between the ATCo and the pilot over the target communications. As a potential solution, we propose an alternative approach that utilizes a speaker role detection (SRD) system based on SC. The system receives text as an input, and it returns as output a category where the communication falls, either uttered by the ATCo or the pilot. In recent years, there has been a growing interest in using deep learning techniques, such as the transformer-based models [69], to improve the performance of SC for SRD in ATC communications. Here, we ablate three types of such models, (i) BERT [70], (ii) RoBERTa [71], and (iii) DEBERTA [80]. These models have been shown to achieve state-of-the-art performance on a wide range of sequence classification tasks, including SRD for ATC. The proposed SRD is illustrated in Figure 10b.

Overall, the SRD and speaker diarization (see Section 5.3.3) tasks can leverage the fact that ATC dialogues follow a well-defined lexicon and dictionary with simple grammar. This standard phraseology has been defined by the ICAO [25] for ATCos. The main idea is to guarantee safety and reduce miscommunications between the ATCos and pilots. Therefore, previous work has shown the potential in performing SRD in an E2E manner on the text-level, as presented here (see in [8,81]).

Experimental setup: he SRD system is built on top of pretrained models (BERT [70], RoBERTa [71], and DEBERTA [80]), which are downloaded from HuggingFace [72,73]. Here, the experimental setup is exactly the same as the one described for the NER system, including the training hyperparameters. For further details, we redirect the reader to Section 5.3.1. Still, the SRD model is fine-tuned on the SC rather than on the NER task. Further, we define an output layer with two units (classes): one for ATCo and one for pilot.

Results and analysis: we evaluate the SRD system on ATCO2-test-set-4h corpus. Differently from the NER system, here, we have access to two training corpora. (1) The Air Traffic Control Corpus (LDC-ATCC) corpus, see URL: https://catalog.ldc.upenn.edu/LDC94S14A (accessed on 10 October 2023). It consists of audio recordings in the area of ASR for air traffic control communications. We use the metadata along the transcripts to perform research on NLU for ATC, i.e., speaker role detection. The data files are sampled at 8 kHz, 16 bit linear, with continuous monitoring and without squelch or silence elimination. (2) the UWB-ATCC corpus by the University of West Bohemia, which can be downloaded for free at the following URL: https://lindat.mff.cuni.cz/repository/xmlui/handle/11858/00-097C-0000-0001-CCA1-0 (accessed on 10 October 2023). The UWB-ATCC corpus contains recordings of air traffic control communication. The speech is manually transcribed with the speaker information; thus, it can be used for speaker role detection) datasets. We evaluate the SRD under two considerations: (i) ablations of different pretrained models for SRD on ATC communications, and (ii) low-resource and incremental training scenarios.

(i) Analysis of the Impact of Pretrained Models and Training Data Type. In this scenario, we evaluate the impact of pretrained models and training data on the SRD task for ATC data. To this end, we compare the performance of three transformer-based [69] models, including BERT, RoBERTa, and deBERTa-V3, trained on two different corpora, LDC-ATCC and UWB-ATCC, and evaluate them on the ATCO2-test-set-4h corpus. The F1-scores for SRD are reported separately for ATCo and pilot speakers in Table 7. Our results show that all the models achieved comparable F1-scores, ranging from 87–88% for ATCo and 84–85% for pilots. These findings suggest that the SRD task for ATC data is not significantly sensitive to the choice of pretrained models. However, we observe that models trained on UWB-ATCC outperform those trained on LDC-ATCC, with up to 4% absolute improvement in F1-scores. For instance, BERT-model with LDC-ATCC → UWB-ATCC gives a comparison of 82.4% → 86.2% for ATCo and 79.2% → 83.2%, for Pilot. Additionally, we find that combining both datasets leads to a 1% absolute improvement in F1-scores. Overall, our study highlights the importance of selecting appropriate training data for the SRD task in ATC data and suggests that using multiple datasets can lead to improved performance. The findings also suggest that the choice of pretrained models has a relatively minor impact on the SRD task for ATC data.

Table 7. *ATCO/PILOT* F1-scores for speaker role identification based on full ATC utterances for ATCO2-test-set-4 test set. Each utterance represents one sample. Metrics reported with three different transformer-based models (BERT [70], RoBERTa [71], deBERTa-V3 [80]). All models are the "base" version, e.g., `bert-base`. Numbers in **bold** refer to the top performance per split, i.e., ATCO or PILOT. Results are averaged over a fivefold cross-validation scheme on *ATCO2-test-set-4h corpus* in order to mitigate overfitting. Each round of fine-tuning is run five times with different training seeds (2222/3333/4444/5555/6666).

Training Corpora	BERT		DEBERTA		ROBERTA	
	ATCO	PILOT	ATCO	PILOT	ATCO	PILOT
LDC-ATCC	82.4	79.2	82.4	79.6	84.0	80.2
UWB-ATCC	86.2	83.2	86.8	84.0	87.0	82.8
↪ + LDC-ATCC	**87.6**	**85.2**	**88.8**	**85.8**	**88.0**	**84.2**

(ii) Analysis of the Impact of Data Quantity on Speaker Role Detection. In this study, we aim to evaluate the impact of the number of text samples on the performance of SRD. The results of this analysis are illustrated in the left panel of Figure 11, where the F1-score on the ATCO2-test-set-4h is plotted against the number of samples in a logarithmic scale on the x-axis. Interestingly, we found that as few as 100 samples are necessary to achieve a reasonably good F1-score of 60% on SRD. Notably, the UWB-ATCC appears to be more informative for the BERT model, which achieves an F1-score of 71% with only 100 training samples. Increasing the training data to 1000 samples further improves the performance, resulting in F1-scores near 80% (LDC-ATCC + UWB-ATCC). These findings are significant, considering that the gold transcription of ATC communications is generally expensive and time-consuming. In the right panel of Figure 11, we present a box plot that shows the variation of the BERT model's performance when fine-tuned on SRD with different training seeds. Each box represents the variation of the model between the ATCo and pilot subsets, over the fivefold cross-validation scheme. Overall, the results indicate that increasing the training data leads to better performance and more consistent results. These observations highlight the importance of selecting a suitable training set size for speaker role detection tasks.

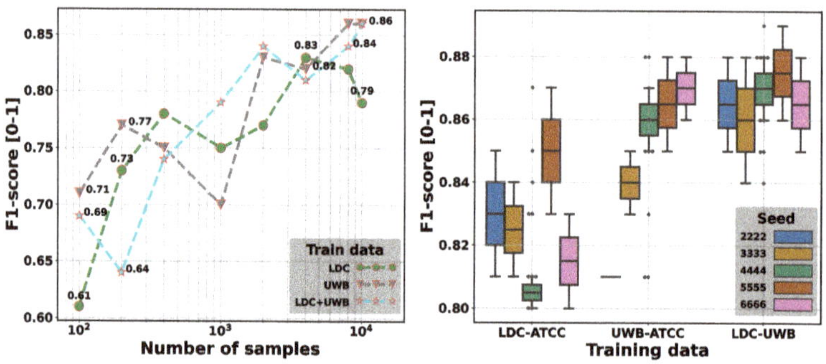

Figure 11. *Metrics for the speaker role detection system (introduced in [7]).* Metrics are reported only on *ATCO2-test-set-1h corpus* with a `bert-base-uncased` model trained with different datasets from Table 1. Left plot: ablation of the F1-score versus the number of samples used to train the system. Right plot: F1-score for models trained with different training seeds. The box plot depicts the performance variability when splitting the test set into ATCo and pilot subsets.

5.3.3. Text-Based Diarization

In addition to only detecting roles in a given ATC communication (e.g., SRD), there are cases where multiple segments end up in the same recording/communication. The task that solves this issue is known as speaker diarization (SD). SD answers the question *"who spoke when?"*. Here, the system receives an audio signal or recording (or text, in our case) and detects the speaker changes or segmentation and the speaker role. The main parts of an SD system are (i) segmentation, (ii) embedding extraction, (iii) clustering, and (iv) labeling (similar to SRD). SD is normally performed on the acoustic level, and previous work based on mel filterbank slope and linear filterbank slope was covered in [82]. Speaker discriminative embeddings such as x-vectors are investigated in [83], and, more recently, a variational Bayesian hidden Markov model (VBx) was investigated in [84], which is the SD system used during the data collection stage of ATCO2 (see Section 3.2). State-of-the-art SD systems are based on the E2E paradigm, named E2E neural diarization (EEND) [85]. This approach was introduced in [86] where an SD model is trained jointly to perform extraction and clustering [87]. Here, differently from SRD, we only used the BERT [70] pretrained model.

Experimental setup: The SD system is built on top of a pretrained BERT model downloaded from HuggingFace [72,73]. As in the NER and SRD system, here, the experimental setup is the same; this also includes the training hyperparameters. For further details we redirect the reader to Section 5.3.1. The SD model is fine-tuned on the NER task, where each speaker role (ATCo or pilot) is a class. Therefore, we have two tags per class, accounting for four classes in total. Readers are directed to our paper on text-based SD presented at The 2022 IEEE Spoken Language Technology Workshop (SLT 2022), see [8].

Evaluation metric: to score the text-based SD system, we use the Jaccard error rate (JER) metric. JER is a recent metric introduced in [88] that aligns with speaker diarization.

JER aims at avoiding the bias that the predominant speaker might cause, i.e., JER evaluates all speakers equally. The JER is defined in Equation (9):

$$JER = 1 - \frac{1}{\#\text{speakers}} \sum_{\text{speaker}} \max_{\text{cluster}} \frac{|\text{speaker} \cap \text{cluster}|}{|\text{speaker} \cup \text{cluster}|}, \quad (9)$$

where (i) speaker is the selected speaker from reference and (ii) $max_{cluster}$ is the cluster from the system with maximum overlap duration with the currently selected speaker.

Results and analysis: we evaluate the SRD system on ATCO2-test-set-4h corpus. Differently from the NER system, but similar to SRD, here, we have access to two training corpora: LDC-ATCC and UWB-ATCC datasets. We evaluate the SD under one consideration: (i) low-resource and incremental training scenario.

(i) Analysis of the Impact of Data Quantity on Text-based Speaker Diarization In this study, we aim to evaluate the impact of the number of text samples on the performance of SD. The results of this analysis are illustrated in the left panel of Figure 12, where the JER (the lower the better) on the ATCO2-test-set-4h is plotted against the number of samples in a logarithmic scale on the x-axis. We found that as few as 100 samples are necessary to achieve a JER score of 45.6% (LDC-UWB). Similar to SRD, the UWB-ATCC dataset seems to be more informative in the SD system. For instance, under the 1000 samples scenario, we noted a 5% absolute JER reduction if UWB-ATCC is used. Furthermore, increasing the training data to 10k samples improved the performance, resulting in JER scores near to 20% (LDC+UWB). A more appropriate comparison of text and acoustic-based SD for ATC communications can be found in our previous work [8]. Additionally, in the right panel of Figure 11, we present a box plot that shows the variation of the BERT-based SD model's performance when fine-tuned with different training seeds. Each box represents the variation of the model between the two proposed classes: ATCo and pilot, over the fivefold cross-validation scheme. The results are listed with F1-scores. Overall, we can conclude that the UWB-ATCC dataset is more informative for the SD model in comparison to the LDC-ATCC dataset.

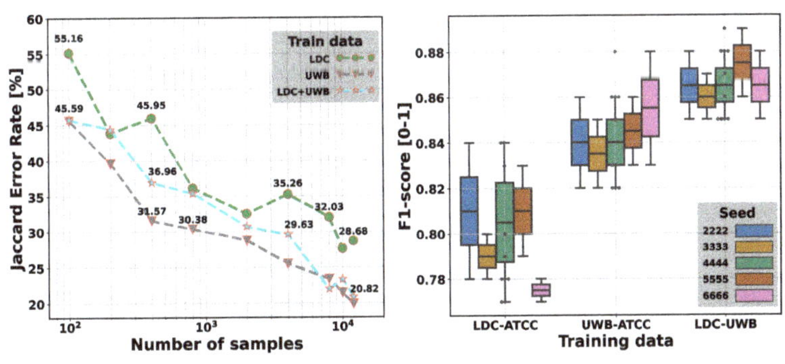

Figure 12. Metrics for the text-based diarization system (introduced in [7,8]). Metrics are reported only on *ATCO2-test-set-4h corpus* with a `bert-base-uncased` model trained with different datasets from Table 1. Left plot: ablation of the Jaccard error rate versus the number of samples used to train the system. Right plot: F1-score for models trained with different training seeds. The box plot depicts the performance variability when splitting the test set by ATCo and pilot subsets.

5.4. Future Work Enabled by ATCO2

In this subsection, we discuss several research directions that can be explored with the ATCO2 corpora. We cover (i) end of communication detection (akin to VAD), (ii) read-back error detection, and (iii) English language detection.

5.4.1. End of Communication Detection

In the ATC domain, it is crucial to detect the end of communications. While push-to-talk (PTT) signals are commonly acquirable in the ATC operations room or in the cockpit, there are cases where PTT is not available, and in such scenarios, the ATCO2 corpora can be leveraged to develop end-of-communication detection systems using either acoustic- or text-based approaches. Acoustic-based systems, known as VAD, perform their task prior to the ATC communication being sent to the ASR system [89], but may require the integration of a new independent module into the recognition pipeline. Text-based systems rely on strong artificial intelligence models like BERT, and previous studies in ATC [8] have shown their effectiveness in detecting callsigns [90], commands [7], and end-of-communication signals from transcripts generated by an ASR system.

5.4.2. Read-Back Error Detection

Pilot read-backs happens when a pilot speaks back the relevant instructions initially uttered by the ATCo. In practice, the ATCo is listening and checking the conformity of each read-back. Therefore, it is important to have a procedure in place, e.g., a read-back error detection (RBED) system. Despite the infrequency of communication errors in ATC, they still have the potential to cause significant safety issues, with some transmissions containing multiple errors. Authors in [91] show that in every hundredth ATC communication, an error may occur, and in [92], the authors show that the error may occur in every sixteenth communication. The possibility to detect such error still remains a challenge, as shown in this recent work [93]. Although, in general, read-back errors are quite rare, preventing even one incident due to automatic RBED can make an important difference in ensuring ATM safety. To support ATCos in this task, previous projects employ ASRU engines to extract high-level information from ATC communications [5]. Previous work in [93] has proposed two approaches for performing RBED. One system is based on rules, while a second system is a data-driven sequence classifier based on a BERT-alike pretrained encoder, named RoBERTa [71]. Here, the input sequence is a concatenation of ATCo and pilot utterance transcriptions with a special separator token `[SEP]` between them. They show that combining these approaches results in an 81% RBED rate in real-life voice recordings from Isavia's en-route airspace. They also cover a proof-of-concept trial with six ATCos producing challenging, artificial read-back error samples.

A main issue with well-known past projects, such as HAAWAII or MALORCA, is that their data cannot be publicly shared. In contrast, ATCO2 corpora are open to the public, e.g., *ATCO2-test-set-1h* set can be accessed for free, and practitioners can follow previous research to implement an RBED module.

5.4.3. English Language Detection

Currently, we have developed and deployed a suitable English language detection system (ELD) to discard non-English utterances in newly collected data. We tested a state-of-the-art acoustic-based system with an x-vector extractor. We also came up with the idea of using an NLP approach that processes ASR output with word confidence for the ELD. Finally, our experiments show that the ELD based on NLP is superior to the acoustic approach in both detection accuracy and computational resources. Moreover, the NLP approach can use outputs from several ASR systems jointly, which further improves the results. For the processing pipeline, we integrated the NLP-based English detector operating on Czech and English ASR. The integrated English detector consists of TF–IDF (term frequency–inverse document frequency) for reweighting the accumulated "soft" word counts and a logistic regression classifier to obtain the English/non-English decision [24].

We created the development and evaluation dataset consisting of data from various airports, data with various English accents, and code-mixing of English and local languages. The data are selected from our ATCO2 corpora introduced in Table 1. The development set is used to estimate the model parameters of our English language detector (the logistic regression classifier). The evaluation set is used for testing. The rules for manually transcribing the utterances are mentioned below. We found several interesting properties of the ATC data during listening and tagging the ELD dataset:

- Various noise conditions. The majority of data are clean, but there are some very noisy segments;
- Strongly accented English. The speakers' English accent varies widely. From native speakers (pilots) to international accents (French, German, Russian, etc.) (pilots and ATCos) and strong Czech accents (pilots and ATCos);
- Mixed words and phrases. For example, the vocabulary of Czech ATCos is a mix of Czech and English words. They use standard greetings in Czech which can be a significant portion of an "English" sentence if a command is short. On the other hand, they use many English words (alphabet, some commands) in "Czech" sentences. Moreover, they use a significant set of "Czenglish" words.

We use the language of spoken numerals as a rule of thumb to decide on the language of a particular ATCo-pilot communication utterance. The language has to be consistent within the audio recording. More detailed information, including experimental results, is covered in our previous work [24].

6. Conclusions

This paper expands upon our previous work [7] and discusses the main lessons learned from the ATCO2 project. The aim of the ATCO2 project was to develop a platform for collecting, preprocessing, and posterior ASR-based transcription generation of ATC communications audio data. With over 5000 h of ASR transcribed audio data, ATCO2 is the largest public ATC dataset to date, thus pushing the research boundaries on robust automatic speech recognition and natural language understanding of ATC communications. The main lessons learned from ATCO2 are sixfold, as follows:

- **Lesson 1:** ATCO2's automatic transcript engine (see Appendix B) and annotation platform (see Appendix C) have proven to be reliable (~20% WER on ATCO2-test-set-4h) for collection of a large-scale audio dataset targeted to ATC communications;
- **Lesson 2:** Good transcription practices for ATC communications have been developed based on ontologies published by previous projects [5]. A cheat sheet (see Appendix E) has been created to provide guidance for future ATC projects and reduce confusion while generating transcripts;
- **Lesson 3:** The most demanding modules of the ATCO2 collection platform are the speaker diarization and automatic speech recognition engines, each accounting for ~32% of the overall system processing time. The complete statistics regarding runtime are covered in the Table 3. In ATCO2, we make these numbers public so they can be used as baselines in future work aligned to reducing the overall memory and runtime footprint of large-scale collection of ATC audio and radar data;
- **Lesson 4:** Training ASR systems purely on ATCO2 datasets (e.g., *ATCO2-T 500h set corpus*) can achieve competitive WERs on ATCO2 test sets (see Table 4). The ASR model can achieve up to 17.9%/24.9% WERs on ATCO2-test-set-1h/ATCO2-test-set-4h, respectively. More importantly, these test sets contain noisy accented speech, which is highly challenging in standard ASR systems;
- **Lesson 5:** ATC surveillance data are an optimal source of real-time information to improve ASR outputs. The integration of air surveillance data can lead to up to 11.8% absolute callsign WERs reduction, which represents an amelioration of 20% (62.6% no boosting → 82.9% GT boosted) absolute callsign accuracy in ATCO2-test-set-4h, as shown in Table 5;
- **Lesson 6:** ATCO2 corpora can be used for natural language understanding of ATC communications. BERT-based NER and speaker role detection modules have been developed based on ATCO2-test-set-4h. These systems can detect callsigns, commands, and values from the textual inputs. Additionally, speaker roles can also be detected based on textual inputs. For instance, as few as 100 samples are necessary to achieve

60% F1-score on speaker role detection. Furthermore, the NLU task is of special interest to the ATC community because this high-level information can be used to assist ATCos in their daily tasks, thus reducing their overall workload.

In addition to these six lessons learned, this paper brings substantial improvements in the domain of automatic speech recognition and understanding for ATC domain, i.e., Tables 5 and 6 show the current best-performing ASR and NLU engines developed on open-source data, and, thus, are replicable by the community. Furthermore, to the authors' knowledge, there is no other research or commercial activity at this moment which would demonstrate a more accurate engine for an ATC domain built on publicly open data.

Author Contributions: Conceptualization, J.Z.-G., S.M., I.S., V.L., M.R. and K.C.; Data curation, J.Z.-G. and I.N.; Formal analysis, J.Z.-G. and I.S.; Funding acquisition, P.M.; Investigation, J.Z.-G.; Methodology, J.Z.-G., I.N., A.P. and A.T.; Project administration, P.M.; Resources, J.Z.-G.; Software, J.Z.-G., S.M., A.T. and I.S.; Supervision, P.M., S.M., I.S. and V.L.; Validation, J.Z.-G., A.P. and I.S.; Visualization, J.Z.-G. and M.R.; Writing—original draft, J.Z.-G. and I.N.; Writing—review and editing, J.Z.-G., I.N., A.P., P.M., D.K., V.L., M.R. and K.C. All authors have read and agreed to the published version of the manuscript.

Funding: This research received no external funding.

Acknowledgments: This paper presents a continuation of our previous work [7], the *ATCO2 Corpora*. ATCO2 is derived from a joint contribution from Clean Sky 2 Joint Undertaking (JU) and EU-H2020, which was supported by Clean Sky 2 Joint Undertaking (JU) and EU-H2020, under Grant Agreement No. 864702—ATCO2 (Automatic collection and processing of voice data from air traffic communications). See our website at https://www.atco2.org/ (accessed on 10 October 2023).

Conflicts of Interest: The authors declare no conflicts of interest.

Nomenclature

AI	Artificial Intelligence
AM	Acoustic Model
ATC	Air Traffic Control
ATM	Air Traffic Management
ASR	Automatic Speech Recognition
ATCo	Air Traffic Controller
ATCC	Air Traffic Control Corpus
ADS-B	Automatic Dependent Surveillance–Broadcast
CTC	Connectionist Temporal Classification
Conformer	Convolution-augmented Transformer
dB	Decibel
DNN	Deep Neural Networks
E2E	End-To-End
ELDA	European Language Resources Association
FST	Finite State Transducer
ICAO	International Civil Aviation Organization
GELU	Gaussian Error Linear Units
LF-MMI	Lattice-Free Maximum Mutual Information
LM	Language Model
ML	Machine Learning
NER	Named Entity Recognition
NLP	Natural Language Processing
NLU	Natural Language Understanding
PTT	Push-To-Talk
SNR	Signal-To-Noise
VAD	Voice Activity Detection
VHF	Very-High Frequency
WER	Word Error Rate
WFST	Weighted Finite State Transducer

RBE Read-back Error
RBED Read-back Error Detection

Appendix A. ATCO2 Project

The ATCO2 project developed an unique platform that allows the collection, organization, and preprocessing of air traffic control (voice communication) data from airspace. The project considers real-time voice communication between air traffic controllers and pilots available either directly through publicly accessible radio frequency channels, or indirectly from air-navigation service providers (ANSPs). In addition to the voice communication, the contextual information available in a form of metadata (i.e., surveillance data) is exploited.

More specifically, data acquisition was based on off-the-shelf automatic-dependent surveillance-broadcast (ADS-B) technology already exploited by OpenSky Network (OSN), collecting detailed (live) aircraft information over the publicly accessible 1090 MHz radio frequency channel. ADS-B sensors are distributed among volunteers (i.e., community of users) throughout the world (at https://opensky-network.org/network/facts; accessed on 10 October 2023), making it possible to analyze billions of ADS-B messages. The aim of ATCO2 was to extend the current cloud setting (i.e., central server responsible for managing the sensors, collecting the received ADS-B messages, and storing all received information in a database) so that ATC voice communication of both channels (ATCo and pilot's read back) will be captured, time correlated with the surveillance data, and stored in the database for further processing. These data will also be complemented by air traffic voice communication data provided by ANSPs. Below are listed some links that might be of interest to the reader.

- The latest news and blog posts from ATCO2 project are located in the following website: https://www.atco2.org/; accessed on 10 October 2023.
- The ATCO2 corpus can be downloaded for a fee at https://catalog.elra.info/en-us/repository/browse/ELRA-S0484/; accessed on 10 October 2023.
- The ATCO2-test-set-1h can be downloaded for free at https://www.atco2.org/data; accessed on 10 October 2023.
- Stats and voice feeding of ATC data is listed at https://ui.atc.opensky-network.org/set-up; accessed on 10 October 2023.
- ATC training and transcription service is provided by SpokenData at: https://www.spokendata.com/atco2; accessed on 10 October 2023.

Appendix B. Automatic Transcription Engine

This appendix describes in detail how we collected the audio and metadata that brought to life the *ATCO2 corpus*. We mainly rely on the automatic transcription engine, described in more detail in Section 3.2. The automatic transcription engine is implemented as a scalable cloud service. It communicates with other services (or partners) using APIs. This service is designed to process large flows of data produced by data feeders. Data feeders are enthusiasts that act as "feeders" of ATC speech and contextual ATC data (e.g., surveillance), see Section 4.2.

The data are pushed to this service by OSN (OpenSky Network: https://opensky-network.org/; accessed on 10 October 2023) servers by calling an API request and providing a job setting JSON file. After the request is accepted, settings parameters are processed and the job is stored in an internal queue for processing. The user (in this case, OSN) may have an ability to tweak the settings and to affect the processing pipeline and the result, namely:

- Audio input format choices;
- Rejection threshold for too-long audio;
- Rejection threshold for too-short audio;
- Rejection threshold for too-noisy audio;
- Rejection threshold for non-English audios;
- Switching the language of automatic speech recognizer.

Most of these are actually disabled due to security reasons (not to interrupt the processing pipeline), but may be easily enabled on the fly if needed. The overall data flow model is described in Figure A1. Any new job (request for a full automatic transcription of recording) accepted via API on the SpokenData (Industrial partner: https://www.spokendata.com/atco2; accessed on 10 October 2023) side is processed by a master processing node. The job is enqueued into a workload manager queue. Once there is a free processing slot, the job is submitted to a processing server, or worker. The master processing node then informs the OSN server about the state of the job by calling a callback.

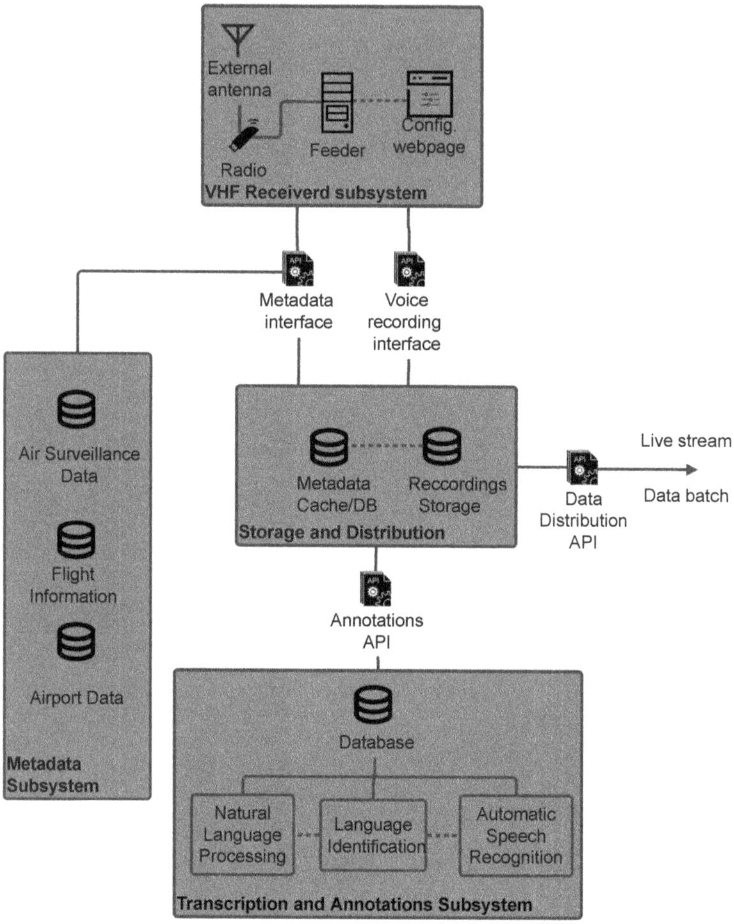

Figure A1. ATCO2 communication schema.

Appendix C. Transcription Platform: Data Flow

The data (the recording for human transcription) life cycle is split into four main states:

The **new recording** state is set as queued and is untouched when the recording is pushed into the transcription platform from the transcription engine. The recording is placed into a queue of transcription jobs and is immediately visible to all annotators. The queue is shown in the open jobs screen. Annotators can interact with the queue—listen to recordings and select some for transcription. Recording in this state may drop off the queue in the case: they are old—no one is interested in annotating them; three annotators marked the recording by thumbs down. The dropped-off recordings are deleted after 7 days.

Once an user selects the recording for transcription, revises the automatic transcription, and saves it, the recording is set as **queued and annotated**. This state prevents the recording from being dropped from the queue and deleted. Also, it is indicated as (to) re-check in the open jobs screen, to inform other annotators that it was modified (annotated) and they should recheck if the transcription is correct rather than annotate from scratch. If any annotator indicates the existence of personal information in the recording (by "Anonymize" label), the recording is dropped off the queue and deleted.

The next state is **annotated**. If the recording is successfully rechecked, then the recording is considered as annotated and the transcription is final. The recording is removed from the open jobs queue and placed on a stack of finished recording transcriptions. The stack is periodically exported to ELRA for further packaging and distribution to the community. This state also triggers a callback to the OSN platform, informing them that the human transcription is completed, and they can download the transcription. After the recording is exported to ELRA, we set the state as **Finished**. Here, the recording can be archived or deleted. The detailed data flow schema is depicted in Figure A2.

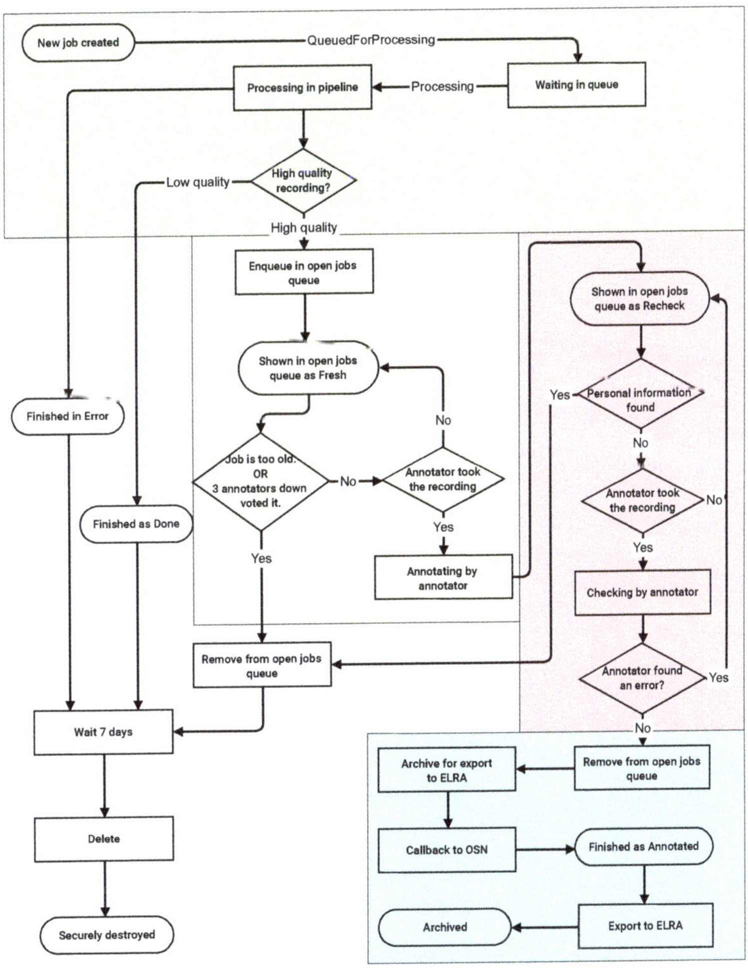

Figure A2. Diagram of the data flow (lifetime) in the transcription platform. Transcription engine in green. Queued and untouched state in yellow. Queued and annotated state in red. Annotated state in blue. The rest (white) is for state, securely destroyed.

Appendix D. Communication Schema

The communication schema developed during ATCO2 project is depicted in Figure A3.

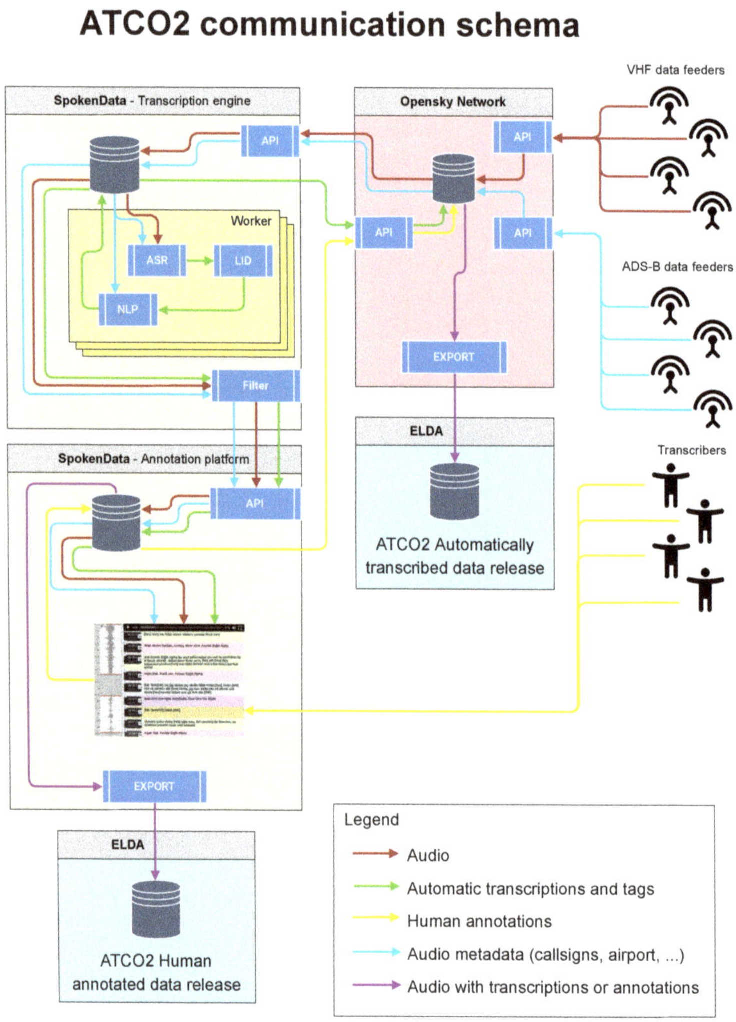

Figure A3. Other view of the ATCO2 communication schema.

Appendix E. Transcription Cheat Sheet

Figure A4 presents the transcription cheat sheet developed by ATCO2 project.

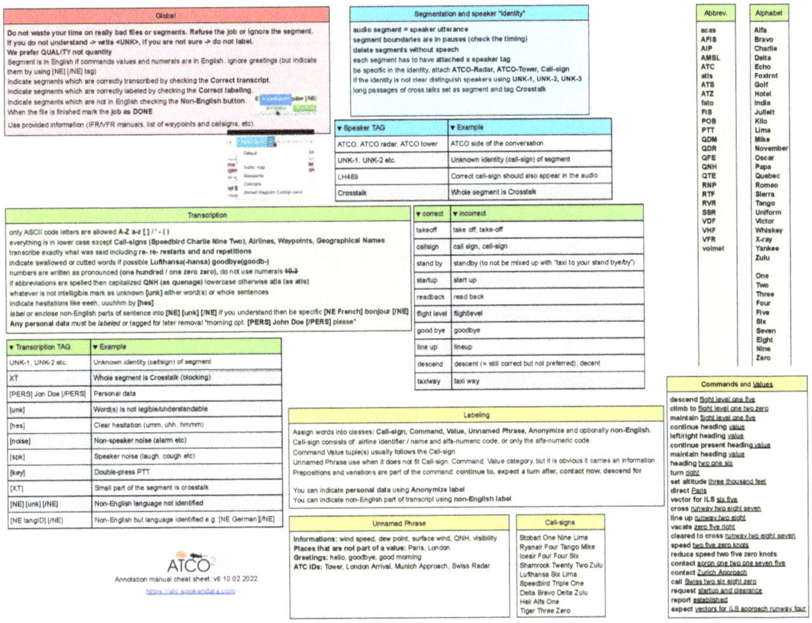

Figure A4. Cheat sheet for ATC communications annotation. This document was created during the transcription process of ATCO2 corpora, which can be used to transcribe air traffic control communications data from different airports.

References

1. Helmke, H.; Ohneiser, O.; Buxbaum, J.; Kern, C. Increasing ATM efficiency with assistant based speech recognition. In Proceedings of the 13th USA/Europe Air Traffic Management Research and Development Seminar, Seattle, WA, USA, 27–30 June 2017.
2. Shetty, S.; Helmke, H.; Kleinert, M.; Ohneiser, O. Early Callsign Highlighting using Automatic Speech Recognition to Reduce Air Traffic Controller Workload. In Proceedings of the International Conference on Applied Human Factors and Ergonomics (AHFE2022), New York, NY, USA, 24–28 July 2022; Volume 60, pp. 584–592.
3. Ohneiser, O.; Helmke, H.; Shetty, S.; Kleinert, M.; Ehr, H.; Murauskas, Š.; Pagirys, T.; Balogh, G.; Tønnesen, A.; Kis-Pál, G.; et al. Understanding Tower Controller Communication for Support in Air Traffic Control Displays. In Proceedings of the SESAR Innovation Days, Budapest, Hungary, 5–8 December 2022.
4. Helmke, H.; Ohneiser, O.; Mühlhausen, T.; Wies, M. Reducing controller workload with automatic speech recognition. In Proceedings of the 2016 IEEE/AIAA 35th Digital Avionics Systems Conference (DASC), Sacramento, CA, USA, 25–29 September 2016; pp. 1–10.
5. Helmke, H.; Slotty, M.; Poiger, M.; Herrer, D.F.; Ohneiser, O.; Vink, N.; Cerna, A.; Hartikainen, P.; Josefsson, B.; Langr, D.; et al. Ontology for transcription of ATC speech commands of SESAR 2020 solution PJ. 16-04. In Proceedings of the IEEE/AIAA 37th Digital Avionics Systems Conference (DASC), London, UK, 23–27 September 2018; pp. 1–10.
6. Guo, D.; Zhang, Z.; Yang, B.; Zhang, J.; Lin, Y. Boosting Low-Resource Speech Recognition in Air Traffic Communication via Pretrained Feature Aggregation and Multi-Task Learning. *IEEE Trans. Circuits Syst. II Express Briefs* **2023**, *70*, 3714–3718. [CrossRef]
7. Zuluaga-Gomez, J.; Veselý, K.; Szöke, I.; Motlicek, P.; Kocour, M.; Rigault, M.; Choukri, K.; Prasad, A.; Sarfjoo, S.S.; Nigmatulina, I.; et al. ATCO2 corpus: A Large-Scale Dataset for Research on Automatic Speech Recognition and Natural Language Understanding of Air Traffic Control Communications. *arXiv* **2022**, arXiv:2211.04054.
8. Zuluaga-Gomez, J.; Sarfjoo, S.S.; Prasad, A.; Nigmatulina, I.; Motlicek, P.; Ondre, K.; Ohneiser, O.; Helmke, H. BERTraffic: BERT-based Joint Speaker Role and Speaker Change Detection for Air Traffic Control Communications. In Proceedings of the IEEE Spoken Language Technology Workshop (SLT), Doha, Qatar, 9–12 January 2023.
9. Kocour, M.; Veselý, K.; Szöke, I.; Kesiraju, S.; Zuluaga-Gomez, J.; Blatt, A.; Prasad, A.; Nigmatulina, I.; Motlíček, P.; Klakow, D.; et al. Automatic processing pipeline for collecting and annotating air-traffic voice communication data. *Eng. Proc.* **2021**, *13*, 8.
10. Ferreiros, J.; Pardo, J.; De Córdoba, R.; Macias-Guarasa, J.; Montero, J.; Fernández, F.; Sama, V.; González, G. A speech interface for air traffic control terminals. *Aerosp. Sci. Technol.* **2012**, *21*, 7–15. [CrossRef]

11. Tarakan, R.; Baldwin, K.; Rozen, N. An automated simulation pilot capability to support advanced air traffic controller training. In Proceedings of the 26th Congress of ICAS and 8th AIAA ATIO, Anchorage, AK, USA, 14–19 September 2008.
12. Cordero, J.M.; Dorado, M.; de Pablo, J.M. Automated speech recognition in ATC environment. In Proceedings of the 2nd International Conference on Application and Theory of Automation in Command and Control Systems, London, UK, 29–31 May 2012; pp. 46–53.
13. Zuluaga-Gomez, J.; Motlicek, P.; Zhan, Q.; Veselý, K.; Braun, R. Automatic Speech Recognition Benchmark for Air-Traffic Communications. In Proceedings of the Interspeech, Shanghai, China, 25–29 October 2020; pp. 2297–2301. [CrossRef]
14. Zuluaga-Gomez, J.; Nigmatulina, I.; Prasad, A.; Motlicek, P.; Veselý, K.; Kocour, M.; Szöke, I. Contextual Semi-Supervised Learning: An Approach to Leverage Air-Surveillance and Untranscribed ATC Data in ASR Systems. In Proceedings of the Interspeech, Brno, Czech Republic, 30 August–3 September 2021; pp. 3296–3300. [CrossRef]
15. Nigmatulina, I.; Zuluaga-Gomez, J.; Prasad, A.; Sarfjoo, S.S.; Motlicek, P. A two-step approach to leverage contextual data: Speech recognition in air-traffic communications. In Proceedings of the IEEE International Conference on Acoustics, Speech and Signal Processing (ICASSP), Virtual, 7–13 May 2022.
16. Chen, S.; Kopald, H.; Avjian, B.; Fronzak, M. Automatic Pilot Report Extraction from Radio Communications. In Proceedings of the 2022 IEEE/AIAA 41st Digital Avionics Systems Conference (DASC), Portsmouth, VA, USA, 18–22 September 2022; pp. 1–8.
17. Godfrey, J. The Air Traffic Control Corpus (ATC0)—LDC94S14A. 1994. Available online: https://catalog.ldc.upenn.edu/LDC94S14A (accessed on 4 September 2023).
18. Šmídl, L.; Švec, J.; Tihelka, D.; Matoušek, J.; Romportl, J.; Ircing, P. Air traffic control communication (ATCC) speech corpora and their use for ASR and TTS development. *Lang. Resour. Eval.* **2019**, *53*, 449–464. [CrossRef]
19. Segura, J.; Ehrette, T.; Potamianos, A.; Fohr, D.; Illina, I.; Breton, P.; Clot, V.; Gemello, R.; Matassoni, M.; Maragos, P. The HIWIRE Database, a Noisy and Non-Native English Speech Corpus for Cockpit Communication. 2007. Available online: http://www.hiwire.org (accessed on 10 October 2023).
20. Delpech, E.; Laignelet, M.; Pimm, C.; Raynal, C.; Trzos, M.; Arnold, A.; Pronto, D. A Real-life, French-accented Corpus of Air Traffic Control Communications. In Proceedings of the Eleventh International Conference on Language Resources and Evaluation (LREC 2018), Miyazaki, Japan, 7–12 May 2018.
21. Pellegrini, T.; Farinas, J.; Delpech, E.; Lancelot, F. The Airbus Air Traffic Control speech recognition 2018 challenge: Towards ATC automatic transcription and call sign detection. *arXiv* **2018**, arXiv:1810.12614.
22. Graglia, L.; Favennec, B.; Arnoux, A. Vocalise: Assessing the impact of data link technology on the R/T channel. In Proceedings of the 24th IEEE Digital Avionics Systems Conference, Washington, DC, USA, 30 October–3 November 2005; Volume 1.
23. Lopez, S.; Condamines, A.; Josselin-Leray, A.; O'Donoghue, M.; Salmon, R. Linguistic analysis of English phraseology and plain language in air-ground communication. *J. Air Transp. Stud.* **2013**, *4*, 44–60. [CrossRef]
24. Szöke, I.; Kesiraju, S.; Novotný, O.; Kocour, M.; Veselý, K.; Černocký, J. Detecting English Speech in the Air Traffic Control Voice Communication. In Proceedings of the Interspeech, Brno, Czech Republic, 30 August–3 September 2021; pp. 3286–3290. [CrossRef]
25. International Civil Aviation Organization. *ICAO Phraseology Reference Guide*; International Civil Aviation Organization: Montreal, QC, Canada, 2020.
26. Panayotov, V.; Chen, G.; Povey, D.; Khudanpur, S. Librispeech: An ASR corpus based on public domain audio books. In Proceedings of the IEEE International Conference on Acoustics, Speech and Signal Processing (ICASSP), South Brisbane, Australia, 19–24 April 2015; pp. 5206–5210.
27. Ardila, R.; Branson, M.; Davis, K.; Henretty, M.; Kohler, M.; Meyer, J.; Morais, R.; Saunders, L.; Tyers, F.M.; Weber, G. Common voice: A massively-multilingual speech corpus. *arXiv* **2019**, arXiv:1912.06670.
28. Godfrey, J.J.; Holliman, E.C.; McDaniel, J. SWITCHBOARD: Telephone speech corpus for research and development. In Proceedings of the Acoustics, Speech, and Signal Processing, IEEE International Conference on IEEE Computer Society, San Francisco, CA, USA, 23–26 March 1992; Volume 1, pp. 517–520.
29. Povey, D.; Ghoshal, A.; Boulianne, G.; Burget, L.; Glembek, O.; Goel, N.; Hannemann, M.; Motlicek, P.; Qian, Y.; Schwarz, P.; et al. The Kaldi speech recognition toolkit. In Proceedings of the IEEE Workshop on Automatic Speech Recognition and Understanding IEEE Signal Processing Society, Waikoloa, HI, USA, 11–15 December 2011.
30. Jurafsky, D.; Martin, J.H. *Speech and Language Processing*, 2nd ed.; Prentice-Hall, Inc.: Upper Saddle River, NJ, USA, 2009.
31. Wang, D.; Wang, X.; Lv, S. An Overview of End-to-End Automatic Speech Recognition. *Symmetry* **2019**, *11*, 1018. [CrossRef]
32. Sennrich, R.; Haddow, B.; Birch, A. Neural Machine Translation of Rare Words with Subword Units. In *Proceedings of the 54th Annual Meeting of the Association for Computational Linguistics (Volume 1: Long Papers)*; Association for Computational Linguistics: Berlin, Germany, 2016; pp. 1715–1725. [CrossRef]
33. Kingsbury, B. Lattice-based optimization of sequence classification criteria for neural-network acoustic modeling. In Proceedings of the 2009 IEEE International Conference on Acoustics, Speech and Signal Processing, Taipei, Taiwan, 19–24 April 2009; pp. 3761–3764.
34. Dehak, N.; Kenny, P.J.; Dehak, R.; Dumouchel, P.; Ouellet, P. Front-end factor analysis for speaker verification. *IEEE Trans. Audio Speech Lang. Process.* **2010**, *19*, 788–798. [CrossRef]

35. Snyder, D.; Garcia-Romero, D.; Povey, D. Time delay deep neural network-based universal background models for speaker recognition. In Proceedings of the 2015 IEEE Workshop on Automatic Speech Recognition and Understanding (ASRU), Scottsdale, AZ, USA, 13–17 December 2015; pp. 92–97.
36. Peddinti, V.; Povey, D.; Khudanpur, S. A time delay neural network architecture for efficient modeling of long temporal contexts. In Proceedings of the Interspeech, Dresden, Germany, 6–10 September 2015; pp. 3214–3218. [CrossRef]
37. Povey, D.; Peddinti, V.; Galvez, D.; Ghahremani, P.; Manohar, V.; Na, X.; Wang, Y.; Khudanpur, S. Purely sequence-trained neural networks for ASR based on lattice-free MMI. In Proceedings of the Interspeech, San Francisco, CA, USA, 8–12 September 2016; pp. 2751–2755.
38. Povey, D.; Cheng, G.; Wang, Y.; Li, K.; Xu, H.; Yarmohammadi, M.; Khudanpur, S. Semi-orthogonal low-rank matrix factorization for deep neural networks. In Proceedings of the Interspeech, Hyderabad, India, 2–6 September 2018; pp. 3743–3747.
39. Nassif, A.B.; Shahin, I.; Attili, I.; Azzeh, M.; Shaalan, K. Speech recognition using deep neural networks: A systematic review. *IEEE Access* **2019**, *7*, 19143–19165. [CrossRef]
40. Graves, A.; Jaitly, N. Towards End-to-End Speech Recognition with Recurrent Neural Networks. In Proceedings of the 31th International Conference on Machine Learning, ICML 2014, Beijing, China, 21–26 June 2014; Volume 32, pp. 1764–1772.
41. Karita, S.; Chen, N.; Hayashi, T.; Hori, T.; Inaguma, H.; Jiang, Z.; Someki, M.; Soplin, N.E.Y.; Yamamoto, R.; Wang, X.; et al. A comparative study on transformer vs rnn in speech applications. In Proceedings of the 2019 IEEE Automatic Speech Recognition and Understanding Workshop (ASRU), Singapore, 14–18 December 2019; pp. 449–456.
42. Graves, A.; Fernández, S.; Gomez, F.; Schmidhuber, J. Connectionist temporal classification: Labelling unsegmented sequence data with recurrent neural networks. In Proceedings of the 23rd International Conference on Machine Learning, Pittsburgh, Pennsylvania, 25–29 June 2006; pp. 369–376.
43. Chorowski, J.; Bahdanau, D.; Serdyuk, D.; Cho, K.; Bengio, Y. Attention-Based Models for Speech Recognition. In Proceedings of the Advances in Neural Information Processing Systems, Montreal, QC, Canada, 7–12 December 2015; pp. 577–585.
44. Watanabe, S.; Hori, T.; Kim, S.; Hershey, J.R.; Hayashi, T. Hybrid CTC/attention architecture for end-to-end speech recognition. *IEEE J. Sel. Top. Signal Process.* **2017**, *11*, 1240–1253. [CrossRef]
45. Oord, A.v.d.; Li, Y.; Vinyals, O. Representation learning with contrastive predictive coding. *arXiv* **2018**, arXiv:1807.03748.
46. Baevski, A.; Zhou, Y.; Mohamed, A.; Auli, M. wav2vec 2.0: A Framework for Self-Supervised Learning of Speech Representations. In Proceedings of the Advances in Neural Information Processing Systems, Virtual, 6–12 December 2020.
47. Baevski, A.; Schneider, S.; Auli, M. vq-wav2vec: Self-Supervised Learning of Discrete Speech Representations. In Proceedings of the 8th International Conference on Learning Representations, ICLR 2020, Addis Ababa, Ethiopia, 26–30 April 2020.
48. Chen, S.; Wang, C.; Chen, Z.; Wu, Y.; Liu, S.; Li, J.; Kanda, N.; Yoshioka, T.; Xiao, X.; Wei, F.; et al. Wavlm: Large-scale self-supervised pre-training for full stack speech processing. *arXiv* **2021**, arXiv:2110.13900.
49. Babu, A.; Wang, C.; Tjandra, A.; Lakhotia, K.; Xu, Q.; Goyal, N.; Singh, K.; von Platen, P.; Saraf, Y.; Pino, J.; et al. XLS-R: Self-supervised cross-lingual speech representation learning at scale. *arXiv* **2021**, arXiv:2111.09296.
50. Zuluaga-Gomez, J.; Prasad, A.; Nigmatulina, I.; Sarfjoo, S.; Motlicek, P.; Kleinert, M.; Helmke, H.; Ohneiser, O.; Zhan, Q. How Does Pre-trained Wav2Vec2.0 Perform on Domain Shifted ASR? An Extensive Benchmark on Air Traffic Control Communications. In Proceedings of the IEEE Spoken Language Technology Workshop (SLT), Doha, Qatar, 9–12 January 2023.
51. Gulati, A.; Qin, J.; Chiu, C.C.; Parmar, N.; Zhang, Y.; Yu, J.; Han, W.; Wang, S.; Zhang, Z.; Wu, Y.; et al. Conformer: Convolution-augmented Transformer for Speech Recognition. In Proceedings of the Interspeech, Shanghai, China, 25–29 October 2020; pp. 5036–5040. [CrossRef]
52. Mai, F.; Zuluaga-Gomez, J.; Parcollet, T.; Motlicek, P. HyperConformer: Multi-head HyperMixer for Efficient Speech Recognition. In Proceedings of the Interspeech, Dublin, Ireland, 20–24 August 2023.
53. Radfar, M.; Lyskawa, P.; Trujillo, B.; Xie, Y.; Zhen, K.; Heymann, J.; Filimonov, D.; Strimel, G.; Susanj, N.; Mouchtaris, A. Conmer: Streaming Conformer without self-attention for interactive voice assistants. In Proceedings of the Interspeech, Dublin, Ireland, 20–24 August 2023.
54. Peng, Y.; Dalmia, S.; Lane, I.; Watanabe, S. Branchformer: Parallel mlp-attention architectures to capture local and global context for speech recognition and understanding. In Proceedings of the International Conference on Machine Learning, Guangzhou, China, 18–21 February 2022; pp. 17627–17643.
55. Ravanelli, M.; Parcollet, T.; Plantinga, P.; Rouhe, A.; Cornell, S.; Lugosch, L.; Subakan, C.; Dawalatabad, N.; Heba, A.; Zhong, J.; et al. SpeechBrain: A general-purpose speech toolkit. *arXiv* **2021**, arXiv:2106.04624.
56. Watanabe, S.; Hori, T.; Karita, S.; Hayashi, T.; Nishitoba, J.; Unno, Y.; Enrique Yalta Soplin, N.; Heymann, J.; Wiesner, M.; Chen, N.; et al. ESPnet: End-to-End Speech Processing Toolkit. In Proceedings of the Interspeech, Hyderabad, India, 2–6 September 2018; pp. 2207–2211. [CrossRef]
57. Kuchaiev, O.; Li, J.; Nguyen, H.; Hrinchuk, O.; Leary, R.; Ginsburg, B.; Kriman, S.; Beliaev, S.; Lavrukhin, V.; Cook, J.; et al. Nemo: A toolkit for building ai applications using neural modules. *arXiv* **2019**, arXiv:1909.09577.
58. Zhang, B.; Wu, D.; Peng, Z.; Song, X.; Yao, Z.; Lv, H.; Xie, L.; Yang, C.; Pan, F.; Niu, J. Wenet 2.0: More productive end-to-end speech recognition toolkit. *arXiv* **2022**, arXiv:2203.15455.
59. Hall, K.; Cho, E.; Allauzen, C.; Beaufays, F.; Coccaro, N.; Nakajima, K.; Riley, M.; Roark, B.; Rybach, D.; Zhang, L. Composition-based on-the-fly rescoring for salient n-gram biasing. In Proceedings of the Interspeech, Dresden, Germany, 6–10 September 2015; pp. 1418–1422.

60. Aleksic, P.; Ghodsi, M.; Michaely, A.; Allauzen, C.; Hall, K.; Roark, B.; Rybach, D.; Moreno, P. Bringing Contextual Information to Google Speech Recognition. 2015. Available online: https://research.google/pubs/pub43819/ (accessed on 4 September 2023).
61. Serrino, J.; Velikovich, L.; Aleksic, P.S.; Allauzen, C. Contextual Recovery of Out-of-Lattice Named Entities in Automatic Speech Recognition. In Proceedings of the Interspeech, Graz, Austria, 15–19 September 2019; pp. 3830–3834.
62. Chen, Z.; Jain, M.; Wang, Y.; Seltzer, M.L.; Fuegen, C. End-to-end contextual speech recognition using class language models and a token passing decoder. In Proceedings of the IEEE International Conference on Acoustics, Speech and Signal Processing (ICASSP), Brighton, UK, 12–17 May 2019; pp. 6186–6190.
63. Yadav, V.; Bethard, S. A Survey on Recent Advances in Named Entity Recognition from Deep Learning models. In Proceedings of the 27th International Conference on Computational Linguistics, Santa Fe, NM, USA, 20–26 August 2018; pp. 2145–2158.
64. Sharma, A.; Chakraborty, S.; Kumar, S. Named Entity Recognition in Natural Language Processing: A Systematic Review. In *Proceedings of the Second Doctoral Symposium on Computational Intelligence*; Springer: Berlin/Heidelberg, Germany, 2022; pp. 817–828.
65. Collobert, R.; Weston, J.; Bottou, L.; Karlen, M.; Kavukcuoglu, K.; Kuksa, P. Natural language processing (almost) from scratch. *J. Mach. Learn. Res.* **2011**, *12*, 2493–2537.
66. Birjali, M.; Kasri, M.; Beni-Hssane, A. A comprehensive survey on sentiment analysis: Approaches, challenges and trends. *Knowl.-Based Syst.* **2021**, *226*, 107134. [CrossRef]
67. Qiao, B.; Zou, Z.; Huang, Y.; Fang, K.; Zhu, X.; Chen, Y. A joint model for entity and relation extraction based on BERT. *Neural Comput. Appl.* **2022**, *34*, 3471–3481. [CrossRef]
68. Zaib, M.; Zhang, W.E.; Sheng, Q.Z.; Mahmood, A.; Zhang, Y. Conversational question answering: A survey. *Knowl. Inf. Syst.* **2022**, *64*, 3151–3195. [CrossRef]
69. Vaswani, A.; Shazeer, N.; Parmar, N.; Uszkoreit, J.; Jones, L.; Gomez, A.N.; Kaiser, Ł.; Polosukhin, I. Attention is all you need. *arXiv* **2017**, arXiv:1706.03762
70. Devlin, J.; Chang, M.W.; Lee, K.; Toutanova, K. BERT: Pre-training of Deep Bidirectional Transformers for Language Understanding. *arXiv* **2018**, arXiv:1810.04805.
71. Liu, Y.; Ott, M.; Goyal, N.; Du, J.; Joshi, M.; Chen, D.; Levy, O.; Lewis, M.; Zettlemoyer, L.; Stoyanov, V. RoBERTa: A Robustly Optimized BERT Pretraining Approach. *arXiv* **2019**, arXiv:1907.11692.
72. Wolf, T.; Debut, L.; Sanh, V.; Chaumond, J.; Delangue, C.; Moi, A.; Cistac, P.; Rault, T.; Louf, R.; Funtowicz, M.; et al. Transformers: State-of-the-Art Natural Language Processing. In *Proceedings of the 2020 Conference on Empirical Methods in Natural Language Processing: System Demonstrations*; Association for Computational Linguistics: Berlin, Germany, 2020; pp. 38–45.
73. Lhoest, Q.; del Moral, A.V.; Jernite, Y.; Thakur, A.; von Platen, P.; Patil, S.; Chaumond, J.; Drame, M.; Plu, J.; Tunstall, L.; et al. Datasets: A Community Library for Natural Language Processing. *arXiv* **2021**, arXiv:2109.02846.
74. Srivastava, N.; Hinton, G.; Krizhevsky, A.; Sutskever, I.; Salakhutdinov, R. Dropout: A simple way to prevent neural networks from overfitting. *J. Mach. Learn. Res.* **2014**, *15*, 1929–1958.
75. Hendrycks, D.; Gimpel, K. Gaussian error linear units (gelus). *arXiv* **2016**, arXiv:1606.08415.
76. Pascanu, R.; Mikolov, T.; Bengio, Y. On the difficulty of training recurrent neural networks. In Proceedings of the International Conference on Machine Learning, Atlanta, GA, USA, 16–21 June 2013; pp. 1310–1318.
77. Loshchilov, I.; Hutter, F. Decoupled Weight Decay Regularization. In Proceedings of the International Conference on Learning Representations, New Orleans, LA, USA, 6–9 May 2019.
78. He, Z.; Wang, Z.; Wei, W.; Feng, S.; Mao, X.; Jiang, S. A Survey on Recent Advances in Sequence Labeling from Deep Learning Models. *arXiv* **2020**, arXiv:2011.06727.
79. Zhou, C.; Cule, B.; Goethals, B. Pattern based sequence classification. *IEEE Trans. Knowl. Data Eng.* **2015**, *28*, 1285–1298. [CrossRef]
80. He, P.; Gao, J.; Chen, W. Debertav3: Improving deberta using electra-style pre-training with gradient-disentangled embedding sharing. *arXiv* **2021**, arXiv:2111.09543.
81. Zuluaga-Gomez, J.; Prasad, A.; Nigmatulina, I.; Motlicek, P.; Kleinert, M. A Virtual Simulation-Pilot Agent for Training of Air Traffic Controllers. *Aerospace* **2023**, *10*, 490. [CrossRef]
82. Madikeri, S.; Bourlard, H. Filterbank slope based features for speaker diarization. In Proceedings of the IEEE International Conference on Acoustics, Speech and Signal Processing (ICASSP), Florence, Italy, 4–9 May 2014; pp. 111–115.
83. Sell, G.; Snyder, D.; McCree, A.; Garcia-Romero, D.; Villalba, J.; Maciejewski, M.; Manohar, V.; Dehak, N.; Povey, D.; Watanabe, S.; et al. Diarization is Hard: Some Experiences and Lessons Learned for the JHU Team in the Inaugural DIHARD Challenge. In Proceedings of the Interspeech, Hyderabad, India, 2–6 September 2018; pp. 2808–2812.
84. Landini, F.; Profant, J.; Diez, M.; Burget, L. Bayesian HMM clustering of x-vector seq uences (VBx) in speaker diarization: Theory, implementation and analysis on standard tasks. *Comput. Speech Lang.* **2022**, *71*, 101254. [CrossRef]
85. Fujita, Y.; Watanabe, S.; Horiguchi, S.; Xue, Y.; Nagamatsu, K. End-to-end neural diarization: Reformulating speaker diarization as simple multi-label classification. *arXiv* **2020**, arXiv:2003.02966.
86. Fujita, Y.; Kanda, N.; Horiguchi, S.; Xue, Y.; Nagamatsu, K.; Watanabe, S. End-to-end neural speaker diarization with self-attention. In Proceedings of the 2019 IEEE Automatic Speech Recognition and Understanding Workshop (ASRU), Singapore, 14–18 December 2019; pp. 296–303.
87. Fujita, Y.; Kanda, N.; Horiguchi, S.; Nagamatsu, K.; Watanabe, S. End-to-end neural speaker diarization with permutation-free objectives. *arXiv* **2019**, arXiv:1909.05952.

88. Ryant, N.; Church, K.; Cieri, C.; Cristia, A.; Du, J.; Ganapathy, S.; Liberman, M. The Second DIHARD Diarization Challenge: Dataset, Task, and Baselines. In Proceedings of the Interspeech, Graz, Austria, 15–19 September 2019; pp. 978–982.
89. Ariav, I.; Cohen, I. An end-to-end multimodal voice activity detection using wavenet encoder and residual networks. *IEEE J. Sel. Top. Signal Process.* **2019**, *13*, 265–274. [CrossRef]
90. Lin, Y. Spoken Instruction Understanding in Air Traffic Control: Challenge, Technique, and Application. *Aerospace* **2021**, *8*, 65. [CrossRef]
91. Cardosi, K.M. *An Analysis of en Route Controller-Pilot Voice Communications*; NASA STI/Recon Technical Report N; Federal Aviation Administration: Washington, DC, USA, 1993; Volume 93, p. 30611.
92. Prasad, A.; Zuluaga-Gomez, J.; Motlicek, P.; Sarfjoo, S.; Nigmatulina, I.; Ohneiser, O.; Helmke, H. Grammar Based Speaker Role Identification for Air Traffic Control Speech Recognition. *arXiv* **2021**, arXiv:2108.12175.
93. Helmke, H.; Ondřej, K.; Shetty, S.; Arilíusson, H.; Simiganoschi, T.; Kleinert, M.; Ohneiser, O.; Ehr, H.; Zuluaga-Gomez, J.; Smrz, P. Readback Error Detection by Automatic Speech Recognition and Understanding—Results of HAAWAII Project for Isavia's Enroute Airspace. In Proceedings of the 12th SESAR Innovation Days. Sesar Joint Undertaking, Budapest, Hungary, 5–8 December 2022.

aerospace

MDPI

Article

In-Vehicle Speech Recognition for Voice-Driven UAV Control in a Collaborative Environment of MAV and UAV

Jeong-Sik Park [1,*] and Na Geng [2]

1 Department of English Linguistics and Language Technology, Hankuk University of Foreign Studies, Seoul 02450, Republic of Korea
2 Department of English Linguistics, Hankuk University of Foreign Studies, Seoul 02450, Republic of Korea; gengna0324@gmail.com
* Correspondence: parkjs@hufs.ac.kr; Tel.: +82-02-2173-8814

Abstract: Most conventional speech recognition systems have mainly concentrated on voice-driven control of personal user devices such as smartphones. Therefore, a speech recognition system used in a special environment needs to be developed in consideration of the environment. In this study, a speech recognition framework for voice-driven control of unmanned aerial vehicles (UAVs) is proposed in a collaborative environment between manned aerial vehicles (MAVs) and UAVs, where multiple MAVs and UAVs fly together, and pilots on board MAVs control multiple UAVs with their voices. Standard speech recognition systems consist of several modules, including front-end, recognition, and post-processing. Among them, this study focuses on recognition and post-processing modules in terms of in-vehicle speech recognition. In order to stably control UAVs via voice, it is necessary to handle the environmental conditions of the UAVs carefully. First, we define control commands that the MAV pilot delivers to UAVs and construct training data. Next, for the recognition module, we investigate an acoustic model suitable for the characteristics of the UAV control commands and the UAV system with hardware resource constraints. Finally, two approaches are proposed for post-processing: grammar network-based syntax analysis and transaction-based semantic analysis. For evaluation, we developed a speech recognition system in a collaborative simulation environment between a MAV and an UAV and successfully verified the validity of each module. As a result of recognition experiments of connected words consisting of two to five words, the recognition rates of hidden Markov model (HMM) and deep neural network (DNN)-based acoustic models were 98.2% and 98.4%, respectively. However, in terms of computational amount, the HMM model was about 100 times more efficient than DNN. In addition, the relative improvement in error rate with the proposed post-processing was about 65%.

Keywords: speech recognition; voice-driven control; acoustic model; grammar network; syntax analysis; semantic analysis; unmanned aerial vehicle (UAV); UAV control

Citation: Park, J.-S.; Geng, N. In-Vehicle Speech Recognition for Voice-Driven UAV Control in a Collaborative Environment of MAV and UAV. *Aerospace* **2023**, *10*, 841. https://doi.org/10.3390/aerospace10100841

Academic Editor: Peng Wei

Received: 18 August 2023
Revised: 11 September 2023
Accepted: 26 September 2023
Published: 27 September 2023

Correction Statement: This article has been republished with a minor change. The change does not affect the scientific content of the article and further details are available within the backmatter of the website version of this article.

1. Introduction

Since speech recognition technology has been successfully used in personal assistant devices such as artificial intelligence (AI) speakers and smartphones, various speech recognition applications have been introduced. In particular, many attempts have been made to apply voice control to moving objects such as cars, and the speech recognition function has played a very important role in controlling flying objects such as unmanned aerial vehicles (UAVs). In order to control a moving object through speech recognition in such a special environment, research considering the specificity of the environment is necessary. This study proposes a speech recognition framework for voice-based UAV control in a collaborative environment of manned aerial vehicles (MAVs) and UAVs. In this environment, multiple MAVs and multiple UAVs fly together, and pilots on board the MAVs perform collaborative tasks with UAVs by controlling multiple UAVs with their voices.

Several previous studies introduced systems for controlling UAVs via voice [1–3]. Most conventional studies have focused on a typical speech recognition environment where speech recognition controls a single UAV. In a previous study, we presented an efficient speech recognition architecture and front end for controlling multiple UAVs with voice [4].

If the current speech recognition scheme (that is, server-centric scheme) that processes speech recognition in a remote server is applied to a collaborative environment of a MAV and an UAV, various problems may arise. First, multiple MAV pilots simultaneously submitting voice commands to a single server can place a heavy burden on the server, delaying the sending of commands. The server-centric scheme also manages indirect UAV control by performing three data transmission sequences: the MAV to the recognition server, the server to the MAV, and the MAV to the UAV. This indirect communication can incur communication costs, resulting in misrecognition or dropped commands due to packet loss while the MAV and the UAV are moving. For high-speed moving, special-purpose UAVs (e.g., military UAVs), the packet loss problem can be more serious.

In [4], we proposed an efficient recognition scheme to solve such disadvantages of conventional speech recognition schemes in the multi-UAV control via voice. The proposed scheme is summarized as distributed speech recognition in which the MAV and UAV share speech recognition processes. The MAV system processes the front-end module to extract acoustic features from the input speech uttered by the MAV pilot. When the acoustic features are sent to the UAVs, the UAV's system performs the recognition process that follows the front-end process.

This study introduces a speech recognition process not covered in the previous study. There are few studies considering an efficient speech recognition framework for the environment where a MAV and an UAV cooperate to perform military operations. This study proposes a speech recognition framework suitable for this environment. In particular, we concentrate on an acoustic model suitable for a distributed speech recognition system in which the MAV and UAV share speech recognition tasks and a post-processing method to minimize the risk caused by speech recognition errors in a collaborative environment of a MAV and an UAV.

In recent years, research on speech recognition and understanding in air traffic control (ATC) environments have been conducted through projects such as HAAWAII [5] and SESAR [6]. In a voice communication environment between air traffic controllers (ATCo) and the pilot, the ATCo receives the pilot's voice command and performs recognition. ATC systems with relatively high-performance hardware can handle complex models such as end-to-end ASR models [7–9]. However, in the collaborative environment between an UAV and a MAV targeted in this study, the speech recognition module is operated on the UAV with hardware resource constraints. Thus, the conventional research has somewhat different characteristics from the environment we are targeting in that the ATC system has relatively few hardware resource limitations and is processed at the ground control center.

The remainder of this paper is organized as follows. In Section 2, we propose an efficient speech recognition framework for voice-driven UAV control in a collaborative environment of MAVs and UAVs. In Section 3, several experiments conducted on speech data and their results are reported and discussed. Finally, Section 4 concludes the paper.

2. In-Vehicle Speech Recognition for Voice-Driven UAV Control in a Collaborative Environment of MAV and UAV

Although considerable research on speech recognition has been conducted in various fields, research on speech recognition for UAV control is relatively insufficient [10]. Several studies on multimodality using speech and visual data have been introduced [10,11], and most of the speech recognition studies have considered situations where speech recognition is performed at a ground control station [12–14]. However, most speech recognition processes for UAV control are performed in a similar way to general speech recognition systems.

A traditional speech recognition system consists of several modules, including front-end, recognition, and post-processing modules, as shown in Figure 1 [15,16]. The front-end module performs several processes, such as noise reduction, voice triggering, and acoustic feature extraction. Next, in the recognition module, speech recognition is performed using pre-trained acoustic models using common pattern recognition techniques such as deep neural networks (DNNs) or hidden Markov models (HMMs). Finally, the post-processing module performs syntax and semantic analyses to improve the recognition output's accuracy and clarity.

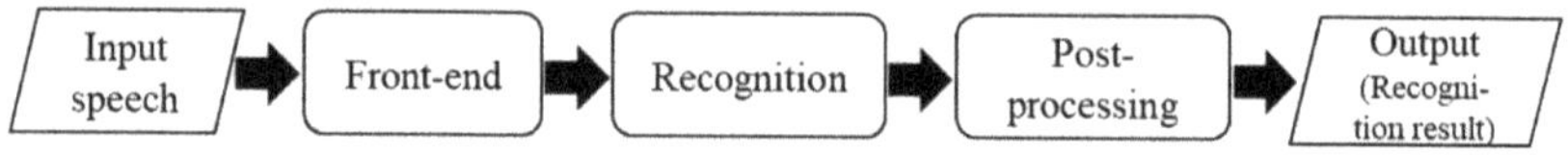

Figure 1. The general procedure of standard speech recognition systems.

As mentioned in Section 1, a speech recognition scheme suitable for the collaborative environment of MAVs and UAVs is a form of distributed speech recognition in which a MAV and an UAV share speech recognition processes. Figure 2 summarizes this scheme in which the MAV processes the front-end module and the UAV performs the recognition process.

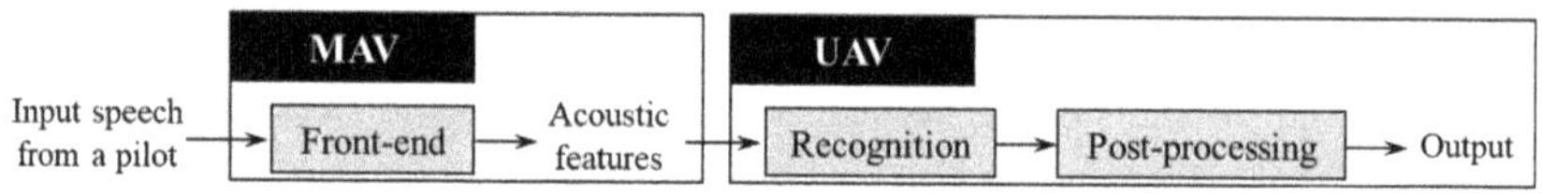

Figure 2. Distributed speech recognition for the collaborative environment of MAVs and UAVs.

In this section, we propose each module suitable for collaborative environments of MAVs and UAVs.

2.1. Front-End of Speech Recognition

The front end of speech recognition consists of four main processes: voice activity detection (VAD), feature extraction, noise reduction, and voice trigger, as shown in Figure 3 [4,17]. The first two processes are essential for speech recognition. VAD is the process of detecting target speech regions to perform speech recognition. Feature extraction is to extract features representing acoustic characteristics in the time or frequency domain from input speech data.

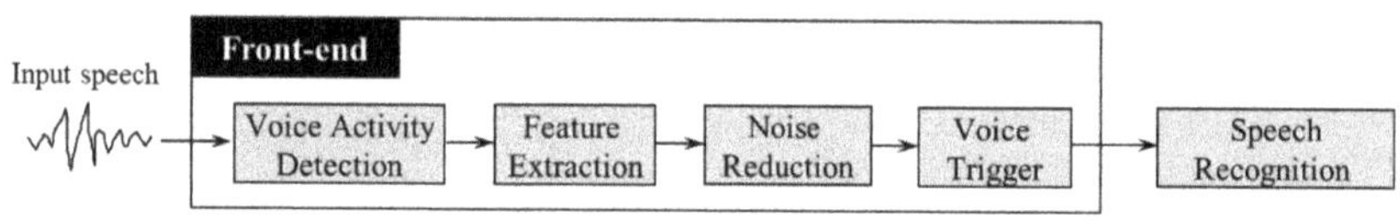

Figure 3. The general procedure of front-end speech recognition.

Noise reduction and voice triggering should be developed according to the system environment. In the previous study, noise reduction and voice-triggering approaches were proposed to handle multi-UAV environments [4]. In particular, to consider multi-UAV control, we proposed a multi-channel voice-trigger approach in which each UAV has a unique name used as a trigger word, and the MAV pilot establishes a connection between the MAV and the target UAV among multiple UAVs. Figure 4 represents the multi-channel voice trigger-based front end and speech recognition procedures for multi-UAV control. When MAV pilots have a conversation and a situation arises where they need to call an UAV among multiple UAVs, they call the name corresponding to the target UAV. Then, the voice-trigger module detects a specific UAV name according to the process shown in the upper block of the figure, and it attempts to connect with the target UAV. When connected to the target UAV, the pilot speaks a command, and the features extracted from the voice are transmitted to the target UAV, and finally, speech recognition proceeds, as shown in the block below in Figure 4.

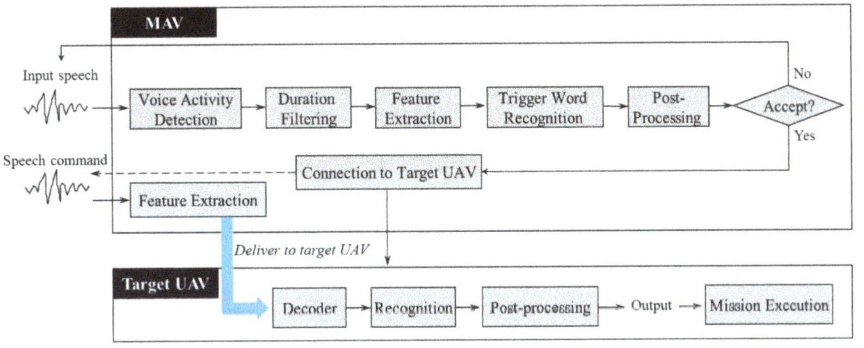

Figure 4. The procedure of multi-channel voice trigger-based front-end and speech recognition.

2.2. Model Construction for Speech Recognition

As described in Figure 4, after the pilot calls the trigger word for the target UAV and a connection is made with the UAV, the pilot delivers a command to the UAV, and the UAV starts speech recognition for this command. This subsection describes the speech recognition process performed in the UAV system.

2.2.1. Definition of Voice Commands for Training Data Collection

For speech recognition, an acoustic model must be constructed in advance (this is called model training), and training data is required in this process. Therefore, for collecting training data, we first define a set of voice commands that the pilot of the MAV delivers to control the UAV. For this work, we conducted expert consultation through several meetings with military aviation officials. The characteristics of commands used for military operations between MAV pilots and UAVs are shown in Table 1.

Table 1. Characteristics of commands used for military operations between MAV pilots and UAVs.

Restrictions	Expert Advice
Structure of commands	Simple and clear commands for precise delivery (1 to 5 connected words)
Vocabulary size	150 to 200 words available to pilots
Language	English is used for communication between military aircraft (International Telecommunications Standard)

In other words, for the voice commands for military operations between MAVs and UAVs, a command structure consisting of 1 to 5 connected words out of approximately 150 to 200 words available to MAV pilots is suitable for accurately delivering commands to the UAV.

The voice command sets must be designed considering various missions the UAV must perform and various collaboration situations between MAV and UAV. In addition, the command sets should be composed of frequently used words for the pilot's convenience and consist of words that are easy for speech recognition. There have been several international cooperation projects related to the collaborative operation of MAVs and UAVs, and various related reports and standards have been published, including the Standardization Agreement (STANAG)-4586, Manned–Unmanned Teaming (MUM-T), and Manned–Unmanned Systems Integration Capability (MUSIC) [18,19]. By analyzing the documents, we investigated the collaboration situations and missions between MAVs and UAVs and specified the division of roles between MAVs and UAVs according to cooperative operation.

In some documents, the core missions of UAVs in the cooperative operation of a MAV and an UAV are known as missions related to reconnaissance, attack, condition monitoring, and location/route management [20,21]. In addition, STANAG-4856, a military standard established by the North Atlantic Treaty Organization (NATO), defines data link interface (DLI) messages between a ground control center and an UAV [18]. Therefore, we derive the voice command sets by linking the UAV core missions with the DLI messages provided by STANAG-4856. Table 2 summarizes the mission command sets configured for several DLI messages.

Table 2. Examples of mission command sets.

DLI Message	Mission Commands
Vehicle Configuration	Check energy storage unit, read back, report energy state, report fuel state, report battery state, ready for launch, acknowledge, are you ready, take off
Vehicle Operating Mode	Set up control mode, request manual control, request automatic control, report control mode
Vehicle Steering	Set up heading point, heading for waypoint (no.), change heading point, report heading point, say heading point, set up altitude, request altitude (no.), maintain altitude, change altitude (no.), say altitude, report altitude, set up speed, reduce speed to (no.), set up loiter position, request loiter position latitude (no.)
Mission Transfer	Set up mission plan, clear route, change route (no.), request route (no.), clear mission, request mission (no.)
AV Loiter Waypoint	Set up loiter type, request loiter type circle, request loiter type racetrack, request loiter radius (no.), report loiter type, report loiter altitude, report loiter speed, request loiter speed (no.), request loiter duration (no.), report loiter duration, request loiter bearing north

Table 3 introduces several scenarios where the UAV executes its mission by passing commands from the MAV pilot to the UAV using the defined command sets. In other words, the phrases in the command scenarios are all examples of commands delivered by the MAV pilot to the UAV, and commands are delivered sequentially according to the phrases in the scenario presented for each mission. It shows that three UAVs (each UAV is named Alpha, Bravo, and Charlie) perform surveillance and reconnaissance under the control of a manned pilot. In a situation where three UAVs are launched simultaneously after the pilot determines basic settings such as route and altitude, Alpha is put into a reconnaissance mission, and Bravo and Charlie are put into surveillance missions. Each mission command starts with calling the UAV to be controlled.

Table 3. Examples of several mission types and command scenarios.

Mission Type	Command Scenario	Mission Type	Command Scenario
Take-off	Agent Alpha Agent Bravo Agent Charlie Set up heading point Heading for waypoint 7 Set up altitude Request altitude 7000 Set up speed Request speed 250 Ready for launch Are you ready Take off Disconnection	Reconnaissance flight instructions	Agent Alpha Request approach Set up altitude Request altitude 3000 Set up area Request vertices number 1 Request area min altitude 2000 Request area max altitude 3000 Request area loop count 10 Disconnection

Table 3. *Cont.*

Mission Type	Command Scenario	Mission Type	Command Scenario
Surveillance flight instructions	Agent Bravo Request activity surveillance Request loiter type circle Request loiter radius 200 Request loiter speed 10 Disconnection Agent Charlie Request activity surveillance Request loiter type figure eight Request loiter speed 20 Disconnection	Return after completing the mission	Agent Alpha Agent Bravo Agent Charlie Clear mission Request flight Change heading point Heading for waypoint 0 Start flight termination Set up control mode Request automatic control Report arrival time Disconnection

2.2.2. Acoustic Model Construction

An acoustic model is a fundamental component of speech recognition [22]. Acoustic model construction is a process of learning to map acoustic features extracted from input speech signals to phonetic units [23,24]. Typical acoustic models are HMM and DNN. HMM is constructed by learning the statistical relationships between the acoustic features and the corresponding phonetic units [25,26]. On the other hand, the DNN-based acoustic model is trained using deep learning techniques to learn a non-linear mapping between the acoustic features and the phonetic units [27].

The HMM is a traditional acoustic model that has been successfully used in many speech recognition systems [24]. It has a simple and interpretable structure and is, therefore, computationally efficient, especially during decoding. However, the HMM has limited ability to model complex non-linear relationships between input features and output phonemes, making it difficult to capture acoustic details [28]. Because of this, the HMM has limitations in recognizing speech with complex structures (e.g., sentence units).

On the other hand, the DNN is capable of modeling complex non-linear relationships between input features and output phonemes, making it possible to capture acoustic details and improve recognition accuracy for sentence-level speech [29,30]. However, it has a more complex structure than HMM, making it more computationally intensive during training and decoding, and it may require specialized hardware to achieve real-time performance. In particular, the DNN requires significant training data to learn many model parameters [31].

Since the HMM and DNN have such conflicting characteristics, selecting and constructing a model suitable for recognizing the mission command sets delivered to the UAV by the MAV pilot and suitable for the system environment driving speech recognition is necessary. As described in the previous section, the pilot's mission commands given to the UAV are relatively short sentences consisting of at most five words. The sentence is considered a series of connected words. The total number of words included in the command sets is only about 400. In terms of the system environment driving speech recognition, UAVs responsible for speech recognition processing typically have limited computing hardware capacity and can perform limited computations.

Based on these characteristics, a speech recognition system that can recognize connected words consisting of a relatively small number of words with medium hardware capacity is appropriate for recognizing UAV mission commands. Therefore, HMM is expected to solve these limitations more effectively than DNN.

For HMM-based connected word recognition, the HMM must be constructed for each word included in the command set. Since the recorded training data are composed of commands in sentence units, it is necessary to segment each speech data sequence into individual words. For each word, a separate HMM is then trained using the segmented speech data. That is, the same number of HMMs as the number of words included in the command set is constructed during training.

After building an HMM for each word, connected word recognition proceeds as follows. First, by detecting the silence regions included in the input speech, the connected word command is divided into sequences of isolated words. Acoustic features are then extracted from each isolated word. Next, as shown in Figure 5, the same decoding process as isolated word recognition is performed using the Viterbi algorithm [32], which computes the likelihood for each HMM ($\lambda_1, \ldots, \lambda_V$) with given acoustic features. Once an HMM representing the maximum likelihood is determined, the word corresponding to the model is regarded as the recognition result.

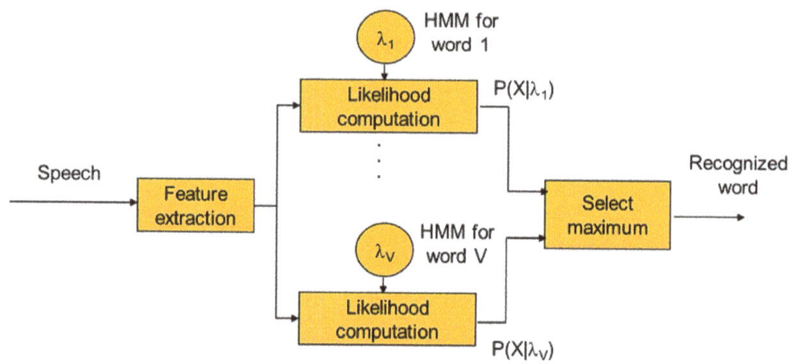

Figure 5. The procedure of HMM-based isolated word recognition.

In the speech recognition process shown in Figure 5, the HMM with the same structure can be used for all words. However, since the amount of information that the model needs to learn varies depending on the length of the word utterance, more effective speech recognition can be performed by modifying the structure of the HMM considering the word length.

As illustrated in Figure 6, if all words have the same HMM structure, inefficient HMMs may be constructed in which one state covers multiple phonemes or several states simultaneously handle one phoneme. In this study, we construct HMMs with different structures for each word to improve the structural problem of HMM. Since we expect that the HMM structure in which one state handles one phoneme is the most effective, we adjust the number of HMM states according to the number of phonemes in each word, as shown in Figure 7.

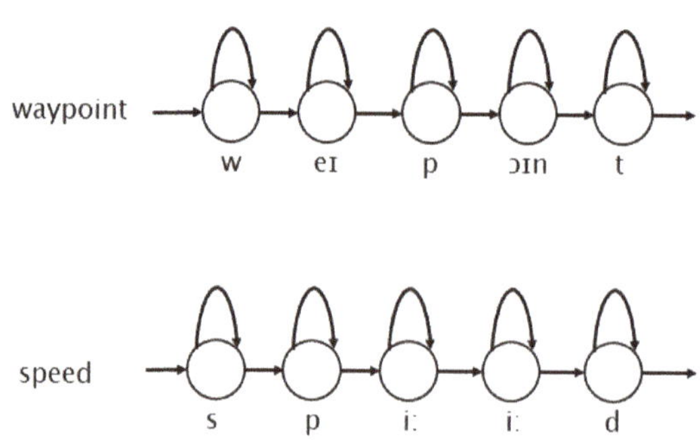

Figure 6. HMM structure with a fixed number of states.

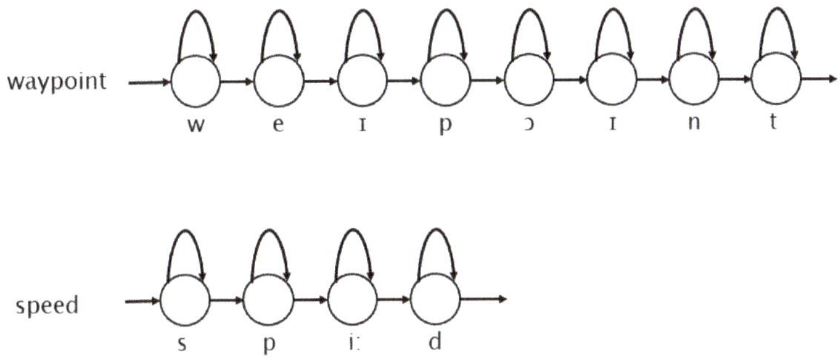

Figure 7. Variable HMM structure considering the number of phonemes in a word.

2.3. Post-Processing with Syntax Analysis and Semantic Analysis

The purpose of post-processing is to improve the accuracy of recognition output through syntax analysis and semantic analysis. In this study, we propose efficient methods for each post-processing task, considering the mission commands established in the collaborative environment of MAVs and UAVs.

2.3.1. Syntax Analysis Based on the Grammar Network

Syntax analysis is the process of analyzing the grammatical structure of a spoken sentence to determine its meaning [33]. It helps clarify sentences with multiple possible meanings. Therefore, this processing is very important in the collaborative environment of MAVs and UAVs where there is a high possibility of misrecognition due to aircraft noise, and the UAV must clarify the pilot's mission commands.

In this study, we organized the grammar structure of mission commands into a tree-type grammar network and performed syntax analysis using this network. That is, the recognition result conforming to this network was determined to have an appropriate grammatical structure; otherwise, it was regarded as misrecognition.

The grammar network has one root node and one terminal node, and the network is formed between the root and the terminal node, with the preceding word becoming the parent node and the succeeding word becoming the child node according to the grammar structure of each command. The starting word of each command becomes the child node of the root node, and the command's last word becomes the terminal node's parent node. When creating a grammar network in this way, several sample commands such as "report loiter altitude", "report loiter duration", and "report battery capacity" can be expressed as a tree, as shown in Figure 8.

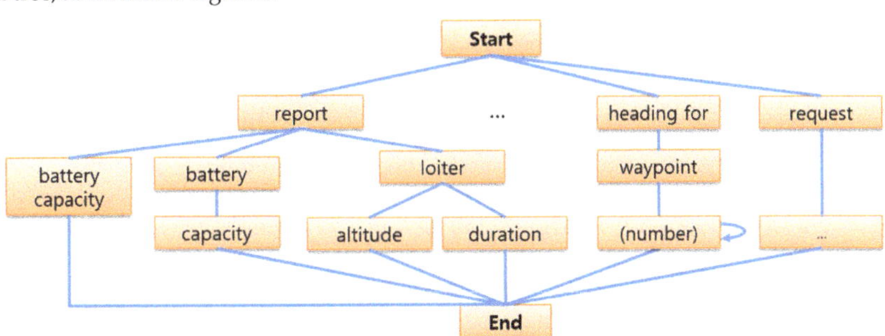

Figure 8. A grammar network created using several sample commands.

In Figure 8, "battery capacity" is handled in two ways: storing two words together in one node or dividing each word into two nodes. The reason for this processing is to recognize it as one word when uttering this command without a pause between two connected words. In addition, some mission commands contain numbers, such as "heading for waypoint 5". Since numbers have various lengths, they are expressed as one node when processing the syntax of such commands.

The principle that the grammar network we built can be used to determine whether the recognition result conforms to the command syntax is as follows. When a sequence of words included in the recognition result has a path starting from the root node and arriving at a terminal node, the word sequence is determined to conform to the command syntax. If the sequence of words starts from the root node and does not reach the terminal node, the word sequence may not conform to the command syntax.

If it is determined that the word sequence of the recognition result does not conform to the command syntax, the result may be regarded as completely incorrect. Nevertheless, there is a possibility that only one or two words in the word sequence might be incorrectly recognized. Therefore, rather than concluding that the recognition result is completely misrecognized, correcting the misrecognized words may help improve overall performance. In this study, we propose a method to correct such misrecognition of several words by combining candidate recognition results with a grammar network.

In general, when word recognition is performed on connected words of an input command, the similarity between given acoustic features and each word model is calculated. Then, the top several word models selected in order of similarity become candidate recognition results for the given features. At this time, the similarity of each candidate's result is also stored.

Figure 9 shows an example of candidate recognition results for the input command "report loiter altitude". If only the first-rank result of each word is accepted, "report route altitude" becomes the recognition result, which cannot pass through our grammar network. However, as shown in this figure, if the second-rank result of each word is also accepted, "report loiter altitude" can be obtained as a recognition result. In other words, after candidate results for each word are selected, among all possible word sequence combinations constructed from the candidates, a word sequence having the highest similarity while passing through the grammar network becomes the final recognition result of the input command. The grammar network is used in this process to verify whether each word sequence constructed from the candidates matches well with the command syntax.

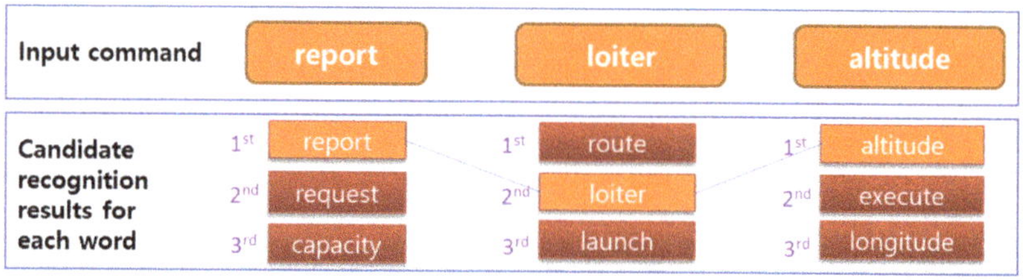

Figure 9. Example of candidate recognition results for the input command "report loiter altitude" and correction of the misrecognition result based on word sequence matching using a grammar network.

As the number of candidate results for each word increases, the amount of computation also increases, so we set the number of candidate results to three, which is considered the most appropriate. If none of all possible word sequence combinations constructed from the candidate results pass through the grammar network, the given input command's recognition result is considered entirely incorrect.

If the first-ranked word sequence does not match the grammar network, it is combined with the next ranked results and attempts to match the grammar network again. If the command consists of five words, the total number of combinations will be 243 = 3^5. There are rarely situations where all 243 combinations are considered, because usually the first or second ranked results match the grammar network and the matching process stops.

2.3.2. Semantic Analysis Based on Transaction Scheme

Semantic analysis is the process of analyzing the meaning of words and phrases contained in recognition results to extract the intended semantic content [34]. This involves understanding the context of recognition results. It is an important process in speech recognition because it allows systems to accurately transcribe speech into its intended meaning rather than simply recognizing sounds or phonemes. As such, this process is particularly important in applications such as natural language processing and virtual assistants, where understanding the meaning of spoken language is essential for providing accurate and useful responses.

Semantic analysis plays a key role in a system that recognizes mission commands in the collaborative environment of MAVs and UAVs. During an important military operation, if a command transmitted by a pilot is recognized as a command that contradicts the current state of the UAV due to a recognition error or a pilot's ignition mistake, it can encounter a very dangerous situation.

For example, in a situation where the UAV receives the command "Request automatic control", meaning to switch from manual flight to automatic flight and performs the mission, if "Ready for launch" to prepare for take-off is recognized as the next command, the UAV should consider it as a recognition error or a pilot's ignition mistake and report it as an unacceptable command. In this study, we utilize semantic analysis to block dangerous situations caused by recognition errors or pilot ignition mistakes.

The proposed method implements semantic analysis using the transaction scheme used in data management. A transaction refers to a sequence of operations in data management that are treated as a single task unit [35]. It is used to ensure data consistency and integrity through key properties referred to as ACID, which represent atomicity, consistency, isolation, and durability.

Atomicity means that a transaction must be treated as a single indivisible operation, and all operations must succeed or fail as a unit. In other words, if any part of a transaction fails, the entire transaction is rolled back to its previous state. Consistency is the property that the data must be in a consistent state before and after a transaction is executed. Isolation means that the transaction must be executed in isolation from other concurrent transactions. That is, the results of one transaction should not be visible to other transactions until it is committed. Finally, durability is the property that once a transaction is committed, its effect on the data must be permanent. The system can provide reliable and robust data management by ensuring that transactions are ACID-compliant, especially in mission-critical applications where data consistency and integrity are essential.

Because of the characteristics of the transaction, transaction-based semantic analysis of speech recognition results can be effectively used to control UAVs performing critical missions. The process of transaction-based semantic analysis is as follows.

First, critical command sets are defined, such as safety-critical commands, mission-critical commands, and flight-critical commands; then, commands corresponding to each set are selected. Next, each mission is classified as a transaction type, such as a take-off transaction, landing transaction, or reconnaissance transaction. Furthermore, as shown in Figure 10, each critical command set is mapped to a mission transaction, allowing more than one command set to be mapped to a single transaction. Although illustrated here, the mapping information between the critical command set and the mission transaction is managed as a kind of mapping table.

Figure 11 shows the process of making a final decision on whether to accept or reject the command recognition result based on the status of the transaction. We apply the concept of transaction status, commonly used in data management, to this study. A transaction has five statuses: active, partially committed, committed, failed, and aborted. When a transaction starts, it becomes "active". When the transaction ends, it becomes "partially committed", and when it is completely finished, it becomes "committed". On the other hand, if the transaction fails to complete in the active status, it goes into the "failed" status and eventually changes to the "aborted" status. Sometimes, it is partially committed and becomes a failed status.

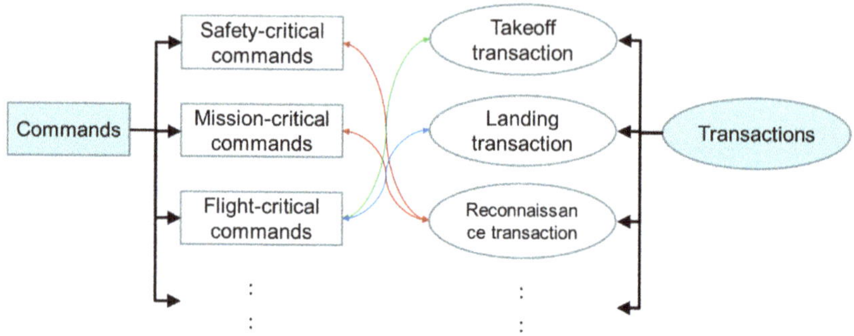

Figure 10. Definition of critical command sets and transaction types and their mapping.

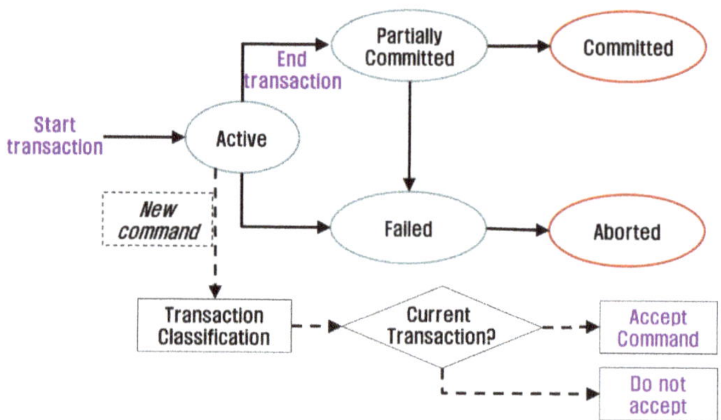

Figure 11. The process of deciding whether to accept the command recognition result based on the transaction status.

The proposed method determines whether to accept or reject the command recognition result based on the transaction status. Assume that a UAV starts a transaction, and a new command is recognized while this transaction is active. At this time, a transaction that the new command corresponds to is found in the mapping table. The new command is accepted if it corresponds to the currently executing transaction; otherwise, it is rejected. As a result, this verification prevents the start of another new transaction before the currently executing transaction becomes committed or aborted.

The proposed method guarantees the independence of individual missions the UAV performs using the transaction scheme. The process for validating recognition results utilizes the ACID characteristics of the transaction discussed above to ensure that the UAV can safely carry out its mission.

Figure 12 shows an example of securing the mission's independence via a UAV by rejecting a disallowed command through transaction-based semantic analysis. In this figure, when a MAV pilot delivers the command "Request activity surveillance" related to the reconnaissance mission to the UAV, the UAV starts the reconnaissance transaction. While performing the reconnaissance transaction, if the UAV recognizes another voice message from the pilot as "Ready for launch" that is related to the take-off mission, the UAV understands that this command is not related to the reconnaissance transaction and informs the MAV that the command is not allowed.

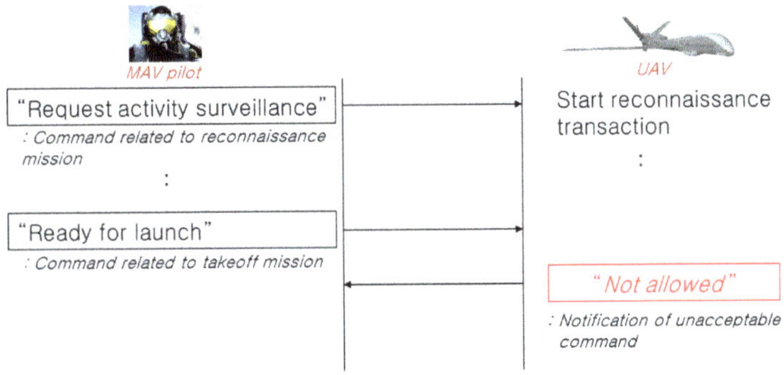

Figure 12. Example of a mission control situation through transaction-based semantic analysis.

3. Evaluation

To validate the efficiency of the proposed speech recognition framework, we conducted several experiments, including speech recognition, syntax analysis, and semantic analysis.

3.1. Validation of Speech Recognition for Mission Command Set in a Collaborative Environment of MAVs and UAVs

Speech recognition experiments were performed to verify the performance of the acoustic models introduced in Section 2.2.2. Training data are required to build acoustic models. As described in Section 2.2.1, we constructed voice command sets related to voice-driven UAV control in a collaborative environment of MAV and UAV, and as a result, obtained about 300 different commands and about 400 different words. We then recorded 50 speakers pronouncing each command and word three times in a clean environment. As a result, 105,000 pieces of voice data (45,000 data for command units and 60,000 for word units) were collected. These data were divided into 10 groups, and speech recognition experiments were conducted using a 10-fold cross-validation method. That is, the data of 5 speakers in the first group were used for testing, and the data of the remaining 45 speakers were used for model training. In this way, the experiments were conducted 10 times by changing the test and training data groups, and the average of each experiment result was calculated.

As explained in Section 2.2.2, we considered the HMM a more efficient model than the DNN in recognizing mission commands in the form of connected words composed of a small number of words. Therefore, an HMM model was built for each word, and connected word recognition experiments were conducted using this model.

In addition, a DNN model was also constructed to compare performance with HMM. However, there is a limit to building a DNN model with about 100,000 voice data we collected, so we built a model using the DARPA Resource Management (RM) speech corpus [36,37]. The DNN used in the experiment is a model with a five-layer structure built based on Kaldi. Kaldi is an automatic speech recognition (ASR) toolkit with many ASR algorithms [38]. It has been released in various versions and has provided various training recipes such as the Wall Street Journal Corpus (wsj), TIMIT (timit), and Resource Management (rm). Since the speech recognition target covered in this study is commands composed of several word sequences, we tried to use a DNN model that shows stable performance while minimizing the amount of computation compared to complex DNN models. For this reason, we trained a TDNN-based triphone model using the Kaldi S5 version by following the recipe using the RM corpus [39,40]. Speech recognition in the two models was performed on a laptop with relatively low specifications (Intel i5 (quad-core, 3.4 GHz), 4 GB RAM) considering the UAV system environment, and the average recognition time was also investigated along with the recognition rate.

Table 4 shows the results. In this experiment, we investigated the command recognition results of sentence units, word error rate, and average recognition time of commands.

In Section 2.2.2, we proposed a method to change the structure of HMM so that each state of HMM processes one phoneme. To verify this method's validity, the performance of the fixed HMM, which consists of a fixed number of states for all words, and the variable HMM, which has a different number of states for each word, were compared. In order to examine the results more elaborately, recognition experiments were conducted according to the length of the command, from a command consisting of two words to a command consisting of five or more words.

Table 4. Performance of acoustic models (HMM and DNN): recognition rate by sentence (%), word error rate (%), and average recognition time (sec) in command units.

Model	Measure	2 Words	3 Words	4 Words	5 or More
Fixed HMM	Rec. Rate (sent.)	100	97.4	95.3	94.2
	Word Error Rate	0	0.87	1.49	2.15
	Avg. Rec. Time *	0.03	0.05	0.06	0.09
Variable HMM (proposed)	Rec. Rate (sent.)	100	98.5	97.4	96.9
	Word Error Rate	0	0.54	0.86	1.13
	Avg. Rec. Time *	0.03	0.04	0.05	0.10
DNN	Rec. Rate (sent.)	100	98.8	97.6	97.2
	Word Error Rate	0	0.49	0.80	1.09
	Avg. Rec. Time *	2.0	4.5	6.5	9.0
	Avg. Rec. Time **	0.20	0.38	0.55	0.85

* Experiment with a low-spec laptop/** Experiment with a high-spec laptop.

As shown in this table, the proposed variable HMM improved performance compared to the fixed HMM. The longer the command length, the more noticeable the performance improvement. However, there was no significant difference in recognition time between the two models. Therefore, this result indicates that the variable HMM configures the number of states differently for each word and is more efficient in recognizing connected words.

Next, we compared the performance of the proposed variable HMM and DNN models. In the experimental results, there was not much difference between the two models in the recognition accuracy, while showing a good performance of 97% or more. As a result of recognizing whole command units consisting of two to five words, the proposed variable HMM and DNN showed an average recognition rate of 98.2% and 98.4%, respectively, derived from four types of sentence recognition rates (ranging from two words to five or more words). For commands composed of two words, both models showed 100% accuracy, and for other commands, the performance of HMM was slightly lower than that of DNN. Among the two measures of recognition accuracy, the word error rate showed a much smaller difference between the two models, and for commands consisting of five or more words, the performance difference was only 0.04%. The reason is that recognition errors in sentence units are mostly caused by the misrecognition of only one word included in a sentence.

On the other hand, the two models showed a big difference in average recognition time. The HMM showed a recognition time shorter than 0.1 s for command sets of all lengths, while the DNN showed a significantly longer recognition time as the command length increased, ranging from 2 s (two words) to 9 s (five or more words). This result is because the DNN has a more complex model structure than the HMM.

Considering that the non-ideally high recognition time of the DNN model was due to the influence of the laptop used in the experiment, we additionally measured the recognition time of the DNN model using a high-spec laptop (Intel i7 (12-core, 2.1 GHz), 16 GB RAM). This result is shown in the last row of Table 4. Because the same program was evaluated on both laptops, the recognition results did not change, but the average recognition time showed a difference. For commands of each length, the second laptop showed recognition times ranging from 0.2 to 0.9 s, reducing the average recognition time by about 10 times compared to the results in the first laptop. However, even with high-spec hardware, the HMM model still showed about 10 times lower recognition time. Therefore, it can be said that HMM, which has a relatively simple structure, is a suitable model to recognize the UAV control commands targeted in this study.

We measured the amount of computation using the number of model parameters to examine the theoretical difference in recognition time between the HMM model and the DNN model. The amount of computation required to process one speech frame in the DNN model can be calculated through the following equation.

$$N^{(DNN)} \approx (L-1) * N^2 + N * D + N * S, \quad (1)$$

where L is the number of layers, N is the number of nodes in each layer, D is the dimension of the feature vector, and S represents the number of nodes in the output layer. The values of the Kaldi model parameter used in our experiment are as follows: $L = 5$, $N = 650$, $D = 39$, and $S = 2000$. Applying these values to (1), about 3 million operations are required to process one frame during the recognition process.

On the other hand, in the case of the HMM model, the amount of computation required to process one speech frame is calculated as follows.

$$N^{(HMM)} \approx S * M * D * 2, \quad (2)$$

where S is the number of HMM states, M is the number of GMM mixtures, and D represents the dimension of the feature vector. The values of the HMM model parameters used in our experiment are as follows: $S = 100$, $M = 8$, and $D = 39$. Accordingly, about 30,000 operations are required to process one frame during the recognition process.

That is, the DNN model requires about 100 times the amount of computation of the HMM model. Furthermore, this theoretical difference is similarly shown in Table 4, with the average recognition time of the DNN model showing a value about 100 times higher than that of the HMM model

3.2. Verification of the Proposed Syntax Analysis Method for the Post-Processing of Mission Command Speech Recognition

We verified the validity of the proposed grammar network-based syntax analysis method through speech recognition experiments. As explained in Figure 9, only the first-rank recognition result of each word is accepted in a general speech recognition framework. The speech recognition results in Table 4 are the recognition accuracy considering only the first-rank results.

However, as mentioned in Section 2.3.1, when the correct answer of a specific word among connected words is ranked second or third, the recognition accuracy can be increased if these candidates are also considered. To implement this, we proposed a syntax analysis method using a grammar network. If the first-rank recognition result is found to be incorrect in the command syntax through the grammar network, the optimal result that matches the command syntax is obtained by combining the second or third-rank candidate results.

Table 5 represents the speech recognition results after applying the proposed syntax analysis method. The proposed variable HMM, which was determined to be the most efficient model in terms of recognition rate and average recognition time in Table 4, was set as the baseline. Furthermore, the recognition rate was investigated after performing grammar network-based syntax analysis for the three best candidate results (that is, the recognition results up to the third rank in the order of output values calculated in HMMs).

Since the commands composed of two words had all correct answers at the first rank in the baseline, the combination of the first-rank words conformed to the command syntax, and therefore, all were recognized as correct answers after applying the syntax analysis. In the case of commands composed of three or more words, the recognition rate increased in all lengths of command sets after applying the proposed syntax analysis method. These results demonstrate that when some of the connected words constituting a command are incorrect, the baseline treats the command as an error, but in the proposed method, many of these data are corrected as the correct answers. In particular, the longer the length of the command, the higher the improvement in the recognition rate by the syntax analysis, which explains that the longer the length of the command, the more errors occur, and the proposed method corrects the errors.

Table 5. Speech recognition results (%) of the whole command unit (not word unit) for validation of the proposed grammar network-based syntax analysis method.

	2 Words	3 Words	4 Words	5 or More
Baseline	100	98.5	97.4	96.9
Applying syntax analysis (proposed)	100	99.7	99.0	98.8

3.3. Verification of the Proposed Semantic Analysis Method for the Post-Processing of Mission Command Speech Recognition

The evaluation of speech recognition models or syntax analysis can be quantitatively evaluated through speech recognition experiments, but the quantitative method is not suitable for evaluating semantic analysis. Accordingly, we implemented a real-time recognition system and a collaborative simulation environment of MAVs and UAVs and attempted to verify the validity of the proposed semantic analysis method.

Figure 13a shows a program for real-time command recognition, and Figure 13b represents an experimental environment created to simulate the collaboration of three UAVs named Alpha, Bravo, and Charlie with a MAV pilot. In this simulation environment, when the pilot issues a command to a device indicating the MAV, the device transmits the command to the target UAV, and then, the UAV recognizes the command. All these processes run in real-time.

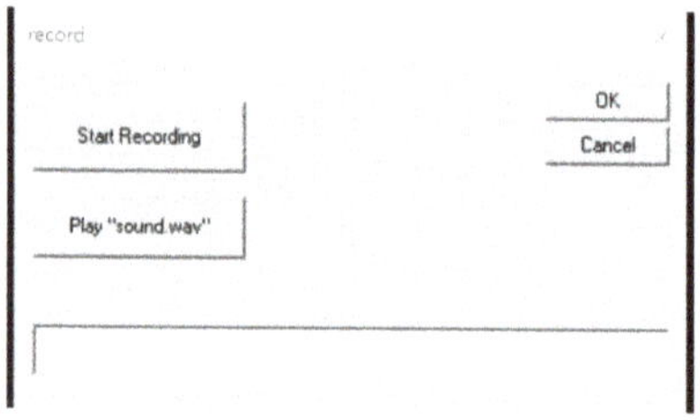

(**a**)

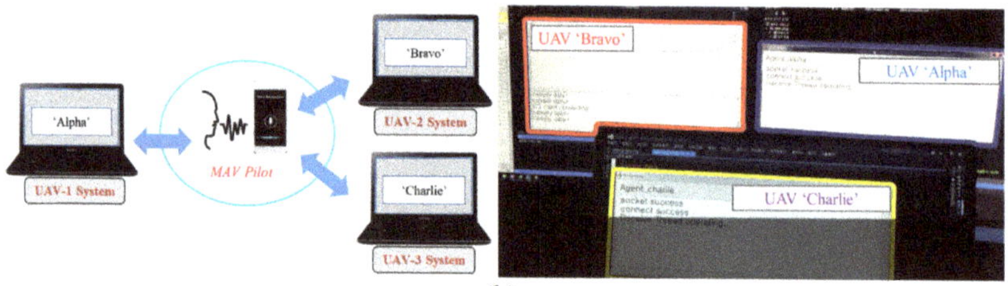

(**b**)

Figure 13. Program for real-time command recognition (**a**) and experimental environment for simulating the collaboration of MAVs and multi-UAVs (**b**).

We created several collaboration scenarios between MAVs and UAVs in this simulation environment to verify whether the proposed transaction-based semantic analysis method works properly while MAVs and UAVs communicate. Figure 14 shows flow charts configured for collaboration scenarios between a MAV and an UAV. The two figures represent collaboration examples: (a) is a collaboration scenario related to condition monitoring and location/route management, and (b) is a scenario representing reconnaissance and attack. We made several such scenarios and tested whether the transaction-based semantic analysis works properly while communicating between the MAV device and the UAV system in real-time in the MAV and UAV collaborative simulation environment presented in Figure 13b.

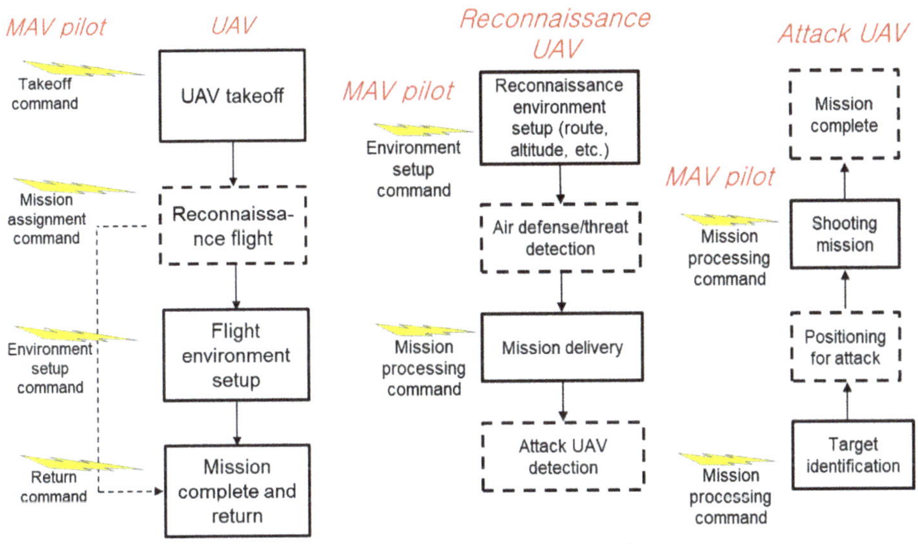

Figure 14. Flow chart for collaboration scenarios between MAVs and UAVs. (**a**) shows the takeoff scenario, and (**b**) illustrates the reconnaissance and attack scenario.

In Figure 14, the arrows indicate the transaction flow, and the yellow signs indicate the situation where the MAV pilot delivers a command to the UAV. The solid boxes represent the transactions the UAV is performing, and the dotted boxes represent the status of the UAV or the situation the UAV is in. For example, in Figure 14a, the pilot issues a take-off command, and the UAV enters a transaction called take-off. Afterward, when the pilot issues a command related to reconnaissance flight, the UAV enters a reconnaissance flight state. Then, when the pilot issues an environment setup command, the UAV enters the Flight environment setup transaction. Finally, when the return command is delivered to the UAV, it changes the UAV into a transaction called mission complete and return. It is impossible for a transaction to start in an order different from the direction of the transactions indicated by the arrow in the figure. For example, after the flight environment setup transaction, the UAV can only enter the mission complete and return transaction and cannot proceed to the take-off transaction. With this process based on a transaction concept, we implement flow control of UAVs collaborating with MAVs and utilize this for semantic analysis of commands to prevent serious situations caused by incorrect speech recognition results.

We conducted an experiment to verify that the proposed transaction-based semantic analysis works properly. For example, in the scenario shown in Figure 14a, let us consider a situation in which a command related to a take-off transaction is entered while the MAV pilot delivers the environment setup command and the UAV performs a transaction of flight environment setup. At this time, we checked whether the UAV system correctly rejected this command by considering it as a context error. As another example, in the attack scenario of Figure 14b, when a command related to the landing transaction was entered while performing a transaction of shooting mission, we checked whether the UAV rejected this command correctly.

This way, we checked whether 100 normal commands that matched any given transaction were correctly accepted and 100 abnormal commands that violated the transaction were accurately rejected. As a result of the experiment, it was found that all normal commands were correctly accepted, and all abnormal commands were rejected by the proposed transaction-based semantic analysis method.

4. Conclusions

In this study, we proposed a speech recognition framework for voice-driven UAV control in a collaborative environment of MAVs and UAVs. The previous study proposed an efficient noise-cancellation method in an aerial vehicle environment and a multi-channel voice-triggering method for controlling multiple UAVs for front-end speech recognition. In this study, we focused on constructing acoustic models for speech recognition and post-processing to perform syntax analysis and semantic analysis. In a collaborative environment between MAVs and UAVs, typical commands that a MAV pilot sends to UAVs are in the form of connected words consisting of at most five words. This study investigated model construction and post-processing methods suitable for recognizing such connected words in a UAV system with low hardware capacity.

First, we explored an efficient acoustic model for recognizing connected words, targeting the HMM, known as the statistical modeling method, and the DNN model using deep learning techniques. In particular, instead of the traditional method using the same HMM structure for each word, we proposed a HMM structure that reflects the number of phonemes in a word. In the average recognition rate of four types of sentence recognition rates (from commands consisting of two words to commands of five or more words), the DNN-based acoustic model showed higher performance than the traditional HMM, while it did not show much difference from the proposed HMM. However, in terms of the amount of computation and recognition time, it was analyzed that the HMM model performs fast recognition with about 100 times less computation than the DNN model. Furthermore, it can be concluded that the proposed HMM model is suitable for recognizing connected words in a UAV system with low hardware capacity.

Naturally, among the various DNN models currently in use, there are models with relatively low computational complexity. Although this study highlighted the fact that the HMM model is less computationally intensive than DNN, this is not a claim that the HMM is the optimal model, and it can be used as an alternative in constrained environments. If the UAV system has high-performance hardware capacity and can allocate many resources to driving speech recognition, the DNN model will also be available.

Next, a grammar network-based syntax analysis method was proposed for post-processing. We configured the structure of the commands that the MAV pilot delivers to the UAV as a grammar network, and if the connected words obtained as a recognition result do not pass through this network, it is determined as a syntax error. In addition, when some of the connected words contain errors, instead of treating the corresponding command as an error, we corrected the recognition error by reflecting the results in the upper rank among the candidate results of each word using the proposed syntax analysis method. As a result of the speech recognition experiment conducted to verify the validity of this method, it was confirmed that the speech recognition performance was remarkably improved after applying the proposed syntax analysis method. As a result of recognizing whole command units consisting of two to five words, the average recognition rates of the baseline approach and the syntax analysis-based approach were 98.2% and 99.4%, respectively, which means that the relative improvement in error rate by the syntax analysis reaches 65%.

Finally, we proposed a semantic analysis approach applying the transaction scheme used in data management. In a very important situation, such as a military operation, misrecognition of a MAV pilot's command may lead to serious danger. To handle this situation, we categorized cooperation missions between MAVs and UAVs as transactions and mapped each command set to related transactions. Then, while the UAV is performing a transaction corresponding to a specific mission when the recognition result of a command delivered by the MAV pilot does not belong to the transaction, the UAV regards it as a recognition error and sends a response indicating that the command cannot be accepted.

To verify the validity of this method, we implemented a real-time recognition system and a collaborative simulation environment of MAVs and UAVs and created several collaboration scenarios between a MAV and an UAV. Then, real-time communication between the MAV and the UAV was performed using the scenarios to confirm that the proposed semantic analysis works properly. In experiments conducted with about 200 commands, it was confirmed that normal commands that match a given transaction are correctly accepted, and commands that do not match are properly rejected.

As described so far, in this study, we introduced a speech recognition framework for voice-based UAV control in a collaborative environment between MAVs and UAVs, proposed useful methods in each process, and successfully verified the validity of each module through various speech recognition experiments. The proposed framework consists of speech database construction, front-end, acoustic model construction, and post-processing and focuses on minimizing the amount of computation so that each module can be directly driven in the UAV system. Therefore, the framework is expected to be efficiently applied in an environment where speech recognition is directly driven in a device with limited hardware resources.

In future research, we plan to expand this research by studying an efficient speech recognition framework for voice-driven communication between the ground control center and an UAV and between the ground control center and a MAV.

Author Contributions: Conceptualization, J.-S.P.; methodology, J.-S.P.; software, J.-S.P. and N.G.; validation, J.-S.P. and N.G.; formal analysis, J.-S.P.; writing—original draft preparation, J.-S.P.; writing—review and editing, J.-S.P. and N.G.; supervision, J.-S.P.; project administration, J.-S.P.; funding acquisition, J.-S.P. All authors have read and agreed to the published version of the manuscript.

Funding: This research was supported by the Hankuk University of Foreign Studies Research Fund, the National Research Foundation of Korea (NRF) grant funded by the Korean Government (MSIT) (No. 2020R1A2C1013162), and the Institute of Information and Communications Technology Planning and Evaluation (IITP) grant funded by the Korean Government (MSIT) (RS-2023-00232949).

Data Availability Statement: Data is contained within the article.

Conflicts of Interest: The authors declare no conflict of interest.

References

1. Oneata, D.; Cucu, H. Kite: Automatic speech recognition for unmanned aerial vehicles. *arXiv* **2019**, arXiv:1907.01195.
2. Lavrynenko, O.Y.; Konakhovych, G.F.; Bakhtiiarov, D.I. Protected voice control system of unmanned aerial vehicle. *Electr. Control Syst.* **2020**, *1*, 92–98. [CrossRef]
3. Anand, S.S.; Mathiyazaghan, R. Design and fabrication of voice controlled unmanned aerial vehicle. *IAES Int. J. Robot. Autom.* **2016**, *5*, 205–212. [CrossRef]
4. Park, J.S.; Na, H.J. Front-end of vehicle-embedded speech recognition for voice-driven multi-UAVs control. *Appl. Sci.* **2020**, *10*, 6876. [CrossRef]
5. Helmke, H.; Kleinert, M.; Shetty, S.; Ohneiser, O.; Ehr, H.; Arilíusson, H.; Simiganoschi, T.S.; Prasad, A.; Motlicek, P.; Veselý, K.; et al. Readback error detection by automatic speech recognition to increase ATM safety. In Proceedings of the Fourteenth USA/Europe Air Traffic Management Research and Development Seminar (ATM2021), Virtual Event, 20–23 September 2021; pp. 20–23.
6. Helmke, H.; Kleinert, M.; Ahrenhold, N.; Ehr, H.; Mühlhausen, T.; Ohneiser, O.; Klamert, L.; Motlicek, P.; Prasad, A.; Zuluaga-Gomez, J.; et al. Automatic speech recognition and understanding for radar label maintenance support increases safety and reduces air traffic controllers' workload. In Proceedings of the Fifteenth USA/Europe Air Traffic Management Research and Development Seminar (ATM2023), Savannah, GA, USA, 5–9 June 2023; pp. 1–11.
7. Guo, D.; Zhang, Z.; Fan, P.; Zhang, J.; Yang, B. A context-aware language model to improve the speech recognition in air traffic control. *Aerospace* **2021**, *8*, 348. [CrossRef]
8. Zhang, S.; Kong, J.; Chen, C.; Li, Y.; Liang, H. Speech GAU: A single head attention for Mandarin speech recognition for air traffic control. *Aerospace* **2022**, *9*, 395. [CrossRef]
9. Lin, Y. Spoken instruction understanding in air traffic control: Challenge, technique, and application. *Aerospace* **2021**, *8*, 65. [CrossRef]

10. Oneață, D.; Cucu, H. Multimodal speech recognition for unmanned aerial vehicles. *Comput. Electr. Eng.* **2021**, *90*, 106943. [CrossRef]
11. Xiang, X.; Tan, Q.; Zhou, H.; Tang, D.; Lai, J. Multimodal fusion of voice and gesture data for UAV control. *Drones* **2022**, *6*, 201. [CrossRef]
12. Galangque, C.M.J.; Guirnaldo, S.A. Speech recognition engine using ConvNet for the development of a voice command controller for fixed wing unmanned aerial vehicle (UAV). In Proceedings of the 12th International Conference on Information & Communication Technology and System (ICTS), Surabaya, Indonesia, 18 July 2019; pp. 93–97. [CrossRef]
13. Zhou, Y.; Hou, J.; Gong, Y. Research and application of human-computer interaction technology based on voice control in ground control station of UAV. In Proceedings of the IEEE 6th International Conference on Computer and Communications (ICCC), Chengdu, China, 11–14 December 2020; pp. 1257–1262. [CrossRef]
14. Contreras, R.; Ayala, A.; Cruz, F. Unmanned aerial vehicle control through domain-based automatic speech recognition. *Computers* **2020**, *9*, 75. [CrossRef]
15. Trivedi, A.; Pant, N.; Shah, P.; Sonik, S.; Agrawal, S. Speech to text and text to speech recognition systems-a review. *IOSR J. Comput. Eng.* **2018**, *20*, 36–43.
16. Karpagavalli, S.; Chandra, E. A review on automatic speech recognition architecture and approaches. *Int. J. Signal Process. Image Process. Pattern Recognit.* **2016**, *9*, 393–404. [CrossRef]
17. Desai, N.; Dhameliya, K.; Desai, V. Feature extraction and classification techniques for speech recognition: A review. *Int. J. Emerg. Technol. Adv. Eng.* **2013**, *3*, 367–371.
18. Marques, M.M. STANAG 4586—Standard interfaces of UAV control system (UCS) for NATO UAV interoperability. *NATO Stand. Agency Afeite Port.* **2012**, *3*, 1–14.
19. Kim, S.; Kim, Y. Development of an MUM-T integrated simulation platform. *IEEE Access.* **2023**, *11*, 21519–21533. [CrossRef]
20. Jameson, S.; Franke, J.; Szczerba, R.; Stockdale, S. Collaborative autonomy for manned/unmanned teams. In Proceedings of the Annual Forum American Helicopter Society, Grapevine, TX, USA, 1–3 June 2005; Volume 61, p. 1673.
21. Alicia, T.J.; Hall, B.T.; Terman, M. Synergistic Unmanned Manned Intelligent Teaming (SUMIT). In *Technical Report*; U.S. Army: Madison County, NY, USA, 2020; pp. 1–92.
22. Juang, B.H.; Rabiner, L.R. Hidden Markov models for speech recognition. *Technometrics* **1991**, *33*, 251–272. [CrossRef]
23. Woodland, P.C.; Odell, J.J.; Valtchev, V.; Young, S.J. Large vocabulary continuous speech recognition using HTK. In Proceedings of the ICASSP'94, IEEE International Conference on Acoustics, Speech and Signal Processing, Adelaide, Australia, 19–22 April 1994; Volume 2, pp. II/125–II/128. [CrossRef]
24. Mor, B.; Garhwal, S.; Kumar, A. A systematic review of hidden Markov models and their applications. *Arch. Comput. Methods Eng.* **2021**, *28*, 1429–1448. [CrossRef]
25. Gales, M.; Young, S. The application of hidden Markov models in speech recognition. *Found. Trends Signal Process.* **2007**, *1*, 195–304. [CrossRef]
26. Mustafa, M.K.; Allen, T.; Appiah, K. A comparative review of dynamic neural networks and hidden Markov model methods for mobile on-device speech recognition. *Neural Comput. Appl.* **2019**, *31*, 891–899. [CrossRef]
27. Hinton, G.; Deng, L.; Yu, D.; Dahl, G.E.; Mohamed, A.; Jaitly, N.; Senior, A.; Vanhoucke, V.; Nguyen, P.; Sainath, T.N.; et al. Deep neural networks for acoustic modeling in speech recognition: The shared views of four research groups. *IEEE Signal Process. Mag.* **2012**, *29*, 82–97. [CrossRef]
28. Shahin, M.A.; Ahmed, B.; McKechnie, J.; Ballard, K.J.; Gutierrez-Osuna, R. A comparison of GMM-HMM and DNN-HMM based pronunciation verification techniques for use in the assessment of childhood apraxia of speech. *Interspeech* **2014**, *1*, 1583–1587.
29. Fohr, D.; Mella, O. New paradigm in speech recognition: Deep neural networks. In Proceedings of the International Conference on Information Systems and Economic Intelligence, Marrakech, Morocco, 13 April 2017.
30. Këpuska, V.; Bohouta, G. Comparing speech recognition systems (Microsoft API, Google API and CMU Sphinx). *Int. J. Eng. Res. Appl.* **2017**, *7*, 20–24. [CrossRef]
31. Deshmukh, A.M. Comparison of hidden Markov model and recurrent neural network in automatic speech recognition. *Eur. J. Eng. Res. Sci.* **2020**, *5*, 958–965. [CrossRef]
32. Lou, H.L. Implementing the Viterbi algorithm. *IEEE Signal Process. Mag.* **1995**, *12*, 42–52. [CrossRef]
33. Arora, S.J.; Singh, R.P. Automatic speech recognition: A review. *Int. J. Comput. Appl.* **2012**, *60*, 34–44. [CrossRef]
34. Tur, G.; DeMori, R. *Spoken Language Understanding: Systems for Extracting Semantic Information from Speech*; John Wiley and Sons: Hoboken, NJ, USA, 2011.
35. Bernstein, P.A.; Newcomer, E. *System Recovery, In Principles of Transaction Processing*; Morgan Kaufmann: San Francisco, CA, USA, 2009; pp. 185–222, ISBN 9781558606234.
36. Hain, T.; Woodland, P.C. Dynamic HMM selection for continuous speech recognition. In Proceedings of the 6th European Conference on Speech Communication and Technology (EUROSPEECH 1999), Budapest, Hungary, 5–9 September 1999.
37. Pallett, D.S.; Fiscus, J.G.; Garofolo, J.S. DARPA resource management benchmark test results June 1990. In Proceedings of the Workshop on Speech and Natural Language, Hidden Valley, PA, USA, 24–27 June 1990.
38. Povey, D.; Ghoshal, A.; Boulianne, G.; Burget, L.; Glembek, O.; Goel, N.; Hannemann, M.; Motlicek, P.; Qian, Y.; Schwarz, P.; et al. The Kaldi speech recognition toolkit. In Proceedings of the IEEE 2011 Workshop on Automatic Speech Recognition and Understanding (ASRU 2011), Waikoloa, HI, USA, 11–15 December 2011.

39. Kaldi Tutorial. Available online: https://kaldi-asr.org/doc/tutorial.html (accessed on 10 January 2023).
40. GitHub: Kaldi Speech Recognition Toolkit. Available online: https://github.com/kaldi-asr/kaldi (accessed on 10 January 2023).

aerospace

MDPI

Article

Analyzing Multi-Mode Fatigue Information from Speech and Gaze Data from Air Traffic Controllers

Lin Xu [1], Shanxiu Ma [2], Zhiyuan Shen [2,*], Shiyu Huang [2] and Ying Nan [1]

1 College of Astronautics, Nanjing University of Aeronautics and Astronautics, Nanjing 211106, China; xulin19851116@163.com (L.X.); nanying@nuaa.edu.cn (Y.N.)
2 College of Civil Aviation, Nanjing University of Aeronautics and Astronautics, Nanjing 211106, China; mashanx@nuaa.edu.cn (S.M.); hsiwoo@nuaa.edu.cn (S.H.)
* Correspondence: shenzy@nuaa.edu.cn

Abstract: In order to determine the fatigue state of air traffic controllers from air talk, an algorithm is proposed for discriminating the fatigue state of controllers based on applying multi-speech feature fusion to voice data using a Fuzzy Support Vector Machine (FSVM). To supplement the basis for discrimination, we also extracted eye-fatigue-state discrimination features based on Percentage of Eyelid Closure Duration (PERCLOS) eye data. To merge the two classes of discrimination results, a new controller fatigue-state evaluation index based on the entropy weight method is proposed, based on a decision-level fusion of fatigue discrimination results for speech and the eyes. The experimental results show that the fatigue-state recognition accuracy rate was 86.0% for the fatigue state evaluation index, which was 3.5% and 2.2%higher than those for speech and eye assessments, respectively. The comprehensive fatigue evaluation index provides important reference values for controller scheduling and mental-state evaluations.

Keywords: fatigue recognition; air traffic controller; feature fusion; multi-mode

Citation: Xu, L.; Ma, S.; Shen, Z.; Huang, S.; Nan, Y. Analyzing Multi-Mode Fatigue Information from Speech and Gaze Data from Air Traffic Controllers. *Aerospace* **2024**, *11*, 15. https://doi.org/10.3390/aerospace11010015

Academic Editor: Hartmut Helmke

Received: 16 October 2023
Revised: 21 December 2023
Accepted: 21 December 2023
Published: 24 December 2023

1. Introduction

Rapidly growing flight volumes have resulted in an increasing workload for air traffic controllers. Research shows that an excessive workload contributes to controller fatigue and negligence, leading to flight accidents [1]. Therefore, fatigue detection in controllers is an important means of preventing air traffic safety accidents.

The current research related to fatigue detection can be divided into two categories: subjective detection and objective detection. Compared to subjective detection methods, objective detection methods do not require subjects to stop their work, making objective detection methods more practical. Among them, objective detection can be divided into contact detection methods such as the ECG signal detection method [2] and the EEG signal detection method [3]. The contact detection method requires the controller to wear detection equipment, which is intrusive and may affect the controller's control work. Therefore, the non-contact fatigue detection method is more suitable for the controller's work scenario.

Controllers need to frequently use radiotelephony in their daily work. Previous studies have shown that the frequency of instruction errors in radio telephony are important fatigue characteristics [4]. Therefore, radio telephony reflects the current fatigue status of the controller. The radio telephony fatigue feature detection method is a very practical and effective detection method. Li [5] conducted a speech-fatigue test on members of an American bombing crew, and found that the pitch and bandwidth of their speech signals showed significant changes after a long flight. When the human body is in a state of fatigue, speech signals will exhibit a decrease in pitch frequency and changes in the position and bandwidth of the spectral resonance peak [6]. Wu [7] extracted MFCC features from radiotelephony communications and proposed a self-adaption quantum genetic algorithm, but their ability to describe audio with only MFCC is limited. A multi-angle description of speech should use multiple features. Kouba [8] discussed the use of a contact method combined with speech, which can more comprehensively present the current fatigue state of the controller through two modalities. However, the collection of EEG information is time-consuming and can affect the progress of control work. Vasconcelos's research [9] shows that low elocution and articulation rates mean that the pilots are in a fatigue state, and it shows that recognizing fatigue states through speech recognition is feasible. Studies have shown that various types of sound quality characteristics reflect the sound quality changes before and after fatigue.

Traditional speech signal processing usually relies on linear theory. Based on their short-term resonance characteristics, speech signals can be simulated as a time series composed of an excitation source and a filter to construct a classic speech signal excitation-source–filter model [10]. These signals can be described using various model parameters for further procession and analyses. Recently, speech-signal research has led to scholars concluding that the process of generating speech signals is a nonlinear process; that is, it is neither a deterministic linear sequence nor a random sequence, but rather a nonlinear sequence with chaotic components. Therefore, traditional linear filter models cannot fully represent the information contained in speech signals [11]. A nonlinear dynamic model of a speech signal is generally constructed using a delay-phase diagram that is obtained by reconstructing the time series of a one-dimensional speech signal in the phase space [12]. Previous studies also showed that the fatigue state of the human body influences the phase-space track of speech, especially the degree of chaos in its phase space [13]. Considering the respective characteristics of both linear and non-linear speech features, this paper extracts linear features and non-linear features for air talk at the same time, and fuses the two types of features to create the features of controller's fatigue speech.

In the field of fatigue recognition, the face is considered an important information system which contains the fatigue status of the subject. Many scholars have considered information about the eye in fatigue detection research. Wierwille [14] analyzed the relationship between a driver's eye closure time and the collision probability in experiments on driving fatigue, and found that eye closure time can be used to characterize fatigue. The US Transportation Administration also conducted experiments investigating nine parameters including blink frequency, closure time, and eyelid closure, and demonstrated that these characteristics can also be used to characterize the degree of fatigue. Jo et al. [15] proposed an algorithm for determining the eye position of drivers based on blob features. Some scholars applied principal-components analysis [16] and linear discriminant analysis [17] to extract ocular features, and finally judged the fatigue state of a driver using a support vector machine (SVM) classifier. It can be seen that eye features have important reference value in fatigue recognition. When the controller stops sending radio telephony, we use his eye features to determine the current fatigue state. To this end, we perform feature-level fusion and decision-level fusion on the controller's voice features and eye features.

Multi-source data fusion can provide the model with richer detailed features. Compared with previous research that relied entirely on radio telephony or facial data, a multi-source feature fusion approach is adopted in this study by extracting various speech features from radio telephony and integrating speech features with facial features, proposing a method for controller fatigue recognition through multi-source feature fusion. In our model, different speech features are fused at the feature layer and passed through a classifier for speech fatigue feature recognition. The recognition results for eye fatigue characteristics and the recognition results for voice fatigue characteristics are fused at the decision-making level via a weighting method to obtain the best fatigue recognition results. In order to verify the effectiveness and robustness of our model, in the experimental part we collected data from a total of six controllers in tower control and approach control positions. In addition, our data collection experiment lasted for one week, and the data from each licensed controller can be divided into six periods, so the data obtained are sufficiently representative.

The main contributions of this work are listed as follows:

1. This paper performs linear feature extraction and nonlinear feature extraction, respectively, for radio telephony, and uses FSVM to perform fatigue recognition for speech fusion features.
2. In order to detect the fatigue status of the controller when he stops sending radio telephony, this paper extracts eye-fatigue-feature *PERCLOS*, and evaluates the fatigue status of the controller using a threshold method.
3. This paper uses a weighting method to perform decision-making fusion based on the fatigue recognition results for voice features and facial features, and conducts experiments to verify the effectiveness and robustness of the model.

This paper is structured as follows: In Section 2, we introduced the multi-level information fusion model used in this paper, including the model's feature extraction from data sources, feature fusion, fusion feature recognition, and the decision layer fusion process based on two types of recognition results. In Section 3, we introduced the controller fatigue feature extraction recognition experiment carried out in this paper, including the process of collecting fatigue data and the processing results according to the model used in this paper. The last part is our summary of this paper. Section 4 draws together our conclusions, demonstrating that the accuracy of our multi-level information fusion model is enhanced when processed using the FSVM classifier.

2. Methods

Currently, in the field of speech processing recognition, there are various speech features. We can divide them into three categories: spectral features, statistical features, and nonlinear features. We choose one from each category so that our data fusion can depict speech from three different perspectives. Here we extract the Mel frequency cepstral coefficient (MFCC), short-time-average zero-crossing rate, and maximum Lyapunov index. Based on the use of the percentage of eyelid closure over time (PERCLOS) as the eye features, feature fusion was applied to the features of the two types of data sources, and an evaluation index system was established.

The structure of the fatigue evaluation index is shown in Figure 1.

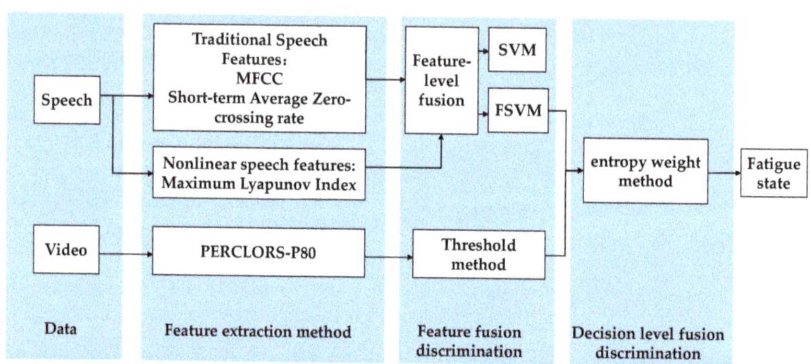

Figure 1. Fatigue-detection scheme based on multilevel information fusion.

2.1. Speech Feature Classification Algorithm Based on the FSVM

In fatigue detection, the fuzzy samples in the feature space often lead to reduction of the classification interval and affect the classification performance of the classifier [18]. In practical applications, training samples are often affected by some noise. This noise may have adverse effects on the decision boundary of the SVM, leading to a decrease in accuracy. To solve this problem, this paper introduces the idea of a fuzzy system and combines it with an FSVM algorithm.

Here, the membership function is first introduced. A membership function is a concept in fuzzy sets that represents the degree to which each element belongs to a certain fuzzy set, with the degree value ranging between 0 and 1.The principle behind the algorithm in the FVSM is to add a membership function $\mu(x_i)$ to the original sample set $(x_i, y_i), i = 1,2,\ldots,n$, so as to change the sample set to $\{x_i, y_i\mu_i(x_i)\}, i = 1,2,\ldots,n$. Membership function $\mu(x_i)$ refers to the probability that the ith sample belongs to the y_i category; this is referred to as the reliability, where $0 < \mu(x_i) \le 1$, and a larger $\mu(x_i)$, indicates a higher reliability. After introducing the membership function, the SVM formula can be expressed as

$$min\ \Phi(w) = \frac{1}{2}\|w\|^2 + C\sum_{i=1}^{n}\mu(x_i)\xi_i$$

$$s.t. y_i(w \cdot x_i + b) \ge 1 - \xi_i, \xi_i \ge 0, 0 < \mu(x_i) \le 1, i = 1,2,\ldots,n \quad (1)$$

where w is the classification interface vector, ξ_i is the relaxation factor, C is the penalty factor, and b is the classification threshold. The original relaxation variable ξ_i is replaced by $\mu(x_i)\xi_i$ since the original description of the error in the sample does not meet the conditions for the relaxation variable presented in the form of the weighted error term of $\mu(x_i)$. Reducing the membership degree of sample $\mu(x_i)$ weakens the influence of error term $\mu(x_i)\xi_i$ on the objective function. Therefore, when locating the optimal segmentation surface, x_i can be regarded as a secondary (or even negligible) sample feature, which can exert a strong inhibitory effect on isolated samples in the middle zone, and yield a relatively good optimization effect in the establishment of the optimal classification surface. The quadratic programming form of Formula (1) is dually transformed to

$$max \sum_{i=1}^{n}\alpha_i - \frac{1}{2}\sum_{i=1}^{n}\sum_{j=1}^{n}\alpha_i\alpha_j y_i y_j K(x_i, x_j)$$

$$s.t. 0 \le \alpha_i \le \mu(x_i)C, \sum_{i=1}^{n}\alpha_i y_i = 0, i = 1,2,\ldots,n \quad (2)$$

In this formula, α is the Lagrange multiplier. The constraint condition is changed from $0 \le \alpha_i \le C$ to $0 \le \alpha_i \le C\mu(x_i)$, and the weight coefficient is added to penalty coefficient C. For samples x_i with different membership degrees, the added penalty coefficients are also different: $\mu(x_i) = 1$ indicates the ordinary SVM, while when $\mu(x_i) \to 0$ sample α_i becomes too small, its contribution to the optimal classification plane will also become smaller.

In the optimal solution $\alpha_i^* = (\alpha_1^*, \alpha_2^*, \ldots, \alpha_n^*)^T$, the support vector is the nonzero solution of $\alpha_i^* > 0$. After introducing the membership function, α_i^* is divided into two parts: (1) the effective support vector distributed on the hyperplane; that is, the sample corresponding to $0 \le \alpha_i \le C\mu(x_i)$, which meets the constraint condition; and (2) the isolated sample that needs to be discarded according to the rules; that is, the sample corresponding to $\alpha_i > C\mu(x_i)$. Therefore, the classification ability of each SVM is tested by the membership function, and finally the FSVM decision model is obtained. For any given test sample $\{(x_i, y_i, s_i)\}_{i=1}^{n}$, perform the following calculation:

$$f(x) = sgn\left[\sum_{i \in SV}\alpha_i^* y_i K(x_i, x) + b^*\right] \quad (3)$$

where

$$b^* = y_i - \sum_{j=1}^{n} y_j a_j K(x_i, x_i), i \in \{i | 0 < a_i^* < \mu(x_i)C\}$$

However, the traditional membership function (see Figure 2) usually only considers the distance between the sample and the center point. Therefore, when there are isolated samples or noise points close to the center point, they will be incorrectly assigned a higher membership.

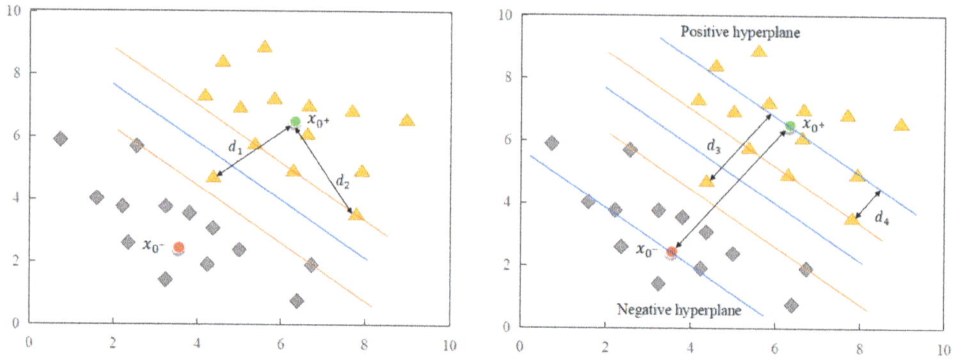

Figure 2. Traditional membership functions (**left**) and normal-plane membership functions (**right**).

In Figure 2, two colors of shapes represent two classes of samples. In traditional membership functions, the yellow line represents SVM, and the blue straight line represents the hyperplane of SVM. In normal-plane membership functions, the blue straight line represents the classification hyperplane, while the yellow straight line represents the degree to which each sample point belongs to a certain class, and the area between the yellow lines is the fuzzy region. As shown in the traditional membership functions in the left part of Figure 2, in the triangle samples of the positive class, the distance d_1 between the isolated sample and the sample center is basically the same as the distance d_2 between the support sample and the sample center, and so the two samples will be prescribed the same membership value, which will adversely affect the determination of the optimal classification plane.

In order to avoid the above situation, a membership function of the normal plane is proposed, which involves connecting the center points of positive and negative samples to determine the normal plane, and synchronously determining the hyperplane of the sample attribute. In this approach, the sample membership is no longer determined by the distance between the sample and the center point, but by the distance from the sample to the attributed hyperplane. As shown in the right part of Figure 2, the distance between the isolated sample and the sample center becomes d_3, and the distance between the supporting sample and the sample center becomes d_4, which is better for eliminating the influence of isolated samples.

In the specific calculation, first set the central point of positive and negative samples as x_{o+} and x_{o-}, respectively, and n_+ and n_- as the number of positive and negative samples, and calculate their center coordinates according to the mean value:

$$x_{o+} = \frac{1}{n_+}\sum_{i=1}^{n_+} x_i, x_{o-} = \frac{1}{n_-}\sum_{i=1}^{n_-} x_i, n_+ + n_- = n \tag{4}$$

$\vec{w} = x_{o+} - x_{o-}$ is the vector connecting the centers of the two samples, and normal vector $\vec{w}^T$ is obtained by transposing $\vec{w} = x_{o+} - x_{o-}$. Then the hyperplane to which the two samples belong is

$$\begin{cases} \vec{w}^T(x - x_{o+}) = 0, \text{Positive sample hyperplane} \\ \vec{w}^T(x - x_{o-}) = 0, \text{Negative sample hyperplane} \end{cases} \tag{5}$$

Then the distance from any sample x_i to its category hyperplane is

$$\begin{cases} d_{i+} = \dfrac{\left|\vec{w}^T(x_i - x_{o+})\right|}{\left\|\vec{w}^T\right\|}, if\ y_i = +1 \\ d_{i-} = \dfrac{\left|\vec{w}^T(x_i - x_{o-})\right|}{\left\|\vec{w}^T\right\|}, if y_i = -1 \end{cases} \tag{6}$$

If the maximum distance between positive samples and the hyperplane is set to $D_+ = max\ (d_{i+})$, and the maximum distance between negative samples and the hyperplane is set to $D_- = max\ (d_{i-})$, then the membership of various inputs is

$$\mu(x_i) = \begin{cases} 1 - \frac{d_{i+}}{D_+ + \delta},\ if\ y_i = +1 \\ 1 - \frac{d_{i-}}{D_- + \delta},\ if\ y_i = -1 \end{cases} \tag{7}$$

where δ takes a small positive value to satisfy $0 < \mu(x_i) < 1$.

In the case of nonlinear classification, kernel function $K(x_i, x_j)$ is also used to map the sample space to the high-dimensional space. According to mapping relationship $\varphi(x)$, the center points of the positive $\varphi(x_{o^+})$ and negative $\varphi(x_{o^-})$ samples become

$$\varphi(x_{o^+}) = \frac{1}{n_+}\sum_{i=1}^{n_+} \varphi(x_i),\ \varphi(x_{o^-}) = \frac{1}{n_-}\sum_{i=1}^{n_-} \varphi(x_i), n_+ + n_- = n \tag{8}$$

The distances from the input sample to the hyperplane can then be expressed as

$$\begin{cases} d_{i^+} = [\varphi(x_{o^+}) - \varphi(x_{o^-})]^T[\varphi(x_i) - \varphi(x_{o^+})],\ if\ y_i = +1 \\ d_{i^-} = [\varphi(x_{o^+}) - \varphi(x_{o^-})]^T[\varphi(x_i) - \varphi(x_{o^-})],\ if\ y_i = -1 \end{cases} \tag{9}$$

This formula can be calculated using inner product function $K(x_i, x_j)$ in the original space rather than $\varphi(x)$. Then, the distance of positive sample x from the positive hyperplane is

$$\begin{aligned} d_{i^+} &= [\varphi(x_{o^+}) - \varphi(x_{o^-})]^T[\varphi(x) - \varphi(x_{o^+})] \\ &= \frac{1}{n_+}\sum_{i=1}^{n_+} \varphi(x_i)\cdot\varphi(x) - \frac{1}{n_-}\sum_{j=1}^{n_-} \varphi(x_j)\cdot\varphi(x) - \\ &\left[\frac{1}{n_+}\sum_{i=1}^{n_+}\varphi(x_i)\right]\cdot\left[\frac{1}{n_+}\sum_{j=1}^{n_+}\varphi(x_j)\right] + \left[\frac{1}{n_+}\sum_{i=1}^{n_+}\varphi(x_i)\right]\cdot\left[\frac{1}{n_-}\sum_{j=1}^{n_-}\varphi(x_j)\right] \\ &= \frac{1}{n_+}\sum_{i=1}^{n_+} K(x_i, x) - \frac{1}{n_-}\sum_{j=1}^{n_-} K(x_j, x) - \frac{1}{n_+^2}\sum_{i=1}^{n_+}\sum_{j=1}^{n_+} K(x_i, x_j) + \frac{1}{n_+ n_-}\sum_{i=1}^{n_+}\sum_{j=1}^{n_-} K(x_i, x_j) \end{aligned} \tag{10}$$

Similarly, the distance of negative sample x from the negative hyperplane is

$$d_{i^-} = \frac{1}{n_+}\sum_{i=1}^{n_+} K(x_i, x) - \frac{1}{n_-}\sum_{j=1}^{n_-} K(x_j, x) - \frac{1}{n_-^2}\sum_{i=1}^{n_-}\sum_{j=1}^{n_-} K(x_i, x_j) + \frac{1}{n_+ n_-}\sum_{i=1}^{n_+}\sum_{j=1}^{n_-} K(x_i, x_j) \tag{11}$$

The parameters in Formula (7) can be obtained from Formulas (10) and (11). Therefore, by combining Formulas (10), (11), and (7), we can solve for $\mu(x_i)$, which gives us the proportion of various input quantities.

2.2. Eye-Fatigue Feature Extraction Algorithm Based on PERCLOS

This section describes eye feature extraction. PERCLOS refers to the proportion of time that the eye is closed and is widely used as an effective evaluation parameter in the field of fatigue discrimination. P80 in PERCLOS, which corresponds to the pupil being covered by more than 80% of the eyelid, was the best parameter for fatigue detection, and so we selected this as the judgment standard for eye fatigue [19].

Before extracting PERCLOS, we first need to identify the closed state of the eyes. This is addressed here by analyzing the aspect ratio of the eyes. Figure 3 shows images of a human eye in different states. When the eye is fully closed, the eye height is 0 and aspect ratio λ is the smallest. Conversely, λ is largest when the eye is fully open.

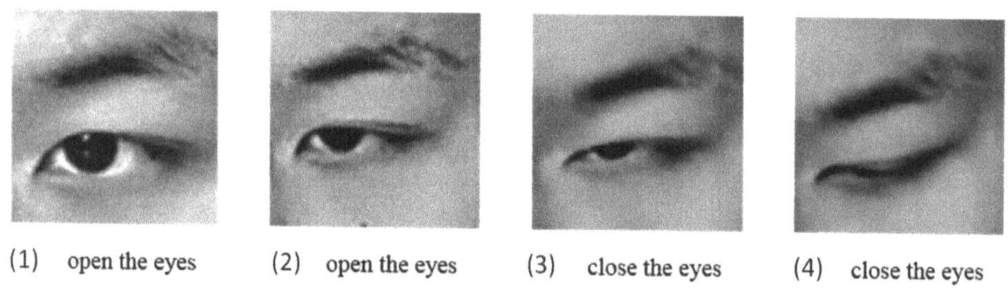

Figure 3. Images of a human eye in different closure states: (1) fully open, (2) partially open, (3) almost closed, and (4) fully closed.

We calculate the PERCLOS value by applying the following steps:

1. Histogram equalization

Histogram equalization of an eye image involves adjusting the grayscale used to display the image to increase the contrast between the eye, eyebrow, and skin color areas, and thereby make these structures more distinct, which will improve the extraction accuracy of the aspect ratio [20]. Figure 3 (1) and (2) show images before and after histogram equalization processing, respectively.

2. Image binarization

Considering the color differences in various parts of the eye image, it is possible to set a grayscale threshold to binarize each part of the image [21]. Through experiments, it was concluded that the processing effect was optimal when the grayscale threshold was between 115 and 127, and so the threshold was set to 121. Figure 4 (3) shows the binarized image of the eye, and the negative image in Figure 4 (4) is used for further data processing since the target area is the eye.

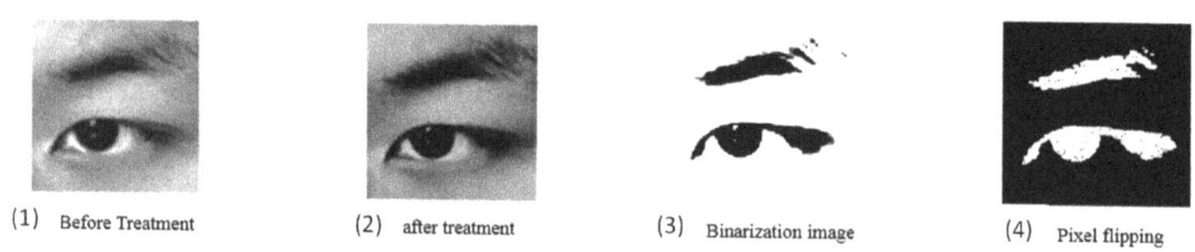

Figure 4. Eye image processing: (1) before histogram equalization, (2) after histogram equalization, (3) after binarization, and (4) the negative binarized image.

3. Calculating eye aspect ratio

Determining the minimum bounding rectangle for the binary image, as shown in Figure 5, yields two rectangles of the eyebrow and eye. The lower one is the circumscribed rectangle of the eye, and the height and width of the rectangle correspond to the height and width of the eye.

Figure 5. Minimum eye-bounding rectangle.

Height h and width w of the rectangle are obtained by calculating the pixel positions of the four vertices of the rectangle, with the height-to-width ratio of the eye being calculated as $\lambda = h/w$. P80 was selected as our fatigue criterion, and so when the pupil is covered by more than 80% of the eyelid, we consider this to indicate fatigue. The height-to-width-ratio statistics for a large number of human eye images were used to calculate the λ values of eyes in different states. We defined $0.12 \leq \lambda \leq 0.23$ as a closed-eye state, and $\lambda \geq 0.23$ as an open-eye state [22]. Figure 6 shows the aspect ratio of the eye for different closure states.

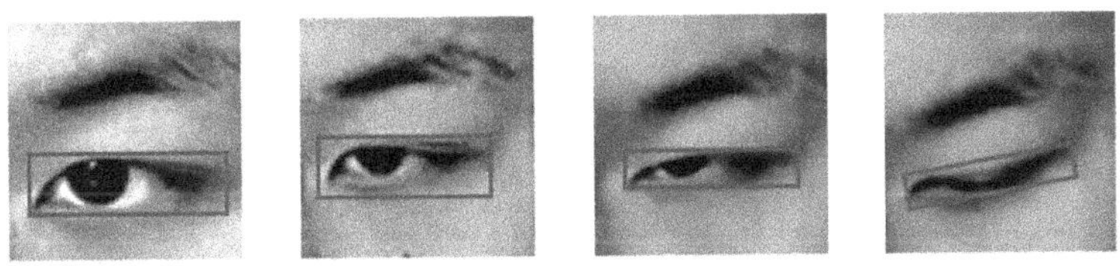

(1) open $\lambda = 0.35$ (2) open $\lambda = 0.32$ (3) close $\lambda = 0.17$ (4) close $\lambda = 0.13$

Figure 6. Eye aspect ratios for the closure states shown in Figure 6 (1) λ = 0.35, (2) λ = 0.32, (3) λ = 0.17, and (4) λ = 0.13.

4. Calculating the PERCLOS value

The principle for calculating the PERCLOS value based on P80 is as follows: assuming that an eye blink lasts for $t_1 - t_4$, where the eyes are open at moments t_1 and t_4, and the time when the eyelid covers the pupil for more than 80% is $t_2 - t_3$, then the value of PERCLOS is

$$PERCLOS = \frac{t_3 - t_2}{t_4 - t_1} \times 100\% \tag{12}$$

Considering that the experiments involved analyzing eye-video data, continuous images can be obtained after frame extraction, and so the timescale can be replaced by fps = 30 with fixed image frames; that is, by analyzing the video of the controller's eye control over a certain period of time, the total number of images collected in the data is M, and the number of closed eyes is N, then the value of PERCLOS is

$$PERCLOS = \frac{N}{M} \times 100\% \tag{13}$$

Figure 7 shows that the PERCLOS value (for P80) is positively correlated with the degree of fatigue. Therefore, we can take the PERCLOS value of controllers as the indicator of their eye fatigue, and quantitatively describe the fatigue state of controllers through PERCLOS (P80).

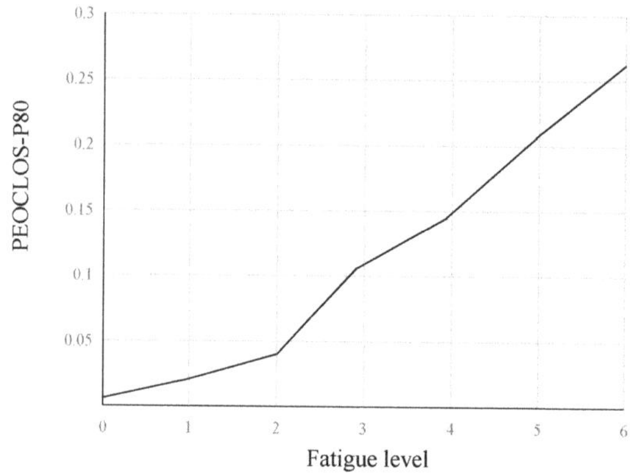

Figure 7. Relationship between PERCLOS value (for P80) and degree of fatigue.

2.3. Decision-Level Fatigue-Information Fusion Based on the Entropy Weight Method

The concept of entropy is often used in information theory to characterize the degree of dispersion in a system. According to the theory behind the entropy weight method, a higher information–entropy index indicates data with a smaller degree of dispersion, and the smaller the impact of the index on the overall evaluation, the lower the weight [23]. According to this theory, the entropy weight method determines the weight according to the degree of discreteness in the data, and the calculation process is more objective. Therefore, the characteristics of information entropy are often used in comprehensive evaluations of multiple indicators to reasonably weight each indicator.

The main steps of the entropy weight method are as follows:

1. Normalization of indicators. Assuming that there are n samples and m indexes, x_{ij} represents the value of the jth indicator ($j = 1, 2, \ldots, m$) corresponding to the ith sample ($i = 1, 2, \ldots, n$), then the normalized value $x_{ij}{}^{*}$ can be expressed as

$$x_{ij}{}^{*} = \frac{x_{ij} - x^{j}_{min}}{x^{j}_{max} - x^{j}_{min}} \tag{14}$$

where x^{j}_{min} and x^{j}_{max} represent the minimum and maximum values of the indicators, respectively.

2. Calculate the proportion of the ith sample value under index j:$p_{ij} = \frac{x^{*}_{ij}}{\sum_{i=1}^{n} x^{*}_{ij}}$.

3. Calculate the entropy of index j. According to the definition of information entropy, the information entropy of a set of data values is

$$E_j = -ln\ (n)^{-1} \sum_{i=1}^{n \sum_{ij}} P_{ij} ln \tag{15}$$

where, when $p_{ij} = 0$, the information entropy is

$$\lim_{p_{ij} \to 0} p_{ij} ln\ p_{ij} = 0 \tag{16}$$

4. Calculate the weight of each indicator.

$$W_j = \frac{1 - E_j}{\sum_{j=1}^{m} (1 - E_j)} \tag{17}$$

After obtaining the information entropy of each index, the weight of each index W_j can be calculated using Formula (17).

3. Experiments and Verification

3.1. Data Acquisition Experiment

The experimental data were collected from the air traffic control simulation laboratory. We used the same type of simulator equipment as the Air Traffic Control Bureau at the experiment scenario, and invited six licensed tower controllers to participate in data collection with the cooperation of the Air Traffic Control Bureau.

Specifically, the radar control equipment in the experiment is shown in Figure 8. Key data acquisition equipment includes controller radio telephony recording equipment and facial data acquisition equipment. Speech data were collected directly from the radio telephony communication system, and eye-video data were collected using a camera with a resolution of 1280 × 720 at 30 fps. The participants included three male and three female certificated air traffic controllers from the Air Traffic Control Bureau. They are all from East China and have more than three years of controller work experience. All subjects were required to have adequate rest (>7 h) every night on rest days, and no food, alcohol, or drinks that might affect the experimental results. All experimental personnel were fully familiar with the control simulator system. All subjects were informed of the experiment's content and had the right to stop the experiment at any time.

Figure 8. Experimental environment.

The experiment lasted for one week, and was designed according to the scheduling system in the actual control work. Severe fatigue is less common during the work process of air traffic controllers. We increased the workload appropriately to avoid a significant difference between the amount of data collected in the non-fatigued and fatigued states. During the week, the six controllers were conducted 24 h of experiment every day on Monday, Wednesday, Friday, and Sunday, and the controllers were prescribed to have complete rest on Tuesday, Thursday and Saturday, and they were required to sleep for more than seven hours. Each experimental day contained six periods of work and six periods of rest of two hours each. Each controller worked for two hours, rested for two hours, and completed the experiment for 24 h in turn. This experimental design was fully consistent with the characteristics of the actual controller post. On the experimental days, the first set of work experiments were conducted from 0:00 to 2:00 for each controller. After two hours of work, the first period of rest was started from 2:00 to 4:00. At the end of each period of work, each controller was asked to fill in the Karolinska Sleepiness Scale (KSS) [24], which took 10 min. They then rested for 110 min. After two hours of work, the controller is scheduled to rest for two hours, which is entirely consistent with the actual work schedule of the controller. The controlled work experiments alternated with rest periods, with up to 12 h of work and 12 h of rest periods within each working day.

KSS is a reliable fatigue detection scale, with scores ranging from one to ten reflecting the subject's state from alert to almost unable to maintain clarity. According to the usual practice of KSS, when the questionnaire score is greater than or equal to seven, we determine that the controller is in a state of fatigue. At the end of the experiment, fatigue was determined according to the fatigue scale filled out by each controller, which was used as the label for the test data. After the experiment, we obtained a total of 462 min of valid voice data and 29 h of valid facial-video data.

3.2. Experiments on Speech-Fatigue Characteristics

We first combined traditional speech features including the MFCC, short-time-average zero-crossing rate, and maximum Lyapunov index, and verified the effectiveness of the FSVM using the classification accuracy of these features. We extracted the largest Lyapunov exponent λ_{max} in the speech signal using data near-fitting and the characteristics of small-sample data. Figure 9 shows the fusion detection process for multiple speech features.

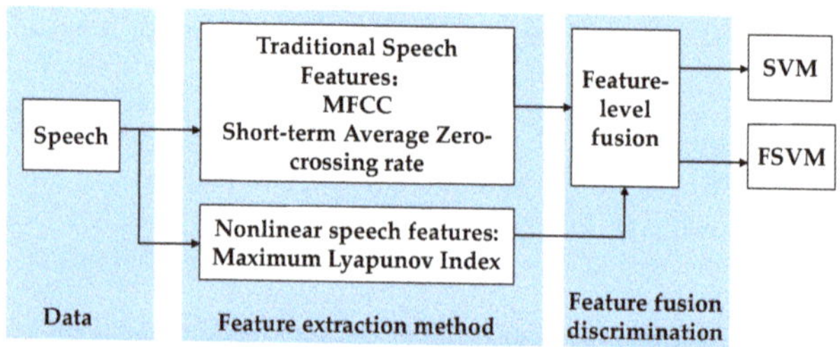

Figure 9. Speech-fatigue feature detection process.

In order to obtain a clearer understanding of the classification performance of each classifier, we input relevant parameters, as presented in Table 1.

Table 1. Internal classifier parameters.

Classification Algorithm	Penalty Parameter C	Kernel-Function Scaling Parameter σ	Number of Support Vectors
Standard SVM	64	0.0625	62
Improved FSVM	128	0.125	39

We selected 2010 speech data (for three males and three females) from the controller speech database as the training set (comprising 1030 fatigue samples and 980 normal samples) for target feature extraction and analysis. Data collected in the past has been processed as necessary and divided into short frames with a window function length of 25 ms. The step size between successive windows was set to 10 ms, and the number of filters used was 40. According to the feature extraction method, the 12th-order MFCC, short-time-average zero-crossing rate, and maximum Lyapunov exponent were extracted from the speech samples under fatigue and normal conditions. The extraction results were tested statistically to identify which of the above features differed significantly from the fatigue state; the results are presented in Table 2.

Table 2. Differences in speech features between different states.

Phonetic Feature	Normal Sample	Fatigue Sample	t	Significance ($p = sig$)
Short-time-average zero-crossing rate (times/frame)	71.4 ± 6.2	66.9 ± 7.3	3.67	$p = 0.041 < 0.05$
12th-order MFCC	-0.91 ± 0.47	0.14 ± 0.51	7.62	$p < 0.01$
Maximum Lyapunov exponent	0.34 ± 0.07	0.25 ± 0.05	10.2	$p < 0.01$

We performed an Analysis of Variance (ANOVA) on three speech features. Our hypotheses are: (1) the short-time-average zero-crossing rate speech feature is related to the level of fatigue; (2) the 12th-order MFCC speech feature is related to the level of fatigue;

(3) the maximum Lyapunov exponent speech feature is related to the level of fatigue. The mean ± standard deviation values are shown in Table 2.

In Table 2, the "Normal sample" represents the mean ± standard deviation of the speech features under normal conditions, while the "Fatigue sample" represents the mean ± standard deviation of the speech features under fatigue conditions. The "t" represents the t-test, which is the difference between the sample mean and the population mean. The significance level "p" represents the probability of the previous three hypotheses being true. Since the p-values in Table 2 are all less than 0.05, the three types of speech features have changed significantly under fatigue conditions. The significance level p-values of MFCC and maximum Lyapunov exponent are less than 0.01, so the differences in these two features before and after the controller's fatigue can be considered to be significant, and these two features are particularly good at representing the fatigue state of the human body. Although the significance level p-value of the short-time average zero-crossing rate is slightly higher than that of MFCC and maximum Lyapunov exponent, its value is still less than 0.05, so this type of speech feature can also be used for fatigue detection.

In order to verify the usefulness of various speech features in detecting controller fatigue, we combined the features in different ways and used the standard SVM and the improved FSVM to classify and detect each combination. We use the data from the previous databases as the training data and the data collected in this experiment as the test data. The two types of speech data are strictly separate for classifiers. The test results after 50% cross-validation are presented in Table 3.

Table 3. Test results for speech-fatigue characteristics.

Group	SVM Accuracy	FSVM Accuracy
MFCC + short-time-average zero-crossing rate	75.4%	77.1%
MFCC + maximum Lyapunov index	78.3%	79.5%
Short-time-average zero-crossing rate + maximum Lyapunov index	73.1%	73.8%
MFCC + short-time-average zero-crossing rate + maximum Lyapunov index	80.4%	82.5%

Table 3 indicates that the fatigue classification performance was worse for the short-time-average zero-crossing rate than for the other two features. Combining the three features produced the best classification performance, which was due to the traditional features and nonlinear features complementing each other. Moreover, the table indicates that the improved FSVM classifier had higher accuracy than the traditional SVM, which was due to the optimization of the membership determination method.

3.3. *Experiments on Eye-Fatigue Characteristics*

The OpenCV software library was used to detect eyes in the images. The human-eye-detection algorithm adopted in this study can accurately determine the eye area in both the open- and closed-eye states.

This study analyzed sub-images of the right eyes of the experimenters, and calculated the aspect ratio λ of the eye by applying equalization, binarization, and other processing methods to the images. The calculation process of the length-width ratio of the eye is shown in Figure 10.Among the 3783 eye images extracted, 748 images had λ values of <0.23, resulting in an eye PERCLOS value of 0.20 for experimenter 1 during this period.

In order to determine the fatigue value of controllers at different time periods more objectively, the Maximum Continuous Eye Closure Duration (MCECD) characteristic values of the experimenters during that time period were synchronously calculated according to the longest continuous eye-closure time within a single time period. As the degree of fatigue increases, the eyes become more difficult to open from a closed state, or more difficult to keep open when in an open state, leading to an increase in MCECD [25]. Considering that the radar signals used for air traffic control are updated every 4 s, we used a 40 s time window to detect the MCECD value of the controller during each time period, and included it in the calculation of the eye-fatigue value. We considered that a subject was in a state of fatigue when their eye-fatigue value was >0.3. Therefore, 0.3 is set as the threshold value. When MCECD is greater than 0.3, the subject is considered to be in a state of fatigue; when MCECD is less than or equal to 0.3, the subject is considered to be in a state of no fatigue. The fatigue state of the six air traffic controllers was judged using the threshold method, and the average value was taken. The accuracy of the discrimination results compared with the results of the KSS scale was shown in the following table. The corresponding accuracy values are listed in Table 4.

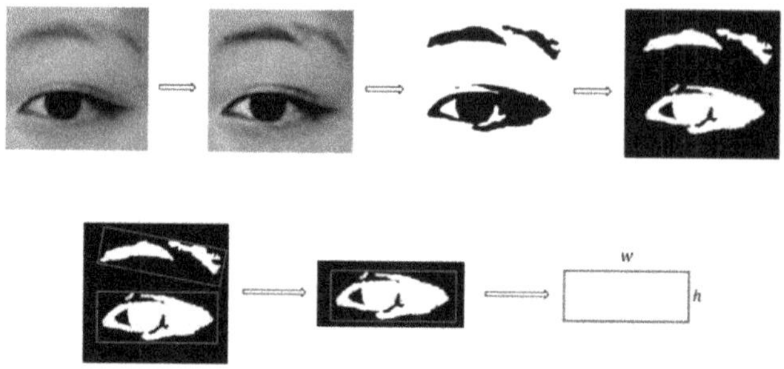

Figure 10. Extraction and processing of eye-fatigue features.

Table 4. Eye-fatigue values calculated for experimenter 1.

Time Interval	MCECD (s)	Eye-Fatigue Value	Accuracy
0:00–2:00	0.48	0.34	83.2%
4:00–6:00	0.08	0.34	82.2%
8:00–10:00	0.82	0.33	79.5%
12:00–14:00	0.77	0.18	78.9%
16:00–18:00	0.97	0.23	88.1%
20:00–22:00	0.82	0.19	86.6%

The experimental results in Table 4 indicate that the average accuracy in identifying fatigue state based on eye characteristics was 83.08%. The experimental results show that the eye features have a high average accuracy in recognizing the fatigue characteristics of different controllers at different time periods, therefore, the eye fatigue features have good application values.

3.4. Multisource Information-Fusion Fatigue-Detection Experiment

The speech features were also evaluated numerically. By combining the trained decision model with $f(x)$, we applied sentence-by-sentence detection to the control speech data of each experimenter during each time period. This yielded the proportion of fatigued-speech epochs to the total number of speech epochs during the entire time period, which was used as the characteristic value for that experimenter's fatigued speech during this time period. The speech-fatigue detection results are presented in Table 5.

According to the theory of the entropy weight method, this study first assigned weights to the speech-fatigue value and eye-fatigue value at the decision level. We calculated the weights of the eye indicators and speech indicators of the six experimenters by normalizing the data and calculating the information entropy. The results are presented in Table 6.

Table 5. Speech-fatigue detection results for experimenter 1.

Time Period	Total Number of Fatigued-Speech Epochs	Total Number of Speech Epochs	Fatigue Value
0:00–2:00	23	308	0.07
4:00–6:00	50	290	0.17
8:00–10:00	27	289	0.09
12:00–14:00	53	309	0.17
16:00–18:00	56	274	0.20
20:00–22:00	56	259	0.22

Table 6. Sample weights of the experimenters.

Experimenter	Eye Indicators	Speech Indicators
1	0.55	0.45
2	0.58	0.42
3	0.55	0.45
4	0.53	0.47
5	0.59	0.41
6	0.57	0.43

Based on the sample data from all experimenters, we averaged the weights of the indicators and used them as comprehensive indicator weights. The final indicator weights for the eye- and speech-fatigue values were 0.56 and 0.44, respectively. Based on the weights of the comprehensive indicators, we could obtain the comprehensive fatigue values of controllers at different time periods. We considered that a comprehensive fatigue value of >0.2 indicated a state of fatigue. Table 7 presents the comprehensive fatigue results for all experimenters.

Table 7. Recognition accuracy of the comprehensive fatigue value.

Time Period	Comprehensive Fatigue Value	Accuracy
0:00–2:00	0.22	84.3%
4:00–6:00	0.25	85.2%
8:00–10:00	0.21	81.7%
12:00–14:00	0.18	89.6%
16:00–18:00	0.22	84.9%
20:00–22:00	0.20	90.3%

The experimental results show that the recognition accuracy of the fatigue state based on integrated features was 86.0%, which was 2.2% and 3.5% higher than those for eye and speech features, respectively, indicating that fatigue recognition after fusion was more robust. In the experiment, some samples could not be correctly classified. The reason is that different controllers have different vocal characteristics and facial expression characteristics. It is undeniable that some people's voice quality, timbre, and features such as eyes are similar to the performance of others when they are fatigued, which makes it difficult for the model to classify correctly. It can be seen that fatigue recognition still has great research potential.

4. Conclusions

This paper aims to solve the controller fatigue detection problem by first collecting linear and nonlinear radio telephony features. Then, the fused radio telephony features are classified and identified. Finally, the radio telephony recognition results and the eye fatigue feature recognition results are combined for decision-making to achieve a comprehensive assessment of the controller's fatigue status.

The experimental results show that different speech feature combinations have different recognition accuracies, among which Mel Frequency Cepstral Coefficient + short-time-average zero-crossing rate + maximum Lyapunov index has the highest recognition accuracy. In the case of the fuzzy support vector machine classifier, the accuracy reached 82.5% compared to the evaluation results using the fatigue scale KSS. The accuracy in identifying fatigue state based on eye characteristics was 83.8%. For voice features and facial features, the initial fatigue results for the speech and eye feature layers were weighted using the entropy weight method, and finally comprehensive controller fatigue characteristic values were obtained through decision-level fusion. The accuracy of the fatigue detection method combined with eye and speech features was 86.0%, which was 2.2% and 3.5% higher than those for eye and speech features, respectively. The present research findings can provide theoretical guidance for air traffic management authorities to detect ATC fatigue. They might also be useful as a reference for controller scheduling improvement.

If the controllers' clearances often need to be corrected, the controller may be fatigued. In the future, we hope to study the relationship between the number of corrections in the controller's radiotelephony and the multiple levels of fatigue.

Author Contributions: Conceptualization, L.X., Z.S. and S.H.; methodology, L.X. and S.M.; formal analysis, L.X., S.M., S.H. and Y.N.; validation, S.H. and Y.N.; software Y.N.; resources, supervision, project administration, funding acquisition, Z.S. All authors have read and agreed to the published version of the manuscript.

Funding: The authors acknowledge the financial support from the National Natural Science Foundation of China (grant no. U2233208).

Data Availability Statement: These data in this study are available from the corresponding author upon request.

Conflicts of Interest: The author(s) declare(s) that there are no conflicts of interest regarding the publication of this paper.

References

1. Terenzi, M.; Ricciardi, O.; Di Nocera, F. Rostering in air traffic control: A narrative review. *Int. J. Environ. Res. Public Health* **2022**, *19*, 4625. [CrossRef] [PubMed]
2. Butkevičiūtė, E.; Michalkovič, A.; Bikulčienė, L. Ecg signal features classification for the mental fatigue recognition. *Mathematics* **2022**, *10*, 3395. [CrossRef]
3. Lei, J.; Liu, F.; Han, Q.; Tang, Y.; Zeng, L.; Chen, M.; Ye, L.; Jin, L. Study on driving fatigue evaluation system based on short time period ECG signal. In Proceedings of the 2018 21st International Conference on Intelligent Transportation Systems (ITSC), Maui, HI, USA, 4–7 November 2018; pp. 2466–2470.
4. Ahsberg, E.; Gamberale, F.; Gustafsson, K. Perceived fatigue after mental work: An experimental evaluation of a fatigue inventory. *Ergonomics* **2000**, *43*, 252–268. [CrossRef] [PubMed]
5. Li, X.; Tan, N.; Wang, T.; Su, S. Detecting driver fatigue based on nonlinear speech processing and fuzzy SVM. In Proceedings of the 2014 12th International Conference on Signal Processing (ICSP), Hangzhou, China, 19–23 October 2014; pp. 510–515.
6. Schuller, B.; Steidl, S.; Batliner, A.; Schiel, F.; Krajewski, J.; Weninger, F.; Eyben, F. Medium-term speaker states—A review on intoxication, sleepiness and the first challenge. *Comput. Speech Lang.* **2014**, *28*, 346–374. [CrossRef]
7. Wu, N.; Sun, J. Fatigue Detection of Air Traffic Controllers Based on Radiotelephony Communications and Self-Adaption Quantum Genetic Algorithm Optimization Ensemble Learning. *Appl. Sci.* **2022**, *12*, 10252. [CrossRef]
8. Kouba, P.; Šmotek, M.; Tichý, T.; Kopřivová, J. Detection of air traffic controllers' fatigue using voice analysis—An EEG validation study. *Int. J. Ind. Ergon.* **2023**, *95*, 103442. [CrossRef]
9. de Vasconcelos, C.A.; Vieira, M.N.; Kecklund, G.; Yehia, H.C. Speech Analysis for Fatigue and Sleepiness Detection of a Pilot. *Aerosp. Med. Hum. Perform.* **2019**, *90*, 415–418. [CrossRef]
10. Reddy, M.K.; Rao, K.S. Inverse filter based excitation model for HMM-based speech synthesis system. *IET Signal Process.* **2018**, *12*, 544–548. [CrossRef]
11. Liang, H.; Liu, C.; Chen, K.; Kong, J.; Han, Q.; Zhao, T. Controller Fatigue State Detection Based on ES-DFNN. *Aerospace* **2021**, *8*, 383. [CrossRef]
12. Shen, Z.; Pan, G.; Yan, Y. A High-Precision Fatigue Detecting Method for Air Traffic Controllers Based on Revised Fractal Dimension Feature. *Math. Probl. Eng.* **2020**, *2020*, 4563962. [CrossRef]
13. Shintani, J.; Ogoshi, Y. Detection of Neural Fatigue State by Speech Analysis Using Chaos Theory. *Sens. Mater.* **2023**, *35*, 2205–2213. [CrossRef]

14. McClung, S.N.; Kang, Z. Characterization of Visual Scanning Patterns in Air Traffic Control. *Comput. Intell. Neurosci.* **2016**, *2016*, 8343842. [CrossRef] [PubMed]
15. Jo, J.; Lee, S.J.; Park, K.R.; Kim, I.-J.; Kim, J. Detecting driver drowsiness using feature-level fusion and user-specific classification. *Expert Syst. Appl.* **2014**, *41*, 1139–1152. [CrossRef]
16. Zyśk, A.; Bugdol, M.; Badura, P. Voice fatigue evaluation: A comparison of singing and speech. In *Innovations in Biomedical Engineering*; Springer International Publishing: Berlin/Heidelberg, Germany, 2019; pp. 107–114.
17. Chan, R.W.; Lee, Y.H.; Liao, C.-E.; Jen, J.H.; Wu, C.-H.; Lin, F.-C.; Wang, C.-T. The Reliability and Validity of the Mandarin Chinese Version of the Vocal Fatigue Index: Preliminary Validation. *J. Speech Lang. Hear. Res.* **2022**, *65*, 2846–2859. [CrossRef]
18. Naz, S.; Ziauddin, S.; Shahid, A.R. Driver Fatigue Detection using Mean Intensity, SVM, and SIFT. *Int. J. Interact. Multimed. Artif. Intell.* **2019**, *5*, 86–93. [CrossRef]
19. Sommer, D.; Golz, M. Evaluation of PERCLOS based current fatigue monitoring technologies. In Proceedings of the 2010 Annual International Conference of the IEEE Engineering in Medicine and Biology, Buenos Aires, Argentina, 31 August–4 September 2010; pp. 4456–4459.
20. Zhuang, Q.; Kehua, Z.; Wang, J.; Chen, Q. Driver Fatigue Detection Method Based on Eye States with Pupil and Iris Segmentation. *IEEE Access* **2020**, *8*, 173440–173449. [CrossRef]
21. Zhang, F.; Wang, F. Exercise Fatigue Detection Algorithm Based on Video Image Information Extraction. *IEEE Access* **2020**, *8*, 199696–199709. [CrossRef]
22. Ji, Y.; Wang, S.; Zhao, Y.; Wei, J.; Lu, Y. Fatigue State Detection Based on Multi-Index Fusion and State Recognition Network. *IEEE Access* **2019**, *7*, 64136–64147. [CrossRef]
23. Sahoo, M.M.; Patra, K.; Swain, J.; Khatua, K. Evaluation of water quality with application of Bayes' rule and entropy weight method. *Eur. J. Environ. Civ. Eng.* **2017**, *21*, 730–752. [CrossRef]
24. Laverde-López, M.C.; Escobar-Córdoba, F.; Eslava-Schmalbach, J. Validation of the Colombian version of the Karolinska sleepiness scale. *Sleep Sci.* **2022**, *15*, 97–104. [CrossRef]
25. Dziuda, Ł.; Baran, P.; Zieliński, P.; Murawski, K.; Dziwosz, M.; Krej, M.; Piotrowski, M.; Stablewski, R.; Wojdas, A.; Strus, W.; et al. Evaluation of a Fatigue Detector Using Eye Closure-Associated Indicators Acquired from Truck Drivers in a Simulator Study. *Sensors* **2021**, *21*, 6449. [CrossRef] [PubMed]

aerospace

MDPI

Article

Ensuring Safety for Artificial-Intelligence-Based Automatic Speech Recognition in Air Traffic Control Environment

Ella Pinska-Chauvin [1,*], Hartmut Helmke [2], Jelena Dokic [1], Petri Hartikainen [1], Oliver Ohneiser [2] and Raquel García Lasheras [3]

[1] Integra Consult A/S, Staktoften 20, 1., 2950 Vedbaek, Denmark; jdj@integra.dk (J.D.); pha@integra.dk (P.H.)
[2] German Aerospace Center (DLR), Institute of Flight Guidance, Lilienthalplatz 7, 38108 Braunschweig, Germany; hartmut.helmke@dlr.de (H.H.); oliver.ohneiser@dlr.de (O.O.)
[3] ATM Research and Development Reference Centre (CRIDA A.I.E.), Las Mercedes Business Park, C/de Campezo 1, 28022 Madrid, Spain; rglasheras@e-crida.enaire.es
* Correspondence: epc@integra.dk

Abstract: This paper describes the safety assessment conducted in SESAR2020 project PJ.10-W2-96 ASR on automatic speech recognition (ASR) technology implemented for air traffic control (ATC) centers. ASR already now enables the automatic recognition of aircraft callsigns and various ATC commands including command types based on controller–pilot voice communications for presentation at the controller working position. The presented safety assessment process consists of defining design requirements for ASR technology application in normal, abnormal, and degraded modes of ATC operations. A total of eight functional hazards were identified based on the analysis of four use cases. The safety assessment was supported by top-down and bottom-up modelling and analysis of the causes of hazards to derive system design requirements for the purposes of mitigating the hazards. Assessment of achieving the specified design requirements was supported by evidence generated from two real-time simulations with pre-industrial ASR prototypes in approach and en-route operational environments. The simulations, focusing especially on the safety aspects of ASR application, also validated the hypotheses that ASR reduces controllers' workload and increases situational awareness. The missing validation element, i.e., an analysis of the safety effects of ASR in ATC, is the focus of this paper. As a result of the safety assessment activities, mitigations were derived for each hazard, demonstrating that the use of ASR does not increase safety risks and is, therefore, ready for industrialization.

Keywords: safety assessment; air traffic control; automatic speech recognition; workload; situational awareness; en-route sector; approach sector

Citation: Pinska-Chauvin, E.; Helmke, H.; Dokic, J.; Hartikainen, P.; Ohneiser, O.; Lasheras, R.G. Ensuring Safety for Artificial-Intelligence-Based Automatic Speech Recognition in Air Traffic Control Environment. *Aerospace* **2023**, *10*, 941. https://doi.org/10.3390/aerospace10110941

Academic Editor: Michael Schultz

Received: 11 September 2023
Revised: 25 October 2023
Accepted: 29 October 2023
Published: 3 November 2023

1. Introduction

Automatic speech recognition (ASR) in the air traffic management (ATM) domain is seen as a promising technology for improving efficiency and safety [1]. Use of ASR technology in ATC environments consists of three conceptual steps. First, a speech-to-text conversion is performed, i.e., an ATC utterance such as "lufthansa two seven victor descend flight level two hundred" is transcribed from the speech signal into a sequence of words. This is followed by the text-to-concepts transformation, i.e., the semantics of the transcription are extracted as machine-readable ontology-conforming annotations with aircraft callsigns and various command elements such as "DLH27V DESCEND 200 FL". In the third step, the output of the two preceding steps is directly presented on the air traffic controllers' (ATCO) human machine interface (HMI) enabling, amongst other benefits, the replacement of manual HMI inputs by the ATCOs. The following ASR functionalities were covered by the safety assessment relevant to this paper:

1. Recognition of relevant aircraft callsigns from ATCO and pilot utterances as well as highlighting the callsigns on the controller working position (CWP) HMI display.

2. Recognition of ATCO commands and input of the command contents into the aircraft radar data labels displayed at the ATCO CWP HMI.

As the ASR technology is relatively new, it is not yet deployed in the various ATM systems developed by the industry. This is due in part to the fact that European regulation [2] requires any new technology introduced into the operational environment undergoes a rigorous safety assessment conducted at the design phase, providing relevant evidence that the physical design satisfies the design requirements. The safety of two ASR prototypes for air traffic control (ATC) purposes supported by artificial intelligence (AI) has been assessed by Integra in accordance with the Single European Sky Air Traffic Management Research (SESAR) safety reference materials [3] in the course of SESAR2020 project PJ.10-W2-96-ASR [4,5]. The safety assessment includes several steps, starting at the design level with the assessment of the introduction of a new system—or a change to an existing system—for the identification of any hazards introduced by the new system elements, and possible associated increase in risks. The system must be proven to be safe in a specific environment by demonstrating that the level of safety is not degraded, and that at least the same level of safety can be achieved as prior to the introduction of the change.

The considered ASR prototypes were designed to improve ATCOs' situational awareness, reduce their workload and increase their productivity. The scope of the safety assessment considered application of ASR supporting ATCOs with the aforementioned goals in approach and en-route sectors of medium traffic complexity.

The main goal of this article is to evaluate possible safety risks introduced by the implementation of ASR and the possible impact of these risks on ATC operations, in order to derive mitigations formulated as system design requirements. In the next step, these system design requirements are implemented into two pre-industrial prototype platforms to demonstrate the feasibility of the evaluated design and expected safety levels of system performance.

In the next section, we provide background work of ASR applications in the ATM domain, including previous safety-related work performed on the topic. In Section 3, the safety assessment performed in the context of ASR application developed in project PJ.10-W2-96 to derive the safety requirements for the design is described. In this section, we also describe the setup of the human-in-the-loop simulations that were conducted to gather the evidence required for the safety assessment. Section 4 presents the results of the two simulations and the safety assessment related results and also the limitations of this study, which is based on only two validation experiments. Concluding remarks are given in Section 5. Lastly, two appendices are included containing additional information for the hazard analysis performed.

2. Background

The SESAR2020 project PJ.16-04 "CWP HMI" (*Controller Working Position Human Machine Interface*) [6] investigated the feasibility of ASR with early prototypes applied in the air traffic control domain. Those prototypes were validated in laboratory-like environments equivalent to technology readiness level (TRL) 4 as per industrial research development standards [7]. Technology readiness levels (TRLs) are a method for estimating the maturity of technologies during research and development phase, that enables consistent and uniform assessment. TRLs are based on a scale from 1 to 9 with 9 being the most mature technology.

The basis of this paper, i.e., the follow-up project PJ.10-W2-96 [4,5] continued developing the application with the aim of demonstrating the technology feasibility in relevant operational environment (TRL6). In parallel, the project PJ.05-W2-97 *HMI Interaction modes for Airport Tower* aimed to develop an ASR system for an aerodrome control tower environment [8]. The ATCO2 platform [9] aims at collecting, pre-processing, and pseudo-anonymizing ATC communications' audio databases of more than 5000 h of audio data with the objective of increasing robustness of speech recognition in the air traffic management domain. The ATCO2 corpus has also been used to detect speaker roles in voice communication, i.e., pilot or ATCO, and clustering speakers [10]. Given enough training data, automatic speech recognition and understanding systems also build the base to train ATCOs [11]. A common goal of prior presented ASR research projects was to define an initial ontology for the annotation of ASR recognized ATC concepts such as command types, values, units, and qualifiers, to be later coordinated and agreed between major European ATM stakeholders enabling industrialization of the technology [12].

Early results demonstrated that ASR facilitated safety in operational environments by detection of read-back errors from comparison of controller and pilot radio communication at aerodrome control towers [13]. ASR together with deep-learning-based methods were applied as safety monitoring function by translating the pilot–controller voice communications into texts, which were then converted to contextual data to be analyzed for flight conformance verification, and potential conflict detection [14]. The Venture capital funded project AcListant® [15] and AcListant®-Strips for ATC approach areas focused on ASR with the aim to significantly reduce controllers' workload [1] and increase ATM efficiency [15]. The exercise of DLR and Austro Control of SESAR2020 project PJ.10-W2-96-ASR, described in detail in Section 3.4, has the same objectives, using Vienna approach and not Dusseldorf as validation airspace. The main difference is that the focus of project PJ.10-W2-96 is on investigating safety aspects regarding the number of erroneous recognitions of ASR that are undetected by the ATCO [5]. These results are summarized in Section 4.2.2.

The STARFiSH (Safety and Artificial Intelligence Speech Recognition) [16] project integrated AI-based speech recognition into an A-SMGCS (Advanced Surface Movement Guidance and Control System) for ground traffic control and monitoring at Frankfurt Airport. The joint application of ASR and A-SMGCS recognized the instructions given by apron controllers to pilots, extracted the commands contained therein and integrated the outputs to the user interface of the A-SMGCS. An additional safety net for AI applications in ASR is intended to ensure that errors in AI-based speech recognition do not have any negative effects on the overall system [16]. The benefits of callsign recognition from flight crew utterances and highlighting the callsign at an en-route CWP HMI were investigated in [4]. The study demonstrated feasibility of the integrated ASR system for the identification of callsigns from flight crew utterances which provide benefits in terms of workload and situational awareness. The papers also highlighted the importance of further work on recognition and timeliness of outputs.

Incorporating ASR into ATC specifically as a safety enhancing feature has been researched by various practitioners, especially in conjunction with integrating ASR into various other safety features contained in an ATM system such as conformance monitoring and trajectory prediction. Karlsson et al. previously hypothesized in 1990 on this basis that the introduction of ASR technology into ATC could result in a reduced occurrence of human-generated errors enabling in turn increased safety of the overall system [17]. More recently in 2023, the use of ASR as a safety enhancing application in ATC operations was investigated and noted that the solutions investigated can improve the safety of ATC operations and can contribute to the reduction in ATCO workload [18]. Zhou et al. argued that ASR represents a gateway between the ATM system and the ATCO in converting speech signal to text inputs and that after spoken instruction understanding (SIU) is applied to the converted text the output information can be used to support safety-critical applications (SCA), enabling safety and reducing possible human errors [19]. The European Union Aviation Safety Agency recently developed a roadmap for the approval and deployment of safety-related AI systems for end-user support (pilots and ATCOs) [20]. In following guidance [21], the process of safety assurance for AI level 1 (assistance to human) and AI level 2 (human machine teaming) is developed with the further classification for different level of safety analysis depending on the AI application level. According to the guidance, AI-supported ASR application can be classified as AI Level 2A: human/machine teaming representing human and AI-based system cooperation. The work presented in this paper did not explicitly address the trustworthiness of AI application as the elements of the safety analysis. This paper concentrates on the initial part of safety assessment in the design phase for such an application. The safety assessment must be continued "in-service" via a data-driven AI safety risk assessment based on operational data and occurrences.

3. Materials and Methods

The validated applications demonstrated ASR capability to effectively support ATCOs by showing evidence appropriate at the pre-industrial feasibility level. This section describes the safety assessment process conducted in accordance with the SESAR Safety Reference Material [3] and its guidance [22] at the design phase to ensure that the proposed implementation of ASR in ATM operations is capable in satisfying the performance requirements as stipulated by European regulation [2]. The SESAR safety assessment process has to demonstrate that the design is safe by using two different approaches:

1. A *success* approach, in which the effectiveness of the new concepts and technologies is assessed, when they are working as intended, i.e., how much the pre-existing risks that are inherent and already present in aviation will be reduced by the changes to the ATM system under assessment, i.e., defining the positive contribution to aviation safety that the ATM changes under assessment may deliver in the absence of failure.
2. A *failure* approach, in which the ATM system generated risks, induced by the ATM changes under assessment are evaluated. This approach defines the negative contribution to the risk of an accident that the ATM changes under assessment may induce in the event of failure(s), however caused.

This paper focuses on the process of deriving the performance requirements for the failure approach based on identification of potential hazards presented by the introduction of ASR, thus ensuring safe implementation of ASR technology to ATC operations.

3.1. Selected Use Cases

The safety assessment covered TRL 5 system development phase, representing technology validated in relevant operational environment, and TRL 6—technology demonstrated in relevant operational environment. For this reason, the operational use cases were selected by a group of subject matter experts from the field who represent the possible users of the technology. The scope of the assessment described was limited to the following uses cases:

3.1.1. Use Case "Highlight of Callsigns (Aircraft Identifier) on the CWP Based on the Recognition of Pilot Voice Communications"

In the scope of this use case, the pilot's voice signal was extracted, processed by ASR for callsign recognition and further verified against contextual flight plan data. This type of use of ASR technology supports the ATCO by identifying new flights entering the sector and making initial contact on the ATC VHF channel, and flight crews requesting actions from ATCOs, e.g., trajectory change, flight level change or information.

3.1.2. Use Case "Highlight of Callsigns on the CWP Based on the Recognition of ATCO Voice Communications"

The ATCO voice signal was extracted, processed by ASR for callsign recognition and further verified against contextual flight plan data. This type of ASR application, where the aircraft callsign is highlighted and is based on the recognition of ATCO voice communications, provides a safety check to the ATCO who will be able to detect, whether there is a difference between the aircraft callsign mentioned and the flight radar data label on the CWP HMI, for which commands are being input.

3.1.3. Use Case "Annotation of ATCO Commands"

The ATCO voice command was extracted, processed, and verified against contextual data to provide the annotation of a specific command on the CWP HMI. This type of ASR application, where annotation of given commands is made available to ATCOs for consultation, enables increased situational awareness and provides a safety check of clearances and instructions given to flights. This use case is an intermediate step prior to the semi-automatic/automatic input of commands in the CWP using ASR.

3.1.4. Use Case "Pre-Filling of Commands in the CWP"

The recognized (and validated) commands were presented to the ATCOs together with the command values in the CWP. ATCOs were able to accept, reject or correct the commands.

Two validation exercises were selected, which jointly address all of the use cases noted above:

1. The exercise performed by CRIDA, Indra and ENAIRE places emphasis on a very low callsign recognition error rate (approx. 0%). Consequently, a lower callsign recognition rate (between 50% and 85%) is foreseen. This exercise will be referred to as the *Callsign Highlighting* exercise in the rest of the text.
2. The second exercise performed by DLR and Austro Control attempts to identify a compromise between low callsign recognition error rate (<1%) and acceptable callsign recognition rate (>97%). This exercise will be referred to as the *Radar Label Maintenance* exercise in the rest of the text.

Both approaches are different with regard to solving the callsign highlighting use cases and are, therefore, very interesting from the perspective of safety considerations. More details of the two validation exercises are provided in Section 3.3 for the *Callsign Highlighting* exercise and in Section 3.4 for the *Radar Label Maintenance* exercise. The safety assessment methodology is addressed before the exercise descriptions.

3.2. Safety Assessment Methodology

The focus of this paper is to present the "failure" part of the assessment, thus the contribution of ASR to the risk of an accident that the ATM changes under assessment may induce in the event of failure(s). The process starts with a hazard identification based on the analysis of the use cases using the walk-through technique supported by sequence diagrams as shown in Figure 1.

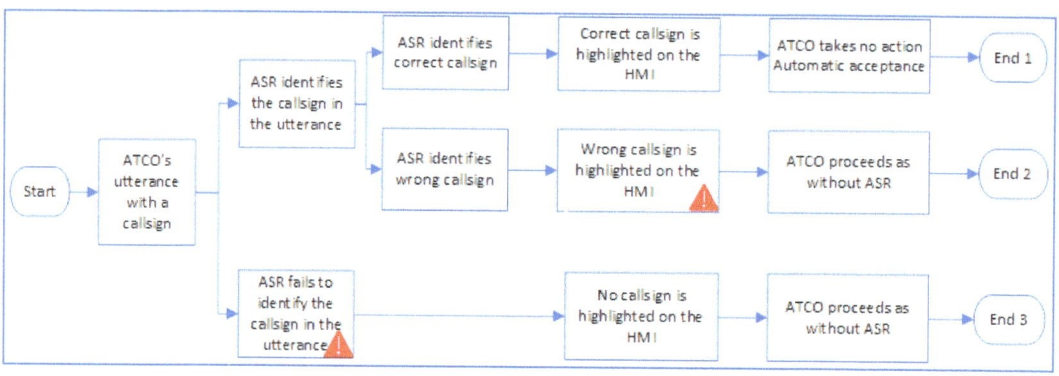

Figure 1. Sequence diagram of use case 2—"Highlight of callsigns on the CWP from ATCO utterances" with indications of the hazard's occurrence, (indicated by the exclamation marks).

Sequence diagrams were produced and analyzed for each use case. The sequence diagrams were used for the walkthrough with subject matter experts to identify potential hazards, meaning each situation that could trigger the unsafe situation. The identified hazards were further assessed according to Functional Hazard Assessment as per Safety Assessment Methodology [23] by applying the following steps:

1. Identification of hazards' effects on operations, including the effect on aircraft operations.
2. Assessment of the severity of each hazard effect.

3. Specification of target performance (safety objectives), i.e., determination of the maximum tolerable frequency of the hazard's occurrence.

A top-down causal analysis was performed for each functional hazard, their causes and associated mitigations. The identified mitigations refer to preventive mitigations for a functional hazard, which either prevent a basic cause from occurring or protect against the propagation of the basic cause effect up to the functionality hazard occurrence.

A complementary bottom-up analysis of the failure modes of the ASR elements/element-to-element interfaces and of their effects was performed in order to determine potential common cause failures.

3.3. Callsign Highlighting Exercise with Focus on Low Callsign Recognition Error Rates

ENAIRE, Indra and CRIDA, conducted an exercise to validate the performance of the pre-industrial ASR prototype covering the following use cases [4]:

- Use Case 1. Highlight of callsigns . . . based on the recognition of pilot voice,
- Use Case 2. Highlight of callsigns . . . based on the recognition of ATCO voice,
- Use Case 3. Annotation of ATCO commands.

This validation exercise used two complementary approaches aiming at providing evidence of ASR applications' performance by providing the following outputs:

1. Collection of subjective operational feedback from ATCOs gathered by means of questionnaires, debriefings and observations. This was achieved through a real-time human-in-the-loop simulation.
2. Collection of statistically significant objective data regarding ASR performance. This was achieved through the analysis of operational recordings of real-life communications between ATCOs and flight crew. Audios from different Spanish en-route sectors were processed by the ASR system to obtain the accuracy on callsign identification and command annotation.

The human-in-the-loop validation exercise simulated two en-route sectors of Madrid Flight information Region (FIR) during the nighttime. The sectors are presented in Figure 2 obtained from Enaire's Aeronautical Information web application [24], each sector in a different color. The sectors are quite wide and have several entry points where flight crew performs their first call (related to use case 1). There are very different traffic flows that require different type of control commands (related to use case 3) and facilitate the creation of situations where the traffic is focused in one area or dispersed along the whole sector (related to use cases 1 and 2).

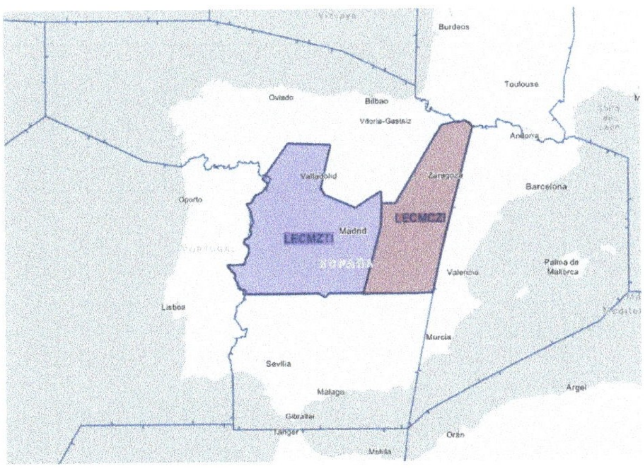

Figure 2. Madrid FIR Simulated sectors in the real-time human-in-the-loop simulation.

The validation exercises were performed in an integrated sector, where one ATCO performs both the executive controller and planning controller roles as is typical in many ACCs during the nighttime. One simulation pilot was assigned for each sector [4]. The exercises were designed with medium-to-high traffic load. A total of two ATCOs from Enaire took part in the simulations in November 2021.

To overcome the limitations of the validation activity (the low number of scheduled runs and participating ATCOs and (simulation) pilots, and the locations specificity to the validated operational environment) a statistical approach was applied. The statistical approach included the analysis of operational recordings from different types of sectors and several actors, both controllers and flight crew. The operational data also serve as a reference to compare performance between laboratory data with real-life data.

During the real-time simulation, it was possible to enable only ATCO speech recognition, pilot speech recognition or both. The different ASR functionalities (callsign recognition and command history window) were activated or deactivated to assess the three use cases separately.

Communications between ATCO and pilot were performed using COMETA, the communication system that ENAIRE has deployed in Spanish ATC units. COMETA uses VoIP and the version used for the exercise was the latest available.

When a radio voice communication is performed the ASR is triggered. The ASR system identifies the callsign in the communication and highlights the corresponding aircraft radar track symbol on the CWP. The ASR also extracts relevant information from ATCO utterances and proceeds to annotate them in a separate window that the ATCO is able to consult.

Context information, i.e., information regarding flight plans and their updates, were sent to the ASR prototype by the simulation platform to reduce callsign recognition error rate. Only callsigns that were completely recognized and present in the flight plans were displayed to the ATCOs, i.e., wrong or partial callsign recognition was considered as not identified, and no flight was highlighted on the screen.

As presented in Figure 3, the callsign recognition is indicated by displaying a white circle around the radar track symbol. The circle flashed for five seconds before disappearing. The functionality allowed highlighting several aircraft at the same time by flashing the indicator around their respective radar track symbols simultaneously.

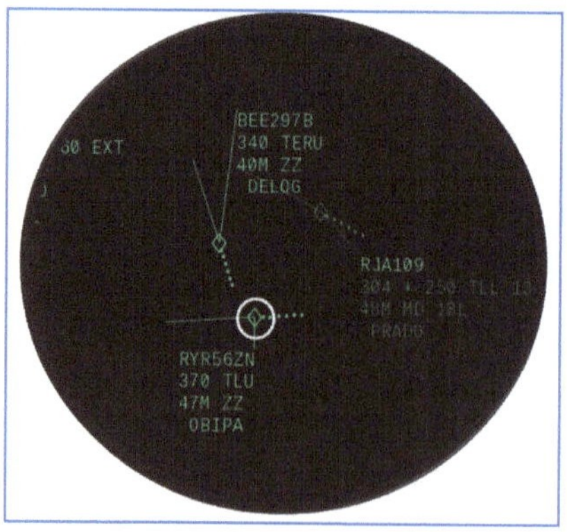

Figure 3. Callsign illumination by flashing.

The annotation window, shown in Figure 4, contains information regarding the commands provided by the ATCO. It includes the callsign of the addressed aircraft, the issuing time, the command annotation in accordance with the standard agreed between the SESAR partners, and an action column. The text in the action column and the colors in the annotation window are coherent with other elements in the CWP. As presented in the figure, if a callsign was not identified by the ASR system, it appeared as NO_IND but the transcription was available in the text field. An annotation window per flight, where only the communications exchanged with the selected flight appeared, was also available.

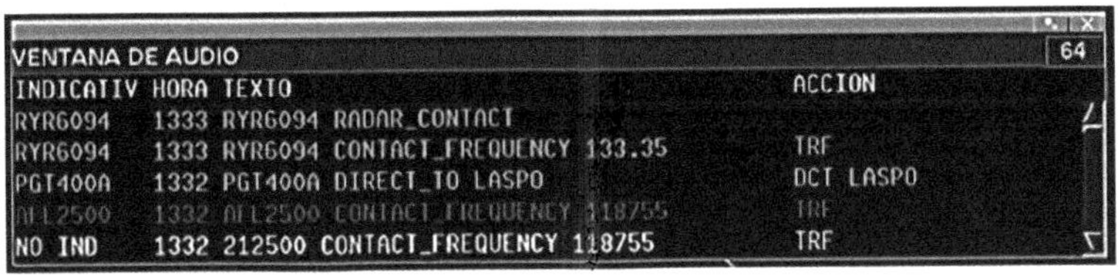

Figure 4. Annotation window with callsign, time, and command content.

The annotation window was displayed only for consultation. ATCOs did not update the information displayed but were able to navigate and sort it.

A mixture of subjective and objective data was used to assess the achievement of the objectives of the exercise. Subjective data were collected via:

- Individual questionnaires: standard and specific questionnaires were developed to assess the validation objectives. The questionnaires were agreed with the subject matter experts participating in dedicated Safety and Human performance workshops.
- Debriefing sessions: after each simulation run the findings, i.e., opportunities, difficulties, general findings observed during the exercise were discussed among all participants (operational and simulation staff).
- Over the shoulder observations: direct and non-intrusive over-the-shoulder observation were carried out by human factors expert, during the runs. This non-intrusive observation had the purpose of providing detailed, complete and reliable information on the way the activity is carried out, especially, if further commented and discussed with the observed users during the debriefing.

Additionally, objective data were obtained from system data recorded during each session by the replay and post-analysis tools. These data contained information on callsign transcription and command annotation generated during the simulation. Data on system reaction times were also recorded. Further quantitative data were obtained from the analysis of operational recordings. The recognized callsigns were compared to the correct callsigns resulting from manual annotations (gold standard annotations).

Two statistical analyses of the outputs were performed: The first one used objective data collected from the real-time simulation (RTS) screen and audio recordings. The second one used operational recordings from different Spanish en-route ATC sectors. These sectors were selected taking into account their complementary characteristics that provided a wide sample of technical (i.e., signal-to-noise ratio, native speakers origin) and operational (i.e., type of commands) characteristics. The statistical analyses were obtained by manually transcribing the recordings, creating the callsign and command annotation standards, and then comparing them against the ASR outcome.

Further details can be found in the project final report [25].

3.4. Radar Label Maintenance Exercise with Focus on High Callsign Recognition Rates

DLR together with Austro Control conducted a real-time human-in-the-loop simulation at DLR's premises in Braunschweig to validate the performance of the pre-industrial ASR prototype covering the following use cases:

- Use Case 2. Highlight of callsigns ... based on the recognition of ATCO voice,
- Use Case 3. Annotation of ATCO commands,
- Use Case 4. Pre-filling of commands in the CWP.

The focus of the simulation was to quantify the benefits of ASR with respect to operational safety and ATCO workload. The traffic scenarios consider inbound flights to

Vienna airport runway 34. Departures and overflights were not modelled. The ATCO, however, was responsible for the four adjacent approach sectors BALAD, MABOD, PESAT and NERDU plus the terminal maneuvering area (TMA) including the landing clearance roughly 6 to 10 miles before touch down, see Figure 5 taken from [5].

Figure 5. Approach chart of Vienna TMA with the four sectors BALAD, NERDU, PESAT, MABOD around the four metering fixes with the same names, taken from [5].

Simulation pilots managed flights and interacted with the ATCO via voice communication. Subjective feedback was gathered by means of questionnaires, debriefings and observations. Objective data regarding system performance were recorded (e.g., flown trajectory length and command recognition rates).

Two different scenarios were created: a medium-density traffic scenario with 30 arriving aircraft per hour and a heavy-density traffic scenario with 42 arriving aircraft per hour. A total of 12 ATCOs from Austro Control took part in the simulations lasting from September to November 2022.

In the baseline scenario, the ATCO was not supported by ASR, but was working and inputting the various commands manually using the current operating method consisting of mouse inputs. The ATCO had to click on one of the nine underlined data fields of the radar labels shown in Figure 6, taken from [5]. The click opens a drop-down menu and the ATCO needs to manually enter the given clearance values, e.g., for altitude, speed, heading, waypoint, etc.

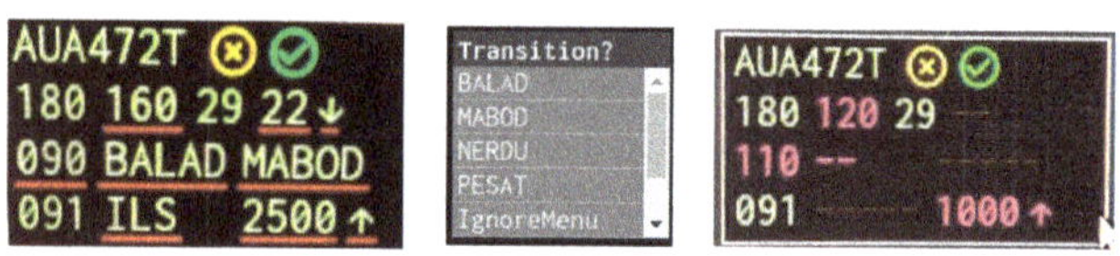

Figure 6. Left: Interactive radar label cells (red underlined); Middle: Drop-down menu to enter given transition names. Right: Radar label showing recognized command values (purple).

In solution runs, the values of the ATCO commands are extracted from the radio telephony utterance by ASR and are automatically input to the radar label cells appearing in purple color. The right part in Figure 6 shows the appearance when a flight level of 120, a heading of 110 and a descent rate of 1000 feet per minute or greater were extracted by the ASR. The dotted line "—" in waypoint field means that a recognized heading value overwrites a previously recognized waypoint value. Thus, the ATCO only needs to check and confirm the automatically generated input with a mouse click on the green checkmark in the first radar label line, or alternatively correct any values in cases of misrecognition. Accepted cell values turned into light green as soon as the ATCO accepted them. The command values are automatically accepted after ten seconds, if the ATCO does not reject or correct them. More details with respect to the HMI design used are described in [5].

Each ATCO participated in four simulation runs of 35 min duration each. Two runs were conducted in baseline mode and two in solution mode with ASR support, so that all combinations of heavy and medium traffic with baseline and solution modes were simulated by each ATCO. To compensate for sequence effects, i.e., training effects, five ATCOs started with the baseline run. In addition, seven ATCOs started with the solution runs. After the baseline run, two solution runs followed or two baseline runs followed, if the ATCO started with the solution runs. Nevertheless, there were sequence effects. A technique to compensate for the sequence effects by subtracting or adding the mean value of all first, second, third and fourth simulation runs was implemented [5].

After each simulation run, the ATCOs filled out several questionnaires, and after the last validation exercise the ATCOs completed an additional final questionnaire. An informal semi-structured debriefing with the ATCOs followed the final validation simulation runs. Table 1 shows the 10 questions, which are safety related and taken from [5].

Table 1. Questions gathering feedback related to safety issues.

Question ID	Content
1	How insecure, discouraged, irritated, stressed, and annoyed were you? (Stress annoyed)
2	What was your peak workload? (Peak workload)
3	In the previous run I ... started to focus on a single problem or a specific aircraft. (Single aircraft)
4	In the previous run there ... was a risk of forgetting something important (such as inputting the spoken command values into the labels). (Risk to Forget)
5	In the previous run, how much effort did it take to evaluate conflict resolution options against the traffic situation and conditions? (Conflict resolution)
6	In the previous run, how much effort did it take to evaluate the consequences of a plan? (Consequences)
7	In the previous working period, I felt that ... the system was reliable. (Reliable)
8	In the previous working period, I felt that ... I was confident when working with the system. (Confidence)
9	I ... found the system unnecessarily complex. (Complexity)
10	Please read the descriptors and score your overall level of user acceptance experienced during the run. Please check the appropriate number. (User Acceptance)

The results from the two validation exercises with focus on safety are presented in the next section.

4. Results

The first part of this section provides the hazards derived from the assessment and the requirements to mitigate the hazards. The second part describes two validation activities conducted to demonstrate the completeness of the design.

4.1. Hazard, Severity and Corresponding Design Requirements

A total of eight functional hazards (FHz) were identified based on the analysis of the use cases.

1. FHz#01: Significant delay in ASR callsign/command recognition and/or display (relevant for use cases 3 and 4)
2. FHz#02: ASR fails to identify an aircraft callsign from pilot's utterance, i.e., no aircraft is highlighted (relevant for use case 1)
3. FHz#03: ASR fails to identify an aircraft callsign from controller's utterance, i.e., no aircraft is highlighted (relevant for use case 2)
4. FHz#04: ASR erroneously identifies an aircraft callsign from pilot's utterance, i.e., the wrong aircraft is highlighted (relevant for use case 1)
5. FHz#05: ASR erroneously identifies an aircraft callsign from controller's utterance, i.e., the wrong aircraft is highlighted (relevant for use case 2)
6. FHz#06: ASR fails to identify a command from controller's utterance, i.e., no given command is shown to the ATCO (relevant for use cases 3, 4)
7. FHz#07: ASR erroneously identifies a command from controller's utterance, i.e., a wrong command or a command never given is shown to the ATCO (relevant for use cases 3, 4)
8. FHz#08: ASR recognizes an incorrect aircraft callsign, and the (correct or wrong) command is displayed for the incorrect flight in the CWP HMI (relevant for all use cases)

Safety assessment requires that the operational effects of identified hazards are classified in accordance with a Risk Classification Scheme (RSC) based on the severity of the operational effect the hazard may trigger [22,26]. The RSC classifies the hazards and provides the safety target, i.e., the maximum tolerable frequency (MToF) for each hazard's occurrence per flight hour in a specific unit.

- Severity Class 1: Accidents (max safety target with a probability of less than 10^{-9}, i.e., one catastrophic accident per one billion flight hours attributable to ATM.
- Severity Class 2: Serious Incidents (max safety target with a probability of less than 10^{-6})
- Severity Class 3: Major Incidents (max safety target with probability of less than 10^{-5})
- Severity Class 4: Significant Incidents (max safety target with a probability of less than 10^{-3})
- Severity Class 5: No Immediate Effect on Safety (no target).

Table 2 provides a list of the identified hazards with their causes, the assessed operational effect, and mitigations considered in defining the operational effect protecting against the functional hazard's effects propagation. The severity classification for each hazard was derived during a workshop with three ATCOs, concept designers and safety experts. The severity of the hazard determines the tolerable frequency of hazard occurrence.

As demonstrated in Table 1, based on the discussion with ATCOs participating in the session, it was recognized that the impacts of the ASR functional hazards are not significant from a safety perspective according to the Risk Classification Scheme, (RCS) [26]. Therefore, the requirements set as a mitigation do not require the safety target as such derived from the RCS and can be derived from operational needs, ensuring performance acceptable for ATCOs and ensuring no degradation in the execution of the ATCOs' tasks.

The impact of never attaining a perfect ASR may lead to situations, which are also present in the current operating method and working procedures impacting human performance negatively, i.e., increased workload and decreased situational awareness. With the support of various tools already used in current operations (such as monitoring aids), these events will in the current mode and with ASR support not escalate to safety relevant events.

Table 2. Hazards' causes, operational effect, possible mitigation and severity of hazards derived during Functional Hazard Assessment workshop.

Functionality Hazard & Severity	Potential Causes & Operational Effect	Mitigations Protecting against Propagation of Effects
FHz#01 Significant delay in ASR callsign/command recognition and/or display **No Immediate Effect on Safety**	-Design issue -ASR provides delayed output If the use of ASR introduces delays in the usage of speech information (display of inputs, identification of aircraft, etc.) this may cause the ATCOs to focus on specific flight/area of the Area of Responsibility, until they can verify that the action induced by ASR has been correctly processed and displayed. This may have a negative impact on ATCO situational awareness.	Contingency measure to switch off ASR.
FHz#02 ASR fails to identify an aircraft from pilot's utterance—no aircraft is highlighted **No Immediate Effect on Safety**	-Pilot utters a non-understandable callsign or noise environment -Pilot utters a legal and understandable callsign, but ASR fails to recognize it. If the pilot performs the radio call and the flight is not highlighted, ATCO may have to scan the area of responsibility (AoR) to locate the aircraft. However, if ASR functionality to highlight the callsign is defined in the new operating method, there is a default expectation by the ATCO that it is functional and assisting in locating aircraft, resulting in minor workload increase and situational awareness reduction.	ATCO may have to scan the Area of Responsibility to locate the aircraft. No difference to current operating method.
FHz#03 ASR fails to identify an aircraft from controller's utterance—no aircraft is highlighted **No Immediate Effect on Safety**	-ATCO utters an illegal/non-understandable callsign -ATCO utters a legal and understandable callsign, but ASR fails to recognize it. If ATCO performs the radio call, it is assumed the impact is minor, because the ATCO's attention is focused on the aircraft being called and the impact is negligible.	No difference to current operating method.
FHz#04 ASR erroneously identifies an aircraft from pilot's utterance—wrong aircraft is highlighted **No Immediate Effect on Safety**	-If pilot performs the radio call and erroneous flight is highlighted, the ATCO may focus on the highlighted aircraft and issue the clearance intended for the calling aircraft to the wrong flight. The difference to the current operating method is that while occasional callsign confusion may occur between similar callsigns, now the ASR system is enforcing the ATCO's perception of issuing the clearance to what is expected to be the correct flight. If the confusion is not clarified through read-back and hear-back procedure or with the assistance of the planning controller, issued clearance to the wrongly highlighted aircraft may result in an unintended trajectory change. From a safety perspective, this is not significantly different from the current operating method, when ATCO enters a clearance into the radar label for the wrong callsign.	If the confidence level of the callsign recognition is not sufficiently high, it is not highlighted. For lower confidence levels to highlight with different color to emphasize the uncertainty of correct recognition. If the erroneous recognition persists, ATCO switches off the ASR and continues working as in today's operations.
FHz#05 ASR erroneously identifies an aircraft from controller's utterance—wrong aircraft is highlighted. **No Immediate Effect on Safety**	If ATCO performs the radio call and erroneous flight is highlighted, it is assumed the impact is minor as the ATCO's attention is on the aircraft being called and the impact of erroneous highlight is negligible.	If the confidence level of the callsign recognition is not sufficiently high, it is not highlighted. For lower confidence levels to highlight with different color to emphasize the uncertainty of correct recognition.
FHz#06 ASR fails to identify a command from controller's utterance. **No Immediate Effect on Safety**	-ATCO utters an illegal/non-understandable command -ATCO utters a legal and understandable command, but ASR fails to recognize it The failure of ASR to identify a complete command force ATCO to manually make the input. In such cases the negative impact on ATCO workload and situational awareness is expected, as in the new operating method there is a default expectation by the ATCO that ASR is functional and assisting in inputting commands in the labels.	ATCO inputs command manually. If the failure of ASR to recognize commands persists, ATCO switches off the ASR and continues working as in today's operations.

Table 2. *Cont.*

Functionality Hazard & Severity	Potential Causes & Operational Effect	Mitigations Protecting against Propagation of Effects
FHz#07 ASR erroneously identifies a command from controller's utterance **No Immediate Effect on Safety**	-ATCO utters an illegal/non-understandable command -ATCO utters a legal and understandable command, but ASR recognizes the incorrect command, and wrong command is displayed in the CWP HMI. In cases where inputs are provided but are erroneous, the ATCO will have to recognize the error and change information already put into the system. Depending on the ATM system, some parts of correcting the clearance, route change, etc., may require manipulation of the flight plan route data to input the correction. In such cases the impact on ATCO workload and potential disruption to the ATCO workflow may be higher than in cases where only missing data need to be input to complete the clearance.	ATCO corrects ASR input manually for the intended callsign. If the failure of ASR to recognize commands correctly persists, ATCO switches off the ASR and continues working as in today's operations.
FHz#08 ASR recognizes an incorrect callsign, and the command is displayed for the incorrect flight in the CWP HMI **No Immediate Effect on Safety**	ATCO utters a legal and understandable command, but ASR recognizes an incorrect callsign, which is being considered by the system, and the command is displayed for the incorrect flight in the CWP HMI In cases where inputs are provided but are erroneous (i.e., command input for wrong aircraft), the ATCO will have to recognize the error and change information already put into the system. If the error is not recognized by the controller, the contacted pilot will nevertheless follow the clearance issued by controller on the frequency. The erroneous input in the label of another aircraft will soon be detected by clearance monitoring aids.	If ATCO recognizes the error, he/she rejects and either repeats the clearance or inputs it manually directly into the label of the correct aircraft radar label. If ATCO does not recognize the error, monitoring aids will detect the discrepancy between the flown trajectory and the command inserted in the label of the erroneous aircraft.

Considering that the hazards identified did not directly impact safety, the requirement for design are derived from operational and functional needs. The following list of requirements was, therefore, developed as preventive mitigations for functional hazards. The hazards were further analysed via top-down and bottom-up techniques with the support of fault trees to identify all possible causes for the hazards to occur, and to limit the propagation of the effects of hazards. The details of the derivation for the hazard are presented in Appendix A: Top-down analysis and Appendix B: Bottom-up analysis. The full list of the requirements can be found in the CORDIS portal of the European Commission [27]. To facilitate understanding of the requirements, they are listed here divided into two subcategories: those related to callsign recognition (as a mitigation for hazards FHz#02, FHz#03, FHz#04, FHz#05 and FHz#08) and those related to command recognition (mitigation to hazards FHz#01, FHz#02, FHz#06, FHz#07 and FHz#08):

4.1.1. Safety and Performance Requirements Concerning Callsign

ASR should send a recognized callsign to the cooperating ATC system, no later than one second after the ATCO has ended the radio transmission.

- For 99.9% of the ATCO utterances (except callsign), the system shall be able to give the output in less than 2 s after the ATCO ended the radio transmission.
- If the confidence level of the callsign recognition is not sufficiently high, it shall not be highlighted. Confidence level corresponds to a plausibility value derived by ASR. If the plausibility value is below a given threshold, the callsign is set to 'not recognized'.
- The HMI shall highlight the track label or part of it (or the track symbol) after recognizing the corresponding callsign.

4.1.2. Safety and Performance Requirements Concerning Commands

- The ASR shall recognize commands of different command categories (such as descend, reduce, heading).
- The Command Recognition Rate of ASR for ATCOs should be higher than 85%.
- The Command Recognition Error Rate of ASR should be less than 2.5% for ATCOs.
- The Command Recognition Error Rate of ASR should be less than 5% for pilots.

- The HMI should present the recognized (and validated) command types together with the command values in the radar label.
- The HMI shall enable manual correction/update of automatically proposed command value/type.
- The ASR system shall have no significant differences in the recognition rates of different command types, if the command types are often used (e.g., more than 1% of the time).

The set of proposed requirements satisfying "failure approach" enables achieving sufficient assurances for safety in the design of the system. The next step of the assessment consisted of demonstrating the evidence for the mitigation of each hazard and achievability of the safety requirements in the validation activities: real time, (human-in-the-loop simulations described in Sections 3.3 and 3.4) conducted in operationally representative environment. The evidence for each hazard was collected via objective metrics and subjective feedback from the ATCOs as shown in Table 3.

Table 3. Specific metrics and measures for collecting evidence from demonstration activities.

Hazard	Objective Metrics	Subjective Feedback on the Statement
FHz#01 Significant delay in ASR command recognition and/or display (all use cases).	Timeliness (processing time)	Applicable to all hazards The accuracy of the information provided by the ASR system is adequate for the accomplishment of operations. Command Recognition Error Rate stays in the acceptable limits.
FHz#02: ASR fails to identify an aircraft from pilot's utterance—no aircraft is highlighted (use case 1)	Pilot's callsign recognition rate (no callsign highlighted)	
FHz#04: ASR erroneously identifies an aircraft from pilot's utterance—wrong aircraft is highlighted (use case 1).	Pilot's callsign recognition error rate	
FHz#03: ASR fails to identify an aircraft from controller's utterance—no aircraft is highlighted (use case 2).	Controller's callsign recognition rate (no callsign highlighted) Controller's callsign recognition error rate (wrong callsign highlighted)	The number and/or severity of errors resulting from the introduction of the ASR system is within tolerable limits, considering error type and operational impact.
FHz#05: ASR erroneously identifies an aircraft from controller's utterance—wrong aircraft is highlighted (use case 2).	Controller's callsign recognition rate (no callsign highlighted) Controller's callsign recognition error rate (wrong callsign highlighted)	The level of ATCO's situational awareness is not reduced with the introduction of the ASR system (ATCO is able to perceive and interpret task relevant information and anticipate future events/actions).
FHz#06: ASR fails to identify a command from controller's utterance (use case 3, 4).	Controller's command recognition rate	The level of ATCO's workload is maintained or decreased with the introduction of the ASR system.
FHz#07: ASR erroneously identifies a command from controller's utterance (use case 3 and 4)	Controller's command recognition error rate	The number and/or severity of errors resulting from the introduction of the ASR system is within tolerable limits, considering error type and operational impact.
FHz#08: ASR recognizes an incorrect callsign, and the command is displayed for the incorrect flight in the CWP HMI (use case 3, 4)	Controller's callsign recognition error rate	

4.2. Results of the Validation Activities

The results of the two real-time human-in-the-loop simulations described in Sections 3.3 and 3.4 are presented in the following two subsections.

4.2.1. Validation Activity "Callsign Highlighting"

Evidence Based on the Objective Metrics

Table 4 presents the total number of callsigns present in the simulation audio logs and the number of callsigns that were correctly detected by the ASR. The percentage of correctly detected callsigns is higher for ATCOs than for flight crew in both cases as the algorithm is optimized for the ATCO locutions.

Table 4. Callsign recognition rates for ATCOs and Flight Crew.

	ATCO			Flight Crew		
Analysis Type	**N° of Callsigns**	**N° Callsigns Detected**	**Percentage**	**N° of Callsigns**	**N° Callsigns Detected**	**Percentage**
RTS recordings	859	721	84%	457	687	67%
Operational recordings	143	127	87%	158	77	49%

Regarding the comparison between simulation and operational recordings, the percentage for ATCOs are similar but the percentage for flight crew is better in the simulation. This was already expected as the quality of the recording (signal-to-noise ratio) is better in the simulation and the accent (mother tongue) of the simulation pilots is unique (Spanish), while the one from the operational recordings is very diverse with 29 airlines from 18 different countries.

No callsign was wrongly recognized as only complete callsigns were detected. Feedback from ATCOs indicated that they would like to have higher recognition rates even if some callsigns were incorrectly detected and highlighted. The error allowance is something to be further investigated in follow-up research.

Table 5 presents the number of commands that were present/detected and the callsign + command that were correctly detected for each analysis. Only commands that fall within the five categories, for which the prototype was optimized, are presented. There were several other commands that ATCOs used during the simulation such as squawk change, standard terminal arrival route (STAR) assignment, and information (traffic information, barometric pressure setting). In the first column "Only Command Recognition", only the type of the command to classify whether the command was detected or not are considered. In the last three columns "Callsign + Command Recognition", the command type plus the callsign must be correctly extracted to count as a detected command. The results show that the lower callsign recognition rate resulting from the very low callsign error rate also results in a lower command recognition rate, if both callsign and command type must be correct.

Table 5. Detected command types, when considering only command type.

	Only Command Recognition			Callsign + Command Recognition		
Analysis Type	**Commands**	**Detected Commands**	**%**	**Commands**	**Detected Callsign + Commands**	**%**
RTS recordings	695	619	89%	695	523	75%
Operational recordings	182	167	92%	182	146	80%

The performance for operational data is better than for RTS recordings. There is a 3% difference between the RTS and the operational recognition percentages regarding command recognition, and a 5% difference in callsign and command recognition percentages. During the exercise, the participating ATCOs were encouraged to test the recognition system. They thus issued longer, and more complex authorizations than usually issued in operational environments. This together with the fact that the ASR prototype was trained and optimized using operational communications explains the difference between both percentages.

If not only the command type, but also the information contained in the command are considered (i.e., values, units, qualifiers), the extraction performance is lower, as shown in Table 6.

Table 6. Detected commands, i.e., command, when callsign plus command information, i.e., values, units, qualifiers are considered.

	Complete Command Recognition			Callsign + Complete Command Recognition		
Analysis Type	**Commands**	**Detected Commands**	**%**	**Commands**	**Detected Commands**	**%**
RTS recordings	695	498	72%	695	416	60%

Evidence Based on Subjective Feedback

Accuracy of ASR was collected through tailor-made questionnaires, debriefings, and data logs. The ATCO feedback was that the tool that needed improvement in the recognition rates to be able to effectively support them in the execution of their tasks. ATCOs indicated that they would prefer some occasional false positive callsign recognized if that would mean higher recognition rates.

Timeliness was collected through tailor-made questionnaires, debriefings, and data logs. Data logs indicated that when the callsigns were located at the end of an utterance the radar track and command information was presented in 0.9 s, but when it was located at the beginning of a sentence it took up to 3.0 s. Controllers subjective feedback indicated that timeliness was rated as adequate for the callsigns at the end of the utterance but inadequate when the callsign was at the begging of the utterance.

ATCOs' situational awareness, measured with SASHA [28], slightly improved with the use of ASR (score 4.0 in the reference questionnaire and 4.4 in the solution scenario). During the debriefings, ATCOs stated that situational awareness was improved but they considered that the ASR recognition rate was not high enough to allow them to completely confide and exploit the tool. They consider that higher callsign recognition rates would further improve their situational awareness.

ATCO workload was collected through Nasa-TLX [29] questionnaire, tailor-made questionnaires, and debriefings. The Nasa-TLX scored 9.1 (out of 20) for the baseline scenario and 7.9 (out of 20) for the solution scenario questionnaire. The tailor-made questionnaire and debriefings indicated that workload slightly decreased in the solution scenario.

4.2.2. Validation Activity 2: Radar Label Maintenance

Evidence Based on the Objective Metrics

The validation activities were performed between September 2022 and November 2022 as described in Section 3.4.

Table 7 provides the speech recognition and understanding performance taken from [5]. A word error rate (WER) of 3.1% was achieved, i.e., only every 33rd word was wrongly recognized. This is extremely good considering that humans usually achieve a WER of 4 to 11%, depending on the noise level and the option to listen more than once [30].

Table 7. Performance at the semantic level quantified as recognition and error rates.

Level of Evaluation	WER	Cmd-Recog-Rate	Cmd-Error-Rate	Csgn-Recog-Rate	Csgn-Error-Rate
Full Command	3.1%	92.1%	2.8%	97.8%	0.6%
Only Label		92.5%	2.4%		

These results are based on 118,800 manually transcribed words resulting in 17,100 commands from 8850 utterances. The word error rate of the used speech recognizer is 3.1%.

Out of all the given commands, 92.1% were recognized and 2.8% were wrongly recognized. The difference to 100% means rejections, i.e., nothing was recognized for this command. A command is only correctly recognized, if the callsign of the command, the command type (descend, reduce, heading...), the values, the unit, the qualifier (left, right, etc.) and the conditions are all correct. Therefore, the callsign recognition rate (column "Csgn-Recog-Rate") is always better than the recognition rate of the total command. A callsign error rate of 0.6% in the last column corresponds to only one of 165 callsigns being wrongly recognized. The last row "Only Label" shows the results when we do not consider all 17,100 recognized commands, but only the 12,600 commands which are also shown in the radar label cells; e.g., a QNH or a squawk command are not shown in the radar label.

A recognition rate of 92.5% means that 7.5% of the given commands were not correctly pre-filled by the ASR support functionality, i.e., the remaining 7.5% of the commands need to be manually input by the ATCOs [5]. Only 50% of them were manually corrected or inputted, respectively. Out of the 6400 commands given in the solution runs, 219, i.e., 3.4% remained incorrect in the radar labels [5]. One can then pose the question whether safety has now decreased, when pre-filling of radar label cells is supported by speech recognition? It can be argued that no, the contrary is the case. Helmke et al. [5] also showed that in the baseline runs without ASR support, 617 of the given 6320 radar cell label relevant commands were not correct or remained missing in the radar label cells, i.e., 11.6% versus 3.4%.

Evidence Based on the Subjective Metrics

Table 8 shows the mean values of the normalized answer differences of the 12 ATCOs after having compensated for sequences effects. The answers were scaled into the interval [1..10]; 1 meaning very good performance and 10 meaning bad performance. Negative values in column "Diff" mean that the ATCO judged the safety aspect relevant to this question higher with ASR than without ASR. The p-value is the statistical significance of a performed t-test. The cells are shaded in green for 0% $\leq$ p-value < 5%, in light green for 5% $\leq$ p-value < 10%, and in yellow for no real evidence (absolute p-value $\geq$ 10%). There were no single cases which would have provided evidence that working without ASR is safer than with ASR, i.e., −10% $\leq$ p-value < 0% was not measured; see [5] for more details.

Table 8. Subjective feedback of ATCOs to safety-related questions.

Question	*Diff*	*p-Value*
Stress annoyed	−0.16	34%
Peak workload	−0.32	9.9%
Single aircraft	0.04	−41%
Risk to forget	−0.64	0.7%
Conflict resolution	−0.26	24%
Consequences	0.30	−21%
Reliable	−0.24	30%
Confidence	−1.59	1.1%
Complexity	−1.98	2.0×10^{-4}
User Acceptance	−1.01	6.3%
Total	−0.56	0.4%

It should be noted that the performance achieved was higher than the minimum performance as defined by the safety/performance requirements [27]; the command recognition rate was expected to be higher than 85%, whereas 92.5% was achieved. The command Recognition Error Rate of ASR showed slightly better performance with 2.4% against 2.5% set by the requirements.

The callsign recognition rate and error rate, although not quantified by the requirements, showed high performance. The subjective data based on the SHAPE Automation Trust Index (SATI) [28] questionnaire confirmed that the level and quality of information provided by the system (as displayed in the radar labels) was acceptable with an average score of 8.8 on a scale from 1 to 10, i.e., 10 indicating the best rating option. With the support of ASR, the value was 0.8 units better than without ASR support.

Timeliness of the information provided by the ASR.

The design recommendation set an expectation of 99.9% for the ATCO utterances, except the callsign, being available in less than two seconds after the ATCO has released the push-to-talk button.

The ATCO subjective feedback demonstrated that the timeliness of ASR output in the aircraft radar labels was considered adequate with an average score of 8.5 on a scale from 1 to 10, i.e., 10 indicating the best rating option.

Number and type (nature) of human errors.

ATCOs confirmed that ASR did not increase the potential for human errors with an average score of 3 on a scale from 1 to 10, i.e., 10 indicating the worst rating option.

An objective analysis actually confirms that the number of errors in the radar label cells, i.e., missing input, is much less if ATCOs are supported by ASR compared to entering everything manually with a mouse ($\alpha < 10^{-7}\%$).

Level of ATCO's situational awareness.

ATCOs confirmed that their situational awareness is maintained at an acceptable level with ASR with an average score of 8.9 on a scale from 1 to 10, i.e., 10 indicating the best rating option.

ATCO's workload.

The secondary validation objective regarding objective workload measurement showed a statistically significant decrease (p-value = 0.3%) in the workload when ATCOs are supported by ASR. The ATCO-self-rated Instantaneous Self-Assessment (ISA) [31] score confirmed this with the same statistical significance (p-value = 3.1%, see Table IX in [5]). ATCOs confirmed that ASR supported them in maintaining the workload at an acceptable level with an average score of 7.9 on a scale from 1 to 10, i.e., 10 indicating the best rating option. More meaningful, however, is the clicking time. In all baseline runs together, the ATCOs need 12,700 s for maintaining the radar label contents. In the solution runs with ASR support, only 405 s were needed, i.e., an improvement by a factor 31.

4.3. Limitations

The safety assessment performed as part of the SESAR2020 PJ.10-W2-96.2 ASR research and development activities covers specific use cases in specific ATC environments and the extrapolation of the results of the safety assessment may, therefore, not be applicable to all operational environments and other ASR applications. The safety analysis described in this paper focuses on the generic application of ASR technology in the pre-industrial phase. Thus, the research results achieved may not be fully transferable to live operations and any local implementation requires further investigation to satisfy the safety requirements as defined in relevant regulation and/or the local competent authority. The results achieved in these validations did not contain a long-term assessment of ASR functionality and potential impacts to safety thereof. Likewise, the impact of external factors such as background noise, various ATCO accents, and radio transmission quality and interference were not assessed beyond their possible occurrence in the research environment. Implementation of ASR capabilities may introduce the ATM system to new cyber security vulnerabilities which would need to be evaluated through a local cyber security assessment. Further research may be required for the implementation of ASR in different operational environments with different traffic demands and complexity characteristics in ATC facilities applying different ATM platforms. If ASR is used to supplement other safety tools present in an ATM system (e.g., conformance monitoring, conflict detection) as suggested by some studies [17–19], the safety considerations of ASR may require a detailed assessment of the various system components' interactions as opposed to the comparison between current and new operating methods focused solely on availability of ASR as presented in this paper. Furthermore, the acceptance of ASR by ATCOs—and subsequent impact on workload and situational awareness—may vary between different organizational cultures and a holistic assessment of ASR suitability to a specific environment would be required.

5. Discussion of Results

Overall, two different approaches for callsign extraction from spoken utterances have been validated. The first one emphasized a very low error rate. The prize is a lower callsign recognition rate. As a result, ATCOs reported preference for higher recognition rates with an occasional false positive callsign. This was addressed by the second approach trying a compromise between low error rate and recognition rate. The analysis shows that it is up to the user, i.e., ATCO, to decide what is preferred on a daily basis. An error rate of 0% will not be possible, if a reasonable recognition rate is needed. Human actors do not achieve an error rate of 0% either.

In general, ATCOs confirmed that they are able to perform their ATC tasks when working with ASR support. The positive results achieved in situational awareness and workload measurements in both validation exercises indicate the potential for further benefits in ATCO performance in an operational environment. Timeliness of ASR output in the first validation—radiotelephony utterances with aircraft callsign at the end only—was found acceptable. In the second validation the timeliness was found to be adequate. ATCOs from both validations confirmed that the application of ASR did not introduce any additional risks for errors.

A recognition rate of 92.5% is still far away from 100%. It means, however, that the time spent by the ATCOs manually updating the radar labels with clearance information could be reduced by a factor of 31 from 12,700 s, i.e., 25% of the total simulation time, without ASR support down to 405 s with ASR support. These numbers are based on a very heavy traffic scenario, in which ATCO plus simulation pilots blocked the frequency for 70% of the time.

The safety and hazard analysis of this research work has shown that no severe hazards exist, when using ASR applications as callsign highlighting or pre-filling radar labels in an operational environment. In heavy traffic scenarios, 3.4% of the given commands are not correctly entered into the radar label cells, when ASR support is available. Without ASR support the missing command rate increases to 11.6%. ASR support does not decrease safety, but rather increases safety, when ATCO and ASR work as a team.

Research, therefore, should not concentrate on increasing command recognition rate from 92.5% to 95% or to 99.9%, which is not expected to happen in the near future. Research needs to focus on attention guidance, which gives hints to the ATCO, when something might be wrong or missing in the radar labels. This can be done by integration of ASR with other assistant systems already available in the ops room. Comparing downlinked mode-S data with radar label cell entries can even further reduce the number of erroneous label value entries.

6. Conclusions

In this article, we focused on demonstrating the safety of ASR application in ATC operational environments.

The safety assessment showed that the eight ASR functional hazards have no significant effect on overall ATM safety. Mitigations were derived from operational needs, to ensure acceptable ATCO performance without degrading ATCO's task execution. A potential decrease in situational awareness or increase in workload in the case of insufficient ASR performance were already present in the current operating method, but can be further mitigated through the use of clearance monitoring tools to prevent the escalation of these events to safety relevant occurrences.

The requirements developed as part of the safety assessment for the application of ASR technology in ATM were achieved in operational environments reflecting real-life ATC centers for en-route and approach control. The technical system, in terms of accuracy and timeliness, outperformed expectations required by the design and associated targets. The subjective feedback of ATCOs from two different validation setups was encouraging and confirmed that ASR application not only generated benefits, but also showed to be feasible for implementation in currently deployed ATM systems.

Author Contributions: Conceptualization, E.P.-C., J.D., P.H. and H.H.; data curation, H.H., R.G.L. and O.O.; formal analysis, E.P.-C., J.D. and H.H.; funding acquisition, P.H.; investigation, E.P.-C., J.D. and H.H.; methodology, E.P.-C. and J.D.; project administration, P.H.; resources, H.H. and O.O.; software, E.P.-C. and H.H.; supervision, P.H. and H.H.; validation, E.P.-C.; J.D., R.G.L. and P.H. visualization, E.P.-C. and O.O.; writing—original draft, E.P.-C., J.D. and H.H.; writing—review and editing: J.D., P.H., R.G.L. and O.O. All authors have read and agreed to the published version of the manuscript.

Funding: This work was funded by SESAR Joint Undertaking under grant agreement No. 874464 under European Union's Horizon 2020 research and innovation programme.

Data Availability Statement: Not applicable.

Conflicts of Interest: The authors declare no conflict of interest. The funding sponsors had no role in the design of this study, in the collection, analyses, or interpretation of the data, in the writing of the manuscript, and in the decision to publish the results.

Appendix A. Top-Down Analysis

The hazards were further analyzed by top-down technique with the support of fault trees to identify all possible causes of the hazards to occur, and to limit the propagation of the effects of hazards. The details of the derivation for the hazard are presented in this appendix.

Cause ID (in Fault Tree)	Cause	Detailed Description	Mitigation/Safety Requirement
FHz#01 Significant delay in ASR callsign/command recognition and/or display	-ASR provides delayed output	One or more of the ASR components is not performing as expected and causing the delay in the ASR output display.	The ASR system should provide the functionality to be switched off and switched on when necessary. ASR should send a recognized callsign to the cooperating ATC system when the controller ends the radio transmission within a maximum of 1.0 s. For 99.9% of the ATCO utterances except callsign itself, the system shall be able to give the output in less than two seconds after the ATCO has released the push to talk button. The ASR system shall have no significant differences in the recognition rates of different command types, if the command types are not very seldom use (e.g., less than 1% of the time).
FHz#02 ASR fails to identify an aircraft from pilot's utterance—no aircraft is highlighted	-Pilot utters a non-understandable callsign -Pilot utters a legal and understandable callsign, but ASR fails to recognize it	Pilot performs the radio call and the flight is not highlighted in the CWP HMI.	The HMI shall highlight the Track Label or part of it (or the track symbol) after recognizing the corresponding callsign. If the confidence level of the callsign recognition is not sufficiently high, it shall not be highlighted.
FHz#03 ASR fails to identify an aircraft from controller's utterance—no aircraft is highlighted	-ATCO utters an illegal/non-understandable callsign -ATCO utters a legal and understandable callsign, but ASR fails to recognize it	ATCO performs the radio call and the flight is not highlighted in the CWP HMI.	The HMI shall highlight the Track Label or part of it (or the track symbol) after recognizing the corresponding callsign. If the confidence level of the callsign recognition is not sufficiently high, it shall not be highlighted.
FHz#04 ASR erroneously identifies an aircraft from pilot's utterance—wrong aircraft is highlighted	-Pilot utters a non-understandable callsign Pilot utters a legal and understandable callsign, but ASR recognizes an existing wrong callsign -Pilot utters a legal and understandable callsign, but ASR recognizes a wrong callsign not matching to callsigns considered by the system	ATCO focuses on the highlighted aircraft and issues the clearance intended for the calling aircraft to the wrong flight. If the ATCO issues the clearance to the wrongly highlighted aircraft, it may result in an unintended trajectory change.	The HMI shall highlight the Track Label or part of it (or the track symbol) after recognizing the corresponding callsign. The Command Recognition Error Rate of ASR should be less than 5% for pilots. The Command Recognition Rate of ASR for pilots should be higher than 75%. If the confidence level of the callsign recognition is not sufficiently high, it shall not be highlighted.

Cause ID (in Fault Tree)	Cause	Detailed Description	Mitigation/Safety Requirement
FHz#05 ASR erroneously identifies an aircraft from controller's utterance—wrong aircraft is highlighted	-ATCO utters an illegal/non-understandable callsign -ATCO utters a legal and understandable callsign, but -ASR recognizes an existing wrong callsign -ATCO utters a legal and understandable callsign, but -ASR recognizes a wrong callsign not matching to callsigns considered by the system	ATCO may get confused and issue a wrong clearance. If the ATCO issues the clearance to the wrongly highlighted aircraft, it may result in an unintended trajectory change.	ATCOs will use standard phraseology as per ICAO Doc.4444 [32]. The ASR shall recognize commands of different command categories (such as descend, reduce, heading). The Command Recognition Error Rate of ASR should be less than 2.5% for ATCOs. The Command Recognition Rate of ASR of ATCOs should be higher than 85%. If the confidence level of the callsign recognition is not sufficiently high, it shall not be highlighted.
FHz#06 ASR fails to identify a command from controller's utterance	-ATCO utters an illegal/non-understandable command ATCO utters a legal and understandable command, but ASR fails to recognize it	ATCO manually makes the input resulting in workflow disruptions and workload increase and situational awareness reduction.	The HMI shall enable manual correction/update of automatically proposed command value/type. The HMI should present the recognized (and validated) command types together with the command values in the radar label. The ASR system should provide the functionality to be switched off and switched on when necessary. The Command Recognition Rate of ASR of ATCOs should be higher than 85%.
FHz#07 ASR erroneously identifies a command from controller's utterance	-ATCO utters an illegal/non-understandable command -ATCO utters a legal and understandable command, but ASR recognizes an incorrect callsign, which is being considered by the system, and the command is displayed for the incorrect flight in the CWP HMI -ATCO utters a legal and understandable command, but ASR recognizes the incorrect command, and wrong command is displayed in the CWP HMI	ATCO will have to change information already input into the system. Depending on the ATM system, some parts of correcting the clearance, route change, etc., may require manipulation of the FPL route data to input the correction.	The HMI shall enable manual correction/update of automatically proposed command value/type. The HMI associated with ASR shall enable the ATCO to reject recognized command values for pre-filling radar label values by clicking on a rejection button. The ASR system should provide the functionality to be switched off and switched on when necessary. The Command Recognition Error Rate of ASR should be less than 2.5% for ATCOs.
FHz#08 ASR recognizes an incorrect callsign, and the command is displayed for the incorrect flight in the CWP HMI	-ATCO utters an illegal/non-understandable callsign followed by a command ATCO utters a legal and understandable callsign and a command, but ASR recognizes the incorrect callsign	ATCO utters a legal and understandable command, but ASR recognizes an incorrect callsign, which is being considered by the system, and the command is displayed for the incorrect flight in the CWP HMI.	The Command Recognition Error Rate of ASR should be less than 2.5% for ATCOs. The Command Recognition Rate of ASR of ATCOs should be higher than 85%. ATCO is supported by the clearance monitoring aids as in today's operations.

Appendix B. Bottom-Up Analysis

In view of complementing the fault tree findings, the bottom-up analysis of the failure modes of the ASR system elements and element interfaces and of their effects was conducted to determine potential common cause failures and to allow a more in-depth causal analysis of certain parts of the functional system design. The details of the derivation for the hazard of the bottom-up analysis are presented in this appendix.

Technical System Element	Failure Mode	Effects	Mitigation/Safety Requirement
Command Prediction	Fails to forecast possible future controller commands. Failure to receive external data required for forecast of future controller commands (external data can be radar data, flight plan data, weather data, airspace data, and also historic data of those types).	The speech recognizer relies on the input of the predicted commands. Commands which are not predicted (normally) cannot be recognized. So, if command prediction accuracy is worse than recognition accuracy itself, the command prediction functionality might have no benefits for the recognition engine any more.	If ASR is used, the Command Prediction Error Rate should not be higher than 10% and also not be higher than 50% of the opposite command recognition rate (i.e., 100% minus the command recognition rate), without using a plausibility checker.
Recognize Voice Words	Fails to analyze the voice flow and to transform into a text string. Does not receive the Voice Flow.	No callsign or command is recognized by ASR and displayed on the CWP HMI.	If ASR does not provide an input, ATCO proceeds as in current operations (manual input and with no highlight of the callsign).
Apply Ontology and Logical check	Fails to analyze the text string and to transform into a set of predefined commands to discard incoherent commands.	The ASR output is erroneous and incoherent.	The ASR shall recognize commands of different command categories.

References

1. Helmke, H.; Ohneiser, O.; Mühlhausen, T.; Wies, M. Reducing Controller Workload with Automatic Speech Recognition. In Proceedings of the 35th Digital Avionics Systems Conference (DASC), Sacramento, CA, USA, 25–29 September 2016. [CrossRef]
2. European Commission. Commission Implementing Regulation (EU) 2017/373 of 1 March 2017 Laying down Common Requirements for Providers of Air Traffic Management/Air Navigation Services and Other Air Traffic Management Network Functions and Their Oversight Repealing Regulation (EC) No 482/2008, Implementing Regulations (EU) No 1034/2011, (EU) No 1035/2011 and (EU) 2016/1377 and Amending Regulation (EU) No 677/2011. 2017. Available online: https://eur-lex.europa.eu/legal-content/EN/TXT/?uri=CELEX%3A32017R0373 (accessed on 23 October 2023).
3. SESAR. SESAR Safety Reference Materials Ed 4.1. 2019. Available online: https://www.sesarju.eu/sites/default/files/documents/transversal/SESAR2020%20Safety%20Reference%20Material%20Ed%2000_04_01_1%20(1_0).pdf (accessed on 23 October 2023).
4. García, R.; Albarrán, J.; Fabio, A.; Celorrio, F.; Pinto de Oliveira, C.; Bárcena, C. Automatic Flight Callsign Identification on a Controller Working Position: Real-Time Simulation and Analysis of Operational Recordings. *Aerospace* **2023**, *10*, 433. [CrossRef]
5. Helmke, H.; Kleinert, M.; Ahrenhold, N.; Ehr, H.; Mühlhausen, T.; Ohneiser, O.; Klamert, L.; Motlicek, P.; Prasad, A.; Zuluaga-Gómez, J.; et al. Automatic Speech Recognition and Understanding for Radar Label Maintenance Support Increases Safety and Reduces Air Traffic Controllers' Workload. In Proceedings of the 15th USA/Europe Air Traffic Management Research and Development Seminar, ATM 2023, Savannah, GA, USA, 5–9 June 2023.
6. Kleinert, M.; Helmke, H.; Moos, S.; Hlousek, P.; Windisch, C.; Ohneiser, O.; Ehr, H.; Labreuil, A. Reducing Controller Workload by Automatic Speech Recognition Assisted Radar Label Maintenance. In Proceedings of the 9th SESAR Innovation Days, Athens, Greece, 2–5 December 2019.
7. European Space Agency. Technology Readiness Levels Handbook for Space Applications. September 2008. TEC-SHS/5551/MG/ap. Available online: https://connectivity.esa.int/sites/default/files/TRL_Handbook.pdf (accessed on 23 October 2023).
8. Santorini, R.; SESAR Digital Academy—Innovation in Airspace Utilization, 29 April 2021. SESAR Joint Undertaking | Automated Speech Recognition for Air Traffic Control. Available online: https://www.sesarju.eu/node/3823 (accessed on 6 October 2023).
9. Zuluaga-Gomez, J.; Nigmatulina, I.; Prasad, A.; Motlicek, P.; Khalil, D.; Madikeri, S.; Tart, A.; Szoke, I.; Lenders, V.; Rigault, M.; et al. Lessons Learned in Transcribing 5000 h of Air Traffic Control Communications for Robust Automatic Speech Understanding. *Aerospace* **2023**, *10*, 898. [CrossRef]
10. Khalil, D.; Prasad, A.; Motlicek, P.; Zuluaga-Gomez, J.; Nigmatulina, I.; Madikeri, S.; Schuepbach, C. An Automatic Speaker Clustering Pipeline for the Air Traffic Communication Domain. *Aerospace* **2023**, *10*, 876. [CrossRef]
11. Zuluaga-Gomez, J.; Prasad, A.; Nigmatulina, I.; Motlicek, P.; Kleinert, M. A Virtual Simulation-Pilot Agent for Training of Air Traffic Controllers. *Aerospace* **2023**, *10*, 490. [CrossRef]

12. Kleinert, M.; Helmke, H.; Shetty, S.; Ohneiser, O.; Ehr, H.; Prasad, A.; Motlicek, P.; Harfmann, J. Automated Interpretation of Air Traffic Control Communication: The Journey from Spoken Words to a Deeper Understanding of the Meaning. In Proceedings of the 40th Digital Avionics Systems Conference (DASC), Hybrid Conference, San Antonio, TX, USA, 3–7 October 2021.
13. Chen, S.; Kopald, H.D.; Chong, R.; Wei, Y.; Levonian, Z. Read back error detection using automatic speech recognition. In Proceedings of the 12th USA/Europe Air Traffic Management Research and Development Seminar (ATM2017), Seattle, WA, USA, 26–30 June 2017.
14. Lin, Y.; Deng, L.; Chen, Z.; Wu, X.; Zhang, J.; Yang, B. A Real-Time ATC Safety Monitoring Framework Using a Deep Learning Approach. *IEEE Trans. Intell. Transp. Syst.* **2020**, *21*, 4572–4581. [CrossRef]
15. Helmke, H.; Ohneiser, O.; Buxbaum, J.; Kern, C. Increasing ATM Efficiency with Assistant Based Speech Recognition. In Proceedings of the 12th USA/Europe Air Traffic Management Research and Development Seminar (ATM2017), Seattle, WA, USA, 26–30 June 2017.
16. Kleinert, M.; Ohneiser, O.; Helmke, H.; Shetty, S.; Ehr, H.; Maier, M.; Schacht, S.; Wiese, H. Safety Aspects of Supporting Apron Controllers with Automatic Speech Recognition and Understanding Integrated into an Advanced Surface Movement Guidance and Control System. *Aerospace* **2023**, *10*, 596. [CrossRef]
17. Karlsson, J. Automatic Speech Recognition in Air Traffic Control: A Human Factors Perspective. In *NASA, Langley Research Center, Joint University Program for Air Transportation Research, 1989–1990*; NASA: Washington, DC, USA, 1990; pp. 9–13.
18. Lin, Y.; Ruan, M.; Cai, K.; Li, D.; Zeng, Z.; Li, F.; Yang, B. Identifying and managing risks of AI-driven operations: A case study of automatic speech recognition for improving air traffic safety. *Chin. J. Aeronaut.* **2023**, *36*, 366–386. [CrossRef]
19. Zhou, S.; Guo, D.; Hu, Y.; Lin, Y.; Yang, B. Data-driven traffic dynamic understanding and safety monitoring applications. In Proceedings of the 2022 IEEE 4th International Conference on Civil Aviation Safety and Information Technology, Dali, China, 12–14 October 2022.
20. European Union Aviation Safety Agency. *EASA Artificial Intelligence Roadmap 2.0; Human-Centric Approach to AI in Aviation*; European Union Aviation Safety Agency: Cologne, Germany, 2023; Available online: https://www.easa.europa.eu/ai (accessed on 23 October 2023).
21. European Union Aviation Safety Agency. EASA Concept Paper: Guidance for Level 1 & 2 Machine Learning Applications—Proposed Issue 02, Cologne, Germany. 2023. Available online: https://www.easa.europa.eu/en/downloads/137631/en (accessed on 23 October 2023).
22. SESAR. Guidance to Apply SESAR Safety Reference Material, Ed. 3.1. 2018. Available online: https://www.sesarju.eu/sites/default/files/documents/transversal/SESAR%202020%20-%20Guidance%20to%20Apply%20the%20SESAR2020%20Safety%20Reference%20Material.pdf (accessed on 23 October 2023).
23. EUROCONTROL. *Safety Assessment Methodology Ed2.2*; EUROCONTROL: Brussels, Belgium, 2006.
24. Insignia. Available online: https://insignia.enaire.es (accessed on 28 February 2023).
25. SESAR. D4.1.100—PJ.10-W2-96 ASR-TRL6 Final TVALR—Part I. V 01.00.00; SESAR Joint Undertaking, Brussels, Belgium, May 2023. Available online: https://cordis.europa.eu/project/id/874464/results (accessed on 23 October 2023).
26. European Organization for Civil Aviation Equipment. *EUROCAE ED-125, Process for Specifying risk Classification Scheme and Deriving Safety Objectives in ATM*; EUROCAE: Malakoff, France, 2010.
27. SESAR. D4.1.020—PJ.10-W2-96 ASR-TRL6 Final TS/IRS—Part I. V 01.00.00; SESAR Joint Undertaking, Brussels, Belgium, May 2023. Available online: https://cordis.europa.eu/project/id/874464/results (accessed on 23 October 2023).
28. Dehn, D.M. Assessing the Impact of Automation on the Air Traffic Controller: The SHAPE Questionnaires. *Air Traffic Control Q.* **2008**, *16*, 127–146. [CrossRef]
29. Hart, S. NASA-task load index (NASA-TLX); 20 years later. In Proceedings of the Human Factors and Ergonomics Society Annual Meeting, Los Angeles, CA, USA, 16–20 October 2006; pp. 904–908.
30. Stolcke, A.; Droppo, J. Comparing Human and Machine Errors in Conversational Speech Transcription. In Proceedings of the Proc. Interspeech 2017, Stockholm, Sweden, 20–24 August 2017; pp. 137–141. Available online: https://www.isca-speech.org/archive/interspeech_2017/stolcke17_interspeech.html (accessed on 23 October 2023).
31. Jordan, C.S.; Brennen, S.D. *Instantaneous Self-Assessment of Workload Technique (ISA)*; Defence Research Agency: Portsmouth, UK, 1992.
32. ICAO. *Procedures for Air Navigation Services (PANS)—Air Traffic Management Doc 4444*, 16th ed.; ICAO: Montreal, QC, Canada, 2016.

MDPI
St. Alban-Anlage 66
4052 Basel
Switzerland
www.mdpi.com

Aerospace Editorial Office
E-mail: aerospace@mdpi.com
www.mdpi.com/journal/aerospace

www.ingramcontent.com/pod-product-compliance
Lightning Source LLC
Chambersburg PA
CBHW061621220926
51850CB00005B/571
9783725803163